T0093104

Self-dual Partial Differential Systems and Their Variational Principles

Nassif Ghoussoub

Self-dual Partial Differential Systems and Their Variational Principles

 Springer

Nassif Ghoussoub
University of British Columbia
Department of Mathematics
Vancouver BC V6T 1Z2
Canada
nassif@math.ubc.ca

ISSN: 1439-7382
ISBN: 978-0-387-84896-9   e-ISBN: 978-0-387-84897-6
DOI: 10.1007/978-0-387-84897-6

Library of Congress Control Number: 2008938377

Mathematics Subject Classification (2000): 46-xx, 35-xx

Printed on acid-free paper

springer.com

To Mireille.

Preface

How to solve partial differential systems by completing the square. This could well have been the title of this monograph as it grew into a project to develop a systematic approach for associating suitable nonnegative energy functionals to a large class of partial differential equations (PDEs) and evolutionary systems. The minima of these functionals are to be the solutions we seek, not because they are critical points (i.e., from the corresponding Euler-Lagrange equations) but from also being zeros of these functionals. The approach can be traced back to Bogomolnyi's trick of "completing squares" in the basic equations of quantum field theory (e.g., Yang-Mills, Seiberg-Witten, Ginzburg-Landau, etc.,), which allows for the derivation of the so-called self (or antiself) dual version of these equations. In reality, the *"self-dual Lagrangians"* we consider here were inspired by a variational approach proposed – over 30 years ago – by Brézis and Ekeland for the heat equation and other gradient flows of convex energies. It is based on Fenchel-Legendre duality and can be used on any convex functional – not just quadratic ones – making them applicable in a wide range of problems. In retrospect, we realized that the "energy identities" satisfied by Leray's solutions for the Navier-Stokes equations are also another manifestation of the concept of self-duality in the context of evolution equations.

The book could have also been entitled *How to solve nonlinear PDEs via convex analysis on phase space.* Indeed, the *self-dual vector fields* we introduce and study here are natural extensions of gradients of convex energies – and hence of self-adjoint positive operators – which usually drive dissipative systems but also provide representations for the superposition of such gradients with skew-symmetric operators, which normally generate conservative flows. Most remarkable is the fact that self-dual vector fields turned out to coincide with *maximal monotone operators*, themselves being far-reaching extensions of subdifferentials of convex potentials. This means that we have a one-to-one correspondence between three fundamental notions of modern nonlinear analysis: maximal monotone operators, semigroups of contractions, and self-dual Lagrangians. As such, a large part of nonlinear analysis can now be reduced to classical convex analysis on phase space, with self-dual Lagrangians playing the role of potentials for monotone vector fields according to a suitable calculus that we develop herein. This then leads to variational formulations and resolutions of a large class of differential systems that cannot otherwise be Euler-Lagrange equations of action functionals.

A note of caution, however, is in order about our chosen terminology. Unlike its use in quantum field theory, our concept of self-duality refers to an invariance under the Legendre transform – up to an automorphism of phase space – of the Lagrangians we consider. It also reflects the fact that many of the functionals we consider here are self-dual in the sense of convex optimization, meaning that the value of the infimum in the primal minimization problem is exactly the negative of the value of the supremum in the corresponding dual problem and therefore must be zero whenever there is no duality gap.

Another note, of a more speculative nature, is also in order, as our notion of self-duality turned out to be also remarkably omnipresent outside the framework of quantum field theory. Indeed, the class of self-dual partial differential systems – as presented here – becomes quite encompassing, as it now also contains many of the classical PDEs, albeit stationary or evolutionary, from gradient flows of convex potentials (such as the heat and porous media equations), Hamiltonian systems, and nonlinear transport equations to Cauchy-Riemann systems, Navier-Stokes evolutions, Schrödinger equations, and many others. As such, many of these basic PDEs can now be perceived as the "self-dual representatives" of families of equations that are still missing from current physical models. They are the absolute minima of newly devised self-dual energy functionals that may have other critical points that correspond – via Euler-Lagrange theory – to a more complex and still uncharted hierarchy of equations.

The prospect of exhibiting a unifying framework for the existence theory of such a disparate set of equations was the main motivating factor for writing this book. The approach is surprising because it suggests that basic convex analysis – properly formulated on phase space – can handle a large variety of PDEs that are normally perceived to be inherently nonlinear. It is also surprisingly simple because it essentially builds on a single variational principle that applies to a deceivingly restrictive-looking class of self-dual energy functionals. The challenges then shift from the analytical issues connected with the classical calculus of variations towards more algebraic/functional analytic methods for identifying and constructing self-dual functionals as well as ways to combine them without destroying their self-dual features.

With this in mind, the book is meant to offer material for an advanced graduate course on convexity methods for PDEs. The generality we chose for our statements definitely puts it under the "functional analysis" classification. The examples – deliberately chosen to be among the simplest of those that illustrate the proposed general principles – require, however, a fair knowledge of classical analysis and PDEs, which is needed to make – among other things – judicious choices of function spaces where the self-dual variational principles need to be applied. These choices necessarily require an apriori knowledge of the expected regularity of the (weak) solutions. We are therefore well aware that this project runs the risk of being perceived as "too much PDEs for functional analysts, and too much functional analysis for PDErs." This is a price that may need to be paid whenever one ventures into any attempt at a unification or classification scheme within PDE theory.

At this stage, I would like to thank Ivar Ekeland for pointing me toward his 1976 conjecture with Haïm Brézis, that triggered my initial interest and eventually led to the development of this program. Most of the results in this book have been obtained in close collaboration with my postdoctoral fellow Abbas Moameni and my former MSc student Leo Tzou. I can certainly say that without their defining contributions – both conceptual and technical – this material would never have reached its present state of readiness.

I would also like to express my gratitude to Yann Brenier, David Brydges, Ivar Ekeland, Craig Evans, Richard Froese, Stephen Gustafson, Helmut Höfer, Robert McCann, Michael Struwe, Louis Nirenberg, Eric Séré, and Tai-peng Tsai for the numerous and fruitful discussions about this project, especially during the foggiest periods of its development. I am also thankful to Ulisse Stefanelli, who made me aware of the large number of related works on evolution equations. Much of this research was done during my visits – in the last five years– to the Centre de Recherches Mathématiques in Montréal, the CEREMADE at l'Université Paris-Dauphine, l'Université Aix-Marseille III, l'Université de Nice-Sophie Antipolis, and the Università di Roma-Sapienza. My gratitude goes to Jacques Hurtubise, Francois Lalonde, Maria Esteban, Jean Dolbeault, Eric Séré, Yann Brenier, Philippe Maisonobe, Michel Rascle, Frédéric Robert, Francois Hamel, PierPaolo Esposito, Filomena Pacella, Italo Capuzzo-Dolcetta, and Gabriella Tarantello, for their friendship and hospitality during these visits. The technical support of my ever reliable assistant Danny Fan has been tremendously helpful. I thank her for it.

Last but not least, "Un Grand Merci" to Louise, Mireille, Michelle, and Joseph for all the times they tried – though often with limited success – to keep me off this project.

Big Bar Lake, British Colombia, Canada   N. A. Ghoussoub
July 2008

Contents

Chapter 1
Introduction

This book is devoted to the development of a calculus of variations that can apply to a large number of partial differential systems and evolution equations, many of which do not fit in the classical Euler-Lagrange framework. Indeed, the solutions of many equations involving nonlinear, nonlocal, or even linear but non self-adjoint operators are not normally characterized as critical points of functionals of the form $\int_\Omega F(x, u(x), \nabla u(x)) dx$. Examples include *transport equations* on a smooth domain Ω of $\mathbf{R}^n$ such as

$$\begin{cases} \Sigma_{i=1}^n a_i \frac{\partial u}{\partial x_i} + a_0 u = |u|^{p-1}u + f & \text{on} \quad \Omega \subset \mathbf{R}^n, \\ \qquad\qquad u(x) = 0 & \text{on} \quad \Sigma_-, \end{cases} \tag{1.1}$$

where $\mathbf{a} = (a_i)_i : \overline{\Omega} \to \mathbf{R}^n$ is a given vector field, $p > 1$, $f \in L^2(\Omega)$, and Σ_- is the entrance set $\Sigma_- = \{x \in \partial\Omega; \mathbf{a}(x) \cdot \mathbf{n}(x) < 0\}$, $\mathbf{n}$ being the outer normal on $\partial\Omega$.
   Another example is the equation

$$\begin{cases} \mathrm{div}(T(\nabla f(x))) = g(x) & \text{on} \quad \Omega \subset \mathbf{R}^n, \\ \qquad\qquad f(x) = 0 & \text{on} \quad \partial\Omega, \end{cases} \tag{1.2}$$

where T is a monotone vector field on $\mathbf{R}^n$ that is not derived from a potential.
   Similarly, dissipative initial-value problems such as the *heat equation, porous media*, or the *Navier-Stokes evolution*

$$\begin{cases} \frac{\partial u}{\partial t} + (u \cdot \nabla)u + f = \alpha\Delta u - \nabla p & \text{on } \Omega \subset \mathbf{R}^n, \\ \qquad\qquad \mathrm{div}\, u = 0 & \text{on } [0, T] \times \Omega, \\ \qquad\qquad u = 0 & \text{on } [0, T] \times \partial\Omega, \end{cases} \tag{1.3}$$

where $\alpha > 0$ and $f \in L^2([0, T] \times \Omega)$, cannot be solved by the standard methods of the calculus of variations since they are not Euler-Lagrange equations of action functionals of the form $\int_0^T L(t, x(t), \dot{x}(t)) dt$. Our goal here is to describe how these examples and many others can still be formulated and resolved variationally by means of a *self-dual variational calculus* that we develop herein.
   The genesis of our approach can be traced to physicists who have managed to formulate – if not solve – variationally many of the basic nonself-adjoint equations of quantum field theory by minimizing their associated action functionals. Indeed,

N. Ghoussoub, *Self-dual Partial Differential Systems and Their Variational Principles*,
Springer Monographs in Mathematics, DOI 10.1007/978-0-387-84897-6_1,
© Springer Science+Business Media, LLC 2009

most equations arising on the interface between Riemannian geometry and quantum field theory (e.g., Yang-Mills, Chern-Simon, Seiberg-Witten, and Ginzburg-Landau) have self-dual and/or antiself-dual versions that enjoy very special features: They are obtained variationally as minima of their action functionals, yet they are not derived from being stationary states (i.e., from the corresponding Euler-Lagrange equations) but from the fact that they are zeros of certain derived nonnegative Lagrangians obtained by Bogomolnyi's trick of completing squares. But this is possible only if the action functional attains a natural and – a priori – known minimum. The identities thus obtained are usually called the *self (or antiself) -dual* equations, which are often lower-order factors of the more complicated Euler-Lagrange equations. Our main premise here is that this phenomenon is remarkably prevalent in the equations originating from geometry, physics, and other applied mathematical models. We shall see that many of the basic partial differential equations, whether stationary or evolutionary, can be perceived as the "self-dual representatives" of a less obvious and more complicated family of equations. They are the absolute minima of appropriately devised new energy functionals that may have other critical points via Euler-Lagrange theory that correspond to more complex hierarchies of equations.

This volume has been written with two objectives in mind:

- First, we develop a general framework, in which solutions of a large class of partial differential equations and evolutionary systems – many of which are not of Euler-Lagrange type – can be identified as the minima of appropriately devised self-dual energy functionals.
- Our second objective is to show how to use such self-dual features to develop a systematic approach for a variational resolution of these equations.

The general framework relies on the observation that a large number of partial differential equations can be written in the form

$$(\Lambda x, Ax) \in \partial L(Ax, \Lambda x), \tag{1.4}$$

where $A : D(A) \subset X \to X$ and $\Lambda : D(\Lambda) \subset X \to X^*$ are – possibly nonlinear – operators on a reflexive Banach space X and ∂L is the subdifferential (in the sense of convex analysis) of a Lagrangian on phase space $X \times X^*$ satisfying the following duality property:

$$L^*(p,x) = L(x,p) \text{ for all } (p,x) \in X^* \times X. \tag{1.5}$$

Here L^* is the Legendre transform of L in both variables, that is,

$$L^*(p,x) = \sup\left\{\langle p,y \rangle + \langle x,q \rangle - L(y,q); (y,q) \in X \times X^*\right\}. \tag{1.6}$$

These equations will be called *self-dual partial differential systems*, while those that correspond to when $\Lambda = 0$ will be called *completely self-dual systems*. This class is remarkably encompassing since besides the equations of quantum field theory mentioned above it also includes many of the classical PDEs: gradient flows of convex potentials (such as the heat and porous media equations), Hamiltonian systems, transport equations, nonlinear Laplace equations with advection, Cauchy-Riemann systems, Navier-Stokes evolutions, Schrödinger equations, and many others.

As for our second objective, developing a systematic approach for a variational resolution of these equations, it consists of noting that such equations can be resolved by simply minimizing the *self-dual energy functional*

$$I(u) = L(Au, \Lambda u) - \langle Au, \Lambda u \rangle. \tag{1.7}$$

However, besides ensuring that the minimum is attained, one needs to establish that the infimum is actually the "natural lower bound", which in our case has been tailor-made to be always *zero*. For that we establish general *self-dual variational principles* that will achieve both goals and will therefore allow for the variational formulation and resolution of these equations. Our principles cover lots of ground and apply to all linear and nonlinear equations mentioned above, though ironically they do not yet cover most equations of quantum field theory because of the prohibitive lack of compactness inherent in these problems.

As mentioned above, a typical example[1] is the Yang-Mills functional on the space of smooth connections over a principal $SU(2)$-bundle P of an oriented closed 4-manifold M. To any connection $A \in \Omega_1(\mathrm{Ad}P)$ on M, one associates a curvature tensor $F_A = dA + \frac{1}{2}[A \wedge A] \in \Omega_2(\mathrm{Ad}P)$, and an exterior differential on k-forms $d_A w = dw + [A \wedge w]$. After completing the square, the Yang-Mills functional on the space of connections looks like

$$I(A) := \int_M \|F_A\|^2 = \frac{1}{2} \int_M \|F_A + *F_A\|^2 - \langle *F_A, F_A \rangle \geq - \int_M \langle F_A \wedge F_A \rangle = 8\pi^2 c_2(P),$$

where $*$ is the Hodge operator and the inner product is the negative of the trace of the product of the matrices. The last term on the right is a topological invariant of the bundle $P \to M$, with $c_2(P)$ being the second Chern class. If now the infimum of the functional I is actually equal to $8\pi^2 c_2(P)$, and if it is attained at some A, it then follows immediately that such a connection satisfies

$$F_A = - *F_A, \tag{1.8}$$

which are then called the *antiself-dual Yang-Mills equations*. Indeed, the Bianchi identities ensure that we then have $d_A^* F_A = 0$, which are the corresponding Euler-Lagrange equations (see, for example, Jost [80]). We note again that, even though equations (1.8) were obtained variationally as absolute minima, they are not derived from the Euler-Lagrange equations of the Yang-Mills functional. In this case, the self-dual Lagrangian is nothing but the "true square" $L(x, p) = \frac{1}{2}(\|x\|^2 + \|p\|^2)$.

From a totally different perspective, Brézis and Ekeland [29] formulated about 30 years ago an intriguing minimization principle that can be associated to the heat equation and other gradient flows of convex energy functionals. It is based on Fenchel-Legendre duality, which can be seen as a more general procedure for "com-

---

[1] Which could/should be skipped by those not familiar with the basics of differential geometry.

pleting squares" that can be used on any convex functional and not just quadratic ones. More recently, Ghoussoub and Tzou eventually demonstrated in [67] the usefulness of this formulation in proving existence results, by showing that one can indeed prove the existence of a gradient flow

$$\begin{cases} \dot{u}(t) \in -\partial \varphi(u(t)) \quad \text{a.e.} \quad \text{on} \quad [0,T] \\ u(0) = u_0, \end{cases} \tag{1.9}$$

for a convex energy φ by minimizing the nonnegative functional

$$I(u) = \int_0^T [\varphi(u(t)) + \varphi^*(-\dot{u}(t))]\,dt + \frac{1}{2}(|u(0)|^2 + |u(T)|^2) - 2\langle u(0), u_0 \rangle + |u_0|^2$$

on an appropriate path space $\mathscr{A}$ and by showing that it has a minimizer $\bar{u}$ in $\mathscr{A}$ such that $I(\bar{u}) = \inf_{u \in \mathscr{A}} I(u) = 0$. The self-dual Lagrangian here is an appropriate "lifting" of $L(x,p) = \varphi(x) + \varphi^*(p)$ to path space.

In [9] and [10], Auchmuty proposed a framework, in which he formalizes and generalizes the Brézis-Ekeland procedure in order to apply it to operator equations of nonpotential type. However, the applicability of these variational principles remained conditional on evaluating the minimum value and – in most cases – could not be used to establish the existence – and sometimes uniqueness – of solutions.

The basic ideas are simple. Starting with a functional equation of the form

$$-Au = \partial \varphi(u), \tag{1.10}$$

on a Banach space X, it is well known that it can be formulated – and sometimes solved – variationally whenever $A : X \to X^*$ is a self-adjoint bounded linear operator and φ is a differentiable or convex functional on X. Indeed, it can be reduced in this case to the equation $\partial \psi(u) = 0$, where ψ is the functional

$$\psi(u) = \varphi(u) + \frac{1}{2}\langle Au, u \rangle. \tag{1.11}$$

A solution can then be obtained, for example, by minimization whenever φ is convex and A is positive. But this variational procedure fails when A is not self-adjoint or when A is a nonpotential operator (i.e., when A is not a gradient vector field), and definitely when A is not linear. In this case, the Brézis-Ekeland procedure – as formalized by Auchmuty – consists of simply minimizing the functional

$$I(u) = \varphi(u) + \varphi^*(-Au) + \langle u, Au \rangle, \tag{1.12}$$

where φ^* is the Fenchel-Legendre dual of φ defined on X^* by $\varphi^*(p) = \sup\{\langle x, p \rangle - \varphi(x); x \in X\}$. The basic Legendre-Fenchel inequality states that

$$\varphi(x) + \varphi^*(p) \geq \langle x, p \rangle \text{ with equality if and only if } p = \partial \varphi(x). \tag{1.13}$$

This clearly yields that $\alpha := \inf_{u \in X} I(u) \geq 0$, and the following simple observation was made by several authors: If the infimum $\alpha = 0$ and if it is attained at $\bar{u} \in X$, then we are in the limiting case of the Fenchel-Legendre duality, $\varphi(\bar{u}) + \varphi^*(-A\bar{u}) = \langle \bar{u}, -A\bar{u} \rangle$, and therefore $-A\bar{u} = \partial\varphi(\bar{u})$.

Note that the procedure does not require any assumption on A, and very general coercivity assumptions on φ often ensure the existence of a minimum. However, the difficulty here is different from that of standard minimization problems in that besides the problem of existence of a minimum one has to ensure that the infimum is actually zero. This is obviously not the case for general operators A, though one can always write (and many authors did) the variational principle (1.12) for the operator equation (1.10).

In this volume, we tackle the real difficulty of deciding when the infimum α is actually zero, and we identify a large and structurally interesting class of *self-dual vector fields* F, for which equations such as

$$0 \in F(u) \quad \text{and} \quad \Lambda u \in F(u), \tag{1.14}$$

with Λ being a suitable linear or nonlinear operator, can be formulated and solved variationally. Such vector fields will be derived from self-dual Lagrangians L and will be denoted by $F = \bar{\partial}L$. Equations of the form $0 \in \bar{\partial}L(u)$ coincide with the *completely self-dual systems* described above and will be dealt with in Part II of this book. The more general class of *self-dual systems* will contain equations of the form $\Lambda u \in \bar{\partial}L(u)$ and will be tackled in Parts III and IV.

For the convenience of the reader, we now give a summarized description of the contents of each chapter.

Part I: Convex analysis on phase space

A large class of PDEs and evolution equations, which we call *completely self-dual differential systems*, can be written in the form

$$(p, x) \in \partial L(x, p), \tag{1.15}$$

where ∂L is the subdifferential of a self-dual Lagrangian on phase space $L : X \times X^* \to \mathbf{R} \cup \{+\infty\}$, and X is a reflexive Banach space. We therefore start in Part I by recalling the classical basic concepts and relevant tools of convex analysis that will be used throughout the text. We then introduce the key notions of convex analysis on phase space and focus on their calculus.

Chapter 2: Legendre-Fenchel duality on phase space

We review basic convex analysis and in particular Fenchel-Legendre duality and its relationship with subdifferentiability. As mentioned before, our approach is based on convex analysis on "phase space", and we shall therefore consider Lagrangians L on $X \times X^*$ that are convex and lower semicontinuous in both variables. All elements of convex analysis will apply, but the calculus on $X \times X^*$ becomes much richer for many reasons, not the least of which being the variety of automorphisms that act on such phase space, as well as the ability of associating Hamiltonians, which are the Legendre transforms of L in one of the two variables.

Chapter 3: Self-dual Lagrangians on phase space

At the heart of the theory is the interplay between certain automorphisms and Legendre transforms. The class of Lagrangians L satisfying the duality conditions (1.5) on phase space is introduced and analyzed in this chapter. Its remarkable permanence properties are also established, in particular, their stability under various operations, such as convolutions, direct sum, regularizations, and superpositions with other Lagrangians and operators.

Chapter 4: Skew-adjoint operators and self-dual Lagrangians

If L is a self-dual Lagrangian on a reflexive Banach space X and $\Gamma : X \to X^*$ is a skew-adjoint operator, then the Lagrangian defined by $L_\Gamma(x, p) = L(x, \Gamma x + p)$ is also self-dual on $X \times X^*$. Here, we deal with the more interesting cases of unbounded antisymmetric operators and with the nonhomogeneous case where operators may be skew-adjoint modulo certain boundary terms. This is normally given by a Green-Stokes type formula of the type

$$\langle x, \Gamma y \rangle + \langle y, \Gamma x \rangle = \langle \mathscr{B}x, R\mathscr{B}y \rangle \text{ for every } x, y \in D(\Gamma), \tag{1.16}$$

where $\mathscr{B} : D(\mathscr{B}) \subset X \to H$ is a boundary operator into a Hilbert space H and R is a self-adjoint automorphism on the "boundary" space H. In other words, the symmetric part of Γ is conjugate to a self-adjoint operator R on the boundary space. In this case, a suitable R-self-dual function ℓ on H is added so as to restore self-duality to the whole system. More precisely, one needs a function on the boundary space that satisfies

$$\ell^*(Rx) = \ell(x) \text{ for all } x \in H, \tag{1.17}$$

so that the Lagrangian on $X \times X^*$

$$L_{\Gamma, \ell}(x, p) = \begin{cases} L(x, -\Gamma x + p) + \ell(\mathscr{B}x) & \text{if } x \in D(\Gamma) \cap D(\mathscr{B}), \\ +\infty & \text{if } x \notin D(\Gamma) \cap D(\mathscr{B}), \end{cases} \tag{1.18}$$

becomes self-dual on $X \times X^*$.

Chapter 5: Self-dual vector fields and their calculus

We introduce here the concept of *self-dual vector fields* and develop their calculus. The starting point is that self-dual Lagrangians on phase space necessarily satisfy

$$L(x, p) \geq \langle p, x \rangle \text{ for all } (p, x) \in X^* \times X, \tag{1.19}$$

and solutions for equation (1.15) can then be found for a given p by simply minimizing the functional $I_p(x) = L(x, p) - \langle x, p \rangle$ and proving that the minimum is actually zero. In other words, by defining the *self-dual vector field of L* at $x \in X$ to be the possibly empty sets

$$\overline{\partial}L(x) := \{p \in X^*; L(x, p) - \langle x, p \rangle = 0\} = \{p \in X^*; (p, x) \in \partial L(x, p)\}, \quad (1.20)$$

one can then find variationally the zeros of those set-valued maps $T : X \to 2^{X^*}$ of the form $T(x) = \overline{\partial}L(x)$, where L is a self-dual Lagrangian on phase space $X \times X^*$. These *self-dual vector fields* are natural extensions of subdifferentials of convex lower semicontinuous energy functionals. Indeed, the most basic self-dual Lagrangians are of the form $L(x, p) = \varphi(x) + \varphi^*(p)$, where φ is such a function on X and φ^* is its Legendre conjugate on X^*, in which case

$$\overline{\partial}L(x) = \partial\varphi(x).$$

More interesting examples of self-dual Lagrangians are of the form $L(x, p) = \varphi(x) + \varphi^*(-\Gamma x + p)$, where φ is a convex and lower semicontinuous function on X and $\Gamma : X \to X^*$ is a skew-symmetric operator. The corresponding self-dual vector field is then,

$$\overline{\partial}L(x) = \Gamma x + \partial\varphi(x).$$

The examples above are typical – possibly multivalued – nonlinear operators T that are *monotone*, meaning that their graphs $G(T) = \{(x, p) \in X \times X^*; p \in T(x)\}$ satisfy

$$\langle x - y, p - q \rangle \geq 0 \text{ for every } (x, p) \text{ and } (y, q) \text{ in } G(T). \quad (1.21)$$

Their graphs are actually *maximal* in the order of set inclusion among monotone subsets of $X \times X^*$, and the theory of such *maximal monotone operators* has been developed extensively over the last 30 years because of its prevalence in both parabolic and elliptic PDEs. Most remarkable is the fact – shown in this chapter – that one can associate to any maximal monotone operator T a self-dual Lagrangian L such that

$$\overline{\partial}L = T, \quad (1.22)$$

so that equations involving such operators can be resolved variationally. The advantages of identifying maximal monotone operators as self-dual vector fields are numerous. Indeed, all equations, systems, variational inequalities, and dissipative initial-value parabolic problems that traditionally involve maximal monotone operators, can now be formulated and resolved variationally. In effect, self-dual Lagrangians play the role of potentials for maximal monotone vector fields in a way similar to how convex energies are the potentials of their own subdifferentials, and in particular how the Dirichlet integral is the potential of the Laplacian. These problems can therefore be analyzed with the full range of methods – computational or not – that are available for variational settings.

Furthermore, while issues around the superposition of, and other operations on, maximal monotone operators are often delicate to prove, the class of self-dual

Lagrangians possesses remarkable permanence properties that are relatively easy to establish. It reflects most variational aspects of convex analysis and is stable under similar types of operations making the calculus of self-dual Lagrangians (and consequently, of maximal monotone operators) as manageable as, yet much more encompassing than, the one for convex functions.

Part II: Completely self-dual systems and their Lagrangians

This part of the book deals with the variety of boundary value problems and evolution equations that can be written in the form of a completely self-dual system

$$0 \in \overline{\partial} L(x) \tag{1.23}$$

and can therefore be solved by minimizing functionals of the form

$$I(x) = L(x, 0) \tag{1.24}$$

on a Banach space X, where L is a self-dual Lagrangian on $X \times X^*$. Such functionals I are always nonnegative, and their main relevance to our study stems from the fact that – under appropriate conditions on L – their infimum is equal to 0. This property allows variational formulations and resolutions of several basic differential systems, which – often for lack of self-adjointness or linearity – cannot be expressed as Euler-Lagrange equations but can, however, be written in the form (1.23).

Chapter 6: Variational principles for completely self-dual functionals and first applications

The fact that the infimum of a completely self-dual functional I is zero follows from the basic duality theory in convex analysis, which in our particular "self-dual case" leads to a situation where the value of the *dual problem* is exactly the negative of the value of the *primal problem*. This value is zero as soon as there is no duality gap, which is normally a prerequisite for the attainment of these extrema.

Several immediate applications follow from this observation coupled with the remarkable fact that the Lagrangian $L_\Gamma(x, p) = L(x, \Gamma x + p)$ remains self-dual on $X \times X^*$, provided L is and as long as Γ is a skew-symmetric operator. One then quickly obtains variational formulations and resolutions of the Lax-Milgram theorem, of variational inequalities, of nonself-adjoint semilinear Dirichlet problems, as well as several other nonpotential operator equations, such as (1.2).

Chapter 7: Semigroups of contractions associated to self-dual Lagrangians

A variational theory for dissipative initial-value problems can be developed via the theory of self-dual Lagrangians. We consider here semilinear parabolic equations with homogeneous state-boundary conditions of the form

$$\begin{cases} -\dot{u}(t) + \Gamma u(t) + \omega u(t) \in \partial\varphi(t,u(t)) & \text{on} \quad [0,T] \\ u(0) = u_0, \end{cases} \tag{1.25}$$

where Γ is an antisymmetric, possibly unbounded, operator on a Hilbert space H, φ is a convex lower semicontinuous function on H, $\omega \in \mathbf{R}$, and $u_0 \in H$. Assuming for now that $\omega = 0$, the framework proposed above for the stationary case yields a formulation of (1.25) as a time-dependent self-dual equation on state space,

$$\begin{cases} -\dot{u}(t) \in \overline{\partial}L(t,u(t)) \\ u(0) = u_0, \end{cases} \tag{1.26}$$

where the self-dual Lagrangian $L(t,\cdot,\cdot)$ on $H \times H$ is associated to the convex functional φ and the operator Γ in the following way:

$$L(t,u,p) = \varphi(t,u) + \varphi^*(t, \Gamma u + p). \tag{1.27}$$

We shall then see that a (time-dependent) self-dual Lagrangian $L : [0,T] \times H \times H \rightarrow \mathbf{R}$ on state space H, "lifts" to a self-dual Lagrangian $\mathscr{L}$ on path space $A_H^2 = \{u : [0,T] \rightarrow H; \dot{u} \in L_H^2\}$ via the formula

$$\mathscr{L}(u,(p,a)) = \int_0^T L(t,u(t) - p(t), -\dot{u}(t))dt + \ell_{u_0}(u(0) - a, u(T)), \tag{1.28}$$

where $(p(t),a) \in L_H^2 \times H$, which happens to be a convenient representation for the dual of A_H^2. Here ℓ_{u_0} is the boundary Lagrangian

$$\ell_{u_0}(x,p) = \frac{1}{2}|x|_H^2 - 2\langle u_0,x\rangle + |u_0|_H^2 + \frac{1}{2}|p|_H^2 \tag{1.29}$$

that is suitable for the initial-value problem (1.26), which can then be formulated as a stationary equation on path space of the form

$$0 \in \overline{\partial}\mathscr{L}(u). \tag{1.30}$$

Its solution $\bar{u}(t)$ can then be obtained by simply minimizing the completely self-dual functional $I(u) = \mathscr{L}(u,0)$ on the path space A_H^2 since it can also be written as the sum of two nonnegative terms:

$$I(u) = \int_0^T \{L(t,u(t), -\dot{u}(t)) + \langle \dot{u}(t), u(t)\rangle\} dt + \|u(0) - u_0\|_H^2.$$

This provides a variational procedure for associating to a self-dual Lagrangian L a semigroup of contractive maps $(S_t)_t$ on H via the formula $S_t u_0 := \bar{u}(t)$, yielding another approach to the classical result associating such semigroups to maximal monotone operators. This chapter is focused on the implementation of this approach

with a minimal set of hypotheses and on the application of this variational approach
to various standard parabolic equations.

Worth noting is the fact that we now have a one-to-one correspondence between three fundamental notions of nonlinear analysis: maximal monotone operators, semigroups of contractions, and self-dual Lagrangians.

Chapter 8: Iteration of self-dual Lagrangians and multiparameter evolutions

Nonhomogeneous boundary conditions reflect the lack of antisymmetry in a differential system. The iteration of a self-dual Lagrangian on phase space $X \times X^*$ with
an operator that is skew-adjoint modulo a boundary triplet $(H, \mathscr{B}, R)$ needs to be
combined with an R-self-dual function ℓ on the boundary H in order to restore self-
duality to the whole system. This is done via the Lagrangian $L_{\Gamma, \ell}$ defined in (1.18),
which is then self-dual, and as a consequence one obtains solutions for the boundary
value problem

$$\begin{cases} \Gamma x \in \bar{\partial} L(x) \\ R\mathscr{B}x \in \partial \ell(\mathscr{B}x), \end{cases} \tag{1.31}$$

by inferring that the infimum on X of the completely self-dual functional $I(x) :=$
$L_{\Gamma, \ell}(x, 0) = L(x, \Gamma x) + \ell(\mathscr{B}x)$ is attained and is equal to zero. Moreover, the addition
of the R-self-dual boundary Lagrangian required to restore self-duality often leads
to the natural boundary conditions.

The latter Lagrangian can then be lifted to path space, provided one adds a suitable self-dual time-boundary Lagrangian. This iteration can be used to solve initial-
value parabolic problems whose state-boundary values are evolving in time such
as

$$\begin{cases} -\dot{x}(t) + \Gamma_t x(t) \in \bar{\partial} L(t, x(t)) & \text{for } t \in [0, T] \\ R_t \mathscr{B}_t(x(t)) \in \partial \ell_t(\mathscr{B}_t x(t)) & \text{for } t \in [0, T] \\ x(0) = x_0, \end{cases} \tag{1.32}$$

where L is a time-dependent self-dual Lagrangian on a Banach space X anchored
on a Hilbert space H (i.e., $X \subset H \subset X^*$), x_0 is a prescribed initial state in X, $\Gamma_t :$
$D(\Gamma_t) \subset X \to X^*$ is antisymmetric modulo a boundary pair $(H_t, R_t, \mathscr{B}_t)$ with $\mathscr{B}_t :$
$D(\mathscr{B}_t) \subset X \to H_t$ as a boundary operator, R_t a self-adjoint automorphism on H_t,
and ℓ_t an R_t-self-dual function on the boundary space H_t. The corresponding self-
dual Lagrangian on $L_X^2[0, T] \times L_{X^*}^2[0, T]$ is then

$$\mathscr{L}(u, p) = \int_0^T L_{\Gamma_t, \ell_t}(t, u(t), p(t) - \dot{u}(t)) dt + \ell_{x_0}(u(0), u(T)).$$

This process can be iterated again by considering the path space $L_X^2[0, T]$ as a new
state space for the newly obtained self-dual Lagrangian, leading to the construction
of multiparameter flows such as

$$\begin{cases} -\frac{\partial x}{\partial t}(s,t) - \frac{\partial x}{\partial s}(s,t) \in \overline{\partial} L\left((s,t), x(s,t), \frac{\partial x}{\partial t}(s,t) + \frac{\partial x}{\partial s}(s,t)\right) & \text{on } [0,S] \times [0,T], \\ x(0,t) = x_0 \text{ for } t \in [0,T], \\ x(s,0) = x_0 \text{ for } s \in [0,S]. \end{cases}$$

(1.33)

This method is quite general and far-reaching but may be limited by the set of conditions needed to accomplish the above-mentioned iterations. This chapter focuses on cases where this can be done.

Chapter 9: Direct sum of completely self-dual functionals

If $\Gamma : X \to X^*$ is an invertible skew-adjoint operator on a reflexive Banach space X, and if L is a self-dual Lagrangian on $X \times X^*$, then $M(x,p) = L(x + \Gamma^{-1}p, \Gamma x)$ is also a self-dual Lagrangian. By minimizing the completely self-dual functional $I(x) = \varphi(x) + \varphi^*(\Gamma x)$ over X, one can then find solutions of $\Gamma x \in \partial \varphi(x)$ as long as φ is convex lower semicontinuous and bounded above on the unit ball of X. In other words, the theory of self-dual Lagrangians readily implies that if the linear system $\Gamma x = p$ is uniquely solvable, then the semilinear system $\Gamma x \in \partial \varphi(x)$ is also solvable for slowly growing convex nonlinearities φ.

Self-dual variational calculus allows us to extend this observation in the following way. Consider bounded linear operators $\Gamma_i : Z \to X_i^*$ for $i = 1, ..., n$. If for each $(p_i)_{i=1}^n \in X_1^* \times X_2^* ... \times X_n^*$ the system of linear equations

$$\Gamma_i x = p_i \qquad (1.34)$$

can be uniquely solved, then one can solve – variationally – the semilinear system

$$\Gamma_i x \in \partial \varphi_i(A_i x) \quad \text{for } i = 1, ..., n,, \qquad (1.35)$$

provided $A_i : Z \to X_i$ are bounded linear operators that satisfy the identity

$$\sum_{i=1}^n \langle A_i x, \Gamma_i x \rangle = 0 \text{ for all } x \in Z. \qquad (1.36)$$

The solution is then obtained by minimizing the functional

$$I(z) = \sum_{i=1}^n \varphi_i(A_i z) + \varphi_i^*(\Gamma_i z), \qquad (1.37)$$

which is then completely self-dual. This result is then applied to derive variational formulations and resolutions for various evolution equations.

Chapter 10: Semilinear evolutions with self-dual boundary conditions

One may use self-dual variational principles to construct solutions of evolution equations that satisfy certain nonlinear boundary conditions. More specifically, we consider evolutions of the form

$$\begin{cases} \dot{u}(t) \in -\overline{\partial}L(t,u(t)) & \forall t \in [0,T] \\ \frac{u(0)+u(T)}{2} \in -\overline{\partial}\ell(u(0)-u(T)), \end{cases} \tag{1.38}$$

where both L and ℓ are self-dual Lagrangians. The novelty here is that the self-dual time-boundary equation we obtain is very general and includes – with judicious choices for ℓ – the more traditional ones such as:

- initial-value problems: $x(0) = x_0$;
- periodic orbits: $x(0) = x(T)$;
- antiperiodic orbits: $x(0) = -x(T)$;
- periodic orbits up to an isometry: $x(T) = e^{-T(\omega I + A)}x(0)$, where $w \in R$ and A is a skew-adjoint operator.

Solutions are obtained by minimizing the self-dual functional

$$I(x) = \int_0^T L\big(t, x(t), \dot{x}(t)\big)\, dt + \ell\left(x(0) - x(T), \frac{x(0)+x(T)}{2}\right).$$

Worth noting here is that many choices for L are possible when one formulates parabolic equations such as

$$-\dot{u}(t) + \Gamma_1 u(t) + \Gamma_2 u(t) + \omega u(t) \in \partial\varphi(t, u(t)) \quad \text{on} \quad [0,T]$$

in a self-dual form as in (1.38). The choices depend on the nature of the skew-adjoint operators $\Gamma_i, i = 1, 2$, on whether the equation contains a diffusive factor or not, or whether ω is a nonnegative scalar or not.

Part III: Self-dual systems and their antisymmetric Hamiltonians

Many more nonlinear boundary value problems and evolution equations can be written in the form

$$0 \in \Lambda x + \overline{\partial}L(x) \tag{1.39}$$

on a Banach space X, where L is a self-dual Lagrangian on $X \times X^*$ and $\Lambda : D(\Lambda) \subset X \to X^*$ is an appropriate linear or nonlinear operator. They can be solved by showing that functionals of the form

$$J(x) = L(x, -\Lambda x) + \langle x, \Lambda x \rangle \tag{1.40}$$

attain their infimum and that the latter is equal to zero. These are very important examples in the class of – what we call – *self-dual equations*. To understand the connection to the systems studied in the previous part, we note that *completely self-dual functionals* can be written as

$$I(x) = L(x,0) = \sup_{y \in X} H_L(y,x) \quad \text{for all } x \in X, \tag{1.41}$$

where L is the corresponding self-dual Lagrangian on $X \times X^*$ and where H_L is the Hamiltonian associated with L (i.e., the Legendre transform of L but only in the second variable). These Hamiltonians are concave-convex functions on state space $X \times X$ and verify the antisymmetry property

$$H_L(x,y) = -H_L(y,x), \tag{1.42}$$

and in particular $H_L(x,x) = 0$. One can easily see that

$$J(x) = L(x,-\Lambda x) + \langle x, \Lambda x \rangle = \sup_{y \in X} \langle x - y, \Lambda x \rangle + H_L(y,x), \tag{1.43}$$

where $M(x,y) = \langle x - y, \Lambda x \rangle + H_L(y,x)$ is again zero on the diagonal of $X \times X$.

Chapter 11: The class of antisymmetric Hamiltonians

We are then led to the class of *antisymmetric Hamiltonians* M on $X \times X$, which – besides being zero on the diagonal – are weakly lower semicontinuous in the first variable and concave in the second. Functionals of the form

$$I(x) = \sup_{y \in X} M(x,y), \tag{1.44}$$

with M being antisymmetric , will be called *self-dual functionals* as they turn out to have many of the variational properties of completely self-dual functionals. They are, however, much more encompassing since they are not necessarily convex, and they allow for the variational resolution of various nonlinear partial differential equations. Indeed, the class of antisymmetric Hamiltonians is a convex cone that contains – Maxwellian – Hamiltonians of the form

$$M(x,y) = \varphi(x) - \varphi(y),$$

with φ being convex and lower semicontinuous, as well as their sum with terms of the form

$$M(x,y) = \langle \Lambda x, x - y \rangle,$$

provided $\Lambda : X \to X^*$ is a not necessarily linear *regular operator* that is, a weak-to-weak continuous operator such that the diagonal map

$$u \to \langle u, \Lambda u \rangle \text{ is weakly lower semicontinuous.} \tag{1.45}$$

Examples include of course all linear positive operators, but also certain linear but not necessarily positive operators such as $\Lambda u = J\dot{u}$, which is regular on the Sobolev space $H^1_{per}[0,T]$ of $\mathbf{R}^{2N}$-valued periodic functions on $[0,T]$, where J is the symplectic matrix. They also include important nonlinear operators such as the Stokes operator $u \rightarrow u \cdot \nabla u$ acting on the subspace of $H_0^1(\Omega, \mathbf{R}^n)$ consisting of divergence-free vector fields (up to dimension 4).

Chapter 12: Variational principles for self-dual functionals and first applications

Here we establish the basic variational principle for self-dual functionals, which again states that under appropriate coercivity conditions, the infimum is attained and is equal to zero. Applied to functionals $J(x) = L(x, -\Lambda x) + \langle x, \Lambda x \rangle$, one then gets solutions to equations of the form $0 \in \Lambda x + \overline{\partial} L(x)$, provided we have the following *strong coercivity condition*:

$$\lim_{\|x\| \rightarrow +\infty} H_L(0, x) + \langle x, \Lambda x \rangle = +\infty. \tag{1.46}$$

This allows for the variational resolution of a large class of PDEs, in particular nonlinear Lax-Milgram problems of the type:

$$\Lambda u + Au + f \in -\partial \varphi(u) \tag{1.47}$$

where φ is a convex lower semicontinuous functional, Λ is a nonlinear regular operator, and A is a linear – not necessarily bounded – positive operator. Immediate applications include a variational resolution to various equations involving nonlinear operators, such as nonlinear transport equations, and the stationary Navier-Stokes equation:

$$\begin{cases} (u \cdot \nabla)u + f = v\Delta u - \nabla p & \text{on } \Omega \subset \mathbf{R}^3, \\ \operatorname{div} u = 0 & \text{on } \Omega, \\ u = 0 & \text{on } \partial\Omega, \end{cases} \tag{1.48}$$

where $v > 0$ and $f \in L^p(\Omega; \mathbf{R}^3)$. The method is also applicable to nonlinear equations involving nonlocal terms such as the generalized Choquard-Pekar Schrödinger equation

$$-\Delta u + V(x)u = (w * f(u))g(u), \tag{1.49}$$

where V and w are suitable real functions.

Chapter 13: The role of the co-Hamiltonian in self-dual variational principles

Self-dual functionals of the form $I(x) = L(Ax, -\Lambda x) + \langle Ax, \Lambda x \rangle$ have more than one antisymmetric Hamiltonian associated to them. In this chapter, we shall see that the one corresponding to the co-Hamiltonian $\tilde{H}_L$ can be more suitable not only when the operator A is nonlinear but also in situations where we need a constrained minimization in order to obtain the appropriate boundary conditions. Furthermore, and even if both A and Λ are linear, the co-Hamiltonian representation can be more suitable for ensuring the required coercivity conditions. Applications are given to solve Cauchy problems for Hamiltonian systems, for doubly nonlinear evolutions, and for gradient flows of non-convex functionals.

Chapter 14: Direct sum of self-dual functionals and Hamiltonian systems

This chapter improves on the results of Chapter 9. The context is similar, as we assume that a system of linear equations

$$\Gamma_i x = p_i, \ i = 1, ..., n, \tag{1.50}$$

with each Γ_i being a linear operator from a Banach space Z into the dual of another one X_i, can be solved for any $p_i \in X^*$. We then investigate when one can solve variationally the semilinear system of equations

$$-\Gamma_i x \in \partial \varphi_i(A_i x), \tag{1.51}$$

where each A_i is a bounded linear operator from Z to X_i. Unlike Chapter 9, where we require $\sum_{i=1}^{n} \langle A_i z, \Gamma_i z \rangle$ to be identically zero, here we relax this assumption considerably by only requiring that the map

$$z \rightarrow \sum_{i=1}^{n} \langle A_i z, \Gamma_i z \rangle \text{ is weakly lower semicontinuous,} \tag{1.52}$$

as long as we have some control of the form

$$\left| \sum_{i=1}^{n} \langle A_i z, \Gamma_i z \rangle \right| \leq \sum_{i=1}^{n} \alpha_i \|\Gamma_i z\|^2, \tag{1.53}$$

for some $\alpha_i \geq 0$. In this case, the growth of the potentials φ_i should not exceed a quadratic growth of factor $\frac{1}{2\alpha_i}$. The existence result is then obtained by minimizing the functional

$$I(z) = \sum_{i=1}^{n} \varphi_i(A_i z) + \varphi_i^*(-\Gamma_i z) + \langle A_i z, \Gamma_i z \rangle, \tag{1.54}$$

which is then self-dual. This is then applied to derive self-dual variational resolutions to Hamiltonian systems with nonlinear boundary conditions of the form

$$\begin{cases} -J\dot{u}(t) & \in \partial \varphi(t, u(t)) \\ -J\frac{u(T)+u(0)}{2} & \in \partial \psi(u(T) - u(0)), \end{cases} \tag{1.55}$$

where for every $t \in [0, T]$, the functions $\varphi(t, \cdot)$ and ψ are convex lower semicontinuous on $\mathbf{R}^{2N}$ and J is the symplectic matrix. By making judicious choices for ψ, these boundary conditions include the traditional ones, such as periodic, antiperiodic and skew-periodic orbits. The method also leads to the construction of solutions that connect two Lagrangian submanifolds associated to given convex lower semicontinuous functions ψ_1 and ψ_2 on $\mathbf{R}^N$; that is,

$$
\begin{cases}
\dot{p}(t) \in \partial_2 \varphi\big(p(t), q(t)\big) & t \in (0, T) \\
-\dot{q}(t) \in \partial_1 \varphi\big(p(t), q(t)\big) & t \in (0, T) \\
q(0) \in \partial \psi_1\big(p(0)\big) \\
-p(T) \in \partial \psi_2\big(q(T)\big).
\end{cases}
\tag{1.56}
$$

In other words, the Hamiltonian path must connect the graph of $\partial \psi_1$ to the graph of $-\partial \psi_2$, which are typical Lagrangian submanifolds in $\mathbf{R}^{2N}$.

Chapter 15: Superposition of interacting self-dual functionals

We consider situations where functionals of the form

$$
I(x) = L_1(A_1 x, -\Lambda_1 x) + \langle A_1 x, \Lambda_1 x \rangle + L_2(A_2 x, -\Lambda_2 x) + \langle A_2 x, \Lambda_2 x \rangle
\tag{1.57}
$$

are self-dual on a Banach space Z, assuming that each $L_i, i = 1, 2$ is a self-dual Lagrangian on the space $X_i \times X_i^*$, and where $(\Lambda_1, \Lambda_2) : Z \to X_1^* \times X_2^*$ and $(A_1, A_2) : Z \to X_1 \times X_2$ are operators on Z that may or may not be linear. Unlike the framework of Chapter 14, the operators $A_1, A_2, \Lambda_1, \Lambda_2$ are not totally independent, and certain compatibility relations between their kernels and ranges are needed for the functional I to become self-dual. One also needs the map $x \to \langle A_1 x, \Lambda_1 x \rangle + \langle A_2 x, \Lambda_2 x \rangle$ to be weakly lower semicontinuous on Z. Under a suitable coercivity condition, I will attain its infimum, which is zero, at a point $\bar{x}$ that solves the system

$$
\begin{cases}
0 \in \Lambda_1 \bar{x} + \overline{\partial} L_1(A_1 \bar{x}) \\
0 \in \Lambda_2 \bar{x} + \overline{\partial} L_2(A_2 \bar{x}).
\end{cases}
\tag{1.58}
$$

This applies for example, to Laplace's equation involving advection terms and nonlinear boundary conditions, but also nonlinear Cauchy-Riemann equations on a bounded domain $\Omega \subset \mathbf{R}^2$, of the type

$$
\begin{cases}
\left(\frac{\partial u}{\partial x} - \frac{\partial v}{\partial y}, \frac{\partial v}{\partial x} + \frac{\partial u}{\partial y} \right) \in \partial \varphi \left(\frac{\partial v}{\partial y} - \frac{\partial u}{\partial x}, -\frac{\partial v}{\partial x} - \frac{\partial u}{\partial y} \right) \\
\qquad u_{|\partial \Omega} \in \partial \psi \left(n_x \frac{\partial v}{\partial y} - n_y \frac{\partial v}{\partial x} \right),
\end{cases}
\tag{1.59}
$$

where φ (resp., ψ) are convex functions on $\mathbf{R}^2$ (resp., $\mathbf{R}$).

Part IV: Perturbations of self-dual systems

Hamiltonian systems of PDEs, nonlinear Schrödinger equations, and Navier-Stokes evolutions can be written in the form

$$
0 \in Au + \Lambda u + \overline{\partial} L(u),
\tag{1.60}
$$

where L is a self-dual Lagrangian on $X \times X^*$, $A : D(A) \subset X \to X^*$ is a linear – possibly unbounded – operator, and $\Lambda : D(\Lambda) \subset X \to X^*$ is a not necessarily linear map. They can be solved by minimizing the functionals

$$I(u) = L(u, -Au - \Lambda u) + \langle Au + \Lambda u, u \rangle$$

on X and showing that their infimum is attained and is equal to zero. However, such functionals are not automatically self-dual functionals on their spaces of definition, as we need to deal with the difficulties arising from the superposition of the operators A and Λ. We are often led to use the linear operator A to strengthen the topology on X by defining a new energy space $D(A)$ equipped with the norm $\|u\|_{D(A)}^2 = \|u\|_X^2 + \|Au\|_X^2$. In some cases, this closes the domain of A and increases the chance for Λ to be regular on $D(A)$, but may lead to a loss of strong coercivity on the new space. We shall present in this part two situations where compactness can be restored without altering the self-duality of the system:

- If Λ is linear and is *almost orthogonal to* A in a sense to be made precise in Chapter 16, one may be able to add to I another functional J in such a way that $\tilde{I} = I + J$ is self-dual and coercive. This is applied in the next chapter when dealing with Hamiltonian systems of PDEs.
- The second situation is when the functional I satisfies what we call *the self-dual Palais-Smale property* on the space $D(A)$, a property that is much weaker than the strong coercivity required in Part III. This method is applied in Chapter 18 to deal with Navier-Stokes and other nonlinear evolutions.

Chapter 16: Hamiltonian systems of PDEs

While dealing with Hamiltonian systems of PDEs, we encounter the standard difficulty arising from the fact that – unlike the case of finite-dimensional systems – the cross product $u \to \int_0^T \langle u(t), J\dot{u}(t) \rangle \, dt$ is not necessarily weakly continuous on the Sobolev space $H_X^1[0,T]$ of all absolutely continuous paths valued in an infinite-dimensional Hilbert space $X := H \times H$. Such systems can often be written in the form

$$J\dot{u}(t) + J\mathscr{A}u(t) \in \overline{\partial}L(t, u(t)),$$

where J is the symplectic operator, $\mathscr{A}$ is an unbounded linear operator on X, and L is a time-dependent self-dual Lagrangian on $[0,T] \times X \times X$. The idea is to use the linear operator $\mathscr{A}$ to strengthen the topology on X by considering the space $D(\mathscr{A})$ equipped with the norm $\|u\|_{D(\mathscr{A})}^2 = \|u\|_X^2 + \|\mathscr{A}u\|_X^2$, and a corresponding path space $\mathscr{W}[0,T]$. The operator $\Lambda u = J\dot{u} + J\mathscr{A}u$ becomes regular on the new path space, but the functional

$$I(u) = \mathscr{L}(u, J\dot{u} + J\mathscr{A}u) - \langle u, J\dot{u} + J\mathscr{A}u \rangle, \tag{1.61}$$

where $\mathscr{L}$ is given by formula (1.28), may cease to be coercive. We propose here a way to restore coercivity by perturbing the functional I without destroying self-duality. It can be used because $J\dot{u}$ is *almost orthogonal to* $J\mathscr{A}$ in a sense described below. In this case, one adds to I another functional J in such a way that $\tilde{I} = I + J$ is self-dual and coercive on $\mathscr{W}[0, T]$. This will be applied to deal with Hamiltonian systems of PDEs such as

$$\begin{cases} -\dot{v}(t) - \Delta(v + u) + b.\nabla v = \partial \varphi_1(t, u), \\ \dot{u}(t) - \Delta(u + v) + a.\nabla u = \partial \varphi_2(t, v), \end{cases} \tag{1.62}$$

with Dirichlet boundary conditions, as well as

$$\begin{cases} -\dot{v}(t) + \Delta^2 v - \Delta v = \partial \varphi_1(t, u), \\ \dot{u}(t) + \Delta^2 u + \Delta u = \partial \varphi_2(t, v), \end{cases}$$

with Navier state-boundary conditions, and where $\varphi_i, i = 1, 2$ are convex functions on some L^p-space.

Chapter 17: The self-dual Palais-Smale condition for noncoercive functionals

We extend the nonlinear variational principle for self-dual functionals of the form $I(x) = L(x, -\Lambda x) + \langle x, \Lambda x \rangle$ to situations where I does not satisfy the strong coercivity condition required in (1.46) but the weaker notion of a *self-dual Palais-Smale property* on the functional I. This says that a sequence $(u_n)_n$ is bounded in X, provided it satisfies

$$\Lambda u_n + \overline{\partial} L(u_n) = -\varepsilon_n D u_n \tag{1.63}$$

for some $\varepsilon_n \to 0$. Here $D : X \to X^*$ is the duality map $\langle Du, u \rangle = \|u\|^2$.

This is often relevant when dealing with the superposition of an unbounded linear operator $A : D(A) \subset X \to X^*$ with the possibly nonlinear map Λ in an equation of the form

$$0 \in Au + \Lambda u + \overline{\partial} L(u) \tag{1.64}$$

by minimizing the functional $I(u) = L(u, -Au - \Lambda u) + \langle u, Au + \Lambda u \rangle$. Unlike in the previous chapter, we consider here the case where A is either positive or skew-adjoint (possibly modulo a boundary operator). This is particularly relevant for the resolution of nonlinear evolution equations and will be considered in detail in the next chapter.

Chapter 18: Navier-Stokes and other nonlinear self-dual evolution equations

The nonlinear self-dual variational principle established in Chapter 12, though sufficient and readily applicable in many stationary nonlinear partial differential equations, does not, however, cover the case of nonlinear evolutions such as those of Navier-Stokes (1.3). One of the reasons is the prohibitive coercivity condition that is not satisfied by the corresponding self-dual functional on the relevant path space. We show here that such a principle still holds for functionals of the form

$$I(u) = \int_0^T \left[L(t, u, -\dot{u} - \Lambda u) + \langle \Lambda u, u \rangle \right] dt + \ell \left(u(0) - u(T), -\frac{u(T) + u(0)}{2} \right),$$

where L (resp., ℓ) is a self-dual Lagrangian on state space (resp., boundary space) and Λ is an appropriate nonlinear operator on path space. As a consequence, we provide a variational formulation and resolution to evolution equations involving nonlinear operators, including those of Navier-Stokes (in dimensions 2 and 3), with various boundary conditions. In dimension 2, we recover the well-known solutions for the corresponding initial-value problem as well as periodic and antiperiodic ones, while in dimension 3 we get Leray weak solutions for the initial-value problems but also solutions satisfying $u(0) = \alpha u(T)$ for any given α in $(-1, 1)$. The approach is quite general and applies to certain nonlinear Schrödinger equations and many other evolutions.

Final remarks: Before we conclude this introduction, we emphasize that this book is focused on questions of existence of solutions for a class of PDEs once they have been formulated as functional equations in suitable energy spaces. It is therefore solely concerned with "weak solutions", which just means here that they belong to whatever apriori function space was considered suitable for our proposed variational setting. This does not preclude the fact – not discussed here – that self-duality may also prove useful in establishing regularity results.

We do not address questions of uniqueness, but it is important to observe that an immediate consequence of our approach is that – apart from very degenerate cases – all completely self-dual systems have unique solutions (at least in the function spaces where they are defined). This is simply because they were obtained as minima of self-dual convex functionals. This is not, however, the case for general self-dual functionals, and a more thorough analysis is needed for each separate case.

Many of the equations in the examples we address here are known to have solutions via other methods. We did not, however, make a serious attempt at tracking their history and therefore could not credit their original authors. This may be regretful, though we did not find it essential to this project, whose objective is simply to establish the efficacy, versatility, and unifying features of a new approach for proving existence results.

Missing from this volume are the following thrusts of current – and potential – research areas related to this self-dual approach to PDEs.

1. The computational advantages of self-duality in problems as basic as those dealing with numerical resolutions of nonsymmetric linear systems of equations (Ghoussoub and Moradifam [66]). We also refer to a somewhat related variational point of view in the case of evolution equations, considered recently by Mielke, Stefanelli, and their collaborators. See for example [71], [99], [102], [108], [112], [145], [150], [152].
2. The relevance of the self-dual approach in the introduction of a *penalty method* in nonlinear inverse problems, as pioneered by Barbu and Kunisch [20] in the case of gradient of convex functions, and extended recently by Ghoussoub [60] and Zaraté [163] to more general monotone nonlinearities.

3. The potential of extending self-duality to certain infinite dimensional manifolds such as the Wasserstein space, so as to give a variational resolution for gradient flows of geodesically convex energies, a topic recently addressed in the ground-breaking book of Ambrosio, N. Gigli, and G. Savare [2].
4. The pertinence of the self-dual approach to second order differential equations [92], and to parabolic equations with measure data [90].
5. The need to develop a self-dual min-max variational approach to deal with the potential of higher critical levels and the multiplicity of solutions for certain semilinear superquadratic elliptic equations involving advection terms.

Finally, we note that while all necessary concepts from functional and convex analysis are spelled out in this book, the same cannot be said unfortunately about the material needed in all of the 53 examples of applications that are included herein. For example, the reader is expected to be somewhat familiar with the basic theory of Sobolev spaces, vector-valued or not, and with their various embeddings into classical L^p-spaces and/or spaces of continuous functions. For this material, we refer the reader to the books of Adams [1], Brézis [26], Evans [48], Gilbarg and Trudinger [70], Mawhin and Willem [96], and Maz'ja [97]. We also refer to the books of Aubin and Cellina [7], Barbu [17] [18], Browder [33], Ekeland [46], Pazy [127], and Showalter [144], for related topics on evolution equations, Hamiltonian systems and general differential inclusions. The books of Dautray and Lions [41], Kinderlehrer and Stampachia [81], Lions and Magenes [88], Roubíček [139], Struwe [153], and Temam [157] are also excellent sources of material related to our chosen examples of applications to partial differential equations.

Glossary of notation

The following list of notation and abbreviations will be used throughout this book. Let Ω be a smooth domain of $\mathbf{R}^n$, and let X be a Banach space.

1. $C^\infty(\Omega, \mathbf{R}^n)$ (resp., $C_0^\infty(\Omega, \mathbf{R}^n)$ will denote the space of infinitely differentiable functions (resp., the space of infinitely differentiable functions with compact support) on Ω.
2. For $1 \le p < +\infty$, $L_X^p(\Omega)$ will be the space of all Bochner-integrable functions $u : \Omega \to X$ with norm

$$\|u\|_{L_X^p} = \left(\int_\Omega \|u(x)\|_X^p dx \right)^{\frac{1}{p}}.$$

3. For $1 \le p < +\infty$, the space $W^{1,p}(\Omega)$ (resp., $W_0^{1,p}(\Omega)$ is the completion of $C^\infty(\Omega, \mathbf{R}^n)$ (resp., $C_0^\infty(\Omega, \mathbf{R}^n)$) for the norm

$$\|u\|_{W^{1,p}(\Omega)} = \|u\|_p + \|\nabla u\|_p \text{ (resp., } \|u\|_{W_0^{1,p}(\Omega)} = \|\nabla u\|_p).$$

We denote the dual of $W^{1,p}(\Omega)$ by $W^{1,-p}(\Omega)$.

4. $W^{1,2}(\Omega)$ (resp., $W_0^{1,2}(\Omega)$ will be denoted by $H^1(\Omega)$ (resp., $H_0^1(\Omega)$) and the dual of $H_0^1(\Omega)$ will be denoted by $H^{-1}(\Omega)$.

5. Suppose now that H is a Hilbert space, $0 < T < \infty$, and $1 < p < +\infty$. We shall consider the space $W^{1,p}([0,T];H)$ of all functions $u : [0,T] \to H$ such that there exists $v \in L_H^p[0,T]$ with the property that, for all $t \in [0,T]$

$$u(t) = u(0) + \int_0^t v(s)\,ds.$$

In this case u is an absolutely continuous function, it is almost everywhere differentiable with $\dot{u} = v$ a.e. on $(0,T)$, and

$$\lim_{h \to 0} \int_0^{T-h} \left\| \frac{u(t+h) - u(t)}{h} - \dot{u}(t) \right\|_H^p dt = 0.$$

See for example the appendix of [25]. The space $W^{1,p}([0,T];H)$ is then equipped with the norm

$$\|u\|_{W^{1,p}([0,T];H)} = (\|u\|_{L_H^p[0,T]}^p + \|\dot{u}\|_{L_H^p[0,T]}^p)^{1/p}.$$

More generally, for any reflexive Banach space X, one can associate the space $W^{1,p}([0,T];X) = \{u : [0,T] \to X; \dot{u} \in L_X^2[0,T]\}$ equipped with the norm

$$\|u\|_{W^{1,p}([0,T];X)} = (\|u\|_{L_X^p}^p + \|\dot{u}\|_{L_X^p}^p)^{1/p}.$$

For simplicity, we shall often denote the space $W^{1,2}([0,T];X)$ by $A_X^2[0,T]$, or simply A_X^2, if there is no ambiguity as to the time interval.

6. An important framework for evolution equations involving PDEs is the so-called *evolution triple* setting. It consists of a Hilbert space H with $\langle , \rangle_H$ as scalar product, a reflexive Banach space X, and its dual X^* in such a way that $X \subset H \subset X^*$, with X being a dense vector subspace of H, while the canonical injection $X \to H$ is continuous. In this case, one identifies the Hilbert space H with its dual H^* and injects it in X^* in such a way that

$$\langle h, u \rangle_{X^*,X} = \langle h, u \rangle_H \quad \text{for all } h \in H \text{ and all } u \in X.$$

This injection is continuous and one-to-one, and H is also dense in X^*. In other words, the dual X^* of X is represented as the completion of H for the dual norm

$$\|h\| = \sup\{\langle h, u \rangle_H; \|u\|_X \leq 1\}.$$

One can then associate the space $W^{1,2}([0,T];X,H)$ of all functions $u \in L_X^2[0,T]$ such that $\dot{u} \in L_{X^*}^2[0,T]$ equipped with the norm

$$\|u\|_{\mathscr{X}_{2,2}[0,T]} = (\|u\|_{L_X^2}^2 + \|\dot{u}\|_{L_{X^*}^2}^2)^{1/2}.$$

For simplicity, we shall often denote this space by $\mathscr{X}_{2,2}[0,T]$, or even $\mathscr{X}_{2,2}$.
7. More generally, we may consider for $1 < p < \infty$ and $\frac{1}{p} + \frac{1}{q} = 1$ the space

$$\mathscr{X}_{p,q}[0,T] := W^{1,p}([0,T];X,H) = \{u;\, u \in L_X^p[0,T], \dot{u} \in L_{X^*}^q[0,T]\}$$

equipped with the norm

$$\|u\|_{W^{1,p}} = \|u\|_{L_X^p[0,T]} + \|\dot{u}\|_{L_{X^*}^q[0,T]},$$

which then leads to a continuous injection $\mathscr{X}_{p,q}[0,T] \subseteq C([0,T];H)$, the latter being the space of continuous functions $u : [0,T] \to H$ equipped with the supremum
norm $\|u\| = \sup_{t \in [0,T]} \|u(t)\|_H$. Moreover, for any pair u,v in $\mathscr{X}_{p,q}[0,T]$, the function
$t \to \langle u(t), v(t) \rangle_H$ is absolutely continuous on $[0,T]$, and for a.e. $t \in [0,T]$ we have

$$\frac{d}{dt} \langle u(t), v(t) \rangle_H = \langle \dot{u}(t), v(t) \rangle_{X^*,X} + \langle u(t), \dot{v}(t) \rangle_{X,X^*}.$$

Finally, if the inclusion $X \to H$ is compact, then $\mathscr{X}_{p,q}[0,T] \to L_H^p[0,T]$ is also
compact. For details, we refer the reader to Evans [48], or Temam [156].

Part I
CONVEX ANALYSIS ON PHASE SPACE

A large class of stationary and dynamic partial differential equations – which will be called *completely self-dual differential systems* – can be written in the form

$$(p,x) \in \partial L(x,p),$$

where ∂L is the subdifferential of a real-valued convex self-dual Lagrangian L on phase space $X \times X^*$, with X being an infinite dimensional reflexive Banach space, and X^* its conjugate. Part I of this volume starts with a recollection of the classical basic concepts and relevant tools of convex analysis that will be used throughout this book, in particular Fenchel-Legendre duality and its relationship with the notion of sub-differentiability of convex functions.

The required notions of convex analysis on phase space are introduced, with a special focus on the basic permanence properties of the class of self-dual Lagrangians and their corresponding self-dual vector fields, including their stability under sums, convolutions, tensor products, iterations, compositions with skew-adjoint operators, superpositions with appropriate boundary Lagrangians, as well as appropriate liftings to various path spaces.

Chapter 2
Legendre-Fenchel Duality on Phase Space

We start by recalling the basic concepts and relevant tools of convex analysis that will be used throughout the book. In particular, we review the Fenchel-Legendre duality and its relationship with subdifferentiability. The material of the first four sections is quite standard and does not include proofs, which we leave and recommend to the interested reader. They can actually be found in most books on convex analysis, such as those of Brézis [26], Ekeland and Temam [47], Ekeland [46], and Phelps [130].

Our approach to evolution equations and partial differential systems, however, is based on convex calculus on "phase space" $X \times X^*$, where X is a reflexive Banach space and X^* is its dual. We shall therefore consider Lagrangians on $X \times X^*$ that are convex and lower semicontinuous in both variables. All elements of convex analysis will apply, but the calculus on state space becomes much richer for many reasons, not the least of which is the possibility of introducing associated Hamiltonians, which are themselves Legendre conjugates but in only one variable.

Another reason for the rich structure will become more evident in the next chapter where the abundance of natural automorphisms on phase space and their interplay with the Legendre transform becomes an essential ingredient of our self-dual variational approach.

2.1 Basic notions of convex analysis

Definition 2.1. A function $\varphi : X \to \mathbf{R} \cup \{+\infty\}$ on a Banach space X is said to be:

1. *lower semicontinuous (weakly lower semicontinuous)* if, for every $r \in \mathbf{R}$, its *epigraph* $\mathrm{Epi}(\varphi) := \{(x,r) \in X \times \mathbf{R}; \varphi(x) \le r\}$ is closed for the norm topology (resp., weak topology) of $X \times \mathbf{R}$, which is equivalent to saying that whenever (x_α) is a net in X that converges strongly (resp., weakly) to x, then $f(x) \le \liminf_\alpha f(x_\alpha)$.
2. *convex* if, for every $r \in \mathbf{R}$, its *epigraph* $\mathrm{Epi}(\varphi)$ is a convex subset of $X \times \mathbf{R}$, which is equivalent to saying that $f(\lambda x + (1-\lambda)y) \le \lambda f(x) + (1-\lambda)f(y)$ for any $x,y \in X$ and $0 \le \lambda \le 1$.

N. Ghoussoub, *Self-dual Partial Differential Systems and Their Variational Principles*,
Springer Monographs in Mathematics, DOI 10.1007/978-0-387-84897-6_2,
© Springer Science+Business Media, LLC 2009

3. *proper* if its effective domain (i.e., the set $\text{Dom}(\varphi) = \{x \in X; \varphi(x) < +\infty\}$) is nonempty, the effective domain being convex whenever φ is convex.

We shall denote by $\mathscr{C}(X)$ the class of convex lower semicontinuous functions on a Banach space X.

Operations on convex lower semicontinuous functions

Consider φ and ψ to be two functions in $\mathscr{C}(X)$. Then,

1. The functions $\varphi + \psi$ and $\lambda \varphi$ when $\lambda \geq 0$ are also in $\mathscr{C}(X)$.
2. The function $x \to \max\{\varphi(x), \psi(x)\}$ is in $\mathscr{C}(X)$.
3. The inf-convolution $x \to \varphi \star \psi(x) := \inf\{\varphi(y) + \psi(x - y); y \in X\}$ is convex. If φ and ψ are bounded below, then $\varphi \star \psi$ is in $\mathscr{C}(X)$ and $\text{Dom}(\varphi \star \psi) = \text{Dom}(\varphi) + \text{Dom}(\psi)$. Moreover, $\varphi \star \psi$ is continuous at a point $x \in X$ if either φ or ψ is continuous at x.
4. If $\rho \in \mathscr{C}(\mathbf{R})$, then $x \to \rho(\|x\|_X)$ is in $\mathscr{C}(X)$.

Convex functions enjoy various remarkable properties that make them agreeable to use in variational problems. We now summarize some of them.

Proposition 2.1. *If $\varphi : X \to \mathbf{R} \cup \{+\infty\}$ is a convex function on a Banach space X, then:*

1. *φ is lower semicontinuous if and only if it is weakly lower semicontinuous, in which case it is the supremum of all continuous affine functions below it.*
2. *If φ is a proper convex lower semicontinuous function on X, then it is continuous on the interior D of its effective domain, provided it is nonempty.*

We shall often use the immediate implication stating that any convex lower semicontinuous function that is finite on the unit ball of X is necessarily continuous. However, one should keep in mind that there exist continuous and convex functions on Hilbert space that are not bounded on the unit ball [130].

2.2 Subdifferentiability of convex functions

Definition 2.2. Let $\varphi : X \to \mathbf{R} \cup \{+\infty\}$ be a convex lower semicontinuous function on a Banach space X. Define the *subdifferential $\partial \varphi$ of φ* to be the following set-valued function: If $x \in \text{Dom}(\varphi)$, set

$$\partial \varphi(x) = \{p \in X^*; \langle p, y - x \rangle \leq \varphi(y) - \varphi(x) \text{ for all } y \in X\}, \qquad (2.1)$$

and if $x \notin \text{Dom}(\varphi)$, set $\partial \varphi(x) = \emptyset$.

The subdifferential $\partial \varphi(x)$ is a closed convex subset of the dual space X^*. It can, however, be empty even though $x \in \text{Dom}(\varphi)$, and we shall write

$$\text{Dom}(\partial\varphi) = \{x \in X; \partial\varphi(x) \neq \emptyset\}. \tag{2.2}$$

An application of the celebrated Bishop-Phelps theorem due to Brondsted and Rock-afellar (see [130]) however yields the following useful result.

Proposition 2.2. *Let φ be a proper convex lower semicontinuous function on X. Then,*

1. $\text{Dom}(\partial\varphi)$ *is dense in* $\text{Dom}(\varphi)$.
2. *Moreover,* $\partial\varphi(x) \neq \emptyset$ *at any point x in the interior of* $\text{Dom}(\varphi)$ *where φ is continuous.*

If $x \in \text{Dom}(\varphi)$, we define the more classical notion $d^+\varphi(x)$ of a "right-derivative" at x as

$$\langle d^+\varphi(x), y\rangle := \lim_{t\to 0^+} \tfrac{1}{t}\big(\varphi(x+ty) - \varphi(x)\big) \text{ for any } y \in X. \tag{2.3}$$

The relationship between the two types of derivatives is given by

$$p \in \partial\varphi(x) \text{ if and only if } \langle p, y\rangle \leq \langle d^+\varphi(x), y\rangle \text{ for any } y \in X. \tag{2.4}$$

Now φ is said to be *Gâteaux-differentiable* at a point $x \in \text{Dom}(\varphi)$ if there exists $p \in X^*$, which will be denoted by $D_G\varphi(x)$ such that

$$\langle p, y\rangle = \lim_{t\to 0} \tfrac{1}{t}\big(\varphi(x+ty) - \varphi(x)\big) \text{ for any } y \in X. \tag{2.5}$$

It is then easy to see the following relationship between the two notions.

Proposition 2.3. *Let φ be a convex function on X.*

1. *If φ is Gâteaux-differentiable at a point $x \in \text{Dom}(\varphi)$, then $\partial\varphi(x) = \{D_G\varphi(x)\}$.*
2. *Conversely, if φ is continuous at $x \in \text{Dom}(\varphi)$, and if the subdifferential of φ at x is single valued, then $\partial\varphi(x) = \{D_G\varphi(x)\}$.*

Subdifferentials satisfy the following calculus.

Proposition 2.4. *Let φ and ψ be in $\mathscr{C}(X)$ and $\lambda \geq 0$. We then have the following properties:*

1. $\partial(\lambda\varphi)(x) = \lambda\partial\varphi(x)$ *and* $\partial\varphi(x) + \partial\psi(x) \subset \partial(\varphi+\psi)(x)$ *for any $x \in X$.*
2. *Moreover, equality $\partial\varphi(x) + \partial\psi(x) = \partial(\varphi+\psi)(x)$ holds at a point $x \in \text{Dom}(\varphi) \cap \text{Dom}(\psi)$, provided either φ or ψ is continuous at x.*
3. *If $A : Y \to X$ is a bounded linear operator from a Banach space Y into X, and if φ is continuous at some point in $R(A) \cap \text{Dom}(\varphi)$, then $\partial(\varphi \circ A)(y) = A^*\partial\varphi(Ay)$ for every point $y \in Y$.*

As a set-valued map, the subdifferential has the following useful properties.

Definition 2.3. A subset G of $X \times X^*$ is said to be

1. *monotone,* provided

$$\langle x - y, p - q\rangle \geq 0 \text{ for every } (x, p) \text{ and } (y, p) \text{ in } G. \tag{2.6}$$

2. *maximal monotone* if it is maximal in the family of monotone subsets of $X \times X^*$ ordered by set inclusion, and
3. *cyclically monotone*, provided that for any finite number of points $(x_i, p_i)_{i=0}^n$ in G with $x_0 = x_n$, we have

$$\sum_{k=1}^n \langle p_k, x_k - x_{k-1} \rangle \geq 0. \tag{2.7}$$

A set-valued map $T : X \to 2^{X^*}$ is then said to be *monotone* (resp., *maximal monotone*) (resp., *cyclically monotone*), provided its graph $G(T) = \{(x, p) \in X \times X^*; p \in T(x)\}$ is *monotone* (resp., *maximal monotone*) (resp., *cyclically monotone*).

The following result was established by Rockafellar. See for example [130].

Theorem 2.1. *Let* $\varphi : X \to \mathbf{R} \cup \{+\infty\}$ *be a proper convex and lower semicontinuous functional on a Banach space X. Then, its differential map $x \to \partial \varphi(x)$ is a maximal cyclically monotone map.*

Conversely, if $T : X \to 2^{X^*}$ is a maximal cyclically monotone map with a nonempty domain, then there exists a proper convex and lower semicontinuous functional on X such that $T = \partial \varphi$.

2.3 Legendre duality for convex functions

Let $\varphi : X \to \mathbf{R} \cup \{+\infty\}$ be any function. Its Fenchel-Legendre dual is the function φ^* on X^* given by

$$\varphi^*(p) = \sup\{\langle x, p \rangle - \varphi(x); x \in X\}. \tag{2.8}$$

Proposition 2.5. *Let* $\varphi : X \to \mathbf{R} \cup \{+\infty\}$ *be a proper function on a reflexive Banach space. The following properties then hold:*

1. φ^* *is a proper convex lower semicontinuous function from X^* to $\mathbf{R} \cup \{+\infty\}$.*
2. $\varphi^{**} := (\varphi^*)^* : X \to \mathbf{R} \cup \{+\infty\}$ *is the largest convex lower semicontinuous function below φ. Moreover, $\varphi = \varphi^{**}$ if and only if φ is convex and lower semicontinuous on X.*
3. *For every $(x, p) \in X \times X^*$, we have $\varphi(x) + \varphi^*(p) \geq \langle x, p \rangle$, and the following are equivalent:*

   i) $\varphi(x) + \varphi^*(p) = \langle x, p \rangle$,
   ii) $p \in \partial \varphi(x)$,
   iii) $x \in \partial \varphi^*(p)$.

Proposition 2.6. *Legendre duality satisfies the following rules:*

1. $\varphi^*(0) = -\inf_{x \in X} \varphi(x)$.
2. *If $\varphi \leq \psi$, then $\varphi^* \geq \psi^*$.*

3. We have $(\inf_{i \in I} \varphi_i)^* = \sup_{i \in I} \varphi_i^*$ and $(\sup_{i \in I} \varphi_i)^* \leq \inf_{i \in I} \varphi_i^*$ whenever $(\varphi_i)_{i \in I}$ is a family of functions on X.

4. For every $\lambda > 0$, $(\lambda \varphi)^*(p) = \lambda \varphi^*(\frac{1}{\lambda} p)$.

5. For every $\alpha \in \mathbf{R}$, $(\varphi + \alpha)^* = \varphi^* - \alpha$.

6. For a fixed $a \in X$, we have, for every $p \in X^*$, $\varphi_a^*(p) = \varphi^*(p) + \langle a, p \rangle$, where $\varphi_a(x) := \varphi(x - a)$.

7. If ρ is an even function in $\mathscr{C}(\mathbf{R})$, then the Legendre transform of $\varphi(x) = \rho(\|u\|_X)$ is $\varphi^*(p) = \rho^*(\|p\|_{X^*})$. In particular, if $\varphi(x) = \frac{1}{\alpha} \|x\|_X^\alpha$, then $\varphi^*(p) = \frac{1}{\beta} \|p\|_{X^*}^\beta$, where $\frac{1}{\alpha} + \frac{1}{\beta} = 1$.

8. If φ and ψ are proper functions, then $(\varphi \star \psi)^* = \varphi^* + \psi^*$.

9. Conversely, if $\mathrm{Dom}(\varphi) - \mathrm{Dom}(\psi)$ contains a neighborhood of the origin, then $(\varphi + \psi)^* = \varphi^* \star \psi^*$.

10. Let $A : D(A) \subset X \to Y$ be a linear operator with a closed graph, and let $\varphi : Y \to \mathbf{R} \cup \{+\infty\}$ be a proper function in $\mathscr{C}(Y)$. Then, the dual of the function φ_A defined on X as $\varphi_A(x) = \varphi(Ax)$ if $x \in D(A)$ and $+\infty$ otherwise, is

$$\varphi_A^*(p) = \inf\{\varphi^*(q); A^* q = p\}.$$

11. Let $h(x) := \inf\{F(x_1, x_2); x_1, x_2 \in X, x = \frac{1}{2}(x_1 + x_2)\}$, where F is a function on $X \times X$. Then, $h^*(p) = F^*(\frac{p}{2}, \frac{p}{2})$ for every $p \in X^*$.

12. Let g be the function on $X \times X$ defined by $g(x_1, x_2) = \|x_1 - x_2\|^2$. Then, $g^*(p_1, p_2) = \frac{1}{4} \|p_1\|^2$ if $p_1 + p_2 = 0$ and $+\infty$ otherwise.

The following lemma will be useful in Chapter 5. It can be used to interpolate between convex functions, and is sometimes called the *proximal average*.

Lemma 2.1. *Let $f_1, f_2 : X \to \mathbf{R} \cup \{+\infty\}$ be two convex lower semicontinuous functions on a reflexive Banach space X. The Legendre dual of the function h defined for $X \in X$ by*

$$h(x) := \inf\left\{\frac{1}{2} f_1(x_1) + \frac{1}{2} f_2(x_2) + \frac{1}{8} \|x_1 - x_2\|^2; x_1, x_2 \in X, x = \frac{1}{2}(x_1 + x_2)\right\}$$

is given by the function h^* defined for $p \in X^*$ by

$$h^*(p) = \inf\left\{\frac{1}{2} f_1^*(p_1) + \frac{1}{2} f_2^*(p_2) + \frac{1}{8} \|p_1 - p_2\|^2; p_1, p_2 \in X^*, p = \frac{1}{2}(p_1 + p_2)\right\}.$$

Proof. Note that

$$h(x) := \inf\left\{F(x_1, x_2); x_1, x_2 \in X, x = \frac{1}{2}(x_1 + x_2)\right\},$$

where F is the function on $X \times X$ defined as $F(x_1, x_2) = g_1(x_1, x_2) + g_2(x_1, x_2)$ with

$$g_1(x_1, x_2) = \frac{1}{2} f_1(x_1) + \frac{1}{2} f_2(x_2) \quad \text{and} \quad g_2(x_1, x_2) = \frac{1}{8} \|x_1 - x_2\|^2.$$

It follows from rules (10) and (7) in Proposition 2.6 that

$$h^*(p) = F^*\left(\frac{p}{2},\frac{p}{2}\right) = (g_1 + g_2)^*\left(\frac{p}{2},\frac{p}{2}\right) = g_1^* \star g_2^*\left(\frac{p}{2},\frac{p}{2}\right).$$

It is easy to see that

$$g_1^*(p_1,p_2) = \frac{1}{2}f_1^*\left(\frac{p_1}{2}\right) + \frac{1}{2}f_2^*\left(\frac{p_2}{2}\right),$$

while rule (11) of Proposition 2.6 gives that

$$g_2^*(p_1,p_2) = 2\|p_1\|^2 \quad \text{if } p_1 + p_2 = 0 \quad \text{and} \quad +\infty \text{ otherwise.}$$

It follows that

$$\begin{aligned}
h^*(p) &= g_1^* \star g_2^*\left(\frac{p}{2},\frac{p}{2}\right) \\
&= \inf\left\{\frac{1}{2}f_1^*\left(\frac{p_1}{2}\right) + \frac{1}{2}f_2^*\left(\frac{p_2}{2}\right) + 2\left\|\frac{p}{2} - \frac{p_1}{4}\right\|^2; p_1,p_2 \in X^*, p = p_1 + p_2\right\} \\
&= \inf\left\{\frac{1}{2}f_1^*(q_1) + \frac{1}{2}f_2^*(q_2) + 2\left\|\frac{p}{2} - \frac{q_1}{2}\right\|^2; q_1,q_2 \in X^*, p = \frac{1}{2}(q_1 + q_2)\right\} \\
&= \inf\left\{\frac{1}{2}f_1^*(q_1) + \frac{1}{2}f_2^*(q_2) + \frac{1}{8}\|q_2 - q_1\|^2; q_1,q_2 \in X^*, p = \frac{1}{2}(q_1 + q_2)\right\}.
\end{aligned}$$

The following theorem can be used to prove rule (8) in Proposition 2.6. It will also be needed in what follows.

Theorem 2.2 (Fenchel and Rockafellar). *Let φ and ψ be two convex functions on a Banach space X such that φ is continuous at some point $x_0 \in \text{Dom}(\varphi) \cap \text{Dom}(\psi)$. Then,*

$$\inf_{x \in X}\{\varphi(x) + \psi(x)\} = \max_{p \in X^*}\{-\varphi^*(-p) - \psi^*(p)\}. \tag{2.9}$$

The theorem above holds, for example, whenever $\text{Dom}(\varphi) - \text{Dom}(\psi)$ contains a neighborhood of the origin or more generally if the set $\text{IntDom}(\varphi) \cap \text{Dom}(\psi)$ is nonempty.

The following simple lemma will be used often throughout this text. Its proof is left as an exercise.

Lemma 2.2. *If $\varphi : X \mapsto \mathbf{R} \cup \{+\infty\}$ is a proper convex and lower semicontinuous functional on a Banach space X such that $-A \le \varphi(y) \le \frac{B}{\alpha}\|y\|_Y^\alpha + C$ with $A \ge 0$, $C \ge 0$, $B > 0$, and $\alpha > 1$, then for every $p \in \partial\varphi(y)$*

$$\|p\|_{X^*} \le \left\{\alpha B^{\frac{\beta}{\alpha}}(\|y\|_X + A + C) + 1\right\}^{\alpha - 1}. \tag{2.10}$$

2.4 Legendre transforms of integral functionals

Let Ω be a Borel subset of $\mathbf{R}^n$ with finite Lebesgue measure, and let X be a separable reflexive Banach space. Consider a bounded below function $\varphi : \Omega \times X \to \mathbf{R} \cup \{+\infty\}$ that is measurable with respect to the σ-field generated by the products of Lebesgue sets in Ω and Borel sets in X. We can associate to φ a functional Φ defined on $L^\alpha(\Omega, X)$ $(1 \le \alpha \le +\infty)$ via the formula

$$\Phi(x) = \int_\Omega \varphi(\omega, x(\omega))\, d\omega,$$

where $x \in L^\alpha(\Omega, X)$. We now relate the Legendre transform and subdifferential of φ as a function of its second variable on X to the Legendre transform and subdifferential of Φ as a function on $L^\alpha(\Omega, X)$. We shall use the following obvious notation. For $\omega \in \Omega$, $x \in X$, and $p \in X^*$,

$$\varphi^*(\omega, p) = \varphi(\omega, \cdot)^*(p) \quad \text{and} \quad \partial\varphi(\omega, x) = \partial\varphi(\omega, \cdot)(x).$$

The following proposition summarizes the relations between the function φ and "its integral" Φ. A proof can be found in [46].

Proposition 2.7. *Assume X is a reflexive and separable Banach space, that $1 \le \alpha \le +\infty$, $\frac{1}{\alpha} + \frac{1}{\beta} = 1$, and that $\varphi : \Omega \times X \to \mathbf{R} \cup \{+\infty\}$ is jointly measurable such that $\int_\Omega |\varphi^*(\omega, \bar{p}(\omega))|\, d\omega < \infty$ for some $\bar{p} \in L^\beta(\Omega, X)$, which holds in particular if φ is bounded below on $\Omega \times X$.*

1. *If the function $\varphi(\omega, \cdot)$ is lower semicontinuous on X for almost every $\omega \in \Omega$, then Φ is lower semicontinuous on $L^\alpha(\Omega, X)$.*
2. *If $\varphi(\omega, \cdot)$ is convex on X for almost every $\omega \in \Omega$, then Φ is convex on $L^\alpha(\Omega, X)$.*
3. *If $\varphi(\omega, \cdot)$ is convex and lower semicontinuous on X for almost every $\omega \in \Omega$, and if $\Phi(\bar{x}) < +\infty$ for some $\bar{x} \in L^\infty(\Omega, X)$, then the Legendre transform of Φ on $L^\beta(\Omega, X)$ is given by*

$$\Phi^*(p) = \int_\Omega \varphi^*(\omega, p(\omega))\, d\omega \quad \text{for all } p \in L^\beta(\Omega, X). \tag{2.11}$$

4. *If $\int_\Omega |\varphi(\omega, \bar{x}(\omega))|\, d\omega < \infty$ and $\int_\Omega |\varphi^*(\omega, \bar{p}(\omega))|\, d\omega < \infty$ for some $\bar{x}$ and $\bar{p}$ in $L^\infty(\Omega, X)$, then for every $x \in L^\alpha(\Omega, X)$ we have*

$$\partial\Phi(x) = \left\{ p \in L^\beta(\Omega, X);\ p(\omega) \in \partial\varphi(\omega, x(\omega)) \text{ a.e.} \right\}. \tag{2.12}$$

Exercises 2.A. Legendre transforms of energy functionals

1. Review and prove all the statements in Sections 2.1 to 2.4.
2. Let Ω be a bounded smooth domain in $\mathbf{R}^n$, and define on $L^2(\Omega)$ the convex lower semicontinuous functional

$$\varphi(u) = \begin{cases} \frac{1}{2} \int_\Omega |\nabla u|^2 & \text{on} \quad H_0^1(\Omega) \\ +\infty & \text{elsewhere.} \end{cases} \tag{2.13}$$

Show that its Legendre-Fenchel conjugate for the L^2-duality is $\varphi^*(v) = \frac{1}{2} \int_\Omega |\nabla(-\Delta)^{-1}v|^2 dx$ and that its subdifferential $\partial \varphi = -\Delta$ with domain $H_0^1(\Omega) \cap H^2(\Omega)$.

3. Consider the Hilbert space $H^{-1}(\Omega)$ equipped with the norm induced by the scalar product $\langle u, v \rangle_{H^{-1}(\Omega)} = \int_\Omega u(-\Delta)^{-1}v dx$. For $m \geq \frac{n-2}{n+2}$, we have $L^{m+1}(\Omega) \subset H^{-1}$, and so we may consider the functional

$$\varphi(u) = \begin{cases} \frac{1}{m+1} \int_\Omega |u|^{m+1} & \text{on} \quad L^{m+1}(\Omega) \\ +\infty & \text{elsewhere.} \end{cases} \tag{2.14}$$

Show that its Legendre-Fenchel conjugate is $\varphi^*(v) = \frac{m}{m+1} \int_\Omega |(-\Delta)^{-1}v|^{\frac{m+1}{m}} dx$ with subdifferential $\partial \varphi(u) = -\Delta(u^m)$ on $D(\partial \varphi) = \{u \in L^{m+1}(\Omega); u^m \in H_0^1(\Omega)\}$.

4. If $0 < m < 1$, then $(-\Delta)^{-1}u$ does not necessarily map $L^{m+1}(\Omega)$ into $L^{\frac{m+1}{m}}$, and so we consider the space X defined as

$$X = \{u \in L^{m+1}(\Omega); (-\Delta)^{-1}u \in L^{\frac{m+1}{m}}(\Omega)\}$$

equipped with the norm $\|u\|_X = \|u\|_{m+1} + \|(-\Delta)^{-1}u\|_{\frac{m+1}{m}}$. Show that the functional $\varphi(u) = \frac{1}{m+1} \int_\Omega |u|^{m+1}$ is convex and lower semicontinuous on X with Legendre-Fenchel transform equal to

$$\varphi^*(v) = \begin{cases} \frac{m}{m+1} \int_\Omega |(-\Delta)^{-1}v|^{\frac{m+1}{m}} dx & \text{if} \quad (-\Delta)^{-1}v \in L^{\frac{m+1}{m}}(\Omega) \\ +\infty & \text{otherwise.} \end{cases} \tag{2.15}$$

2.5 Legendre transforms on phase space

Let X be a reflexive Banach space. Functions $L : X \times X^* \to \mathbf{R} \cup \{+\infty\}$ on phase space $X \times X^*$ will be called *Lagrangians*, and we shall consider the class $\mathscr{L}(X)$ of those Lagrangians that are proper convex and lower semicontinuous (in both variables). The Legendre-Fenchel dual (in both variables) of L is defined at $(q, y) \in X^* \times X$ by

$$L^*(q, y) = \sup\{\langle q, x \rangle + \langle y, p \rangle - L(x, p); x \in X, p \in X^*\}.$$

The (partial) domains of a Lagrangian L are defined as

$$\mathrm{Dom}_1(L) = \{x \in X; L(x, p) < +\infty \text{ for some } p \in X^*\}$$

and

$$\mathrm{Dom}_2(L) = \{p \in X^*; L(x, p) < +\infty \text{ for some } x \in X\}.$$

To each Lagrangian L on $X \times X^*$, we can define its corresponding Hamiltonian $H_L : X \times X \to \bar{\mathbf{R}}$ (resp., co-Hamiltonian $\tilde{H}_L : X^* \times X^* \to \bar{\mathbf{R}}$) by

$$H_L(x, y) = \sup\{\langle y, p \rangle - L(x, p); p \in X^*\} \text{ and } \tilde{H}_L(p, q) = \sup\{\langle y, p \rangle - L(y, q); y \in X\},$$

which is the Legendre transform in the second variable (resp., first variable). Their domains are

$$\text{Dom}_1(H_L) : = \{x \in X; H_L(x,y) > -\infty \text{ for all } y \in X\}$$
$$= \{x \in X; H_L(x,y) > -\infty \text{ for some } y \in X\}$$

and

$$\text{Dom}_2(\tilde{H}_L) : = \{q \in X^*; \tilde{H}_L(p,q) > -\infty \text{ for all } p \in X^*\}$$
$$= \{q \in X^*; \tilde{H}_L(p,q) > -\infty \text{ for some } p \in X^*\}.$$

It is clear that $\text{Dom}_1(L) = \text{Dom}_1(H_L)$ and $\text{Dom}_2(L) = \text{Dom}_2\tilde{H}_L$.

Remark 2.1. To any pair of proper convex lower semicontinuous functions φ and ψ on a Banach space X, one can associate a Lagrangian on state space $X \times X^*$ via the formula $L(x,p) = \varphi(x) + \psi^*(p)$. Its Legendre transform is then $L^*(p,x) = \psi(x) + \varphi^*(p)$. Its Hamiltonian is $H_L(x,y) = \psi(y) - \varphi(x)$ if $x \in \text{Dom}(\varphi)$ and $-\infty$ otherwise, while its co-Hamiltonian is $\tilde{H}_L(p,q) = \varphi^*(p) - \psi^*(q)$ if $q \in \text{Dom}(\psi^*)$ and $-\infty$ otherwise. The domains are then $\text{Dom}_1 H_L := \text{Dom}(\varphi)$ and $\text{Dom}_2(\tilde{H}_L) := \text{Dom}(\psi^*)$. These Lagrangians will be the building blocks of the variational approach developed in this book.

Operations on Lagrangians

We define on the class of Lagrangians $\mathcal{L}(X)$ the following operations:

Scalar multiplication: If $\lambda > 0$ and $L \in \mathcal{L}(X)$, define the Lagrangian $\lambda \cdot L$ on $X \times X^*$ by

$$(\lambda \cdot L)(x,p) = \lambda^2 L\left(\frac{x}{\lambda}, \frac{p}{\lambda}\right).$$

Addition: If $L, M \in \mathcal{L}(X)$, define the sum $L \oplus M$ on $X \times X^*$ by:

$$(L \oplus M)(x,p) = \inf\{L(x,r) + M(x,p-r); r \in X^*\}.$$

Convolution: If $L, M \in \mathcal{L}(X)$, define the convolution $L \star M$ on $X \times X^*$ by

$$(L \star M)(x,p) = \inf\{L(z,p) + M(x-z,p); z \in X\}.$$

Right operator shift: If $L \in \mathcal{L}(X)$ and $\Gamma : X \to X^*$ is a bounded linear operator, define the Lagrangian L_Γ on $X \times X^*$ by

$$L_\Gamma(x,p) := L(x, -\Gamma x + p).$$

Left operator shift: If $L \in \mathcal{L}(X)$ and if $\Gamma : X \to X^*$ is an invertible operator, define the Lagrangian $_\Gamma L$ on $X \times X^*$ by

$$_\Gamma L(x,p) := L(x - \Gamma^{-1} p, \Gamma x).$$

Free product: If $\{L_i; i \in I\}$ is a finite family of Lagrangians on reflexive Banach spaces $\{X_i; i \in I\}$, define the Lagrangian $L := \Sigma_{i \in I} L_i$ on $(\Pi_{i \in I} X_i) \times (\Pi_{i \in I} X_i^*)$ by

$$L((x_i)_i, (p_i)_i) = \Sigma_{i \in I} L_i(x_i, p_i).$$

Twisted product: If $L \in \mathscr{L}(X)$ and $M \in \mathscr{L}(Y)$, where X and Y are two reflexive spaces, then for any bounded linear operator $A : X \to Y^*$, define the Lagrangian $L \oplus_A M$ on $(X \times Y) \times (X^* \times Y^*)$ by

$$(L \oplus_A M)((x,y), (p.q)) := L(x, A^*y + p) + M(y, -Ax + q).$$

Antidualization of convex functions: If φ, ψ are convex functions on $X \times Y$ and if A is any bounded linear operator $A : X \to Y^*$, define the Lagrangian $\varphi \oplus_A \psi$ on $(X \times Y) \times (X^* \times Y^*)$ by

$$\varphi \oplus_A \psi((x,y), (p,q)) = \varphi(x,y) + \psi^*(A^*y + p, -Ax + q).$$

Remark 2.2. The convolution operation defined above should not be confused with the standard convolution for L and M as convex functions in both variables. Indeed, it is easy to see that in the case where $L(x, p) = \varphi(x) + \varphi^*(p)$ and $M(x, p) = \psi(x) + \psi^*(p)$, addition corresponds to taking

$$(L \oplus M)(x, p) = (\varphi + \psi)(x) + \varphi^* \star \psi^*(p),$$

while convolution reduces to

$$(L \star M)(x, p) = (\varphi \star \psi)(x) + (\varphi^* + \psi^*)(p).$$

Proposition 2.8. *Let X be a reflexive Banach space. Then,*

1. *$(\lambda \cdot L)^* = \lambda \cdot L^*$ for any $L \in \mathscr{L}(X)$ and any $\lambda > 0$.*
2. *$(L \oplus M)^* \leq L^* \star M^*$ and $(L \star M)^* \leq L^* \oplus M^*$ for any $L, M \in \mathscr{L}(X)$.*
3. *If M is a basic Lagrangian of the form $\varphi(Ux) + \psi^*(V^*p)$, where ψ is continuous on X and U, V are two automorphisms of X, then $(L \star M)^* = L^* \oplus M^*$ for any $L \in \mathscr{L}(X)$.*
4. *If $L, M \in \mathscr{L}(X)$ are such that $\mathrm{Dom}_2(L^*) - \mathrm{Dom}_2(M^*)$ contains a neighborhood of the origin, then $(L \star M)^* = L^* \oplus M^*$.*
5. *If $L, M \in \mathscr{L}(X)$ are such that $\mathrm{Dom}_1(L) - \mathrm{Dom}_1(M)$ contains a neighborhood of the origin, then $(L \oplus M)^* = L^* \star M^*$.*
6. *If $L \in \mathscr{L}(X)$ and $\Gamma : X \to X^*$ is a bounded linear operator, then $(L_\Gamma)^*(p, x) = L^*(\Gamma^*x + p, x)$.*
7. *If $L \in \mathscr{L}(X)$ and if $\Gamma : X \to X^*$ is an invertible operator, then $(_\Gamma L)^*(p, x) = L^*(-\Gamma^*x, (\Gamma^{-1})^*p + x)$.*
8. *If $\{L_i; i \in I\}$ is a finite family of Lagrangians on reflexive Banach spaces $\{X_i; i \in I\}$, then*

$$(\Sigma_{i \in I} L_i)^*((p_i)_i, (x_i)_i) = \Sigma_{i \in I} L_i^*(p_i, x_i).$$

9. *If $L \in \mathscr{L}(X)$ and $M \in \mathscr{L}(Y)$, where X and Y are two reflexive spaces, then for any bounded linear operator $A : X \to Y^*$, we have*

$$(L \oplus_A M)^*((p,q), (x,y)) = L^*(A^*y + p, x) + M^*(-Ax + q, y).$$

10. If φ and ψ are convex functions on $X \times Y$ and A is any bounded linear operator $A : X \to Y^*$, then the Lagrangian L defined on $(X \times Y) \times (X^* \times Y^*)$ by $L((x,y),(p,q)) = \varphi(x,y) + \psi^*(A^*y + p, -Ax + q)$ has a Legendre transform

$$L^*((p,q),(x,y)) = \psi(x,y) + \varphi^*(A^*y + p, -Ax + q).$$

Proof. (1) is obvious.

To prove (2) fix $(q,y) \in X^* \times X$ and use the formula $(\varphi \star \psi)^* \leq \varphi^* + \psi^*$ in one variable on the functions $\varphi(p) = L(z,p)$ and $\psi(p) = M(v,p)$ to write

$$
\begin{aligned}
(L \star M)^*(q,y) &= \sup\{\langle q,x\rangle + \langle y,p\rangle - L(z,p) - M(x-z,p); (z,x,p) \in X \times X \times X^*\} \\
&= \sup\{\langle q,v+z\rangle + \langle y,p\rangle - L(z,p) - M(v,p); (z,v,p) \in X \times X \times X^*\} \\
&\leq \sup_{(z,v) \in X \times X} \{\langle q,v+z\rangle + \sup\{\langle y,p\rangle - L(z,p) - M(v,p); p \in X^*\}\} \\
&\leq \sup_{(z,v) \in X \times X} \left\{ \langle q,v+z\rangle + \inf_{w \in X} \{ \sup_{p_1 \in X^*} (\langle w,p_1\rangle - L(z,p_1)) \right. \\
&\qquad\qquad\qquad \left. + \sup_{p_2 \in X^*} (\langle y-w,p_2\rangle - M(v,p_2))\} \right\} \\
&\leq \inf_{w \in X} \left\{ \sup_{(z,p_1) \in X \times X^*} \{\langle q,z\rangle + \langle w,p_1\rangle - L(z,p_1))\} \right. \\
&\qquad\qquad\qquad \left. + \sup_{(v,p_2) \in X \times X^*} \{\langle q,v\rangle + \langle y-w,p_2\rangle - M(v,p_2)\} \right\} \\
&= \inf_{w \in X} \{L^*(q,w) + M^*(q,y-w)\} \\
&= (L^* \oplus M^*)(q,y).
\end{aligned}
$$

For (3), assume that $M(x,p) = \varphi(Ux) + \psi^*(V^*p)$, where φ and ψ are convex continuous functions and U and V are automorphisms of X. Fix $(q,y) \in X^* \times X$ and write

$$
\begin{aligned}
(L \star M)^*(q,y) &= \sup\{\langle q,x\rangle + \langle y,p\rangle - L(z,p) - M(x-z,p); (z,x,p) \in X \times X \times X^*\} \\
&= \sup\{\langle q,v+z\rangle + \langle y,p\rangle - L(z,p) - M(v,p); (z,v,p) \in X \times X \times X^*\} \\
&= \sup_{p \in X^*} \left\{ \langle y,p\rangle + \sup_{(z,v) \in X^2} \{\langle q,v+z\rangle - L(z,p) - \varphi(Uv)\} - \psi^*(V^*p) \right\} \\
&= \sup_{p \in X^*} \left\{ \langle y,p\rangle + \sup_{z \in X}\{\langle q,z\rangle - L(z,p)\} \right. \\
&\qquad\qquad \left. + \sup_{v \in X}\{\langle q,v\rangle - \varphi(Uv)\} - \psi^*(V^*p) \right\} \\
&= \sup_{p \in X^*} \left\{ \langle y,p\rangle + \sup_{z \in X}\{\langle q,z\rangle - L(z,p)\} + \varphi^*((U^{-1})^*q) - \psi^*(V^*p) \right\}
\end{aligned}
$$

$$= \sup_{p\in X^*} \sup_{z\in X} \{\langle y,p\rangle + \langle q,z\rangle - L(z,p) - \psi^*(V^*p)\} + \varphi^*((U^{-1})^*q)$$

$$= (L+T)^*(q,y) + \varphi^*((U^{-1})^*q),$$

where $T(z,p) := \psi^*(V^*p)$ for all $(z,p) \in X \times X^*$. Note now that

$$T^*(q,y) = \sup_{z,p}\{\langle q,z\rangle + \langle y,p\rangle - \psi^*(V^*p)\} = \begin{cases} +\infty & \text{if } q \neq 0, \\ \psi((V^{-1}y) & \text{if } q = 0, \end{cases}$$

in such a way that by using the duality between sums and convolutions in both variables, we get

$$(L+T)^*(q,y) = \text{conv}(L^*,T^*)(q,y)$$
$$= \inf_{r\in X^*,z\in X}\{L^*(r,z) + T^*(-r+q,-z+y)\}$$
$$= \inf_{z\in X}\{L^*(q,z) + \psi(V^{-1}(-z+y))\}.$$

Finally,

$$(L\star M)^*(q,y) = (L+T)^*(q,y) + \varphi^*((U^{-1})^*q)$$
$$= \inf_{z\in X}\{L^*(q,z) + \psi(V^{-1}(-z+y))\} + \varphi^*((U^{-1})^*q)$$
$$= \inf_{z\in X}\{L^*(q,z) + (\varphi\circ U)^*(q) + (\psi^*\circ V^*)^*(-z+y)\}$$
$$= (L^*\oplus M^*)(q,y).$$

For (4), again fix $(q,y) \in X^* \times X$, and write

$$(L\star M)^*(q,y) = \sup_{(z,x,p)\in X\times X\times X^*}\{\langle q,x\rangle + \langle y,p\rangle - L(z,p) - M(x-z,p)\}$$
$$= \sup_{(z,v,p)\in X\times X\times X^*}\{\langle q,v+z\rangle + \langle y,p\rangle - L(z,p) - M(v,p)\}$$
$$= \sup_{(z,v,p)\in X\times X\times X^*}\{-\varphi^*(-z,-v,-p) - \psi^*(z,v,p)\}$$

with $\varphi^*(z,v,p) = \langle q,z\rangle + L(-z,-p)$ and $\psi^*(z,v,p) = -\langle y,p\rangle - \langle q,v\rangle + M(v,p)$. Note that now

$$\varphi(r,s,x) = \sup_{(z,v,p)\in X\times X\times X^*}\{\langle r,z\rangle + \langle v,s\rangle + \langle x,p\rangle - \langle q,z\rangle - L(-z,-p)\}$$
$$= \sup_{(z,v,p)\in X\times X\times X^*}\{\langle r-q,z\rangle + \langle v,s\rangle + \langle x,p\rangle - L(-z,-p)\}$$
$$= \sup_{v\in X}\{\langle v,s\rangle + L^*(q-r,-x)\},$$

which is equal to $+\infty$ whenever $s \neq 0$. Similarly, we have

$$\psi(r,s,x) = \sup_{(z,v,p)\in X\times X\times X^*} \{\langle r,z\rangle + \langle v,s\rangle + \langle x,p\rangle + \langle y,p\rangle + \langle v,q\rangle - M(v,p)\}$$

$$= \sup_{(z,v,p)\in X\times X\times X^*} \{\langle r,z\rangle + \langle v,q+s\rangle + \langle x+y,p\rangle - M(v,p)\}$$

$$= \sup_{z\in X}\{\langle z,r\rangle + M^*(q+s,x+y)\},$$

which is equal to $+\infty$ whenever $r \neq 0$. If now $\mathrm{Dom}_2(L^*) - \mathrm{Dom}_2(M^*)$ contains a neighborhood of the origin, then we apply the theorem of Fenchel and Rockafellar to get

$$(L\star M)^*(q,y) = \sup\{-\varphi^*(-z,-v,-p) - \psi^*(z,v,p); (z,v,p)\in X\times X\times X^*\}$$

$$= \inf\{\varphi(r,s,x) + \psi(r,s,x); (r,s,x)\in X^*\times X^*\times X\}$$

$$= \inf_{(r,s,x)\in X^*\times X^*\times X} \left\{\sup_{v\in X}\{\langle v,s\rangle + L^*(q-r,-x)\}\right.$$

$$\left. + \sup_{z\in X}\{\langle z,r\rangle + M^*(q+s,x+y)\}\right\}$$

$$= \inf\{L^*(q,-x) + M^*(q,x+y); x\in X\}$$

$$= (L^*\oplus M^*)(q,y).$$

Assertion (5) can be proved in a similar fashion.

For (6), fix $(q,y)\in X^*\times X$, set $r = \Gamma x + p$ and write

$$(L_\Gamma)^*(q,y) = \sup\{\langle q,x\rangle + \langle y,p\rangle - L(x,-\Gamma x+p); (x,p)\in X\times X^*\}$$

$$= \sup\{\langle q,x\rangle + \langle y,r+\Gamma x\rangle - L(x,r); (x,r)\in X\times X^*\}$$

$$= \sup\{\langle q+\Gamma^*y,x\rangle + \langle y,r\rangle - L(x,r); (x,r)\in X\times X^*\}$$

$$= L^*(q+\Gamma^*y,y).$$

For (7), let $r = x - \Gamma^{-1}p$ and $s = \Gamma x$ and write

$$(_\Gamma L)^*(q,y) = \sup\{\langle q,x\rangle + \langle y,p\rangle - L(x-\Gamma^{-1}p,\Gamma x); (x,p)\in X\times X^*\}$$

$$= \sup\{\langle q,\Gamma^{-1}s\rangle + \langle y,s-\Gamma r\rangle - L(r,s); (r,s)\in X\times X^*\}$$

$$= \sup\{\langle(\Gamma^{-1})^*q+y,s\rangle - \langle\Gamma^*y,r\rangle - L(r,s); (r,s)\in X\times X^*\}$$

$$= L^*(-\Gamma^*y,(\Gamma^{-1})^*q+y).$$

The proof of (8) is obvious, while for (9) notice that if $(\tilde{z},\tilde{r})\in (X\times Y)\times(X^*\times Y^*)$, where $\tilde{z} = (x,y)$ and $\tilde{r} = (p,q)$, we can write

$$L\oplus_A M(\tilde{z},\tilde{r}) = (L+M)(\tilde{z},\tilde{A}\tilde{z}+\tilde{r}),$$

where $\tilde{A}: X\times Y\to X^*\times Y^*$ is the skew-adjoint operator defined by $\tilde{A}(\tilde{z}) = \tilde{A}((x,y)) = (-A^*y,Ax)$. Now apply (6) and (8) to $L+M$ and $\tilde{A}$ to obtain

$$(L\oplus_A M)^*((p,q),(x,y)) = (L+M)^*(\tilde{r}+\tilde{A}^*\tilde{z},\tilde{z})$$

$$= (L^* + M^*)(\tilde{r} - \tilde{A}\tilde{z}, \tilde{z})$$
$$= L^*(A^*y + p, x) + M^*(-Ax + q, y).$$

Assertion (10) follows again from (6) since the Lagrangian $M((x,y),(p,q)) = \varphi(x,y) + \psi^*(-A^*y - p, Ax - q)$ is of the form $L((x,y), \tilde{A}(x,y) + (p,q))$, where $L((x,y),(p,q)) = \varphi(x,y) + \psi^*(p,q)$ and $\tilde{A} : X \times Y \to X^* \times Y^*$ is again the skew-adjoint operator defined by $\tilde{A}((x,y)) = (-A^*y, Ax)$. The Legendre transform is then equal to $L^*((p,q),(x,y)) = \psi(x,y) + \varphi^*(A^*y + p, -Ax + q)$.

2.6 Legendre transforms on various path spaces

Legendre transform on the path space $L^\alpha([0,T],X)$

For $1 < \alpha < +\infty$, we consider the space $L^\alpha_X[0,T]$ of Bochner integrable functions from $[0,T]$ into X with norm

$$\|u\|_{L_\alpha(X)} = \left(\int_0^T \|u(t)\|_X^\alpha dt \right)^{\frac{1}{\alpha}}.$$

Definition 2.4. Let $[0,T]$ be a time interval and let X be a reflexive Banach space. A *time-dependent convex function on* $[0,T] \times X$ (resp., *a time-dependent convex Lagrangian on* $[0,T] \times X \times X^*$) is a function $\varphi : [0,T] \times X \to \mathbf{R} \cup \{+\infty\}$ (resp., $L : [0,T] \times X \times X^* \to \mathbf{R} \cup \{+\infty\}$) such that :

1. φ (resp., L) is measurable with respect to the σ-field generated by the products of Lebesgue sets in $[0,T]$ and Borel sets in X (resp., in $X \times X^*$).
2. For each $t \in [0,T]$, the function $\varphi(t, \cdot)$ (resp., $L(t, \cdot, \cdot)$) is convex and lower semi-continuous on X (resp., $X \times X^*$).

The Hamiltonian H_L of L is the function defined on $[0,T] \times X \times X^*$ by

$$H_L(t, x, y) = \sup\{\langle y, p \rangle - L(t, x, p); p \in X^*\}.$$

To each time-dependent Lagrangian L on $[0,T] \times X \times X^*$, one can associate the corresponding Lagrangian $\mathscr{L}$ on the path space $L^\alpha_X \times L^\beta_{X^*}$, where $\frac{1}{\alpha} + \frac{1}{\beta} = 1$ to be

$$\mathscr{L}(u, p) := \int_0^T L(t, u(t), p(t)) dt,$$

as well as the associated Hamiltonian on $L^\alpha_X \times L^\alpha_X$,

$$H_\mathscr{L}(u, v) = \sup \left\{ \int_0^T (\langle p(t), v(t) \rangle - L(t, u(t), p(t))) dt \; ; \; p \in L^\beta_{X^*} \right\}$$

The Fenchel-Legendre dual of $\mathscr{L}$ is defined for any $(q,v) \in L_{X^*}^\beta \times L_X^\alpha$ as

$$\mathscr{L}^*(q,v) = \sup_{(u,p) \in L_X^\alpha \times L_{X^*}^\beta} \int_0^T \{\langle q(t), u(t)\rangle + \langle p(t), v(t)\rangle - L(t, u(t), p(t))\} dt.$$

Proposition 2.7 immediately yields the following.

Proposition 2.9. *Suppose that L is a Lagrangian on $[0,T] \times X \times X^*$, and let $\mathscr{L}$ be the corresponding Lagrangian on the path space $L_X^\alpha \times L_{X^*}^\beta$. Then,*

1. $\mathscr{L}^*(p,u) = \int_0^T L^*(t, p(t), u(t)) dt.$
2. $H_{\mathscr{L}}(u,v) = \int_0^T H_L(t, u(t), v(t)) dt.$

Suppose now that H is a Hilbert space, and consider the space A_H^2 of all functions in L_H^2 such that $\dot{u} \in L_H^2$ equipped with the norm

$$\|u\|_{A_H^2} = (\|u\|_{L_H^2}^2 + \|\dot{u}\|_{L_H^2}^2)^{1/2}.$$

Theorem 2.3. *Suppose ℓ is a convex lower semicontinuous function on $H \times H$, and let L be a time-dependent Lagrangian on $[0,T] \times H \times H$ such that*

For each $p \in L_H^2$, the map $u \to \int_0^T L(t, u(t), p(t)) dt$ is continuous on L_H^2. (2.16)

The map $u \to \int_0^T L(t, u(t), 0) dt$ is bounded on the unit ball of L_H^2. (2.17)

$$-C \le \ell(a,b) \le \tfrac{1}{2}(1 + \|a\|_H^2 + \|b\|_H^2) \text{ for all } (a,b) \in H \times H. \quad (2.18)$$

Consider the following Lagrangian on $L_H^2 \times L_H^2$:

$$\mathscr{L}(u,p) = \begin{cases} \int_0^T L(t, u(t), p(t) - \dot{u}(t)) dt + \ell(u(0), u(T)) & \text{if } u \in A_H^2 \\ +\infty & \text{otherwise.} \end{cases}$$

The Legendre transform of $\mathscr{L}$ is then

$$\mathscr{L}^*(p,u) = \begin{cases} \int_0^T L^*(t, p(t) - \dot{u}(t), u(t)) dt + \ell^*(-u(0), u(T)) & \text{if } u \in A_H^2 \\ +\infty & \text{otherwise.} \end{cases}$$

Proof. For $(q,v) \in L_H^2 \times A_H^2$, write

$$\mathscr{L}^*(q,v) = \sup_{u \in L_H^2} \sup_{p \in L_H^2} \left\{ \int_0^T (\langle u(t), q(t)\rangle + \langle v(t), p(t)\rangle - L(t, u(t), p(t) - \dot{u}(t))) dt \right.$$
$$\left. -\ell(u(0), u(T)) \right\}$$

$$= \sup_{u \in A_H^2} \sup_{p \in L_H^2} \left\{ \int_0^T (\langle u(t), q(t)\rangle + \langle v(t), p(t)\rangle - L(t, u(t), p(t) - \dot{u}(t))) dt \right.$$
$$\left. -\ell(u(0), u(T)) \right\}.$$

Make a substitution $p(t) - \dot{u}(t) = r(t) \in L_H^2$. Since u and v are both in A_H^2, we have

$$\int_0^T \langle v, \dot{u} \rangle = -\int_0^T \langle \dot{v}, u \rangle + \langle v(T), u(T) \rangle - \langle v(0), u(0) \rangle.$$

Since the subspace $A_H^{2,0} = \{u \in A_H^2; u(0) = u(T) = 0\}$ is dense in L_H^2, and since $u \to \int_0^T L(t, u(t), p(t)) dt$ is continuous on L_H^2 for each p, we obtain

$$
\begin{aligned}
\mathscr{L}^*(q,v) &= \sup_{u \in A_H^2} \sup_{r \in L_H^2} \left\{ \int_0^T \{ \langle u(t), q(t) \rangle + \langle v(t), r(t) + \dot{u}(t) \rangle - L(t, u(t), r(t)) \} dt \right. \\
&\qquad\qquad\qquad - \ell(u(0), u(T)) \Big\} \\
&= \sup_{u \in A_H^2} \sup_{r \in L_H^2} \left\{ \int_0^T \{ \langle u(t), q(t) - \dot{v}(t) \rangle + \langle v(t), r(t) \rangle - L(t, u(t), r(t)) \} dt \right. \\
&\qquad\qquad\qquad + \langle v(T), u(T) \rangle - \langle v(0), u(0) \rangle - \ell(u(0), u(T)) \Big\} \\
&= \sup_{u \in A_H^2} \sup_{r \in L_H^2} \sup_{u_0 \in A_H^{2,0}} \left\{ \int_0^T \{ \langle u, q - \dot{v} \rangle + \langle v, r \rangle - L(t, u(t), r(t)) \} dt \right. \\
&\qquad\qquad\qquad + \langle v(T), (u + u_0)(T) \rangle - \langle v(0), (u + u_0)(0) \rangle \Big\} \\
&\qquad\qquad\qquad - \ell((u + u_0)(0), (u + u_0)(T)) \Big\} \\
&= \sup_{w \in A_H^2} \sup_{r \in L_H^2} \sup_{u_0 \in A_H^{2,0}} \left\{ \int_0^T \langle w(t) - u_0(t), q(t) + \dot{v}(t) \rangle + \langle v(t), r(t) \rangle \, dt \right. \\
&\qquad\qquad\qquad - \int_0^T L(t, w(t) - u_0(t), r(t)) \, dt \Big\} \\
&\qquad\qquad\qquad + \langle v(T), w(T) \rangle - \langle v(0), w(0) \rangle - \ell(w(0), w(T)) \Big\} \\
&= \sup_{w \in A_H^2} \sup_{r \in L_H^2} \sup_{x \in L_H^2} \left\{ \int_0^T \{ \langle x, q - \dot{v} \rangle + \langle v(t), r(t) \rangle - L(t, x(t), r(t)) \} dt \right. \\
&\qquad\qquad\qquad + \langle v(T), w(T) \rangle - \langle v(0), w(0) \rangle - \ell(w(0), w(T)) \Big\}.
\end{aligned}
$$

Now, for each $(a, b) \in H \times H$, there is $w \in A_H^2$ such that $w(0) = a$ and $w(T) = b$, namely the linear path $w(t) = \frac{(T-t)}{T} a + \frac{t}{T} b$. Since ℓ is continuous on H, we finally obtain that

$$
\begin{aligned}
\mathscr{L}^*(q,v) &= \sup_{(a,b) \in H \times H} \sup_{(r,x) \in L_H^2 \times L_H^2} \left\{ \int_0^T \{ \langle x, q - \dot{v} \rangle + \langle v, r \rangle - L(t, x(t), r(t)) \} dt \right. \\
&\qquad\qquad\qquad + \langle v(T), b \rangle - \langle v(0), a \rangle - \ell(a, b) \Big\} \\
&= \sup_{x \in L_H^2} \sup_{r \in L_H^2} \left\{ \int_0^T \{ \langle x(t), q(t) - \dot{v}(t) \rangle + \langle v(t), r(t) \rangle - L(t, x(t), r(t)) \} dt \right\}
\end{aligned}
$$

$$+ \sup_{a \in H} \sup_{b \in H} \left\{ \langle v(T), b \rangle - \langle v(0), a \rangle - \ell(a, b) \right\}$$

$$= \int_0^T L^*(t, q(t) - \dot{v}(t), v(t)) dt + \ell^*(-v(0), v(T)).$$

If now $(q, v) \in L_H^2 \times L_H^2 \setminus A_H^2$, then we use the fact that $u \to \int_0^T L(t, u(t), 0) dt$ is bounded on the unit ball of A_H^2 and the growth condition on ℓ to deduce that

$$\mathcal{L}^*(q, v) \geq \sup_{u \in A_H^2} \sup_{r \in A_H^2} \left\{ \int_0^T \langle u(t), q(t) \rangle + \langle v(t), r(t) \rangle + \langle v(t), \dot{u}(t) \rangle - L(t, u(t), r(t)) dt \right.$$

$$\left. - \ell(u(0), u(T)) \right\}$$

$$\geq \sup_{u \in A_H^2} \sup_{r \in A_H^2} \left\{ -\|u\|_{L_H^2} \|q\|_{L_H^2} - \|v\|_{L_H^2} \|r\|_{L_H^2} + \int_0^T \langle v, \dot{u} \rangle - L(t, u(t), r(t)) dt \right.$$

$$\left. - \ell(u(0), u(T)) \right\}$$

$$\geq \sup_{\|u\|_{A_H^2} \leq 1} \left\{ -\|q\|_2 + \int_0^T \{ \langle v(t), \dot{u}(t) \rangle - L(t, u(t), 0) \} dt - \ell(u(0), u(T)) \right\}$$

$$\geq \sup_{\|u\|_{A_H^2} \leq 1} \left\{ C + \int_0^T \langle v(t), \dot{u}(t) \rangle - L(t, u, 0) dt - \frac{1}{2} (\|u(0)\|^2 + \|u(T)\|^2) \right\}$$

$$\geq \sup_{\|u\|_{A_H^2} \leq 1} \left\{ D + \int_0^T \langle v(t), \dot{u}(t) \rangle dt - \frac{1}{2} (\|u(0)\|_H^2 + \|u(T)\|_H^2) \right\}.$$

Since now v does not belong to A_H^2, we have that

$$\sup_{\|u\|_{A_H^2} \leq 1} \left\{ \int_0^T \langle v(t), \dot{u}(t) \rangle dt - \frac{1}{2} (\|u(0)\|_H^2 + \|u(T)\|_H^2) \right\} = +\infty,$$

which means that $\mathcal{L}^*(q, v) = +\infty$.

Legendre transform on spaces of absolutely continuous functions

Consider now the path space $A_H^2 = \{ u : [0, T] \to H ; \dot{u} \in L_H^2 \}$ equipped with the norm

$$\|u\|_{A_H^2} = \left(\|u(0)\|_H^2 + \int_0^T \|\dot{u}\|^2 dt \right)^{\frac{1}{2}}.$$

One way to represent the space A_H^2 is to identify it with the product space $H \times L_H^2$ in such a way that its dual $(A_H^2)^*$ can also be identified with $H \times L_H^2$ via the formula

$$\langle u, (p_1, p_0)\rangle_{A_H^2, H \times L_H^2} = \langle u(0), p_1\rangle_H + \int_0^T \langle \dot{u}(t), p_0(t)\rangle dt, \tag{2.19}$$

where $u \in A_H^2$ and $(p_1, p_0) \in H \times L_H^2$. With this duality, we have the following theorem.

Theorem 2.4. *Let L be a time-dependent convex Lagrangian on $[0, T] \times H \times H$ and let ℓ be a proper convex lower semicontinuous function on $H \times H$. Consider the Lagrangian on $A_H^2 \times (A_H^2)^* = A_H^2 \times (H \times L_H^2)$ defined by*

$$\mathscr{N}(u, p) = \int_0^T L(t, u(t) - p_0(t), -\dot{u}(t))dt + \ell(u(0) - a, u(T)), \tag{2.20}$$

where $u \in A_H^2$ and $(p_0(t), a) \in L_H^2 \times H$ represents an element p in the dual of A_H^2. Then, for any $(v, q) \in A_H^2 \times (A_H^2)^*$ with q of the form $(q_0(t), 0)$, we have

$$\mathscr{N}^*(q, v) = \int_0^T L^*(t, -\dot{v}(t), v(t) - q_0(t),)dt + \ell^*(-v(0), v(T)). \tag{2.21}$$

Proof. For $(v, q) \in A_H^2 \times (A_H^2)^*$ with q represented by $(q_0(t), 0)$, write

$$\mathscr{N}^*(q, v) = \sup_{p_1 \in H} \sup_{p_0 \in L_H^2} \sup_{u \in A_H^2} \left\{ \langle p_1, v(0)\rangle + \int_0^T \langle p_0(t), \dot{v}(t)\rangle + \langle q_0(t), \dot{u}(t)\rangle\, dt \right.$$
$$\left. - \int_0^T L(t, u(t) - p_0(t), -\dot{u}(t))\, dt - \ell(u(0) - p_1, u(T)) \right\}.$$

Making a substitution $u(0) - p_1 = a \in H$ and $u(t) - p_0(t) = y(t) \in L_H^2$, we obtain

$$\mathscr{N}^*(q, v) = \sup_{a \in H} \sup_{y \in L_H^2} \sup_{u \in A_H^2} \left\{ \langle u(0) - a, v(0)\rangle - \ell(a, u(T)) \right.$$
$$\left. + \int_0^T \left\{ \langle u(t) - y(t), \dot{v}(t)\rangle + \langle q_0(t), \dot{u}(t)\rangle - L(t, y(t), -\dot{u}(t)) \right\}dt \right\}.$$

Since $\dot{u}$ and $\dot{v} \in L_H^2$, we have

$$\int_0^T \langle u, \dot{v}\rangle = -\int_0^T \langle \dot{u}, v\rangle + \langle v(T), u(T)\rangle - \langle v(0), u(0)\rangle,$$

which implies

$$\mathscr{N}^*(q, v) = \sup_{a \in H} \sup_{y \in L_H^2} \sup_{u \in A_H^2} \left\{ \langle -a, v(0)\rangle + \langle v(T), u(T)\rangle - \ell(a, u(T)) \right.$$
$$\int_0^T [-\langle y(t), \dot{v}(t)\rangle + \langle v(t) - q_0(t), -\dot{u}(t)\rangle - L(t, y(t), -\dot{u}(t))]\, dt \bigg\}.$$

Now identify A_H^2 with $H \times L_H^2$ via the correspondence

$$(b,r) \in H \times L^2_H \mapsto b + \int_t^T r(s)\,ds \in A^2_H,$$

$$u \in A^2_H \mapsto \big(u(T), -\dot{u}(t)\big) \in H \times L^2_H.$$

We finally obtain

$$\mathscr{N}^*(q,v) = \sup_{a \in H} \sup_{b \in H} \Big\{ \langle a, -v(0) \rangle + \langle v(T), b \rangle - \ell(a,b) \Big\}$$

$$+ \sup_{y \in L^2_H} \sup_{r \in L^2_H} \Big\{ \int_0^T -\langle y(t), \dot{v}(t) \rangle + \langle v(t) - q_0(t), r(t) \rangle - L(t, y(t), r(t))dt \Big\}$$

$$= \int_0^T L^*(t, -\dot{v}(t), v(t) - q_0(t))dt + \ell^*(-v(0), v(T)).$$

Legendre transform for a symmetrized duality on spaces of absolutely continuous functions

Consider again $A^2_H := \{u : [0,T] \to H; \dot{u} \in L^2_H\}$ equipped with the norm

$$\|u\|_{A^2_H} = \left\{ \left\| \frac{u(0) + u(T)}{2} \right\|_H^2 + \int_0^T \|\dot{u}\|_H^2\,dt \right\}^{\frac{1}{2}}.$$

We can again identify the space A^2_H with the product space $H \times L^2_H$ in such a way that its dual $(A^2_H)^*$ can also be identified with $H \times L^2_H$ via the formula

$$\left\langle u, (p_1, p_0) \right\rangle_{A^2_H, H \times L^2_H} = \left\langle \frac{u(0) + u(T)}{2}, p_1 \right\rangle + \int_0^T \langle \dot{u}(t), p_0(t) \rangle\,dt,$$

where $u \in A^2_H$ and $(p_1, p_0(t)) \in H \times L^2_H$.

Theorem 2.5. *Suppose L is a time-dependent Lagrangian on $[0,T] \times H \times H$ and ℓ is a Lagrangian on $H \times H$. Consider the following Lagrangian defined on the space $A^2_H \times (A^2_H)^* = A^2_H \times (H \times L^2_H)$ by*

$$\mathscr{M}(u,p) = \int_0^T L\big(t, u(t) + p_0(t), -\dot{u}(t)\big)\,dt + \ell\Big(u(T) - u(0) + p_1, \frac{u(0) + u(T)}{2}\Big).$$

The Legendre transform of $\mathscr{M}$ on $A^2_H \times (L^2_H \times H)$ is given by

$$\mathscr{M}^*(p,u) = \int_0^T L^*\big(t, -\dot{u}(t), u(t) + p_0(t)\big)\,dt + \ell^*\Big(\frac{u(0) + u(T)}{2}, u(T) - u(0) + p_1\Big).$$

Proof. For $(q,v) \in A^2_H \times (A^2_H)^*$ with q represented by $(q_0(t), q_1)$, we have

$$\mathscr{M}^*(q,v) = \sup_{p_1 \in H} \sup_{p_0 \in L^2_H} \sup_{u \in A^2_H} \left\langle p_1, \frac{v(0)+v(T)}{2} \right\rangle + \left\langle q_1, \frac{u(0)+u(T)}{2} \right\rangle$$

$$- \int_0^T \left[\langle p_0(t), \dot{v}(t) \rangle + \langle q_0(t), \dot{u} \rangle - L(t, u(t)+p_0(t), -\dot{u}(t)) \right] dt$$

$$- \ell \left(u(T) - u(0) + p_1, \frac{u(0)+u(T)}{2} \right) \Bigg\}.$$

Making a substitution $u(T) - u(0) + p_1 = a \in H$ and $u(t) + p_0(t) = y(t) \in L^2_H$, we obtain

$$\mathscr{M}^*(q,v) = \sup_{a \in H} \sup_{y \in L^2_H} \sup_{u \in A^2_H} \left\langle a - u(T) + u(0), \frac{v(0)+v(T)}{2} \right\rangle + \left\langle q_1, \frac{u(0)+u(T)}{2} \right\rangle$$

$$- \int_0^T \left[\langle y(t) - u(t), \dot{v} \rangle + \langle q_0(t), \dot{u}(t) \rangle - L(t, y(t), -\dot{u}(t)) \right] dt$$

$$- \ell \left(a, \frac{u(0)+u(T)}{2} \right) \Bigg\}.$$

Again, since $\dot{u}$ and $\dot{v} \in L^2_H$, we have

$$\int_0^T \langle u(t), \dot{v}(t) \rangle \, dt = - \int_0^T \langle \dot{u}(t), v(t) \rangle \, dt + \langle u(T), v(T) \rangle - \langle v(0), u(0) \rangle,$$

which implies

$$\mathscr{M}^*(q,v) = \sup_{a \in H} \sup_{y \in L^2_H} \sup_{u \in A^2_H} \left\{ \left\langle a, \frac{v(0)+v(T)}{2} \right\rangle - \left\langle u(T), \frac{v(0)+v(T)}{2} - v(T) \right\rangle \right.$$

$$- \left\langle u(0), v(0) - \frac{v(0)+v(T)}{2} \right\rangle + \left\langle q_1, \frac{u(0)+u(T)}{2} \right\rangle$$

$$- \int_0^T \left[\langle y(t), \dot{v} \rangle + \langle \dot{u}(t), v(t) + q_0(t) \rangle - L(t, y(t), -\dot{u}(t)) \right] dt$$

$$- \ell \left(a, \frac{u(0)+u(T)}{2} \right) \Bigg\}.$$

Hence,

$$\mathscr{M}^*(q,v) = \sup_{a \in H, y \in L^2_H, u \in A^2_H} \left\langle a, \frac{v(0)+v(T)}{2} \right\rangle + \left\langle q_1 + v(T) - v(0), \frac{u(0)+u(T)}{2} \right\rangle$$

$$- \ell \left(a, \frac{u(0)+u(T)}{2} \right)$$

$$- \int_0^T \left[\langle y(t), \dot{v}(t) \rangle + \langle \dot{u}(t), v(t) + q_0(t) \rangle - L(t, y(t), -\dot{u}(t)) \right] dt \Bigg\}.$$

Now identify A_H^2 with $H \times L_H^2$ via the correspondence:

$$\left(b, f(t)\right) \in H \times L_H^2 \longmapsto b + \frac{1}{2}\left(\int_t^T f(s)\,ds - \int_0^t f(s)\,ds\right) \in A_H^2,$$

$$u \in A_H^2 \longmapsto \left(\frac{u(0)+u(T)}{2}, -\dot{u}(t)\right) \in H \times L_H^2.$$

We finally obtain

$$\mathcal{M}^*(q,v) = \sup_{a \in H} \sup_{b \in H}\left\{\langle a, \frac{v(0)+v(T)}{2}\rangle + \langle q_1 + v(T) - v(0), b\rangle - \ell(a,b)\right\}$$

$$+ \sup_{y \in L_H^2, r \in L_H^2}\left\{\int_0^T -\langle y(t), \dot{v}(t)\rangle + \langle v(t) + q_0(t), r(t)\rangle - L\left(t, y(t), r(t)\right) dt\right\}$$

$$= \ell^*\left(\frac{v(0)+v(T)}{2}, q_1 + v(T) - v(0)\right) + \int_0^T L^*\left(t, -\dot{v}(t), v(t) + q_0(t)\right) dt.$$

Exercises 2.B. Legendre transforms on path spaces

1. Prove Proposition 2.9.
2. Establish the identification between the Hilbert spaces $A_H^2[0,T]$ and $H \times L_H^2$ via the isomorphism $u \in A_H^2 \mapsto \left(u(T), -\dot{u}(t)\right) \in H \times L_H^2$.
3. Establish the identification between the Hilbert spaces $A_H^2[0,T]$ and $H \times L_H^2$ via the isomorphism $u \in A_H^2 \mapsto \left(\frac{u(0)+u(T)}{2}, -\dot{u}(t)\right) \in H \times L_H^2$.
4. Show that the Legendre transform of the Lagrangian on $L_H^2 \times L_H^2$

$$\mathscr{L}(u,p) = \begin{cases} \int_0^T L(t, u(t), p(t) - \dot{u}(t))dt + \ell\left(u(T) - u(0), \frac{u(0)+u(T)}{2}\right) & \text{if } u \in A_H^2 \\ +\infty & \text{otherwise} \end{cases}$$

is

$$\mathscr{L}^*(p,u) = \begin{cases} \int_0^T L^*(t, p(t) - \dot{u}(t), u(t))dt + \ell^*\left(\frac{u(0)+u(T)}{2}, u(T) - u(0)\right) & \text{if } u \in A_H^2 \\ +\infty & \text{otherwise,} \end{cases}$$

provided the conditions of Theorem 2.3 are satisfied.

2.7 Primal and dual problems in convex optimization

Consider the problem of minimizing a convex lower semicontinuous function I that is bounded below on a Banach space X. This is usually called the *primal problem*:

$$(\mathscr{P}) \qquad\qquad \inf_{x \in X} I(x). \qquad\qquad (2.22)$$

One can sometimes associate to I a family of perturbed problems in the following way. Let Y be another Banach space, and consider a convex lower semicontinuous

Lagrangian $L : X \times Y \to \mathbf{R} \cup \{+\infty\}$ such that the following holds:

$$I(x) = L(x,0) \quad \text{for all } x \in X. \tag{2.23}$$

For any $p \in Y$, one can consider the perturbed minimization problem

$$(\mathscr{P}_p) \qquad\qquad\qquad \inf_{x \in X} L(x,p) \tag{2.24}$$

in such a way that $(\mathscr{P}_0)$ is clearly the initial primal problem. By considering the Legendre transform L^* of L on the dual space $X^* \times Y^*$, one can consider the so-called *dual problem*

$$(\mathscr{P}^*) \qquad\qquad\qquad \sup_{p^* \in Y^*} -L^*(0,p^*). \tag{2.25}$$

Consider the function $h : Y \to \mathbf{R} \cup \{+\infty\}$ on the space of perturbations Y defined by

$$h(p) = \inf_{x \in X} L(x,p) \quad \text{for every } p \in Y. \tag{2.26}$$

The following proposition summarizes the relationship between the primal problem and the behavior of the value function h.

Theorem 2.6. *Assume L is a proper convex lower semicontinuous Lagrangian that is bounded below on $X \times Y$. Then, the following assertions hold:*

1. *(Weak duality)* $\quad -\infty < \sup_{p^* \in Y^*} \{-L^*(0,p^*)\} \le \inf_{x \in X} L(x,0) < +\infty.$
2. *h is a convex function on Y such that $h^*(p^*) = L^*(0,p^*)$ for every $p^* \in Y^*$, and*

$$h^{**}(0) = \sup_{p^* \in Y^*} \{-L^*(0,p^*)\}.$$

3. *h is lower semicontinuous at 0 (i.e., $(\mathscr{P})$ is normal) if and only if there is no duality gap, i.e., if*

$$\sup_{p^* \in Y^*} \{-L^*(0,p^*)\} = \inf_{x \in X} L(x,0).$$

4. *h is subdifferentiable at 0 (i.e., $(\mathscr{P})$ is stable) if and only if $(\mathscr{P})$ is normal and $(\mathscr{P}^*)$ has at least one solution. Moreover, the set of solutions for $(\mathscr{P}^*)$ is equal to $\partial h^{**}(0)$.*
5. *If for some $x_0 \in X$ the function $p \to L(x_0,p)$ is bounded on a ball centered at 0 in Y, then $(\mathscr{P})$ is stable and $(\mathscr{P}^*)$ has at least one solution.*

Proof. (1) For each $p^* \in Y^*$, we have

$$\begin{aligned}
L^*(0,p^*) &= \sup\{\langle p^*,p\rangle - L(x,p); x \in X, p \in Y\} \\
&\ge \sup\{\langle p^*,0\rangle - L(x,0); x \in X\} \\
&= -\inf\{L(x,0); x \in X\}.
\end{aligned}$$

(2) To prove the convexity of h, consider $\lambda \in (0,1)$ and elements $p, q \in Y$ such that $h(p)$ and $h(q)$ are finite. For every $a > h(p)$ (resp., $b > h(q)$), find $u \in X$ (resp., $v \in X$) such that

$$h(p) \leq L(x, p) \leq a \quad \text{and} \quad h(q) \leq L(v, q) \leq b.$$

Now use the convexity of L in both variables to write

$$
\begin{aligned}
h(\lambda p + (1 - \lambda)q) &= \inf\{L(x, \lambda p + (1 - \lambda)q); x \in X\} \\
&\leq L(\lambda u + (1 - \lambda)v, \lambda p + (1 - \lambda)q) \\
&\leq \lambda L(u, p) + (1 - \lambda)L(v, q) \\
&\leq \lambda a + (1 - \lambda)b,
\end{aligned}
$$

from which the convexity of h follows.

(3) Note first that the Legendre dual of h can be written for $p^* \in Y^*$ as

$$
\begin{aligned}
h^*(p^*) &= \sup\{\langle p^*, p \rangle - h(p); p \in Y\} \\
&= \sup\left\{\langle p^*, p \rangle - \inf_{x \in X}\{L(x, p); p \in Y\}\right\} \\
&= \sup\{\langle p^*, p \rangle - L(x, p); p \in Y, x \in X\} \\
&= L^*(0, p^*).
\end{aligned}
$$

It follows that

$$\sup_{p^* \in Y^*}\{-L^*(0, p^*)\} = \sup_{p^* \in Y^*} -h^*(p^*) = h^{**}(0) \leq h(0) = \inf_{x \in X} L(x, 0). \qquad (2.27)$$

Our claim follows from the fact that h is lower semicontinuous at 0 if and only if $h(0) = h^{**}(0)$.

For claim 4), we start by establishing that the set of solutions for $(\mathscr{P}^*)$ is equal to $\partial h^{**}(0)$. Indeed, if $p^* \in Y^*$ is a solution of $(\mathscr{P}^*)$, then

$$
\begin{aligned}
-h^*(p^*) &= -L^*(0, p^*) \\
&= \sup\{-L^*(0, q^*); q^* \in Y^*\} \\
&= \sup\{-h^*(q^*); q^* \in Y^*\} \\
&= \sup\{\langle 0, q^* \rangle - h^*(q^*); q^* \in Y^*\} \\
&= h^{**}(0),
\end{aligned}
$$

which is equivalent to $p^* \in \partial h^{**}(0)$.

Suppose now that $\partial h(0) \neq \emptyset$. Then, $h(0) = h^{**}(0)$ (i.e., $(\mathscr{P})$ is normal) and $\partial h(0) = \partial h^{**}(0) \neq \emptyset$, and hence $(\mathscr{P}^*)$ has at least one solution. Conversely, if h is lower semicontinuous at 0, then $h(0) = h^{**}(0)$, and if $\partial h^{**}(0) \neq \emptyset$, then $\partial h(0) = \partial h^{**}(0) \neq \emptyset$.

The condition in (5) readily implies that h is bounded above on a neighborhood of zero in Y^*, which implies that h is subdifferentiable at 0.

Further comments

The first four sections summarize the most basic concepts and relevant tools of convex analysis that will be used throughout this text. Proofs are not included, as they can be found in a multitude of books on convex analysis. We refer to the books of Aubin and Ekeland [8], Brézis [26], Ekeland and Temam [47], Ekeland [46], and Phelps [130].

The particularities of convex calculus on phase space were developed in Ghoussoub [55]. Legendre transforms on path space for the basic action functionals of the calculus of variations have already been dealt with by Rockafellar [137]. Theorem 2.4 is due to Ghoussoub and Tzou [68], while the new symmetrized duality for A_H^2 and the corresponding Legendre transform were first discussed in Ghoussoub and Moameni [63].

Chapter 3
Self-dual Lagrangians on Phase Space

At the heart of this theory is the interplay between certain automorphisms and Legendre transforms. The main idea originates from the fact that a large class of PDEs and evolution equations –the *completely self-dual differential systems*– can be written in the form

$$(p,x) \in \partial L(x,p),$$

where ∂L is the subdifferential of a self-dual Lagrangian $L : X \times X^* \to \mathbf{R} \cup \{+\infty\}$ on phase space. This class of Lagrangians is introduced in this chapter, where its remarkable permanence properties are also established, in particular, their stability under various operations such as convolution, direct sum, superposition, iteration, and certain regularizations, as well as their composition with skew-adjoint operators.

3.1 Invariance under Legendre transforms up to an automorphism

Definition 3.1. Given a bounded linear operator R from a reflexive Banach space E into its dual E^*, we say that a convex lower semicontinuous functional $\ell : E \to \mathbf{R} \cup \{+\infty\}$ is R-self-dual if

$$\ell^*(Rx) = \ell(x) \text{ for any } x \in E, \tag{3.1}$$

where here ℓ^* is the Legendre transform of ℓ on E.

The following easy proposition summarizes the properties of R-self-dual functions to be used in what follows.

Proposition 3.1. *Let ℓ be an R-self-dual convex functional on a reflexive Banach space E, where $R : E \to E^*$ is a bounded linear operator. Then,*

1. For every $x \in E$, we have $\ell(x) \geq \frac{1}{2}\langle Rx, x\rangle$.
2. For $\bar{x} \in E$, we have $\ell(\bar{x}) = \frac{1}{2}\langle R\bar{x}, \bar{x}\rangle$ if and only if $R\bar{x} \in \partial \ell(\bar{x})$.

Proof. It is sufficient to combine self-duality with the Fenchel-Legendre inequality to obtain

N. Ghoussoub, *Self-dual Partial Differential Systems and Their Variational Principles*,
Springer Monographs in Mathematics, DOI 10.1007/978-0-387-84897-6_3,
© Springer Science+Business Media, LLC 2009

$$2\ell(x) = \ell^*(Rx) + \ell(x) \geq \langle Rx, x \rangle \text{ with equality if and only if } Rx \in \partial\ell(x).$$

This leads us to introduce the following definition.

Definition 3.2. The R-core of ℓ is the set

$$\mathscr{C}_R\ell = \{x \in E; Rx \in \partial\ell(x)\} = (R - \partial\ell)^{-1}\{0\}.$$

It is easy to see that the only functional satisfying $\varphi^*(x) = \varphi(x)$ (i.e., when R is the identity) is the quadratic function $\varphi(x) = \frac{1}{2}\|x\|^2$. In this case, the I-core of φ is the whole space. On the other hand, by simply considering the operator $Rx = -x$, we can see that the notion becomes much more interesting. The following proposition is quite easy to prove.

Proposition 3.2. *Let E be a reflexive Banach space.*

1. *If R is self-adjoint and satisfies $\langle Rx, x \rangle \geq \delta\|x\|^2$ for some $\delta > 0$, then the only R-self-dual function on E is $\ell(x) = \frac{1}{2}\langle Rx, x \rangle$. In this case, $\mathscr{C}_R\ell = E$.*
2. *On the other hand, for every a in a Hilbert space H, the function*

$$\ell_a(x) = \frac{1}{2}\|x\|^2 - 2\langle a, x \rangle + \|a\|^2$$

satisfies $\ell_a^*(-x) = \ell_a(x)$ for every $x \in H$. In this case, $\overline{\partial}_{-I}\ell = \{a\}$.
3. *If $E = E_1 \times E_2$ is a product space and $R(x_1, x_2) = (R_1x_1, R_2x_2)$, where $R_i : E_i \to E_i^*$ ($i = 1, 2$), then a function on $E_1 \times E_2$ of the form $\ell(x_1, x_2) = \ell_1(x_1) + \ell_2(x_2)$ is R-self-dual as long as ℓ_1 is R_1-self-dual and ℓ_2 is R_2-self-dual. In this case, $\mathscr{C}_R\ell = \mathscr{C}_{R_1}\ell_1 \times \mathscr{C}_{R_2}\ell_2$. In particular, for any a in a Hilbert space E_2, the function*

$$\ell(x_1, x_2) = \frac{1}{2}\|x_1\|^2 + \frac{1}{2}\|x_2\|^2 - 2\langle a, x_2 \rangle + \|a\|^2$$

is $(I, -I)$-self-dual on $E_1 \times E_2$ and $\mathscr{C}_{(I,-I)}\ell = E_1 \times \{a\}$.
4. *If $R(x_1, x_2) = (x_2, x_1)$ and $S(x_1, x_2) = (-x_2, -x_1)$ from a Hilbert space $H \times H$ into itself, then for any convex lower semicontinuous functions ψ on H and any skew-adjoint operator $A : H \to H$, the function defined for $(x_1, x_2) \in E = H^2$ by*

$$\ell_1(x_1, x_2) = \psi(x_1) + \psi^*(Ax_1 + x_2) \text{ (resp., } \ell_2(x_1, x_2) = \psi(x_1) + \psi^*(-Ax_1 - x_2))$$

is R-self-dual (resp., S-self-dual) on $E = H^2$. In this case,

$$\mathscr{C}_R\ell_1 = \{(x_1, x_2) \in H \times H; x_2 \in -Ax_1 + \partial\psi(x_1)\}$$

and

$$\mathscr{C}_S\ell_2 = \{(x_1, x_2) \in H \times H; x_2 \in -Ax_1 - \partial\psi(x_1)\}.$$

Exercises 3.A. R-self-dual functions

1. Establish claims 1- 4 in Proposition 3.2.
2. Give an example of a J-self-dual function where $J(x,p) = (-p,x)$.
3. Are functions of the form ℓ_a above the only convex functions satisfying $\ell^*(x) = \ell(-x)$?
4. Is assertion (1) valid without the condition $\delta > 0$?

3.2 The class of self-dual Lagrangians

A rich class of automorphisms appears when E is a phase space $X \times X^*$, which is of particular interest when studying partial differential systems. One may consider the self-adjoint automorphism $R : X \times X^* \to X^* \times X$ defined by $R(x,p) = (p,x)$, or more generally $R(x,p) = (B^*p, Bx)$, where B is any bounded linear operator on X.

The following classes of self-dual convex functionals on phase space will play a significant role in the sequel.

Definition 3.3. Say that a convex Lagrangian $L : X \times X^* \to \mathbf{R} \cup \{+\infty\}$ on a reflexive Banach space X is a *self-dual (resp., antiself-dual) Lagrangian* on $X \times X^*$ if

$$L^*(p,x) = L(x,p) \quad (\text{resp.}, L^*(p,x) = L(-x,-p)) \quad \text{for all } (p,x) \in X^* \times X. \quad (3.2)$$

Denote by $\mathscr{L}^{\mathrm{sd}}(X)$ (resp., $\mathscr{L}^{\mathrm{asd}}(X)$) the class of self-dual (antiself-dual) Lagrangians. More generally, given a bounded linear operator $B : X \to X$, say that L is a *B-self-dual* Lagrangian if

$$L^*(B^*p, Bx) = L(x,p) \quad \text{for all } (p,x) \in X^* \times X. \quad (3.3)$$

We denote by $\mathscr{L}^{\mathrm{sd}}(X;B)$ the class of B-self-dual Lagrangians.

The basic B-self-dual Lagrangians

Following are the first examples of B-self-dual Lagrangians. More elaborate examples will be devised later, though all constructions will be based on these important building blocks. The proofs are easy and will be left to the interested reader.

1. Any convex lower semicontinuous function on X defines a self-dual (resp., antiself-dual) Lagrangian on $E = X \times X^*$ via the formula

$$L(x,p) = \varphi(x) + \varphi^*(p) \quad (\text{resp.}, L(x,p) = \varphi(x) + \varphi^*(-p).)$$

2. If B is a bounded linear operator on X and φ is a convex lower semicontinuous function, then a typical B-self-dual Lagrangians is given by $L(x,p) = \varphi(Bx) + \varphi^*(p)$, provided that B is either an onto operator on X or has dense range, while φ is continuous.

3. Another way to construct a B-self-dual Lagrangian is to consider $L(x,p) = \varphi(x) + \varphi^*(B^*p)$ again, provided B^* is either an onto operator on X^*, or has dense range and φ^* is continuous.
4. As seen below, one can iterate the procedure above to construct new self-dual Lagrangians from old ones. Indeed, if $B, C : X \to X$ are such that B and C^* are onto, then $N(x,p) = \varphi(Bx) + \varphi^*(C^*p)$ is a $C \circ B$-self-dual Lagrangian.
5. Furthermore, if $\Gamma : X \to X^*$ is a bounded linear operator such that $\Gamma^* \circ C \circ B$ is a skew-adjoint operator, then $N_\Gamma(x,p) = \varphi(Bx) + \varphi^*(C^*\Gamma x + C^*p)$ is again a $C \circ B$-self-dual Lagrangian.

These examples can be checked directly, but they also follow from the following propositions that summarize the permanence properties enjoyed by self-dual Lagrangians and will be frequently used (and extended) in the sequel.

Permanence properties of the class of self-dual Lagrangians

The proof of the following proposition is straightforward.

Proposition 3.3. *Let L be a B-self-dual Lagrangian on a reflexive Banach space X, where B is a bounded linear operator on X.*

1. *If $C : X \to X$ is an onto bounded linear operator or if C has dense range and L is continuous in the first variable, then $M(x,p) = L(Cx,p)$ is a $B \circ C$-self-dual Lagrangian.*
2. *If $D : X \to X$ is a bounded linear operator such that D^* is onto or if D^* has dense range and L is continuous in the second variable, then $N(x,p) = L(x, D^*p)$ is a $D \circ B$-self-dual Lagrangian.*

Proof. Indeed, fix $(q,y) \in X^* \times X$ and write

$$
\begin{aligned}
M^*(C^*B^*q, BCy) &= \sup\{\langle C^*B^*q, x\rangle + \langle BCy, p\rangle - L(Cx, p); (x, p) \in X \times X^*\} \\
&= \sup\{\langle B^*q, Cx\rangle + \langle BCy, p\rangle - L(Cx, p); (x, p) \in X \times X^*\} \\
&= \sup\{\langle B^*q, z\rangle + \langle BCy, p\rangle - L(z, p); (z, p) \in X \times X^*\} \\
&= L^*(B^*q, BCy) \\
&= L(Cy, q) = M(y, q).
\end{aligned}
$$

The proof of the rest is similar.

The following proposition summarizes some of the most useful permanence properties of B-self-dual Lagrangians.

Proposition 3.4. *Let B be a bounded linear operator on a reflexive Banach space X. The following properties hold:*

1. *If L is in $\mathscr{L}^{sd}(X; B)$ and if $\lambda > 0$, then $\lambda \cdot L$ also belongs to $\mathscr{L}^{sd}(X; B)$.*

2. If L is in $\mathscr{L}^{sd}(X;B)$, then for any $y \in X$ and $q \in X^*$, the translated Lagrangians M_y and N_p defined respectively by $M_y(x,p) = L(x+y,p) - \langle By,p \rangle$ and $N_q(x,p) = L(x,p+q) - \langle x,B^*q \rangle$ are also in $\mathscr{L}^{sd}(X;B)$.

3. If $L \in \mathscr{L}^{sd}(X;B)$ for some automorphism B of X and if $M(x,p) = \varphi(Bx) + \varphi^*(p)$ or $\varphi(x) + \varphi^*(B^*p)$, where φ is a convex finite function on X, then the Lagrangians $L \oplus M$ and $L \star M$ also belong to $\mathscr{L}^{sd}(X;B)$.

4. If $L,M \in \mathscr{L}^{sd}(X;B)$ are such that $\mathrm{Dom}_1(L) - \mathrm{Dom}_1(M)$ contains a neighborhood of the origin, then $L \star M$ and $L \oplus M$ are also in $\mathscr{L}^{sd}(X;B)$.

5. If U is a unitary operator $(U^{-1} = U^*)$ on a Hilbert space X that commutes with B and if L is in $\mathscr{L}^{sd}(X;B)$, then $M(x,p) = L(Ux,Up)$ also belongs to $\mathscr{L}^{sd}(X;B)$.

6. If $L_i \in \mathscr{L}^{sd}(X_i;B_i)$, where B_i is a bounded operator on a reflexive Banach space X_i for each $i \in I$, then $\Sigma_{i \in I} L_i$ is in $\mathscr{L}^{sd}(\Pi_{i \in I} X_i, (B_i)_{i \in I})$.

7. If $L \in \mathscr{L}^{sd}(X;B)$ and if $\Gamma : X \to X^*$ is a bounded linear operator such that $\Gamma^* B$ is skew-adjoint, then the Lagrangian L_Γ is also in $\mathscr{L}^{sd}(X;B)$.

8. If $L \in \mathscr{L}^{sd}(X;B)$ and if $\Gamma : X \to X^*$ is an invertible operator such that $B^*\Gamma$ and $\Gamma^{-1}B^*$ are skew-adjoint, then the Lagrangian $_\Gamma L$ is also in $\mathscr{L}^{sd}(X;B)$.

9. If $L \in \mathscr{L}^{sd}(X;B_1)$ and $M \in \mathscr{L}^{sd}(Y;B_2)$, then for any bounded linear operator $A : X \to Y^*$ such that $AB_1 = B_2^*A$, the Lagrangian $L \oplus_A M$ belongs to $\mathscr{L}^{sd}(X \times Y;(B_1,B_2))$ on $X \times Y$.

10. If φ is a convex continuous function on $X \times Y$, $B = (B_1,B_2)$ is a dense range operator on $X \times Y$ and $A : X \to Y^*$ is such that $AB_1 = B_2^*A$, then the Lagrangian L defined on $(X \times Y) \times (X^* \times Y^*)$ by

$$L((x,y),(p,q)) = \varphi(B_1x,B_2y) + \varphi^*(A^*y+p,-Ax+q)$$

is in $\mathscr{L}^{sd}(X \times Y;B)$.

Proof. (1) and (2) are straightforward.

To prove (3), use Proposition 2.8 (3) to get that $(L \star M)^* = L^* \oplus M^*$. On the other hand, note that

$$\begin{aligned} L^* \oplus M^*(B^*q,By) &= \inf\{L^*(B^*q,r) + M^*(B^*q,By-r); r \in X\} \\ &= \inf\{L^*(B^*q,Bt) + M^*(B^*q,By-Bt); t \in X\} \\ &= \inf\{L(t,q) + M(y-t,q); t \in X\} \\ &= L \star M(y,q). \end{aligned}$$

The proof of (4) is similar, provided one uses Proposition 2.8.(4) and (5).

The proofs of (5) and (6) are straightforward, while those of (7) and (8) readily follow from (6) and (7) in Proposition 2.8.

For (9), it is enough to note that for $(\tilde{z},\tilde{r}) \in (X \times Y) \times (X^* \times Y^*)$, where $\tilde{z} = (x,y) \in X \times Y$ and $\tilde{r} = (p,q) \in X^* \times Y^*$, we can write

$$L \oplus_A M(\tilde{z},\tilde{r}) = L(x,A^*y+p) + M(y,-Ax+q) = (L+M)(\tilde{z},\tilde{A}\tilde{z}+\tilde{r}),$$

where $\tilde{A} : X \times Y \to X^* \times Y^*$ is defined by $\tilde{A}(\tilde{z}) = \tilde{A}((x,y)) = (A^*y, -Ax)$. The conditions ensure that the operator $(\tilde{A})^* \circ (B_1, B_2)$ is a skew-adjoint operator, so that the assertion follows from (7) above.

Assertion (10) follows again from Proposition 3.3 and (7) above since again the operator $(\tilde{A})^* \circ (B_1, B_2)$ is skew-adjoint.

Self-duality and orthogonal decompositions

Proposition 3.5. *Let L be a self-dual Lagrangian on a Hilbert space E, and let $E = Y \oplus Y^\perp$ be a decomposition of E into two orthogonal subspaces, with $\pi : E \to Y$ denoting the orthogonal projection onto Y and $\pi^\perp = I - \pi$. The Lagrangian*

$$M(u, p) = L(\pi(u) + \pi^\perp(p), \pi(p) + \pi^\perp(u)) \tag{3.4}$$

is then self-dual on $E \times E$.

Proof. Fix $(v,q) \in E \times E$. Note that if we set $r = \pi(u) + \pi^\perp(p)$ and $s = \pi(p) + \pi^\perp(u)$, then $p = \pi(s) + \pi^\perp(r)$ and $u = \pi(r) + \pi^\perp(s)$ and therefore

$$
\begin{aligned}
M^*(q,v) &= \sup\left\{ \langle q, u \rangle + \langle p, v \rangle - L(\pi(u) + \pi^\perp(p), \pi(p) + \pi^\perp(u)); u \in E, p \in E \right\} \\
&= \sup\left\{ \langle q, \pi(r) + \pi^\perp(s) \rangle + \langle \pi(s) + \pi^\perp(r), v \rangle - L(r,s); r \in E, s \in E \right\} \\
&= \sup\left\{ \langle \pi(q) + \pi^\perp(v), r \rangle + \langle \pi(v) + \pi^\perp(q), s \rangle - L(r,s); r \in E, s \in E \right\} \\
&= L^*(\pi(q) + \pi^\perp(v), \pi(v) + \pi^\perp(q)) \\
&= L(\pi(v) + \pi^\perp(q), \pi(q) + \pi^\perp(v)) \\
&= M(v,q).
\end{aligned}
$$

Exercises 3.C. Fenchelian and subself-dual Lagrangians

1. Show that if L is a B-self-dual Lagrangian on $X \times X^*$, then

$$L^*(B^*p, Bx) \geq L(x,p) \geq \langle Bx, p \rangle \text{ for all } (p,x) \in X^* \times X. \tag{3.5}$$

   Any Lagrangian L that satisfies (3.5) will be called B-*subself-dual*. It is said to be B-*Fenchelian* if it only satisfies

$$L(x,p) \geq \langle Bx, p \rangle \quad \text{for all } (p,x) \in X^* \times X. \tag{3.6}$$

   Denote by $\mathscr{L}_+^{\mathrm{sd}}(X;B)$ the class of B-subself-dual Lagrangians and by $\mathscr{L}_+^{\mathrm{sd}}(X)$ the class corresponding to when B is the identity operator. Similarly, denote by $\mathscr{F}(X,B)$ the class of B-Fenchelian Lagrangians on $X \times X^*$.
2. Let φ be a finite convex lower semicontinuous function on X and $\Gamma : X \to X^*$ be any bounded linear operator, and define the following Lagrangian on $X \times X^*$:

$$L(x,p) = \varphi(x) + \varphi^*(-\Gamma x + p) + \langle \Gamma x, x \rangle.$$

Show that if Γ is nonnegative (i.e., $\langle \Gamma x, x \rangle \geq 0$ for all $x \in X$), then L is Fenchelian on $X \times X^*$, and $M(x, p) = L^*(p, x)$ is subself-dual on $X \times X^*$.

3. Establish the following permanence properties for the classes $\mathscr{L}_+^{sd}(X; B)$ and $\mathscr{F}(X, B)$.

   a. If L is in $\mathscr{L}_+^{sd}(X; B)$ (resp., $\mathscr{F}(X, B)$) and if $\lambda > 0$, then $\lambda \cdot L$ also belongs to $\mathscr{L}_+^{sd}(X, B)$ (resp., $\mathscr{F}(X, B)$).

   b. If L is in $\mathscr{L}_+^{sd}(X; B)$ (resp., $\mathscr{F}(X, B)$), then for any $y \in X$ and $q \in X^*$, the translated Lagrangians M_y and N_p defined respectively by $M_y(x, p) = L(x + y, p) - \langle By, p \rangle$ and $N_q(x, p) = L(x, p + q) - \langle x, B^* q \rangle$ are also in $\mathscr{L}_+^{sd}(X; B)$ (resp., $\mathscr{F}(X, B)$).

   c. If $L \in \mathscr{L}_+^{sd}(X; B)$ (resp., $\mathscr{F}(X, B)$) for some automorphism B of X, and if $M(x, p) = \varphi(Bx) + \varphi^*(p)$ or $\varphi(x) + \varphi^*(B^* p)$, then the Lagrangians $L \oplus M$ and $L \star M$ also belong to $\mathscr{L}_+^{sd}(X; B)$ (resp., $\mathscr{F}(X, B)$).

   d. If $L, M \in \mathscr{L}_+^{sd}(X; B)$ (resp., $\mathscr{F}(X, B)$) are such that $\mathrm{Dom}_1(L) - \mathrm{Dom}_1(M)$ contains a neighborhood of the origin, then $L \star M$ and $L \oplus M$ are also in $\mathscr{L}_+^{sd}(X; B)$ (resp., $\mathscr{F}(X, B)$).

   e. If U is a unitary operator ($U^{-1} = U^*$) on a Hilbert space X that commutes with B, and if L is in $\mathscr{L}_+^{sd}(X; B)$ (resp., $\mathscr{F}(X, B)$), then $M(x, p) = L(Ux, Up)$ also belongs to $\mathscr{L}_+^{sd}(X; B)$ (resp., $\mathscr{F}(X, B)$).

   f. If $L_i \in \mathscr{L}_+^{sd}(X_i, B_i)$ (resp., $\mathscr{F}(X_i, B_i)$), where B_i is a bounded operator on a reflexive Banach space X_i for each $i \in I$, then $\Sigma_{i \in I} L_i$ is in $\mathscr{L}_+^{sd}(\Pi_{i \in I} X_i, (B_i)_{i \in I})$ (resp., $\mathscr{F}(\Pi_{i \in I} X_i, (B_i)_{i \in I})$).

   g. If $L \in \mathscr{L}_+^{sd}(X; B)$ (resp., $\mathscr{F}(X, B)$) and if $\Gamma : X \to X^*$ is a bounded linear operator, then the Lagrangian L_Γ is in $\mathscr{L}_+^{sd}(X; B)$ (resp., $\mathscr{F}(X, B)$), provided $B^* \Gamma$ is skew-adjoint (resp., $B^* \Gamma$ is positive).

   h. If $L \in \mathscr{L}_+^{sd}(X; B_1)$ (resp., $\mathscr{F}(X, B_1)$) and $M \in \mathscr{L}_+^{sd}(Y : B_2)$ (resp., $\mathscr{F}(X, B_2)$), then for any bounded linear operator $A : X \to Y^*$ such that $AB_1 = B_2^* A$ the Lagrangian $L \oplus_A M$ belongs to $\mathscr{L}_+^{sd}(X \times Y; (B_1, B_2))$ (resp., $\mathscr{F}(X \times Y, (B_1, B_2))$).

   i. If L is in $\mathscr{L}_+^{sd}(X; B)$ (resp., $\mathscr{F}(X, B)$), then for any onto operator $C : X \to X$, or if C has dense range and L is continuous in the first variable (resp., for any linear operator C), the Lagrangian $M(x, p) = L(Cx, p)$ is in $\mathscr{L}_+^{sd}(X; B \circ C)$ (resp., $\mathscr{F}(X, B \circ C)$).

   j. If L is in $\mathscr{L}_+^{sd}(X; B)$ (resp., $\mathscr{F}(X, B)$), then for any $D : X \to X$ such that D^* is onto, or if D^* has dense range and L is continuous in the second variable (resp., for any D), the Lagrangian $N(x, p) = L(x, D^* p)$ is in $\mathscr{L}_+^{sd}(X; D \circ B)$ (resp., $\mathscr{F}(X, D \circ B)$).

3.3 Self-dual Lagrangians on path spaces

We now present two different ways to "lift" a B-self-dual Lagrangian from a Hilbertian state space to path space.

Self-dual Lagrangians on $A_H^2 \times (H \times L_H^2)$

Let H be a Hilbert space, and consider again the space A_H^2 equipped with the equivalent norm

$$\|u\|_{A_H^2} = \left\{ \left\| \frac{u(0) + u(T)}{2} \right\|_H^2 + \int_0^T \|\dot{u}\|_H^2 \, dt \right\}^{\frac{1}{2}}.$$

The space A_H^2 can be identified with the product space $H \times L_H^2$ in such a way that its dual $(A_H^2)^*$ can also be identified with $H \times L_H^2$ via the formula

$$\Big\langle u, (p_1, p_0)\Big\rangle_{A_H^2, H \times L_H^2} = \Big\langle \frac{u(0) + u(T)}{2}, p_1 \Big\rangle + \int_0^T \langle \dot{u}(t), p_0(t)\rangle \, dt,$$

where $u \in A_H^2$ and $(p_1, p_0(t)) \in H \times L_H^2$.

Proposition 3.6. *Let B be a self-adjoint bounded linear operator on H, and suppose L is a time-dependent B-self-dual Lagrangian on $[0, T] \times H \times H$ and that ℓ is a B-self-dual Lagrangian on $H \times H$. The Lagrangian defined on $A_H^2 \times (A_H^2)^* = A_H^2 \times (H \times L_H^2)$ by*

$$\mathscr{L}(u, p) = \int_0^T L\big(t, u(t) + p_0(t), -\dot{u}(t)\big) \, dt + \ell\Big(u(T) - u(0) + p_1, \frac{u(0) + u(T)}{2}\Big).$$

is then a $\bar{B}$-self-dual Lagrangian on $A_H^2 \times (A_H^2)^*$, where $\bar{B}$ is defined on A_H^2 by $(\bar{B}u)(t) = B(u(t))$.

Proof. This follows immediately from Theorem 2.5, which states that the Legendre transform of $\mathscr{L}$ on $A_H^2 \times (L_H^2 \times H)$ is given by

$$\mathscr{L}^*(p, u) = \int_0^T L^*\big(t, -\dot{u}(t), u(t) + p_0(t)\big) \, dt + \ell^*\Big(\frac{u(0) + u(T)}{2}, u(T) - u(0) + p_1\Big).$$

Self-dual Lagrangians on $L_H^2 \times L_H^2$

Theorem 3.1. *Let B be a bounded linear operator on a Hilbert space H, and suppose ℓ is a $(-B, B)$-self-dual function on $H \times H$. Let L be a time-dependent B-self-dual Lagrangian on $H \times H$ such that:*

$$\text{For each } p \in L_H^2, \text{ the map } u \to \int_0^T L(t, u(t), p(t)) dt \text{ is continuous on } L_H^2. \quad (3.7)$$

$$\text{The map } u \to \int_0^T L(t, u(t), 0) dt \text{ is bounded on the balls of } L_H^2. \quad (3.8)$$

$$-C \leq \ell(a, b) \leq C(1 + \|a\|_H^2 + \|b\|_H^2) \text{ for all } (a, b) \in H \times H. \quad (3.9)$$

The Lagrangian defined on $L_H^2 \times L_H^2$ by

$$\mathscr{L}(u, p) = \begin{cases} \int_0^T L(t, u(t), p(t) - \dot{u}(t)) dt + \ell(u(0), u(T)) & \text{if } u \in A_H^2 \\ +\infty & \text{otherwise} \end{cases}$$

is then $\bar{B}$-self-dual, where $\bar{B}$ is defined on L_H^2 by $(\bar{B}u)(t) = B(u(t))$.

Proof. This follows immediately from Theorem 2.3, which states that the Legendre transform of $\mathscr{L}$ on $L_H^2 \times L_H^2$ is given by

$$\mathscr{L}^*(p, u) = \begin{cases} \int_0^T L^*(t, p(t) - \dot{u}(t), u(t)) dt + \ell^*(-u(0), u(T)) & \text{if } u \in A_H^2 \\ +\infty & \text{otherwise.} \end{cases}$$

Remark 3.1. Note the differences between the two ways of lifting self-dual Lagrangians to path space. For one we need more boundedness hypotheses on L and ℓ in the case of L_H^2. Moreover, in the case of A_H^2, the boundary Lagrangian ℓ needs to be a B-self-dual Lagrangian (i.e., $\ell^*(B^*p, Bx) = \ell(x,p)$), while for the case of L_H^2, ℓ needs to be a $(-B,B)$-self-dual function (i.e., $\ell^*(-Bx, Bp) = \ell(x,p)$) and B need not be self-adjoint.

3.4 Uniform convexity of self-dual Lagrangians

We shall need the following notion of uniform convexity for Lagrangians since it will yield certain regularity properties for the solutions of evolution equations driven by such Lagrangians.

Definition 3.4. Say that a Lagrangian L on $X \times X^*$ is *uniformly convex in the first variable* (resp., uniformly convex in both variables) if for some $\varepsilon_0 > 0$ (resp., for some $\varepsilon_0 > 0$ and $\delta_0 > 0$), the function

$$M(x,p) := L(x,p) - \varepsilon \|x\|^2 \ (\text{resp.}, M(x,p) := L(x,p) - \varepsilon \|x\|^2 - \delta \|p\|^2)$$

is convex on $X \times X^*$ for all $0 < \varepsilon < \varepsilon_0$ (resp., for all $0 < \varepsilon < \varepsilon_0$ and $0 < \delta < \delta_0$).

Lemma 3.1. *Let $F : H \to \mathbf{R} \cup \{+\infty\}$ be a convex and lower semicontinuous function on a Hilbert space such that its Legendre dual F^* is uniformly convex. Then, for every $x \in H$, the subdifferential $\partial F(x)$ is nonempty and single valued and the map $x \to \partial F(x)$ is Lipschitz on H.*

Proof. Since F^* is uniformly convex, then $F^*(x) = G(x) + \frac{\varepsilon \|x\|^2}{2}$ for some convex lower semicontinuous function G and some $\varepsilon > 0$. It follows that $F^*(x) \geq C + \langle a, x \rangle + \frac{\varepsilon \|x\|^2}{2}$ for some $a \in H$ and $C > 0$, and hence $F(x) = F^{**}(x) \leq C(1 + \|x\|^2)$, which means that F is subdifferentiable for all $x \in H$.

Consider now $p_j \in \partial F(x_j)$ for $j = 1, 2$ in such a way that $x_j \in \partial F^*(p_j) = \partial G(p_j) + \varepsilon p_j$. By monotonicity, we have $0 \leq \langle p_1 - p_2, \partial G(p_1) - \partial G(p_2) \rangle = \langle p_1 - p_2, x_1 - \varepsilon p_1 - x_2 + \varepsilon p_2 \rangle$, which yields that $\varepsilon \|p_1 - p_2\| \leq \|x_1 - x_2\|$ and we are done.

Proposition 3.7. *Assume that $L : H \times H \to \mathbf{R} \cup \{+\infty\}$ is a Lagrangian on a Hilbert space H such that both L and L^* are uniformly convex in both variables. Then, for all $x, u \in H$, there exists a unique $v \in H$, denoted $v = R(u,x)$, such that $x = \partial_2 L(u,v)$. Moreover, the map $(u,x) \to R(u,x)$ is jointly Lipschitz on $H \times H$.*

Proof. Since L is uniformly convex, then $L(x,p) = M(x,p) + \varepsilon \left(\frac{\|x\|^2}{2} + \frac{\|p\|^2}{2} \right)$, where M is convex lower semicontinuous in such a way that $x = \partial_2 L(u,v)$ if and only if $0 \in \partial_2 M(u,v) + \varepsilon v - x$ if and only if v is the solution to the minimization problem

$$\min_p \left\{ M(u,p) + \frac{\varepsilon \|p\|^2}{2} - \langle x,p \rangle \right\}.$$

But for each fixed u and x, the map $p \mapsto M(u,p) - \langle x,p \rangle$ majorizes a linear functional, and therefore the minimum is attained uniquely at v by strict convexity and obviously $x = \partial_2 L(u,v)$.

To establish the Lipschitz property, write

$$R(u_1,x_1) - R(u_2,x_2) = R(u_1,x_1) - R(u_1,x_2) + R(u_1,x_2) - R(u_2,x_2).$$

We first bound $\|R(u_1,x_1) - R(u_1,x_2)\|$ as follows. Since $x_1 = \partial_2 L(u_1, R(u_1,x_1))$, $x_2 = \partial_2 L(u_1, R(u_1,x_2))$, and $L(u_1,v) = M(u_1,v) + \frac{\varepsilon \|v\|^2}{2}$ for some M convex and lower semicontinuous, it follows that $x_j = \partial_2 M(u_1, R(u_1,x_j)) + \varepsilon R(u_1,x_j)$ for $j = 1,2$, so by monotonicity we get

$$0 \le \left\langle R(u_1,x_1) - R(u_1,x_2), \partial M(u_1, R(u_1,x_1)) - \partial M(u_1, R(u_1,x_2)) \right\rangle$$
$$= \left\langle R(u_1,x_1) - R(u_1,x_2), x_1 - \varepsilon R(u_1,x_1) - x_2 + \varepsilon R(u_1,x_2) \right\rangle,$$

which yields that

$$\varepsilon \|R(u_1,x_1) - R(u_1,x_2)\|^2 \le \|R(u_1,x_1) - R(u_1,x_2)\| \|x_1 - x_2\|$$

and therefore

$$\|R(u_1,x_1) - R(u_1,x_2)\| \le \frac{1}{\varepsilon} \|x_1 - x_2\|. \tag{3.10}$$

Now we bound $\|R(u_1,x_2) - R(u_2,x_2)\|$. For that we let $x_2 = \partial_2 L(u_j, R(u_j,x_2)) = \partial_2 M(u_j, R(u_j,x_2)) + \varepsilon R(u_j,x_2)$ for $j = 1,2$, and note that

$$0 \le \left\langle R(u_1,x_2) - R(u_2,x_2), \partial_2 M(u_1, R(u_1,x_2)) - \partial_2 M(u_1, R(u_2,x_2)) \right\rangle$$

by monotonicity. Setting $p_j = R(u_j,x_2)$, we have with this notation

$$\langle p_1 - p_2, \partial_2 M(u_1,p_2) - \partial_2 M(u_2,p_2) \rangle \le \langle p_1 - p_2, \partial_2 M(u_1,p_1) - \partial_2 M(u_2,p_2) \rangle$$
$$= \langle p_1 - p_2, x_2 - \varepsilon p_1 - x_2 + \varepsilon p_2 \rangle$$
$$= -\varepsilon \|p_1 - p_2\|^2,$$

so that $\varepsilon \|p_1 - p_2\|^2 \le \|p_1 - p_2\| \|\partial_2 M(u_1,p_2) - \partial_2 M(u_2,p_2)\|$. Since $\partial_2 M(u_j,p_2) = \partial_2 L(u_j,p_2) - \varepsilon p_2$, we get that

$$\varepsilon \|p_1 - p_2\| \le \|\partial_2 L(u_1,p_2) - \partial_2 L(u_2,p_2)\| \le \|\partial L(u_1,p_2) - \partial L(u_2,p_2)\|$$

since L^* is also uniformly convex. We then apply Lemma 3.1 to get

$$\|\partial L(u,p) - \partial L(u',p')\| \le C(\|u - u'\| + \|p - p'\|),$$

from which follows that $\|p_1 - p_2\| \le \frac{C}{\varepsilon} \|u_1 - u_2\|$, and hence,

$$\|R(u_1,x_2) - R(u_2,x_2)\| \leq \frac{C}{\varepsilon}\|u_1 - u_2\|. \tag{3.11}$$

Combining estimates (3.10) and (3.11), we finally get

$$\|R(u_1,x_1) - R(u_2,x_2)\| \leq \frac{1}{\varepsilon}(1+C)\left(\|u_1 - u_2\| + \|x_1 - x_2\|\right).$$

We can now deduce the following regularity result for certain flows driven by uniformly convex self-dual Lagrangians.

Corollary 3.1. *Assume $L : H \times H \to \mathbf{R} \cup \{+\infty\}$ is a Lagrangian on a Hilbert space H such that both L and L^* are uniformly convex in both variables. Suppose the paths $v, x, u : [0,T] \to H$ are such that $x, u \in C([0,T];H)$ and $x(t) = \partial_2 L\big(u(t), v(t)\big)$ for almost all $t \in [0,T]$. Then, $v \in C([0,T];H)$ and $x(t) = \partial_2 L\big(u(t), v(t)\big)$ for all $t \in [0,T]$.*

Exercises 3.D. Uniform convexity of self-dual Lagrangians

1. Show that the Lagrangian $L(x,p) = \varphi(x) + \varphi^*(p)$ is uniformly convex in the first variable (resp., uniformly convex in both variables) on $X \times X^*$, provided φ is a (resp., φ and φ^* are) uniformly convex function.
2. Prove the following analogue of Corollary 3.7, where the Hilbert space H is replaced by a reflexive Banach space X: If L is a self-dual Lagrangian on $X \times X^*$ that is uniformly convex in both variables, then there exists $C > 0$ and a map $R : X \times X \to X^*$ such that $R(x,y) \leq C(\|x\| + \|y\|)$ and $y = \partial_2 L(x, R(x,y))$ for every $(x,y) \in X \times X$.
   Hint: Use the fact that X then has an equivalent locally uniformly convex norm [42], from which it follows that the duality map between X and X^*, $D(x) = \{p \in X^*; \langle p,x \rangle = \|x\|^2\}$, is single valued and linear.

3.5 Regularization of self-dual Lagrangians

We now describe three ways of regularizing a self-dual Lagrangian. The first one ensures that it becomes bounded on bounded sets in the first variable, while the second guarantees coercivity in that variable. The third regularization is a combination of the first two and leads to both properties being satisfied. What is remarkable is that the regularized Lagrangians remain self-dual. It is worth comparing these regularization procedures to the inf-convolution operations on convex functions but also to the regularization by resolvents in Yosida's theory for operators (see Exercise 5.B). This should not be surprising, as self-dual Lagrangians include convex lower semicontinuous potentials, skew-adjoint operators and their superpositions.

Now it is clear that the most basic and the most regular self-dual Lagrangian on a Banach space X is $M_\lambda(x,p) = \frac{\|x\|^2}{2\lambda} + \frac{\lambda\|p\|^2}{2}$, where $\lambda > 0$. We shall use it to regularize other self-dual Lagrangians.

Lemma 3.2. *For $L : X \times X^* \to \mathbf{R} \cup \{+\infty\}$, we associate the following Lagrangians. For $(x, r) \in X \times X^*$, set*

$$L_\lambda^1(x, r) = \inf\left\{L(y, r) + \frac{\|x - y\|_X^2}{2\lambda} + \frac{\lambda \|r\|_{X^*}^2}{2}; y \in X\right\}$$

$$L_\lambda^2(x, r) = \inf\left\{L(x, s) + \frac{\|r - s\|_{X^*}^2}{2\lambda} + \frac{\lambda \|x\|_X^2}{2}; s \in X^*\right\}$$

and

$$L_\lambda^{1,2}(x, r) = \inf_{y \in X, s \in X^*}\left\{L(y, s) + \frac{1}{2\lambda}\|x - y\|_X^2 + \frac{\lambda}{2}\|r\|_{X^*}^2 + \frac{1}{2\lambda}\|s - r\|_{X^*}^2 + \frac{\lambda}{2}\|y\|_X^2\right\}.$$

If L is a self-dual (or antiself-dual) Lagrangian on $X \times X^*$, then the following properties hold:

1. *L_λ^1, L_λ^2 and $L_\lambda^{1,2}$ are also self-dual (or antiself-dual) Lagrangians on $X \times X^*$.*
2. *L_λ^1 (resp., L_λ^2) (resp., $L_\lambda^{1,2}$) is bounded on bounded sets in the first variable (resp., in the second variable) (resp., in both variables).*
3. *Suppose L is bounded from below. If $x_\lambda \rightharpoonup x$ and $p_\lambda \rightharpoonup p$ weakly in X and X^* respectively, as $\lambda \to 0$, and if $L_\lambda^i(x_\lambda, p_\lambda)$ is bounded from above for $i = 1$ or 2, then $L(x, p) \leq \liminf_{\lambda \to 0} L_\lambda^i(x_\lambda, p_\lambda)$. The same holds for $L_\lambda^{1,2}$.*

Proof. It suffices to note that $L_\lambda^1 = L \star M_\lambda$ and $L_\lambda^2 = L \oplus M_\lambda$, where the Lagrangian $M_\lambda(x, r) = \psi_\lambda(x) + \psi_\lambda^*(r)$ with $\psi_\lambda(x) = \frac{1}{2\lambda}\|x\|^2$. Note that $L_\lambda^{1,2} = (L \oplus M_\lambda) \star M_\lambda$ with $M_\lambda(x, r) = \frac{1}{2\lambda}\|x\|^2 + \frac{\lambda}{2}\|r\|^2$. The fact that they are self-dual Lagrangians follows from the calculus of self-dual Lagrangians developed in Chapter 2 and Proposition 3.4. The rest is straightforward and is left as an exercise.

For $(x, p) \in X \times X^*$, we denote by $J_\lambda(x, p)$ the minimizer in (3.12), that is,

$$L_\lambda^1(x, p) = L(J_\lambda(x, p), p) + \frac{\|x - J_\lambda(x, p)\|^2}{2\lambda} + \frac{\lambda \|p\|^2}{2}$$

$$= \inf\left\{L(z, p) + \frac{\|x - z\|^2}{2\lambda} + \frac{\lambda \|p\|^2}{2}; z \in X\right\},$$

in such a way that for all $(x, p) \in X \times X^*$

$$\partial_1 L_\lambda^1(x, p) = \frac{x - J_\lambda(x, p)}{\lambda} \in \partial_1 L(J_\lambda(x, p), p). \tag{3.12}$$

We shall need the following proposition, which relates the properties of a Lagrangian to those of its λ-regularization.

Lemma 3.3. *Let $L : X \times X^* \to \mathbf{R} \cup \{+\infty\}$ be a convex lower semicontinuous Lagrangian and consider for each $\lambda > 0$ its λ-regularizations L_λ^1 and L_λ^2.*

1. If L is uniformly convex in the first variable (resp., the second variable), then L_λ^1 (resp., L_λ^2) is uniformly convex (in both variables) on $X \times X^*$.
2. If L is self-dual and uniformly convex in the first variable, then L_λ^1 and $(L_\lambda^1)^*$ are self-dual and uniformly convex in both variables, and the map $(x, p) \mapsto J_\lambda(x, p)$ is Lipschitz on $X \times X^*$.

Proof. (1) For each $\lambda > 0$, there exists $\varepsilon > 0$ such that $M(x, p) := L(x, p) - \frac{\varepsilon \|x\|^2}{\lambda^2}$ is convex. Pick $\delta = \frac{1 - \frac{1}{1+\varepsilon}}{\lambda}$ so that $1 + \varepsilon = \frac{1}{1 - \lambda \delta}$ and note that the quantity

$$N_{\lambda, \delta}(x, p) := L_\lambda^1(x, p) - \frac{\lambda \|p\|^2}{2} - \delta \frac{\|x\|^2}{2}$$

can be written as

$$N_{\lambda, \delta}(x, p) = \inf_z \left\{ L(z, p) + \frac{\|x - z\|^2}{2\lambda} - \frac{\delta \|x\|^2}{2} \right\}$$

$$= \inf_z \left\{ L(z, p) + \frac{\|x\|^2}{2\lambda} - \frac{\langle x, z \rangle}{\lambda} + \frac{\|z\|^2}{2\lambda} - \frac{\delta \|x\|^2}{2} \right\}$$

$$= \inf_z \left\{ L(x, p) + \frac{\left\| \sqrt{\frac{1}{\lambda} - \delta}\, x \right\|^2}{2} - \frac{\langle x, z \rangle}{\lambda} + \frac{\|z\|^2}{2\lambda} \right\}$$

$$= \inf_z \left\{ L(z, p) + \frac{\left\| \sqrt{1 - \lambda \delta}\, x \right\|^2}{2\lambda} - \frac{\langle \sqrt{1 - \lambda \delta}\, x, \frac{z}{\sqrt{1 - \lambda \delta}} \rangle}{\lambda} + \frac{\|z\|^2}{2\lambda} \right\}$$

$$= \inf_z \left\{ M(z, p) + \frac{\varepsilon \|z\|^2}{2\lambda} + \frac{\left\| \sqrt{1 - \lambda \delta}\, x \right\|^2}{2\lambda} - \frac{\langle \sqrt{1 - \lambda \delta}\, x, \frac{z}{\sqrt{1 - \lambda \delta}} \rangle}{\lambda} + \frac{\|z\|^2}{2\lambda} \right\}$$

$$= \inf_z \left\{ M(z, p) + \frac{(1 + \varepsilon) \|z\|^2}{2\lambda} - \frac{\langle \sqrt{1 - \lambda \delta}\, x, \frac{z}{\sqrt{1 - \lambda \delta}} \rangle}{\lambda} + \frac{\left\| \sqrt{1 - \lambda \delta}\, x \right\|^2}{2\lambda} \right\}$$

$$= \inf_z \left\{ M(z, p) + \frac{\left\| \frac{z}{\sqrt{1 - \lambda \delta}} - \sqrt{1 - \lambda \delta}\, x \right\|^2}{2\lambda} \right\}.$$

This means that $(z, p, x) \mapsto M(z, p) + \frac{\left\| \frac{z}{\sqrt{1 - \lambda \delta}} - \sqrt{1 - \lambda \delta}\, x \right\|^2}{2\lambda}$ is convex and therefore the infimum in z is convex, which means that $L_\lambda^1(x, p) - \frac{\lambda \|p\|^2}{2} - \delta \frac{\|x\|^2}{2}$ is itself convex and hence, L_λ^1 is uniformly convex in both variables. The same proof shows that if L is uniformly convex in the second variable, then $L_\lambda^2 = L \oplus M_\lambda$ is uniformly convex in both variables.

For (2), apply (1) to L^*, which is then uniformly convex in the second variable, and use Proposition 3.4 (3), to get that both $L_\lambda^1 = L \star M_\lambda$ and $(L_\lambda^1)^* = (L \star M_\lambda)^* = L^* \oplus M_\lambda = (L^*)_\lambda^2$ are self-dual and uniformly convex in both variables. From Lemma

3.1, we get that $(x,p) \mapsto \partial L_\lambda^1(x,p)$ is Lipschitz, which yields – in view of (3.12) above – that $J_\lambda(x,p) = x - \lambda \partial_1 L_\lambda(x,p)$ is Lipschitz as well.

Exercises 3.E. More on regularizations of self-dual Lagrangians

1. Assume $p > 1$ and $\frac{1}{p} + \frac{1}{q} = 1$. For a Lagrangian $L : X \times X^* \to \mathbf{R} \cup \{+\infty\}$, define for every $(x,r) \in X \times X^*$

$$L_{\lambda,p}^1(x,r) = \inf \left\{ L(y,r) + \frac{\|x-y\|^p}{\lambda p} + \frac{\lambda^{q-1} \|r\|^q}{q}; y \in X \right\} \qquad (3.13)$$

and

$$L_{\lambda,p}^2(x,r) = \inf \left\{ L(x,s) + \frac{\|r-s\|^q}{\lambda q} + \frac{\lambda^{p-1} \|x\|^p}{p}; s \in X^* \right\}. \qquad (3.14)$$

Show that $L_{\lambda,p}^1$ and $L_{\lambda,p}^2$ satisfy most of the claims in Lemma 3.2 and 3.3.

2. Consider a self-dual Lagrangian of the form $L(x,p) = \varphi(x) + \varphi^*(Ax+p)$, where φ is a convex lower semicontinuous function on a reflexive Banach space X and $A : X \to X^*$ is a bounded skew-adjoint operator. Compare the self-dual Lagrangian L_λ^1 (resp., L_λ^2) with the self-dual Lagrangians

$$M_\lambda^1(x,p) = \varphi_\lambda(x) + (\varphi_\lambda)^*(Ax+p) \ (\text{resp.}, M_\lambda^2(x,p) = \varphi(x) + \tfrac{\lambda}{2}\|x\|^2 + (\varphi^*)_\lambda(Ax+p)),$$

where, for any $\lambda > 0$, the λ-inf-convolution of a function ψ is defined as

$$\psi_\lambda(x) := \inf \left\{ \psi(y) + \frac{\lambda}{2}\|x-y\|^2; y \in X \right\}.$$

3.6 Evolution triples and self-dual Lagrangians

A common framework for PDEs and evolution equations is the so-called *evolution triple* setting. It consists of a Hilbert space H with $\langle , \rangle_H$ as its scalar product, a reflexive Banach space X, and its dual X^* in such a way that $X \subset H \subset X^*$, with X being a dense vector subspace of H, while the canonical injection $X \to H$ is continuous. In this case, one identifies the Hilbert space H with its dual H^* and injects it in X^* in such a way that

$$\langle h,u \rangle_{X^*,X} = \langle h,u \rangle_H \quad \text{for all } h \in H \text{ and all } u \in X.$$

This injection is continuous and one-to-one, and H is also dense in X^*. In other words, the dual X^* of X is represented as the completion of H for the dual norm $\|h\| = \sup\{\langle h,u \rangle_H; \|u\|_X \le 1\}$.

1. A typical example of an evolution triple is $H_0^1(\Omega) \subset L^2(\Omega) \subset H^{-1}(\Omega)$, where Ω is a smooth bounded domain in $\mathbf{R}^n$.

2. Another example that is relevant for evolution equations is $A_H^2[0,T] \subset L_H^2[0,T] \subset (A_H^2[0,T])^*$, where H is a Hilbert space and $T > 0$.

3. More generally, for any evolution triple $X \subset H \subset X^*$, one can associate an evolution triple of path spaces such as $A_X^2[0,T] \subset L_H^2[0,T] \subset (A_X^2[0,T])^*$, where $A_X^2[0,T] = \{u : [0,T] \to X; \dot{u} \in L_X^2[0,T]\}$ equipped with the norm $\|u\|_{A_X^2} = (\|u\|_{L_X^2}^2 + \|\dot{u}\|_{L_X^2}^2)^{1/2}$.

4. Another choice is $\mathscr{X}_{2,2}[0,T] \subset L_H^2[0,T] \subset \mathscr{X}_{2,2}^*[0,T]$, where $\mathscr{X}_{2,2}[0,T]$ is the space of all functions in $L_X^2[0,T]$ such that $\dot{u} \in L_{X^*}^2[0,T]$, equipped with the norm $\|u\|_{\mathscr{X}_{2,2}} = (\|u\|_{L_X^2}^2 + \|\dot{u}\|_{L_{X^*}^2}^2)^{1/2}$.

5. More generally, we may consider for $1 < p < \infty$ and $\frac{1}{p} + \frac{1}{q} = 1$ the space

$$\mathscr{X}_{p,q} := W^{1,p}(0,T;X,H) = \{u; u \in L^p(0,T:X), \dot{u} \in L^q(0,T:X^*)\}$$

equipped with the norm $\|u\|_{W^{1,p}} = \|u\|_{L^p(0,T:X)} + \|\dot{u}\|_{L^q(0,T:X^*)}$, which leads to a continuous injection $W^{1,p}(0,T;X,H) \subseteq C(0,T:H)$.

The following useful lemma relates self-dual Lagrangians on $X \times X^*$ to those on the intermediate Hilbert space H.

Lemma 3.4. *Let $X \subset H \subset X^*$ be an evolution triple, and suppose $L : X \times X^* \to \mathbf{R} \cup \{+\infty\}$ is a self-dual (resp., antiself-dual) Lagrangian on the Banach space X. Assume the following two conditions:*

1. *For all $x \in X$, the map $L(x,\cdot) : X^* \to \mathbf{R} \cup \{+\infty\}$ is continuous on X^*.*
2. *There exists $x_0 \in X$ such that $p \to L(x_0,p)$ is bounded on the bounded sets of X^*.*

Then, the Lagrangian defined on $H \times H$ by

$$M(x,p) := \begin{cases} L(x,p) & x \in X \\ +\infty & x \in H \backslash X \end{cases}$$

is self-dual (resp., antiself-dual) on the Hilbert space $H \times H$.

In particular, the result holds for any self-dual (resp., antiself-dual) Lagrangian L on $X \times X^*$ that satisfies for some $C_1, C_2 > 0$ and $r_1 \geq r_2 > 1$

$$C_1(\|x\|_X^{r_2} - 1) \leq L(x,0) \leq C_2(1 + \|x\|_X^{r_1}) \text{ for all } x \in X. \tag{3.15}$$

Proof. Assume L is self-dual on $X \times X^*$. For $(\tilde{x}, \tilde{p}) \in X \times H$, write

$$M^*(\tilde{p}, \tilde{x}) = \sup_{\substack{x \in X \\ p \in H}} \{\langle \tilde{x}, p \rangle_H + \langle \tilde{p}, x \rangle_H - L(x,p)\}$$

$$= \sup_{x \in X} \sup_{p \in X^*} \{\langle \tilde{x}, p \rangle_{X,X^*} + \langle x, \tilde{p} \rangle_{X,X^*} - L(x,p)\}$$

$$= L(\tilde{x}, \tilde{p}).$$

If $\tilde{x} \in H \backslash X$, then

$$M^*(\tilde{p},\tilde{x}) = \sup_{\substack{x \in X \\ p \in H}} \{\langle \tilde{x}, p \rangle_H + \langle \tilde{p}, x \rangle_H - L(x,p)\} \geq \langle \tilde{p}, x_0 \rangle + \sup_{p \in H} \{\langle \tilde{x}, p \rangle_H - L(x_0, p)\}.$$

Since $\tilde{x} \notin X$, we have $\sup\{\langle \tilde{x}, p \rangle; p \in H, \|p\|_{X^*} \leq 1\} = +\infty$. Since $p \to L(x_0, p)$ is bounded on the bounded sets of X^*, it follows that

$$M^*(\tilde{p},\tilde{x}) \geq \langle \tilde{p}, x_0 \rangle + \sup_{p \in H} \{\langle \tilde{x}, p \rangle_H - L(x_0, p)\} = +\infty,$$

and we are done.

To prove the second part, one can use Lemma 3.5 below to get that for some $C_1, C_2 > 0$ and for $\frac{1}{r_i} + \frac{1}{s_i} = 1$, $i = 1, 2$,

$$C_1 (\|p\|_{X^*}^{s_1} + \|x\|_X^{r_2} - 1) \leq L(x,p) \leq C_2 (1 + \|x\|_X^{r_1} + \|p\|_{X^*}^{s_2}),$$

and therefore it satisfies the required continuity properties for M to be self-dual (or antiself-dual) on $H \times H$.

The following easy lemma establishes how boundedness in one of the variables of a self-dual Lagrangian, relates to coercivity in the other variable and vice-versa.

Lemma 3.5. *Let L be a self-dual (or an antiself-dual) Lagrangian on a reflexive Banach space $X \times X^*$.*

1. *Assume that for some $r > 1$ we have $L(x,0) \leq C(1 + \|x\|_X^r)$ for all $x \in X$. Then, there exist $D > 0$ such that*

$$L(x,q) \geq D(\|q\|_{X^*}^s - 1) \text{ for every } (x,q) \in X \times X^*, \text{ where } \frac{1}{r} + \frac{1}{s} = 1. \quad (3.16)$$

2. *Assume that for $C_1, C_2 > 0$ and $r_1 \geq r_2 > 1$ such that*

$$C_1 (\|x\|_X^{r_2} - 1) \leq L(x,0) \leq C_2 (1 + \|x\|_X^{r_1}) \text{ for all } x \in X. \quad (3.17)$$

Then, there exist $D_1, D_2 > 0$ such that

$$D_1 (\|p\|_{X^*}^{s_1} + \|x\|_X^{r_2} - 1) \leq L(x,p) \leq D_2 (1 + \|x\|_X^{r_1} + \|p\|_{X^*}^{s_2}), \quad (3.18)$$

where $\frac{1}{r_i} + \frac{1}{s_i} = 1$ for $i = 1, 2$, and L is therefore continuous in both variables.

Proof. For $(x,q) \in X \times X^*$, we have

$$\begin{aligned} L(x,q) &= \sup_{(y,p) \in X \times X^*} \{\langle x, p \rangle + \langle y, q \rangle - L^*(p,y)\} \\ &= \sup_{(y,p) \in X \times X^*} \{\langle x, p \rangle + \langle y, q \rangle - L(y,p)\} \\ &\geq \sup_{y \in X} \{\langle y, q \rangle - L(y,0)\} \\ &\geq \sup_{y \in X} \{\langle y, q \rangle - C(1 + \|y\|_X^r)\} \end{aligned}$$

$$= C_1 \|q\|_{X^*}^s - C_2$$

for some positive constants C_1 and C_2. Part (2) then follows by duality.

Remark 3.2. It is clear that Lemma 3.5 holds for any B-self-dual Lagrangian L, provided B is a bounded linear operator on the Hilbert space H such that its restriction to X is also a bounded linear operator on X that satisfies $Bx \in X$ if and only if $x \in X$.

Remark 3.3. We shall need the following extension of property 5) in Proposition 3.4 above to the setting of an evolution triple $X \subset H \subset X^*$, where X is reflexive and H is a Hilbert space. Also assume that there exists a linear and symmetric duality map D between X and X^* in such a way that $\|x\|^2 = \langle x, Dx \rangle$. We can then consider X and X^* as Hilbert spaces with the following inner products:

$$\langle u, v \rangle_{X \times X} := \langle Du, v \rangle \quad \text{and} \quad \langle u, v \rangle_{X^* \times X^*} := \langle D^{-1}u, v \rangle. \tag{3.19}$$

A typical example is the evolution triple $X = H_0^1(\Omega) \subset H := L^2(\Omega) \subset X^* = H^{-1}(\Omega)$, where the duality map is given by $D = -\Delta$.

If now $\bar{S}$ is an isometry on X^*, then $S = D^{-1}\bar{S}D$ is also an isometry on X in such a way that

$$\langle u, p \rangle = \langle Su, \bar{S}p \rangle \text{ for all } u \in X \text{ and } p \in X^*. \tag{3.20}$$

Indeed, we have

$$\langle Su, \bar{S}p \rangle = \langle DSu, \bar{S}p \rangle_{X^* \times X^*} = \langle \bar{S}Du, \bar{S}p \rangle_{X^* \times X^*} = \langle Du, p \rangle_{X^* \times X^*} = \langle u, p \rangle,$$

from which we can deduce that

$$\|Su\|_X^2 = \langle Su, Su \rangle_{X \times X} = \langle Su, DSu \rangle = \langle Su, \bar{S}Du \rangle = \langle u, Du \rangle = \|u\|_X^2.$$

Moreover, if L is a self-dual Lagrangian on $X \times X^*$, then $L_S := L(Su, \bar{S}p)$ is also a self-dual Lagrangian on $X \times X^*$ since

$$\begin{aligned} L_S^*(p, u) &= \sup\{\langle v, p \rangle + \langle u, q \rangle - L_S(v, q); (v, q) \in X \times X^*\} \\ &= \sup\{\langle Sv, \bar{S}p \rangle + \langle Su, \bar{S}q \rangle - L(Sv, \bar{S}q); (v, q) \in X \times X^*\} \\ &= L^*(\bar{S}p, Su) = L(Su, \bar{S}p) = L_S(u, p). \end{aligned}$$

Further comments

The basic properties of the class of B-self-dual Lagrangians were exhibited in Ghoussoub [55]. Their various regularizations appeared in [68], [64], and [65]. Since self-dual Lagrangians represent extensions of both convex functions and skew-adjoint operators, their λ-regularization can be seen as an extension of both Yosida's resolvent theory for operators and the inf-convolution procedure often used to regularize convex functions.

Chapter 4
Skew-Adjoint Operators and Self-dual Lagrangians

If L is a self-dual Lagrangian on a reflexive Banach space X and if $\Gamma : X \to X^*$ is a skew-adjoint operator, we have seen that the Lagrangian defined by $L_\Gamma(x,p) = L(x, \Gamma x + p)$ is then also self-dual on X. In this chapter, we deal with the more interesting cases of unbounded antisymmetric operators and with nonhomogeneous problems where operators may be skew-adjoint modulo certain boundary terms normally given by a Green-Stokes type formula of the type:

$$\langle x, \Gamma y \rangle + \langle y, \Gamma x \rangle + \langle \mathscr{B}x, R\mathscr{B}y \rangle = 0 \text{ for every } x,y \in D(\Gamma),$$

where $\mathscr{B} : D(\mathscr{B}) \subset X \to H$ is a boundary operator into a Hilbert space H and R is a self-adjoint automorphism on the "boundary" space H. In this case, a suitable self-dual boundary Lagrangian – namely an R-self-dual convex function ℓ on H – should be added to restore self-duality to the whole system. Under appropriate boundedness conditions, the following Lagrangian is then again self-dual on $X \times X^*$:

$$L_{\Gamma,\ell}(x,p) = \begin{cases} L(x, \Gamma x + p) + \ell(\mathscr{B}x) & \text{if } x \in D(\Gamma) \cap D(\mathscr{B}) \\ +\infty & \text{if } x \notin D(\Gamma) \cap D(\mathscr{B}). \end{cases}$$

4.1 Unbounded skew-symmetric operators and self-dual Lagrangians

Definition 4.1. Let B be a bounded linear operator on a reflexive Banach space X, and let $\Gamma : D(\Gamma) \subset X \to X^*$ be a – not necessarily bounded – linear operator with a domain $D(\Gamma)$ that is dense in X. We consider the set

$$D^*(B,\Gamma) = \{x \in X; \sup\{\langle Bx, \Gamma y \rangle; y \in D(\Gamma), \|y\|_X \le 1\} < +\infty\},$$

which – in the case where B is the identity on X – is nothing but the domain of the adjoint operator Γ^* of Γ.

N. Ghoussoub, *Self-dual Partial Differential Systems and Their Variational Principles*, Springer Monographs in Mathematics, DOI 10.1007/978-0-387-84897-6_4,
© Springer Science+Business Media, LLC 2009

1. The pair (B,Γ) is said to be *nonnegative* if

$$\langle Bx, \Gamma x \rangle \geq 0 \text{ for every } x \in D(\Gamma). \tag{4.1}$$

2. (B,Γ) is said to be *antisymmetric* if

$$\langle Bx, \Gamma x \rangle = 0 \text{ for every } x \in D(\Gamma). \tag{4.2}$$

3. The pair (B,Γ) is said to be *skew-adjoint* if in addition to being antisymmetric it satisfies

$$D(\Gamma) = D^*(B,\Gamma). \tag{4.3}$$

When B is the identity, we shall simply say that Γ is nonnegative (resp. antisymmetric , resp., skew-adjoint), provided the pair (I,Γ) is nonnegative (resp. antisymmetric , resp., skew-adjoint).

Exercises 4.A. Examples of antisymmetric operators

1. The Schrödinger operator: Let Ω be an open subset of $\mathbf{R}^n$.
   (i) Consider the Sobolev space $H := H^{-1}(\Omega;\mathbf{C}) = H^{-1}(\Omega) + iH^{-1}(\Omega)$ endowed with its real Hilbert space structure. Show that the operator $\Gamma : D(\Gamma) \subset H^{-1}(\Omega;\mathbf{C}) \to H^{-1}(\Omega;\mathbf{C})$ defined as

$$\Gamma u = i\Delta u \text{ with domain } D(\Gamma) = H_0^1(\Omega, \mathbf{C})$$

is a skew-adjoint operator on $H^{-1}(\Omega;\mathbf{C})$.
   (ii) Consider the Hilbert space $H := L^2(\Omega;\mathbf{C}) = L^2(\Omega) + iL^2(\Omega)$ endowed with its real Hilbert structure. Show that the operator $\Gamma : D(\Gamma) \subset L^2(\Omega;\mathbf{C}) \to L^2(\Omega;\mathbf{C})$ defined as

$$\Gamma u = i\Delta u \text{ with domain } D(\Gamma) = \{u \in H_0^1(\Omega,\mathbf{C}); \Delta u \in L^2(\Omega;\mathbf{C})\}$$

is a skew-adjoint operator on $L^2(\Omega;\mathbf{C})$.
   (iii) On the other hand, if $p \neq 2$, show that the operator $\Gamma_p : D(\Gamma) \subset H^{-1}(\Omega;\mathbf{C}) \to H^{-1}(\Omega;\mathbf{C})$ defined as

$$\Gamma_p u = i\Delta u \text{ with domain } D(\Gamma_p) = W_0^{1,p}(\Omega,\mathbf{C})$$

is antisymmetric but not necessarily skew-adjoint.
2. The transport operator: Let $\mathbf{a} : \Omega \to \mathbf{R}^n$ be a smooth function on a bounded domain Ω of $\mathbf{R}^n$.
   (i) Show that the first-order linear operator $\Gamma : D(\Gamma) \subset H^{-1}(\Omega) \to H^{-1}(\Omega)$

$$\Gamma v = \mathbf{a} \cdot \nabla v = \Sigma_{i=1}^n a_i \frac{\partial v}{\partial x_i} \text{ with domain } D(\Gamma) = H_0^1(\Omega)$$

is skew-adjoint whenever $\mathrm{div}(\mathbf{a}) = 0$.
   (ii) On the other hand, if $p \neq 2$, show that the operator $\Gamma_p : D(\Gamma) \subset H^{-1}(\Omega;\mathbf{C}) \to H^{-1}(\Omega;\mathbf{C})$ defined as

$$\Gamma_p u = \mathbf{a} \cdot \nabla v \text{ with domain } D(\Gamma_p) = W_0^{1,p}(\Omega,\mathbf{C})$$

is antisymmetric but not necessarily skew-adjoint.
   (iii) If $\mathbf{a} : \Omega \to \mathbf{R}^n$ has compact support and $\mathrm{div}(\mathbf{a}) = 0$, show that for $p \geq 1$ the operator $\Gamma : D(\Gamma) \subset L^p(\Omega) \to L^p(\Omega)$ defined by

$$\Gamma v = \mathbf{a} \cdot \nabla v \text{ with domain } D(\Gamma) = \{u \in L^p(\Omega); \mathbf{a} \cdot \nabla v \in L^p(\Omega)\}$$

is also skew-adjoint.

3. The wave operator: Let Ω be an open subset of $\mathbf{R}^n$.

    (i) Show that the operator $\Gamma : D(\Gamma) \subset H := L^2(\Omega) \times H^{-1}(\Omega) \to L^2(\Omega) \times H^{-1}(\Omega)$ defined by

$$\Gamma(u,v) = (v, \Delta u) \text{ with domain } D(\Gamma) = H_0^1(\Omega) \times L^2(\Omega)$$

is skew-adjoint.

    (ii) Similarly, show that the operator $\Gamma : D(\Gamma) \subset H := H_0^1(\Omega) \times L^2(\Omega) \to H_0^1(\Omega) \times L^2(\Omega)$

$$\Gamma(u,v) = (v, \Delta u) \text{ with domain } D(\Gamma) = \{(u,v) \in H; \Delta u \in L^2(\Omega), v \in H_0^1(\Omega)\}$$

is skew-adjoint.

4. The Klein-Gordon operator: Let Ω be an open subset of $\mathbf{R}^n$, and let $\lambda < \lambda_1(\Omega)$, the first eigenvalue of the Laplacian on $H_0^1(\Omega)$. In this case, the operator $-\Delta - \lambda I : H_0^1(\Omega) \to H^{-1}(\Omega)$ is invertible and we set $T_\lambda = (-\Delta - \lambda I)^{-1}$. Now consider $H := L^2(\Omega) \times H^{-1}(\Omega)$, equipped with the equivalent scalar product

$$\langle (u_1, v_1), (u_2, v_2) \rangle = \int_\Omega \nabla(T_\lambda u_1) \cdot \nabla(T_\lambda u_2) \, dx - \lambda \int_\Omega T_\lambda u_1 \cdot T_\lambda u_2 \, dx + \int_\Omega v_1 v_2 \, dx.$$

    (i) Show that the operator $\Gamma : D(\Gamma) \subset H := L^2(\Omega) \times H^{-1}(\Omega) \to L^2(\Omega) \times H^{-1}(\Omega)$ defined by

$$\Gamma(u,v) = (v, \Delta u + \lambda u) \text{ with domain } D(\Gamma) = H_0^1(\Omega) \times L^2(\Omega)$$

is skew-adjoint.

    (ii) Similarly, show that the operator $\Gamma : D(\Gamma) \subset H := H_0^1(\Omega) \times L^2(\Omega) \to H_0^1(\Omega) \times L^2(\Omega)$

$$\Gamma(u,v) = (v, \Delta u + \lambda u) \text{ with domain } D(\Gamma) = \{(u,v) \in H; \Delta u \in L^2(\Omega), v \in H_0^1(\Omega)\}$$

is skew-adjoint.

5. The Airy operator defined on $L^2(\mathbf{R})$ by

$$\Gamma(u) = -\frac{d^3 u}{dx^3} \text{ with domain } D(\Gamma) = H^3(\mathbf{R})$$

is clearly skew-adjoint.

6. The Maxwell operator: Consider the Hilbert space $H = (L^2(\mathbf{R}^3))^3 \times (L^2(\mathbf{R}^3))^3$ and its subspace $H_0 = \{(\mathbf{E}, \mathbf{H}) \in H; \nabla \cdot \mathbf{E} = \nabla \cdot \mathbf{H} = 0\}$, where the differential operator ∇ is defined in the sense of distributions. Show that the Maxwell operator $\Gamma : D(\Gamma) \subset H_0 \to H_0$ defined by

$$\Gamma(\mathbf{E}, \mathbf{H}) = (-c\nabla \times \mathbf{H}, c\nabla \times \mathbf{E}) \text{ with domain } D(\Gamma) = \{(\mathbf{E}, \mathbf{H}) \in H_0; \Gamma(\mathbf{E}, \mathbf{H}) \in H_0\}$$

is skew-adjoint.

Definition 4.2. We say that a Lagrangian L is *B-standard* if for every $y \in X \setminus \text{Dom}_1(L)$, there is $x \in B^{-1}(\text{Dom}_1(L))$ such that $H_L(Bx, y) = +\infty$, where H_L is the Hamiltonian associated to L.

Note that any basic self-dual Lagrangian of the form $L(x, p) = \varphi(x) + \varphi^*(p)$, where φ is a proper convex lower semicontinuous function, is B-standard whenever $B^{-1}(\text{Dom}(\varphi)) \neq \emptyset$. Indeed, we have then $H_L(Bx, y) = \varphi(y) - \varphi(Bx)$ whenever $Bx \in \text{Dom}(\varphi)$.

The following permanence propery is the extension of Proposition 3.4.7) to the setting of unbounded linear operators.

Proposition 4.1. *Let B be a bounded linear operator on a reflexive Banach space X with dense range, $L : X \times X^* \to \mathbf{R} \cup \{+\infty\}$ a Lagrangian in $\mathscr{L}(X)$ that is continuous in the first variable, and $\Gamma : D(\Gamma) \subset X \to X^*$ a linear operator with dense domain. Assume one of the following two conditions:*

1. *The pair (B, Γ) is antisymmetric , L is B-standard, and $B^{-1}(\mathrm{Dom}_1(L)) \subset D(\Gamma)$,*
   or
2. *the pair (B, Γ) is skew-adjoint, and the function $x \to L(x, p_0)$ is bounded on the balls of X for some $p_0 \in X^*$.*

Then, the Legendre dual of the Lagrangian

$$L_\Gamma(x, p) = \begin{cases} L(Bx, \Gamma x + p) & \text{if } x \in D(\Gamma) \\ +\infty & \text{if } x \notin D(\Gamma) \end{cases}$$

satisfies

$$L_\Gamma^*(B^* p, Bx) = \begin{cases} L^*(\Gamma x + p, Bx) & \text{if } x \in D(\Gamma) \\ +\infty & \text{if } x \notin D(\Gamma). \end{cases}$$

It follows that L_Γ is B-self-dual whenever L is a self-dual Lagrangian on $X \times X^*$.

Proof. Assume first that the pair (B, Γ) is antisymmetric and that L is a Lagrangian that is continuous on X in the first variable. We first consider the case where $y \in D(\Gamma)$. Indeed, write

$$L_\Gamma^*(B^* q, By) = \sup \{\langle By, p \rangle + \langle x, B^* q \rangle - L(Bx, \Gamma x + p); x \in D(\Gamma), p \in X^*\}.$$

Substituting $r = \Gamma x + p$ and since for $y \in D(\Gamma)$ we have $\langle By, \Gamma x \rangle = -\langle Bx, \Gamma y \rangle$, and using that L is continuous in the first variable, we obtain

$$
\begin{aligned}
L_\Gamma^*(B^* q, By) &= \sup_{\substack{x \in D(\Gamma) \\ r \in X^*}} \{\langle By, r - \Gamma x \rangle + \langle Bx, q \rangle - L(Bx, r)\} \\
&= \sup_{\substack{x \in D(\Gamma) \\ r \in X^*}} \left\{ \langle Bx, \Gamma y \rangle + \langle By, r \rangle + \langle Bx, q \rangle - L(Bx, r) \right\} \\
&= \sup \left\{ \langle Bx, \Gamma y + q \rangle + \langle By, r \rangle - L(Bx, r); x \in D(\Gamma), r \in X^* \right\} \\
&= \sup \left\{ \langle z, \Gamma y + q \rangle + \langle By, r \rangle - L(z, r); z \in X, r \in X^* \right\} \\
&= L^*(\Gamma y + q, By).
\end{aligned}
$$

If now $y \notin D(\Gamma)$, we shall distinguish the two cases.

Case 1. Under condition (1), we have that $By \notin \mathrm{Dom}_1(L)$. Hence, since L is B-standard, we have for some $Bx_0 \in \mathrm{Dom}_1(L)$ that $H_L(Bx_0, By) = +\infty$ and therefore, since $x_0 \in B^{-1}(\mathrm{Dom}_1(L)) \subset D(\Gamma)$, we have:

$$L_\Gamma^*(B^*q,By) = \sup_{\substack{x\in \mathrm{Dom}_1(L)\\ r\in X^*}} \{\langle By, r-\Gamma x\rangle + \langle Bx, q\rangle - L(Bx,r)\}$$

$$\geq \sup_{r\in X^*} \{\langle By, r\rangle + \langle By, -\Gamma x_0\rangle + \langle Bx_0, q\rangle - L(Bx_0, r)\}$$

$$= H_L(Bx_0, y) + \langle By, -\Gamma x_0\rangle + \langle Bx_0, q\rangle$$

$$= +\infty.$$

Case 2. Under condition (2), write

$$L_\Gamma^*(B^*q,By) = \sup_{\substack{x\in D(\Gamma)\\ r\in X^*}} \{\langle By, r-\Gamma x\rangle + \langle Bx, q\rangle - L(Bx,r)\}$$

$$\geq \sup_{\substack{x\in D(\Gamma)\\ \|Bx\|_X<1}} \{\langle -By, \Gamma x\rangle + \langle Bx, q\rangle + \langle By, p_0\rangle - L(Bx, p_0)\}.$$

Since by assumption $L(Bx, p_0) \leq K$ whenever $\|x\|_X \leq 1$, we obtain since $y \notin D(\Gamma) = D^*(B,\Gamma)$ that

$$L_\Gamma^*(B^*q,By) \geq \sup_{\substack{x\in D(\Gamma)\\ \|x\|_X<1}} \langle By, \Gamma x\rangle - \|B\|\|q\| + \langle By, p_0\rangle - K = +\infty.$$

The formula is then verified for all $(y,q) \in X \times X^*$, and the proposition is proved.

Remark 4.1. Unlike the case of a bounded skew-adjoint operator, the conditions in Theorem 4.1 are essential to conclude that L_Γ is self-dual. Here is a counterexample: Let $H = L^2[0,1]$, and consider the operator $\Gamma = \frac{d}{dt}$ on H with domain

$$D(\Gamma) = \{x \in A_H^2[0,1]; x(0) = x(1)\}.$$

Note that $y \in D(\Gamma)$ if and only if $y \in L^2$ and $x \to \int_0^1 \langle \dot{x}(t), y(t)\rangle\, dt$ is bounded on $D(\Gamma)$, i.e., for some $C > 0$,

$$\left|\int_0^1 \langle \dot{x}(t), y(t)\rangle\, dt\right| \leq C\|x\|_{L^2} \text{ for all } x \in D(\Gamma).$$

Hence, $-\Gamma^* = \Gamma$ and $D(\Gamma^*) = D(\Gamma)$, meaning that Γ is skew-adjoint. Now consider the convex lower semicontinuous function φ on H defined by

$$\varphi(x) = \begin{cases} \frac{1}{2}\|\dot{x}\|_{L^2}^2 & \text{if } x \in A_H^2 \\ +\infty & \text{otherwise,} \end{cases}$$

and let L be the self-dual Lagrangian $L(x,p) = \varphi(x) + \varphi^*(p)$. We claim that the Lagrangian defined by

$$L_\Gamma(x,p) = \begin{cases} L(x, \Gamma x + p) & \text{if } x \in D(\Gamma) \\ +\infty & \text{otherwise} \end{cases}$$

is not self-dual on $H \times H$.

Indeed, if $\tilde{x} \notin D(\Gamma)$ but $\tilde{x} \in A_H^2$, then

$$
\begin{aligned}
L_\Gamma^*(\tilde{p}, \tilde{x}) &= \sup_{\substack{x \in D(\Gamma)\, z \in H}} \sup \left\{ \langle \tilde{p}, x \rangle + \langle z - \Gamma x, \tilde{x} \rangle - \varphi(x) - \varphi^*(z) \right\} \\
&= \sup_{\substack{x \in D(\Gamma) z \in H}} \left\{ \langle \tilde{p}, x \rangle + \langle z - \dot{x}, \tilde{x} \rangle - \frac{\|\dot{x}\|_{L^2}}{2} - \varphi^*(z) \right\} \\
&= \sup_{\substack{x \in D(\Gamma) \\ z \in H}} \left\{ \int_0^1 \int_0^t \tilde{p}(s) \dot{x}(t)\, ds\, dt + \int_0^1 \tilde{p}(s)\, ds + \langle z - \dot{x}, \tilde{x} \rangle - \frac{\|\dot{x}\|_2}{2} - \varphi^*(z) \right\} \\
&\le \sup_{\substack{v \in H \\ z \in H}} \left\{ \left\langle \int_0^t \tilde{p}(s)\, ds - \tilde{x}, v \right\rangle + \langle z, \tilde{x} \rangle - \frac{\|v\|_2^2}{2} - \varphi^*(z) \right\} + \int_0^1 \tilde{p}(s)\, ds \\
&= \frac{1}{2} \left\| \int_0^t \tilde{p}(s)\, ds - \tilde{x} \right\|_{L^2}^2 + \varphi(\tilde{x}) + \int_0^1 \tilde{p}(s)\, ds < \infty,
\end{aligned}
$$

and so L_Γ is not a self-dual Lagrangian.

More generally, suppose $\Gamma : D(\Gamma) \to H$ is any skew-adjoint operator such that $D(\Gamma) \subseteq X \subseteq H \subseteq X^*$ and $\|\Gamma x\|_{X^*} \le C\|x\|_X$ for all $x \in D(\Gamma)$. Consider on H the convex lower semicontinuous function

$$
\varphi(x) = \begin{cases} \frac{\|x\|_X^2}{2} & \text{if } x \in X \\ +\infty & \text{if } x \in H \backslash X \end{cases}
$$

and its Legendre transform φ^* on H. Let L_Γ be the Lagrangian defined by

$$
L_\Gamma(x, p) = \begin{cases} \varphi(x) + \varphi^*(\Gamma x + p) & \text{if } x \in D(\Gamma) \\ +\infty & \text{otherwise.} \end{cases}
$$

We show that if $D(\Gamma) \neq X$, then L_Γ cannot be self-dual on $H \times H$. Indeed, $L_\Gamma(\tilde{x}, p) = +\infty$ for $\tilde{x} \in X \backslash D(\Gamma)$, while

$$
\begin{aligned}
L_\Gamma^*(\tilde{p}, \tilde{x}) &= \sup_{\substack{x \in D(\Gamma) \\ z \in H}} \left\{ (\tilde{p}, x) + (z - \Gamma x, \tilde{x}) - \varphi(x) - \varphi^*(z) \right\} \\
&\le \sup_{\substack{x \in D(\Gamma) \\ z \in H}} \left\{ \|\tilde{p}\|_{X^*} \|x\|_X + \|\Gamma x\|_{X^*} \|\tilde{x}\|_X + \|z\|_{X^*} \|\tilde{x}\|_X - \frac{\|x\|_X^2}{2} - \frac{\|z\|_{X^*}^2}{2} \right\}.
\end{aligned}
$$

Using that $\|\Gamma x\|_{X^*} \le C\|x\|_X$, we get

$$
\begin{aligned}
L_\Gamma^*(\tilde{p}, \tilde{x}) &\le \sup_{x \in D(\Gamma) z \in H} \left\{ \left(\|\tilde{p}\|_{X^*} + C\|\tilde{x}\|_X \right) \|x\|_X + \|\tilde{x}\|_X \|z\|_{X^*} - \frac{\|x\|_X^2}{2} - \frac{\|z\|_{X^*}^2}{2} \right\} \\
&= \frac{\left(\|\tilde{p}\|_{X^*} + C\|\tilde{x}\|_X \right)^2}{2} + \frac{\|\tilde{x}\|_X^2}{2} < \infty.
\end{aligned}
$$

Exercise 4.B. Properties of L_Γ when Γ is not skew-adjoint

1. Assume $L : X \times X^* \to \mathbf{R}$ is a self-dual B-standard Lagrangian that is continuous in the first variable. Suppose $\Gamma : D(\Gamma) \subset X \to X^*$ is such that the pair (B,Γ) is a nonnegative linear operator with dense domain such that $\text{Dom}_1(L) \subset D(\Gamma)$. Show that the following Lagrangian

$$L_\Gamma(x,p) = \begin{cases} L(Bx, -\Gamma x + p) + \langle Bx, \Gamma x \rangle & \text{if } x \in D(\Gamma) \\ +\infty & \text{if } x \notin D(\Gamma) \end{cases}$$

   is B-Fenchelian, while the Lagrangian $M_\Gamma(x,p) = L_\Gamma^*(B^* p, Bx)$ is B-subself-dual on $X \times X^*$.

4.2 Green-Stokes formulas and self-dual boundary Lagrangians

We now consider operators that are antisymmetric modulo boundary terms. The following framework covers many concrete situations.

Definition 4.3. Let X be a Banach space (the "state space"). A *boundary triplet* $(H, R, \mathscr{B})$ associated to X consists of a Hilbert space H (the "boundary space"), a self-adjoint automorphism R on H, and a linear "boundary operator" $\mathscr{B} : D(\mathscr{B}) \subset X \to H$ with a dense range and a dense domain.

Consider a bounded linear operator B on X and a linear operator $\Gamma : D(\Gamma) \subset X \to X^*$.

• The pair (B,Γ) is said to be *antisymmetric modulo the boundary triplet* $(H, R, \mathscr{B})$ if the following properties are satisfied:

1. The image of the space $X_0 := \text{Ker}(\mathscr{B}) \cap D(\Gamma)$ by B is dense in X.
2. The image of the space $X_1 := D(\Gamma) \cap D(\mathscr{B})$ by $\mathscr{B}$ is dense in H.
3. For every $x \in X_1$, we have that $\langle Bx, \Gamma x \rangle + \frac{1}{2}\langle \mathscr{B}x, R\mathscr{B}x \rangle = 0$.

• The pair (B,Γ) is said to be *skew-adjoint modulo the boundary triplet* $(H, R, \mathscr{B})$ if it is *antisymmetric modulo* $(H, R, \mathscr{B})$ and if in addition it satisfies for $C > 0$

4. $X_1 = \left\{ y \in X; \sup\{ \langle By, \Gamma x \rangle - C\|\mathscr{B}x\|^2; x \in X_1, \|x\|_X \leq 1 \} < +\infty \right\}$.

As Theorem 4.2 below will show, skew-adjoint operators modulo boundary triplets allow the building of new self-dual Lagrangians from old ones, provided one superposes a suitable boundary Lagrangian. First, we give a few examples of antisymmetric operators modulo a boundary.

Exercises 4.C. Antisymmetric operators modulo a boundary

1. **The operator $\frac{du}{dt}$:** Let H be a Hilbert space, and consider the space $X = L_H^2[0,T]$ and the operator $\Gamma : D(\Gamma) \subset X \to X$ defined by

$$\Gamma u = \frac{du}{dt} \text{ with domain } D(\Gamma) = A_H^2[0,T].$$

In view of the identity

$$\int_0^T \big(\dot{u}(t)v(t) + u(t)\dot{v}(t) \big)dt + \langle u(0), v(0) \rangle - \langle u(T), v(T) \rangle = 0,$$

which holds for any u, v in $A_H^2[0,T]$, show that the operator Γ is skew-adjoint modulo a boundary triplet $(H \times H, R, \mathscr{B})$ in two different ways:

- The boundary operator $\mathscr{B} : D(\mathscr{B}) \subset L_H^2[0,T] \to H \times H$ can be defined as $\mathscr{B}u = \big(u(0), u(T) \big)$ with domain $A_H^2[0,T]$, while the automorphism R_1 is defined on $H \times H$ by $R_1(a,b) = (a,-b)$. In other words,

$$\int_0^T \dot{u}(t)u(t)dt + \frac{1}{2}\langle \mathscr{B}u, R_1\mathscr{B}u \rangle = \int_0^T \dot{u}(t)u(t)dt + \frac{1}{2}\big\langle \big(u(0), u(T)\big), \big(u(0), -u(T)\big) \big\rangle = 0.$$

- The boundary operator $\mathscr{C} : D(\mathscr{C}) \subset L_H^2[0,T] \to H \times H$ can also be written as $\mathscr{C}u = (\mathscr{C}_1u, \mathscr{C}_2u) = \big(u(0) - u(T), \frac{u(0)+u(T)}{2}\big)$ with domain $A_H^2[0,T]$, while the automorphism R_2 is defined on $H \times H$ by $R_2(a,b) = (b,a)$. In other words,

$$\int_0^T \dot{u}(t)u(t)dt + \frac{1}{2}\langle \mathscr{C}u, R_2\mathscr{C}u \rangle = \int_0^T \dot{u}(t)u(t)dt + \frac{1}{2}\big\langle (\mathscr{C}_1u, \mathscr{C}_2u), (\mathscr{C}_2u, \mathscr{C}_1u) \big\rangle = 0.$$

2. **The transport operator:** We shall adopt the framework of [21], and in particular all conditions imposed there on the smooth vector field $\mathbf{a}$ defined on a neighborhood of a C^∞ bounded open set Ω in $\mathbf{R}^n$. Consider the space $X = L^p(\Omega)$ with $1 \le p < +\infty$, and the operator $\Gamma : D(\Gamma) \subset X \to X^*$ defined by

$$\Gamma u = \mathbf{a} \cdot \nabla u + \frac{(\nabla \cdot \mathbf{a})}{2}u \text{ with domain } D(\Gamma) = \Big\{ u \in L^p(\Omega); \mathbf{a} \cdot \nabla u + \frac{(\nabla \cdot \mathbf{a})}{2}u \in L^q(\Omega) \Big\} \text{ into } L^q(\Omega),$$

where $\frac{1}{p} + \frac{1}{q} = 1$. Observe that $D(\Gamma)$ is a Banach space under the norm

$$\|u\|_{D(\Gamma)} = \|u\|_p + \|\mathbf{a} \cdot \nabla u\|_q.$$

In view of Green's formula

$$\int_\Omega (\mathbf{a} \cdot \nabla u)u + \frac{1}{2}(\mathrm{div}\,\mathbf{a})|u|^2 dx - \frac{1}{2}\int_{\partial\Omega} |u|^2\, \hat{n} \cdot \mathbf{a}d\sigma = 0,$$

which holds for all $u, v \in C^\infty(\bar{\Omega})$ and where $\hat{n}$ is the outer normal to $\partial\Omega$, we define

$$\Sigma_\pm = \{ x \in \partial\Omega; \pm\mathbf{a}(x) \cdot \hat{n}(x) \ge 0 \} \text{ to be the entrance and exit sets of the transport operator } \mathbf{a} \cdot \nabla,$$

the corresponding Hilbert spaces $H_1 = L^2(\Sigma_+; |\mathbf{a} \cdot \hat{n}|d\sigma)$, $H_2 = L^2(\Sigma_-; |\mathbf{a} \cdot \hat{n}|d\sigma)$, and the boundary (trace) operators $\mathscr{B}u = (\mathscr{B}_1u, \mathscr{B}_2u) = (u|_{\Sigma_+}, u|_{\Sigma_-}) : L^p(\Omega) \to H := H_1 \times H_2$, whose domain is

$$D(\mathscr{B}) = \{ u \in L^p(\Omega); (u|_{\Sigma_+}, u|_{\Sigma_-}) \in H_1 \times H_2 \}.$$

Observe that $X_1 := D(\Gamma) \cap D(\mathscr{B})$ is also a Banach space under the norm

$$\|u\|_{X_1} = \|u\|_p + \|\mathbf{a} \cdot \nabla u\|_q + \|u|_{\Sigma_+}\|_{L^2(\Sigma_+; |\mathbf{a}\cdot\hat{n}|d\sigma)}$$

and that under our assumptions on the vector field $\mathbf{a}$ and Ω, we have that

$$C^\infty(\bar{\Omega}) \subset D(\Gamma) \cap D(\mathscr{B}) \subset D(\Gamma)$$

and that $C^\infty(\bar{\Omega})$ is dense in both spaces (see Bardos [21]). Conclude the exercise by using the following lemma.

Lemma 4.1. *The linear operator* Γ *is skew-adjoint modulo the boundary triplet* $(H_1 \oplus H_2, R, \mathscr{B})$, *where* R *is the automorphism on* $H_1 \times H_2$ *defined by* $R(h,k) = (-h,k)$.

Proof. We check the four criteria of Definition 4.3. For (1) it suffices to note that $C_0^\infty(\Omega) \subset \mathrm{Ker}(\mathscr{B}) \cap D(\Gamma)$ in such a way that $\mathrm{Ker}(\mathscr{B}) \cap D(\Gamma)$ is dense in X. Criterion (2) follows by a simple argument with coordinate charts, as it is easy to show that for all $(v_+, v_-) \in C_0^\infty(\Sigma_+) \times C_0^\infty(\Sigma_-)$ there exists $u \in C^\infty(\bar{\Omega})$ such that $(u|_{\Sigma_+}, u|_{\Sigma_-}) = (v_+, v_-)$. The embedding of $C_0^\infty(\Sigma_\pm) \subset L^2(\Sigma_\pm; |\mathbf{a} \cdot \hat{n}| d\sigma)$ is dense, and therefore the image of $C^\infty(\bar{\Omega}) \subset D(\Gamma) \cap D(\mathscr{B})$ under $\mathscr{B}$ is dense in $H_1 \times H_2$.

For (3), note that by Green's theorem we have

$$\int_\Omega (\mathbf{a} \cdot \nabla u)v + \frac{1}{2}(\mathrm{div}\,\mathbf{a})uv dx = -\int_\Omega (\mathbf{a} \cdot \nabla v)u + \frac{1}{2}(\mathrm{div}\,\mathbf{a})uv dx + \int_{\partial\Omega} uv\hat{n} \cdot \mathbf{a} d\sigma$$

for all $u, v \in C^\infty(\bar{\Omega})$, and the identity on X_1 follows since $C^\infty(\bar{\Omega})$ is dense in X_1 for the norm $\|u\|_{X_1}$.

For (4), we need to check that if $u \in X$, then it belongs to $X_1 := D(\Gamma) \cap D(\mathscr{B})$ if and only if

$$\sup\left\{\langle u, \Gamma v\rangle - \frac{1}{2}(\|\mathscr{B}_1(v)\|_{H_1}^2 + \|\mathscr{B}_2(v)\|_{H_2}^2); v \in X_1, \|v\|_X < 1\right\} < \infty. \qquad (4.4)$$

The "only if" direction follows directly from Green's theorem and the fact that $C^\infty(\bar{\Omega})$ is dense in the Banach space X_1 under the norm $\|u\|_{X_1}$.

For the reverse implication, suppose that (4.4) holds. Then, we clearly have that $\sup\{\langle u, \Gamma v\rangle; v \in C_0^\infty(\Omega), \|v\|_X < 1\}$, which means that $\mathbf{a} \cdot \nabla u + \frac{\nabla \cdot \mathbf{a}}{2} u \in L^q(\Omega)$ in the sense of distribution and therefore $u \in D(\Gamma)$. Now, to show that $u \in D(\mathscr{B})$, we observe that if $u \in X$ and $u \in D(\Gamma)$, then $u|_{\Sigma_+} \in L^2_{loc}(\Sigma_+; |\mathbf{a} \cdot \hat{n}| d\sigma)$. To check that $u|_{\Sigma_+} \in L^2(\Sigma_+; |\mathbf{a} \cdot \hat{n}| d\sigma)$, a simple argument using Green's theorem shows that (4.4) implies that

$$\sup\left\{\int_{\Sigma_+} uv|\mathbf{a} \cdot \hat{n}|d\sigma; v \in C_0^\infty(\Sigma_+), \int_{\Sigma_+} |v|^2|\mathbf{a} \cdot \hat{n}|d\sigma \le 1\right\} < +\infty,$$

which means that $u|_{\Sigma_+} \in L^2(\Sigma_+; |\mathbf{a} \cdot \hat{n}| d\sigma)$ and $u \in D(\mathscr{B}_1)$. The same argument works for $u|_{\Sigma_-}$, and (4) is therefore satisfied. Note that

$$\int_{\partial\Omega} uv\hat{n} \cdot \mathbf{a} d\sigma = \int_{\Sigma_+} uv|\hat{n} \cdot \mathbf{a}|d\sigma - \int_{\Sigma_-} uv|\hat{n} \cdot \mathbf{a}|d\sigma$$

$$= \langle(u|_{\Sigma_+}, u|_{\Sigma_-}), (v|_{\Sigma_+}, -v|_{\Sigma_-})\rangle$$

$$= -\langle\mathscr{B}u, R\mathscr{B}v\rangle,$$

where $R(h,k) = (-h,k)$.

Self-dual Lagrangians associated to skew-adjoint operators modulo a boundary

Proposition 4.2. *Let $B : X \to X$ be a bounded linear operator on a reflexive Banach space X with dense range, and let $\Gamma : D(\Gamma) \subset X \to X^*$ be a linear operator such that the pair (B, Γ) is antisymmetric modulo a boundary triplet $(H, R, \mathscr{B})$. Let $L : X \times X^* \to \mathbf{R}$ be a convex Lagrangian that is continuous in the first variable, and let $\ell : H \to \mathbf{R} \cup \{+\infty\}$ be a convex continuous function on H such that one of the following two conditions holds:*

1. *L is B-standard, $B^{-1}(\mathrm{Dom}_1(L)) \subset D(\Gamma) \cap D(\mathscr{B})$, and $\mathscr{B}(\mathrm{Dom}_1(L)) \cap \mathrm{Dom}(\ell)$ is nonempty.*
2. *The pair (B, Γ) is skew-adjoint modulo $(H, R, \mathscr{B})$, the map $x \to L(x, p_0)$ is bounded on the balls of X for some $p_0 \in X^*$, and $\ell(s) \leq C(1 + \|s\|^2)$ for $s \in H$.*

(i) The Lagrangian on $X \times X^*$

$$
L_{\Gamma,\ell}(x,p) = \begin{cases} L(Bx, \Gamma x + p) + \ell(\mathscr{B}x) & \text{if } x \in D(\Gamma) \cap D(\mathscr{B}) \\ +\infty & \text{if } x \notin D(\Gamma) \cap D(\mathscr{B}) \end{cases}
$$

has a Legendre transform $L^*_{\Gamma,\ell}$ that satisfies

$$
L^*_{\Gamma,\ell}(B^* q, By) = \begin{cases} L^*(\Gamma y + q, By) + \ell^*(R\mathscr{B}y) & \text{if } y \in D(\Gamma) \cap D(\mathscr{B}) \\ +\infty & \text{if } y \notin D(\Gamma) \cap D(\mathscr{B}) \end{cases} \qquad (4.5)
$$

(ii) In particular, if L is a self-dual Lagrangian on X and ℓ is an R-self-dual function on H, then $L_{\Gamma,\ell}$ is a B-self-dual Lagrangian on X.

Proof. Assume first that (B, Γ) is antisymmetric modulo the boundary triplet $(H, R, \mathscr{B})$ and that, for every $p \in X^*$, the function $x \to L(x, p)$ is continuous on X. We shall first prove formula (4.5) in the case where $y \in X_1 := D(\Gamma) \cap D(\mathscr{B})$. Indeed, we can then write

$$
L^*_{(\Gamma,\ell)}(B^* q, By) = \sup\{\langle By, p \rangle + \langle x, B^* q \rangle - L(Bx, \Gamma x + p) - \ell(\mathscr{B}x); x \in X_1, p \in X^*\}.
$$

Substituting $r = \Gamma x + p$, and since for $y \in X_1$, $\langle By, \Gamma x \rangle = -\langle Bx, \Gamma y \rangle - \langle \mathscr{B}x, R\mathscr{B}y \rangle$, we obtain

$$
\begin{aligned}
L^*_{(\Gamma,\ell)}(B^* q, By) &= \sup_{\substack{x \in X_1 \\ r \in X^*}} \{\langle By, r - \Gamma x \rangle + \langle Bx, q \rangle - L(Bx, r) - \ell(\mathscr{B}x)\} \\
&= \sup_{\substack{x \in X_1 \\ r \in X^*}} \Big\{ \langle Bx, \Gamma y \rangle + \langle \mathscr{B}x, R\mathscr{B}y \rangle + \langle By, r \rangle + \langle Bx, q \rangle \\
&\qquad\qquad -L(Bx, r) - \ell(\mathscr{B}x) \Big\}
\end{aligned}
$$

$$= \sup\Big\{ \langle Bx, \Gamma y + q \rangle + \langle \mathscr{B}(x + x_0), R\mathscr{B}y \rangle + \langle By, r \rangle - L(Bx, r)$$

$$- \ell(\mathscr{B}(x + x_0)); x \in X_1, r \in X^*, x_0 \in \text{Ker}(\mathscr{B}) \cap D(\Gamma) \Big\}.$$

Since X_1 is a linear space, we may set $w = x + x_0$ and write

$$L^*_{(\Gamma, \ell)}(B^* q, By) = \sup\Big\{ \langle B(w - x_0), \Gamma y + q \rangle + \langle \mathscr{B}w, R\mathscr{B}y \rangle + \langle By, r \rangle - L(B(w - x_0), r)$$

$$- \ell(\mathscr{B}w); w \in X_1, r \in X^*, x_0 \in \text{Ker}(\mathscr{B}) \cap D(\Gamma) \Big\}.$$

Now, for each fixed $w \in X_1$ and $r \in X^*$, the supremum over Bx_0 where $x_0 \in \text{Ker}(\mathscr{B}) \cap D(\Gamma)$ can be taken as a supremum over $z \in X$ since the image of $\text{Ker}(\mathscr{B}) \cap D(\Gamma)$ by B is dense in X and all terms involving Bx_0 are continuous in that variable. Furthermore, since L is continuous in the first variable, we can for each fixed $w \in X_1$ and $r \in X^*$ replace the supremum over $z \in X$ of the terms $w - z$ by a supremum over $v \in X$, where $v = w - z$. We therefore get

$$L^*_{(\Gamma, \ell)}(B^* q, By) = \sup\Big\{ \langle v, \Gamma y + q \rangle + \langle \mathscr{B}w, R\mathscr{B}y \rangle + \langle By, r \rangle - L(v, r)$$

$$- \ell(\mathscr{B}w); v \in X, r \in X^*, w \in X_1 \Big\}$$

$$= \sup_{v \in X} \sup_{r \in X^*} \{ \langle v, \Gamma y + q \rangle + \langle By, r \rangle - L(v, r) \}$$

$$+ \sup_{w \in X_1} \{ \langle \mathscr{B}w, R\mathscr{B}y \rangle - \ell(\mathscr{B}w) \}$$

$$= L^*(\Gamma y + q, By) + \sup_{w \in X_1} \{ \langle \mathscr{B}w, R\mathscr{B}y \rangle - \ell(\mathscr{B}w) \}.$$

Since the range of $\mathscr{B} : X_1 \to H$ is dense and ℓ is continuous, the boundary term can be written as

$$\sup_{a \in H} \{ \langle a, R\mathscr{B}y \rangle - \ell(a) \} = \sup_{a \in H} \{ \langle a, R\mathscr{B}y \rangle - \ell(a) \} = \ell^*(R\mathscr{B}y).$$

If now $y \notin X_1$, we shall distinguish the two cases.

Case 1. Under condition (1), we have that $y \notin B^{-1}(\text{Dom}_1(L))$, and since L is B-standard, we have for some $x_0 \in X$ with $Bx_0 \in \text{Dom}_1(L)$ and $\mathscr{B}x_0 \in \text{Dom}(\ell)$ that $H_L(Bx_0, By) = +\infty$, where H_L is the Hamiltonian associated to L. Since $x_0 \in B^{-1}(\text{Dom}_1(L)) \subset D(\Gamma)$, we can write

$$L^*_{(\Gamma, \ell)}(B^* q, By) = \sup_{\substack{Bx \in \text{Dom}_1(L) \\ r \in X^*}} \{ \langle By, r - \Gamma x \rangle + \langle Bx, q \rangle - L(Bx, r) - \ell(\mathscr{B}x) \}$$

$$\geq \sup_{r \in X^*} \{ \langle By, r \rangle - L(Bx_0, r) \} - \langle By, \Gamma x_0 \rangle + \langle Bx_0, q \rangle - \ell(\mathscr{B}x_0)$$

$$= H_L(Bx_0, By) - C.$$

Case 2. Under condition (2), write

$$L^*_{(\Gamma,\ell)}(B^*q, By) = \sup_{\substack{x \in X_1 \\ r \in X^*}} \left\{ \langle By, r - \Gamma x \rangle + \langle Bx, q \rangle - L(Bx, r) - \ell(\mathscr{B}x) \right\}$$

$$\geq \sup_{\substack{x \in X_1 \\ \|x\|_X < 1}} \left\{ \langle -By, \Gamma x \rangle + \langle Bx, q \rangle + \langle By, p_0 \rangle - L(Bx, p_0) \right.$$

$$\left. -C(1 + \|\mathscr{B}(x)\|^2) \right\}.$$

Since by assumption $L(Bx, p_0) \leq K$ whenever $\|x\|_X \leq 1$, we obtain since $y \notin X_1$ that

$$L^*_{(\Gamma,\ell)}(B^*q, By) \geq \sup_{\substack{x \in X_1 \\ \|x\|_X \leq 1}} \left\{ \langle -By, \Gamma x \rangle - \|B\| \|q\| + \langle By, p_0 \rangle - K - C(1 + \|\mathscr{B}(x)\|^2) \right\}$$

$$= C \sup_{\substack{x \in X_1 \\ \|x\|_X < 1}} \left\{ \langle B\tilde{y}, \Gamma x \rangle - \|\mathscr{B}(x)\|^2 \right\} + C'$$

$$= +\infty$$

since $\tilde{y} = -\frac{y}{C} \notin X_1$. The formula is then verified for all $(y, q) \in X \times X^*$, and the proposition is proved. $\qquad \blacksquare$

4.3 Unitary groups associated to skew-adjoint operators and self-duality

We now apply Proposition 4.2 to lift self-dual Lagrangians on state space to self-dual Lagrangians on the path space L_X^α. For that, we shall consider an evolution triple $X \subset H \subset X^*$, where H is a Hilbert space equipped with $\langle, \rangle$ as scalar product and where X is a dense vector subspace of H that is a reflexive Banach space once equipped with its own norm $\| \cdot \|$. Let $[0, T]$ be a fixed real interval, and consider for $\alpha, \beta > 1$ such that $\frac{1}{\alpha} + \frac{1}{\beta} = 1$ the Banach space $\mathscr{X}_{\alpha,\beta}$ of all functions in L_X^α such that $\dot{u} \in L_{X^*}^\beta$ equipped with the norm

$$\|u\|_{\mathscr{X}_{\alpha,\beta}} = \|u\|_{L_X^\alpha} + \|\dot{u}\|_{L_{X^*}^\beta}.$$

Proposition 4.3. *Consider an evolution triple $X \subset H \subset X^*$, and let B be an automorphism of H whose restriction to X is also an automorphism of X. Suppose ℓ is a $(-B, B)$-self-dual function on $H \times H$, and let L be a time-dependent B-self-dual Lagrangian on $X \times X^*$ such that:*

For each $p \in L_{X^*}^\beta$, the map $u \to \int_0^T L(t, u(t), p(t)) dt$ is continuous on L_X^α. (4.6)

The map $u \to \int_0^T L(t, u(t), 0) dt$ is bounded on the balls of L_X^α. (4.7)

$$-C \leq \ell(a,b) \leq C(1+\|a\|_H^2 + \|b\|_H^2) \text{ for all } (a,b) \in H \times H. \qquad (4.8)$$

The Lagrangian defined by

$$\mathscr{L}(u,p) = \begin{cases} \int_0^T L(t,u(t),p(t)-\dot{u}(t))dt + \ell(u(0),u(T)) & \text{if } u \in \mathscr{X}_{\alpha,\beta} \\ +\infty & \text{otherwise} \end{cases}$$

is then B-self-dual on $L_X^\alpha \times L_{X^*}^\beta$.

Proof. This follows immediately from Proposition 4.2 applied to the B-self-dual Lagrangian $\tilde{L}(u,p) := \int_0^T L(t,u(t),p(t))dt$ on $L_X^\alpha \times L_{X^*}^\beta$, the linear unbounded operator $u \rightarrow -\dot{u}$ that is skew-adjoint modulo the boundary operator $\mathscr{B} : D(\mathscr{B}) : L_X^\alpha \rightarrow H \times H$ defined by $\mathscr{B}u = (u(0),u(T))$, and where the automorphism is $R(a,b) = (-a,b)$ since for u and v in $\mathscr{X}_{\alpha,\beta}$ we have

$$\int_0^T \langle v,\dot{u} \rangle = -\int_0^T \langle \dot{v},u \rangle + \langle v(T),u(T) \rangle - \langle v(0),u(0) \rangle.$$

Note that the subspace $\mathscr{X}_{\alpha,\beta}^0 = \{u \in \mathscr{X}_{\alpha,\beta}; u(0) = u(T) = 0\}$ is dense in L_X^α. Moreover, for each $(a,b) \in X \times X$, there is $w \in \mathscr{X}_{\alpha,\beta}$ such that $w(0) = a$ and $w(T) = b$, namely the linear path $w(t) = \frac{(T-t)}{T}a + \frac{t}{T}b$. Since X is also dense in H and ℓ is continuous on H, all the required hypotheses of Proposition 4.2 are satisfied. $\qquad \square$

There are several possible ways to associate to a skew-adjoint operator Γ : $D(\Gamma) \subset X \rightarrow X^*$, a self-dual Lagrangian on path space, and we shall see later that each choice can be dictated by the hypothesis on the parabolic equation involving such a skew-adjoint operator. Start with a given time-dependent self-dual Lagrangian L on $X \times X^*$.

First choice: Use Proposition 4.1 to deduce that the Lagrangian

$$L_\Gamma(t,x,p) = \begin{cases} L(t,x,\Gamma x + p) & \text{if } x \in D(\Gamma) \\ +\infty & \text{if } x \notin D(\Gamma). \end{cases}$$

is also a time-dependent self-dual Lagrangian on $[0,T] \times X \times X^*$ and then Proposition 4.3 above to conclude that

$$\mathscr{L}(u,p) = \begin{cases} \int_0^T L(t,u(t),\Gamma u(t)+p(t)-\dot{u}(t))dt + \ell(u(0),u(T)) & \text{if } u \in \mathscr{X}_{\alpha,\beta} \\ +\infty & \text{otherwise} \end{cases}$$

is then a self-dual Lagrangian on $L_X^\alpha \times L_{X^*}^\beta$.

Second choice: Skew-adjoint operators on Hilbert spaces can be seen as infinitesimal generators of certain groups of unitary operators. Indeed, we recall the following and we refer to [127] for details.

Definition 4.4. A C_0-group on H is a family of bounded operators $S = \{S_t\}_{t\in\mathbf{R}}$ satisfying

(i) $S_t S_s = S_{t+s}$ for each $t,s \in \mathbf{R}$.
(ii) $S(0) = I$.
(iii)   The function $(t,x) \to S_t x$ is jointly continuous from $\mathbf{R} \times H \to H$.

We recall the following celebrated result of Stone.

Proposition 4.4. *An operator $\Gamma : D(\Gamma) \subset H \to H$ on a Hilbert space H is skew-adjoint if and only if it is the infinitesimal generator of a C_0-group of unitary operators $(S_t)_{t\in\mathbf{R}}$ on H. In other words, we have $\Gamma x = \lim\limits_{t\downarrow 0} \frac{S_t x - x}{t}$ for every $x \in D(\Gamma)$.*

More generally, assume that $X \subset H \subset X^*$ is an evolution triple such that there exists a linear and symmetric duality map D between X and X^* in such a way that $\|x\|^2 = \langle x, Dx \rangle$. As mentioned in Remark 3.3, we can then consider X and X^* as Hilbert spaces with the following inner products:

$$\langle u,v \rangle_{X\times X} := \langle Du, v \rangle \quad \text{and} \quad \langle u,v \rangle_{X^*\times X^*} := \langle D^{-1}u, v \rangle.$$

We then use Stone's theorem on Γ as a skew-adjoint operator on the Hilbert space X^* to associate a C_0-group of unitary operators $\bar{S}_t$ on X^*. Then, by Remark 3.3, $S_t = D^{-1}\bar{S}_t D$ is also a C_0-group of unitary operators on X in such a way that

$$L_S(t,u,p) := L(t, S_t u, \bar{S}_t p)$$

is also a time-dependent self-dual Lagrangian on $[0,T] \times X \times X^*$. We again conclude by Theorem 4.3 above that the Lagrangian

$$\mathscr{L}(u,p) = \begin{cases} \int_0^T L(t, S_t u(t), \bar{S}_t p(t) - \bar{S}_t \dot{u}(t))dt + \ell(u(0), u(T)) & \text{if } u \in \mathscr{X}_{\alpha,\beta} \\ +\infty & \text{otherwise} \end{cases}$$

is then self-dual on $L_X^\alpha \times L_{X^*}^\beta$.

Mixed choice: We may of course use the first approach on a skew-adjoint operator Γ_1 and the second on Γ_2 to associate to the sum $\Gamma_1 + \Gamma_2$, the self-dual Lagrangian

$$\mathscr{L}(u,p) = \begin{cases} \int_0^T L(t, S_t u, \Gamma_1 S_t u + \bar{S}_t p - \bar{S}_t \dot{u})dt + \ell(u(0), u(T)) & \text{if } u \in \mathscr{X}_{\alpha,\beta} \\ +\infty & \text{otherwise,} \end{cases}$$

where S_t is the C_0-unitary group associated to the skew-adjoint operator Γ_2.

Further comments

The stability of self-dual Lagrangians under iteration with bounded skew-adjoint operators was noted by Ghoussoub [55] and was verified in the unbounded case by Ghoussoub and Tzou [68]. The general case of a skew-adjoint operator modulo boundary terms were given in Ghoussoub [58]. Applications to nonhomogeneous boundary value problems and differential systems will be given in Chapter 8.

Chapter 5
Self-dual Vector Fields and Their Calculus

We introduce the concept of *self-dual vector fields* and develop their functional calculus. Given a function on phase space $L : X \times X^* \to \mathbf{R} \cup \{+\infty\}$, its *symmetrized vector field* at $x \in X$ is defined as the - possibly empty - subset of X^* given by

$$\overline{\partial} L(x) = \{p \in X^*; L(x,p) + L^*(p,x) = 2\langle x,p \rangle\}.$$

If L is convex and lower semicontinuous on $X \times X^*$, then

$$\overline{\partial} L(x) = \{p \in X^*; (p,x) \in \partial L(x,p)\}.$$

If now L is a self-dual Lagrangian, then

$$\overline{\partial} L(x) = \{p \in X^*; L(x,p) - \langle x,p \rangle = 0\}.$$

A *self-dual vector field* is any map of the form $\overline{\partial} L$, where L is a self-dual Lagrangian. A key point is that self-dual Lagrangians on phase space necessarily satisfy $L(x,p) \geq \langle p,x \rangle$ for all $(p,x) \in X^* \times X$, and therefore the zeros v of a self-dual vector field (i.e., $0 \in \overline{\partial} L(v)$) can be obtained by simply minimizing the functional $I(x) = L(x,0)$ and proving that the infimum is actually zero.

Self-dual vector fields are natural extensions of subdifferentials of convex lower semicontinuous functions as well as their sum with skew-adjoint operators. Most remarkable is the fact – shown in this chapter – that one can associate to any maximal monotone operator T a self-dual Lagrangian L in such a way that $T = \overline{\partial} L$, so that equations involving these operators can be resolved variationally. In effect, self-dual Lagrangians play the role of potentials of maximal monotone vector fields in a way similar to how convex energies are the potentials of their own subdifferentials. This representation reduces the often delicate calculus of maximal monotone operators to the more manageable standard convex analysis of self-dual Lagrangians on phase space. Moreover, all equations involving maximal monotone operators can now be analyzed with the full range of methods – computational or not – that are available for variational settings.

N. Ghoussoub, *Self-dual Partial Differential Systems and Their Variational Principles*, Springer Monographs in Mathematics, DOI 10.1007/978-0-387-84897-6_5, © Springer Science+Business Media, LLC 2009

5.1 Vector fields derived from self-dual Lagrangians

Definition 5.1. Let B be a bounded linear operator on a reflexive Banach space X and let $L : X \times X^* \to \mathbf{R} \cup \{+\infty\}$ be a proper convex lower semicontinuous function on $X \times X^*$.

1) The *symmetrized B-vector fields of L* at $x \in X$ (resp., $p \in X^*$) are the possibly empty sets

$$\overline{\partial}_B L(x) := \{p \in X^*; (B^* p, Bx) \in \partial L(x, p)\},$$

resp.

$$\tilde{\partial}_B L(p) := \{x \in X; (B^* p, Bx) \in \partial L(x, p)\}.$$

2) The *domains* of these vector fields are the sets

$$\mathrm{Dom}(\overline{\partial}_B L) = \{x \in X; \overline{\partial} L_B(x) \neq \emptyset\}, \quad \text{resp.} \quad \mathrm{Dom}(\tilde{\partial}_B L) = \{p \in X^*; \tilde{\partial}_B L(p) \neq \emptyset\}.$$

3) The *graph* of a symmetrized B-vector field $\overline{\partial}_B L$ is the following subset of phase space $X \times X^*$:

$$\mathcal{M}_B(L) = \{(x, p) \in X \times X^*; (B^* p, Bx) \in \partial L(x, p)\}.$$

Remark 5.1. The vector fields defined above should not be confused with the subdifferential ∂L of L as a convex function on $X \times X^*$. It is only in the case where the Lagrangian is of the form $L(x, p) = \varphi(x) + \varphi^*(p)$, where φ is convex and lower semicontinuous, that the field $x \to \overline{\partial} L(x)$ is identical to the subdifferential map $x \to \partial \varphi(x)$.

Proposition 5.1. *For any convex lower semicontinuous Lagrangian L on $X \times X^*$, the map $x \to \overline{\partial} L(x)$ is monotone.*

Proof. Indeed, if $p \in \overline{\partial} L(x)$ and $q \in \overline{\partial} L(y)$, then

$$\langle p - q, x - y \rangle = \frac{1}{2} \langle (p, x) - (q, y), (x, p) - (y, q) \rangle \geq 0$$

since we have that $(p, x) \in \partial L(x, p)$ and $(q, y) \in \partial L(y, q)$.
We shall denote by $\delta_B L$ the vector field

$$\delta_B L(x) =: \{p \in X^*; L(x, p) - \langle Bx, p \rangle = 0\}. \tag{5.1}$$

It is clear that $\overline{\partial}_B L = \delta_B L$ whenever L is a B-self-dual Lagrangian. The following summarizes their relationship for more general Lagrangians.

Proposition 5.2. *Let L be a B-Fenchelian Lagrangian on $X \times X^*$. Then,*

1. *$\delta_B L(x) \subset \overline{\partial}_B L(x)$ for any $x \in X$.*
2. *If L is B-subself-dual, then $\delta_B L(x) = \overline{\partial}_B L(x)$ for any $x \in X$.*

Proof. (1) Assuming $L(x,p) - \langle Bx, p \rangle = 0$, consider $(y,q) \in X \times X^*$ and write

$$L(x+y, p+q) - L(x,p) \geq t^{-1}\left[L(x+ty, p+tq) - L(x,p)\right]$$
$$\geq t^{-1}\left[\langle B(x+ty), p+tq \rangle - \langle Bx, p \rangle\right]$$
$$\geq \langle Bx, q \rangle + \langle y, B^*p \rangle + t\langle By, q \rangle.$$

Letting $t \to 0^+$, we get that $L(x+y, p+q) - L(x,p) \geq \langle Bx, q \rangle + \langle y, B^*p \rangle$, which means that we have $(B^*p, B^*x) \in \partial L(x,p)$.

For (2), we assume that $p \in \overline{\partial}_B L(x)$ and use Legendre-Fenchel duality to write

$$0 = L(x,p) - \langle Bx, p \rangle + L^*(B^*p, Bx) - \langle Bx, p \rangle.$$

We can conclude since $L(x,p) \geq \langle Bx, p \rangle$ and $L^*(B^*p, Bx) \geq L(x,p) \geq \langle Bx, p \rangle$.

For self-dual Lagrangians, the relationship between the various notions is more transparent.

Proposition 5.3. *Let L be a B-self-dual Lagrangian on phase space $X \times X^*$. Then, the following assertions are equivalent:*

1. *$p \in \overline{\partial}_B L(x)$.*
2. *$x \in \tilde{\partial} L_B(p)$.*
3. *$0 \in \overline{\partial}_B L_p(x)$, where L_p is the B-self-dual Lagrangian $L_p(x,q) = L(x, p+q) + \langle Bx, p \rangle$.*
4. *The infimum of the functional $I_p(u) = L(u,p) - \langle Bu, p \rangle$ is zero and is attained at $x \in X$.*

Proof. Assertions (1) and (2) are readily equivalent. It is clear that (3) is a reformulation of (1) in view of Proposition 3.4 (2). For the equivalence with (3) and (4), use the definition of B-self-duality and Legendre-Fenchel duality to write

$$2\langle Bx, p \rangle = L(x,p) + L^*(B^*p, Bx) \geq \langle x, B^*p \rangle + \langle p, Bx \rangle = 2\langle Bx, p \rangle,$$

and therefore $L(x,p) - \langle Bx, p \rangle = 0$ if and only if $L(x,p) + L^*(B^*p, Bx) = \langle x, B^*p \rangle + \langle p, Bx \rangle$, which is equivalent to $(B^*p, Bx) \in \partial L(x,p)$.

The following vector fields are central to the subject of this book.

Definition 5.2. A set-valued function $F : \text{Dom}(F) \subset X \to 2^{X^*}$ is said to be a B-*self-dual vector field* if there exists a B-self-dual Lagrangian on $X \times X^*$ such that $F(x) = \overline{\partial}_B L(x)$ for every $x \in \text{Dom}(F)$.

5.2 Examples of B-self-dual vector fields

1. Vector fields associated to convex energy functionals

For a basic self-dual Lagrangian of the form $L(x,p) = \varphi(x) + \varphi^*(p)$, where φ is a convex lower semicontinuous function, and φ^* is its Legendre conjugate on X^*, it is clear that

$$\overline{\partial}L(x) = \partial\varphi(x) \quad \text{while} \quad \tilde{\partial}L(p) = \partial\varphi^*(p).$$

The corresponding variational problem reduces to minimizing the convex functional $I(x) = L(x,0) = \varphi(x) + \varphi^*(0)$ in order to solve equations of the form $0 \in \partial\varphi(x)$. The associated self-dual Lagrangian submanifold is then the graph of $\partial\varphi$, which is

$$\mathscr{M}_{I,\varphi} = \{(x,p) \in X \times X^*; p \in \partial\varphi(x)\}.$$

2. Superposition of conservative and dissipative vector fields

More interesting examples of self-dual Lagrangians are of the form $L(x,p) = \varphi(x) + \varphi^*(-\Gamma x + p)$, where φ is a convex and lower semicontinuous function on X and $\Gamma : X \to X^*$ is a skew-symmetric operator. The corresponding self-dual vector fields are then

$$\overline{\partial}L(x) = \Gamma x + \partial\varphi(x) \quad \text{while} \quad \tilde{\partial}L(p) = (\Gamma + \partial\varphi)^{-1}(p).$$

More generally, if the operator Γ is merely nonnegative (i.e., $\langle \Gamma x, x \rangle \geq 0$), then one can still write the vector field $\Gamma + \partial\varphi$ as $\overline{\partial}L$ for some self-dual Lagrangian L on $X \times X^*$. Indeed, denoting by $\mathscr{C}(X)$ the cone of bounded below, proper convex lower semicontinuous functions on X and by $\Gamma \in \mathscr{A}(X)$ the cone of nonnegative linear operators from X into X^*, we can show the following proposition.

Proposition 5.4. *To any pair $(\varphi, \Gamma) \in \mathscr{C}(X) \times \mathscr{A}(X)$, one can associate a self-dual Lagrangian $L := L_{(\varphi,\Gamma)}$ such that for $p \in X^*$ the following are equivalent:*

1. *The equation $\Gamma x + \partial\varphi(x) = p$ has a solution $\bar{x} \in X$.*
2. *The functional $I_p(x) = L(x,p) - \langle x, p \rangle$ attains its infimum at $\bar{x} \in X$ with $I_p(\bar{x}) = 0$.*
3. *$p \in \overline{\partial}L(\bar{x})$.*

Proof. Let $L(x,p) = \psi(x) + \psi^*(-\Gamma^{as}x + p)$, where ψ is the convex function $\psi(x) = \frac{1}{2}\langle \Gamma x, x \rangle + \varphi(x)$, and where $\Gamma^{as} = \frac{1}{2}(\Gamma - \Gamma^*)$ is the antisymmetric part of Γ and $\Gamma^{sym} = \frac{1}{2}(\Gamma + \Gamma^*)$ is its symmetric part. The fact that the minimum of $I(x) = \psi(x) + \psi^*(-\Gamma^{as}x + p) - \langle x, p \rangle$ is equal to 0 and is attained at some $\bar{x} \in X$ means that $\psi(\bar{x}) + \psi^*(-\Gamma^{as}\bar{x} + p) = \langle \Gamma^{as}\bar{x} + p, \bar{x} \rangle$, which yields, in view of Legendre-Fenchel duality, that $-\Gamma^{as}\bar{x} + p \in \partial\psi(\bar{x}) = \Gamma^{sym}\bar{x} + \partial\varphi(\bar{x})$, and hence, $\bar{x}$ solves $p \in \Gamma x + \partial\varphi(x)$.

4. Superposition of dissipative vector fields with skew-adjoint operators modulo a boundary

Now let $\Gamma : D(\Gamma) \subset X \to X^*$ be an antisymmetric operator modulo a boundary triplet $(H, R, \mathscr{B})$. Let $L : X \times X^* \to \mathbf{R}$ be a self-dual Lagrangian and let $\ell : H \to \mathbf{R} \cup \{+\infty\}$ be an R-self-dual convex continuous function on H. Then, the self-dual vector field associated to the Lagrangian $L_{\Gamma, \ell}$ on $X \times X^*$

$$L_{\Gamma, \ell}(x, p) = \begin{cases} L(x, \Gamma x + p) + \ell(\mathscr{B}x) & \text{if } x \in D(\Gamma) \cap D(\mathscr{B}) \\ +\infty & \text{if } x \notin D(\Gamma) \cap D(\mathscr{B}) \end{cases}$$

– which is self-dual, for example under the hypotheses of Proposition 4.2 – is given by

$$p \in \overline{\partial} L_{\Gamma, \ell}(x) \quad \text{if and only if} \quad \begin{cases} p \in \overline{\partial} L(x) - \Gamma x \\ 0 \in \partial \ell(\mathscr{B}x) - R\mathscr{B}x. \end{cases} \tag{5.2}$$

Indeed, if $L_{(\Gamma, \ell)}(x, p) - \langle x, p \rangle = 0$, we write

$$L_{(\Gamma, \ell)}(x, p) - \langle x, p \rangle = L(x, \Gamma x + p) + \ell(\mathscr{B}x) - \langle x, p \rangle - \langle x, \Gamma x \rangle - \frac{1}{2} \langle \mathscr{B}x, R\mathscr{B}x \rangle$$

$$= L(x, \Gamma x + p) - \langle x, \Gamma x + p \rangle - \frac{1}{2} \langle \mathscr{B}x, R\mathscr{B}x \rangle + \ell(\mathscr{B}x).$$

Since $L(x, p) \geq \langle x, p \rangle$ and $\ell(s) \geq \frac{1}{2} \langle s, Rs \rangle$, we get

$$\begin{cases} L(x, \Gamma x + p) = \langle x, \Gamma x + p \rangle \\ \ell(\mathscr{B}x) = \frac{1}{2} \langle \mathscr{B}x, R\mathscr{B}x \rangle, \end{cases} \tag{5.3}$$

and we are done.

4. Maximal monotone operators

The vector fields of the form $\Gamma + \partial \varphi$ above are actually very particular but important examples of maximal monotone operators. It turns out that this latter class of operators coincides with the class of self-dual vector fields. This remarkable result is only of theoretical value since the corresponding self-dual Lagrangians are not given by an explicit formula as in the case of the vector fields $\Gamma + \partial \varphi$. We shall postpone its proof until the next section.

5. AntiHamiltonian vector fields

By considering a basic B-self-dual Lagrangian of the form $L(x, p) = \varphi(Bx) + \varphi^*(p)$, where φ is convex lower semicontinuous and B is onto (or φ continuous and B has dense range), we obtain a B-self-dual vector field of the form

$$\overline{\partial}_B L(x) = \partial \varphi(Bx) \quad \text{while} \quad \tilde{\partial} L(p) = B^{-1} \partial \varphi^*(p).$$

In the case where B is the symplectic matrix $J(x,y) = (-y,x)$, these correspond to vector fields of the form $J\partial H$ that appear in Hamiltonian systems. In other words, the J-self-dual Lagrangian associated to the vector field $J\partial H$, where H is a convex lower semicontinuous function on $\mathbf{R}^n \times \mathbf{R}^n$, is

$$L((x,y),(p,q)) = H(x,y) + H^*(q,-p).$$

If $A : X \to X^*$ is a bounded linear operator, then the operator $J^* \circ (A^*, A)$ is skew-adjoint, which means that the Lagrangian

$$M((x,y),(p,q)) = H(x,y) + H^*(Ay+q, -A^*x-p)$$

is J-self-dual and the vector field $(A^*, A) + J\partial H$ is J-self-dual. This means that if A is self-adjoint, then solving the system

$$(Ax, Ay) \in J\partial H(x,y) \tag{5.4}$$

reduces to minimizing the functional $E(x,y) = H(x,y) + H^*(Ay, -Ax)$. This simple case will be used in Chapter 7 to solve certain convex-concave Hamiltonian systems. This is, however, not the case for Hamiltonian evolutions where $A = \frac{d}{dt}$, which is essentially skew-adjoint. In this case and in order to solve $(Ax, Ay) \in -J\partial H(x,y)$, we are led to minimize the more complicated functional

$$\tilde{E}(x,y) = H(x,y) + H^*(Ay, -Ax) - 2\langle Ay, x \rangle.$$

This will warrant an extension of our results on J-self-dual functionals and will be considered in Part III of this book.

5.3 Operations on self-dual vector fields

The following proposition summarizes the elementary properties of the operator $\overline{\overline{\partial}}_B$ on self-dual Lagrangians.

Proposition 5.5. *Let B be a bounded linear operator on a reflexive Banach space X and let L be a B-self-dual Lagrangian on $X \times X^*$. The following properties hold:*

1. *If $\lambda > 0$, then the vector field associated to the Lagrangian $(\lambda \cdot L)(x,p) := \lambda^2 L(\frac{x}{\lambda}, \frac{p}{\lambda})$ is*

$$\overline{\partial}_B(\lambda \cdot L)(x) = \lambda \overline{\partial}_B L\left(\frac{x}{\lambda}\right).$$

2. *For $y \in X$ and $q \in X^*$, consider M_y and N_p to be the translated Lagrangians defined respectively by $M_y(x,p) = L(x+y,p) - \langle By, p \rangle$ and $N_q(x,p) = L(x,p+q) - \langle x, B^*q \rangle$. Then,*

$$\overline{\partial}_B M_y(x) = \overline{\partial}_B L(x+y) \text{ and } \overline{\partial}_B N_q(x) = \overline{\partial}_B L(x) - q.$$

3. *If X is a Hilbert space, U is a unitary operator ($U^{-1} = U^*$) on X that commutes with B, and if $M(x,p) := L(Ux, Up)$, then*

$$\overline{\partial}_B M(x) = U^* \overline{\partial}_B L(Ux).$$

4. *If B_i is a bounded operator on a reflexive Banach space X_i for each $i \in I$ and if L_i is a convex lower semicontinuous Lagrangian on X_i, then on the product space $\Pi_{i \in I} X_i$, we have*

$$\overline{\partial}_B (\Sigma_{i \in I} L_i) = \Pi_{i \in I} \overline{\partial}_{B_i} L_i.$$

5. *If $\Gamma : X \to X^*$ is any skew-adjoint operator and $L_\Gamma(x, p) = L(x, -\Gamma x + p)$, then*

$$\overline{\partial}_B L_\Gamma = \overline{\partial}_B L + \Gamma.$$

6. *If $\Gamma : X \to X^*$ is an invertible skew-adjoint operator and $_\Gamma L(x, p) = L(x + \Gamma^{-1} p, \Gamma x)$, then*

$$\overline{\partial}_B(_\Gamma L)(x) = \Gamma \tilde{\partial} L_B(\Gamma x) - \Gamma x.$$

7. *If $L \in \mathscr{L}(X)$ and $M \in \mathscr{L}(Y)$, B_1 (resp., B_2) is a linear operator on X (resp., Y) and $A : X \to Y^*$ is such that $AB_1 = B_2^* A$, then setting $B = (B_1, B_2)$ on $X \times Y$, and writing J for the symplectic operator $J(x, y) = (-y, x)$, we have*

$$\overline{\partial}_B(L \oplus_A M) = (\overline{\partial}_{B_1} L, \overline{\partial}_{B_2} M) + (A^*, A) \circ J.$$

8. *If φ is a convex continuous function on $X \times Y$, B_1, B_2, and A are as in (7), and if L is the Lagrangian on $(X \times Y) \times (X^* \times Y^*)$ defined by $L((x,y), (p,q)) = \varphi(B_1 x, B_2 y) + \varphi^*(A^* y + p, -Ax + q)$, then*

$$\overline{\partial}_B L = \partial \varphi(B_1, B_2) + (A^*, A) \circ J.$$

Proposition 5.6. *Let B be a bounded linear operator on a reflexive Banach space X. The following properties hold for any B-self-dual Lagrangians L and M:*

1. $\overline{\partial}_B(L \oplus M) \subset \overline{\partial}_B(L) + \overline{\partial}_B(M)$.
2. *If $\text{Dom}_1(L) - \text{Dom}_1(M)$ also contains a neighborhood of the origin, then $\overline{\partial}_B(L \oplus M) = \overline{\partial}_B(L) + \overline{\partial}_B(M)$.*
3. *If $M(x, p) = \varphi(Bx) + \varphi^*(p)$ with φ convex and finite on X, then $\overline{\partial}_B(L \oplus M)(x) = \overline{\partial}_B L(x) + \partial \varphi(Bx)$.*
4. *If $B = \text{Id}$ and $p \in \overline{\partial} L_\lambda^1(x)$, then $p \in \overline{\partial} L(J_\lambda(x, p))$, where J_λ is defined in (3.12).*
5. *For each $x \in X$, we have $\sup \left\{ \|p\|; p \in \overline{\partial} L_\lambda^1(x) \right\} \leq \inf \left\{ \|q\|; q \in \overline{\partial} L(x) \right\}.$*

Proof. The proofs of (1), (2) and (3) are left as exercises. For (4), we assume $L_\lambda(x, p) = \langle x, p \rangle$ and write

$$2\langle x, p \rangle = L_\lambda(x, p) + L_\lambda(x, p)$$

$$= 2\left(L\big(J_\lambda(x,p),p\big) + \frac{\|x - J_\lambda(x,p)\|_X^2}{2\lambda} + \frac{\lambda\|p\|_{X^*}^2}{2} \right)$$

$$= L^*\big(p,J_\lambda(x,p)\big) + L\big(J_\lambda(x,p),p\big) + 2\left(\frac{\|x - J_\lambda(x,p)\|_X^2}{2\lambda} + \frac{\lambda\|p\|_{X^*}^2}{2} \right)$$

$$\geq 2\langle p,J_\lambda(x,p)\rangle + 2\langle x - J_\lambda(x,p),p\rangle$$

$$= 2\langle x,p\rangle.$$

The second to last inequality is deduced by applying the Legendre-Fenchel inequality to the first two terms and the last two terms. The chain of inequality above shows that all inequalities are equalities. This implies, again by the Legendre-Fenchel inequality, that $\big(p,J_\lambda(x,p)\big) \in \partial L\big(J_\lambda(x,p),p\big)$.

For (5), if $p_\lambda \in \bar\partial L_\lambda^1(x)$, then $(p_\lambda,x) = \partial L_\lambda(x,p_\lambda)$, and we get from (3.12) that

$$p_\lambda = \frac{x - J_\lambda(x,p_\lambda)}{\lambda} \in \partial_1 L\big(J_\lambda(x,p_\lambda),p_\lambda\big),$$

and from property (4) that

$$\big(p_\lambda,J_\lambda(x,p_\lambda)\big) \in \partial L\big(J_\lambda(x,p_\lambda),p_\lambda\big).$$

If now $p \in \bar\partial L(x)$, then setting $v_\lambda = J_\lambda(x,p_\lambda)$ and using that $(p_\lambda,v_\lambda) \in \partial L(v_\lambda,p_\lambda)$, we get from monotonicity and the fact that $p_\lambda = \frac{x - v_\lambda}{\lambda}$

$$0 \leq \big\langle (x,p) - (v_\lambda,p_\lambda), \big(\partial_1 L(x,p), \partial_2 L(x,p)\big) - (p_\lambda,v_\lambda)\big\rangle$$

$$= \left\langle (x,p) - (v_\lambda,p_\lambda),(p,x) - \big(\frac{x - v_\lambda}{\lambda},v_\lambda\big)\right\rangle$$

$$= -\frac{\|x - v_\lambda\|_X^2}{\lambda} + \langle x - v_\lambda,p\rangle + \langle p,x - v_\lambda\rangle - \langle p_\lambda,x - v_\lambda\rangle$$

$$= -2\frac{\|x - v_\lambda\|_X^2}{\lambda} + 2\langle x - v_\lambda,p\rangle,$$

which yields that $\frac{\|x - v_\lambda\|_X}{\lambda} \leq \|p\|_{X^*}$ and finally the desired bound $\|p_\lambda\| \leq \|p\|$ for all $\lambda > 0$.

Exercises 5.A. More on self-dual vector fields

1. Prove claims (1) - (8) in Proposition 5.5 and show that they hold for any convex lower semi-continuous Lagrangian on $X \times X^*$.

2. Prove claims (1), (2) and (3) in Proposition 5.6, and show that it suffices to assume that L and M are Fenchelian.

3. Assume $p > 1$, $\frac{1}{p} + \frac{1}{q} = 1$, $D : X \to X^*$ is a duality map, and let L be a self-dual Lagrangian on $X \times X^*$. Prove that, for every $x \in X$ and $\lambda > 0$, we have $\bar\partial L_{\lambda,p}^2(x) = \bar\partial L(x) + \lambda^{p-1}\|x\|^{p-2}Dx$.

4. Show also that, for every $x \in X$, we have $\bar{\partial}L^1_{\lambda,p}(x) = \bar{\partial}L(x + \lambda^{q-1}\|r\|^{q-2}D^{-1}r)$, where $r = \overline{\partial}L(x)$.

5.4 Self-dual vector fields and maximal monotone operators

We now show that self-duality is closely related to a more familiar notion in nonlinear analysis. The following establishes a one-to-one correspondence between maximal monotone operators and self-dual vector fields.

Theorem 5.1. *If $T : D(T) \subset X \to 2^{X^*}$ is a maximal monotone operator with a nonempty domain, then there exists a self-dual Lagrangian L on $X \times X^*$ such that $T = \overline{\partial}L$. Conversely, if L is a proper self-dual Lagrangian L on a reflexive Banach space $X \times X^*$, then the vector field $x \to \overline{\partial}L(x)$ is maximal monotone.*

We shall need the following two lemmas.

Lemma 5.1. *Let $T : D(T) \subset X \to 2^{X^*}$ be a monotone operator, and consider on $X \times X^*$ the Lagrangian L_T defined by*

$$L_T(x,p) = \sup\{\langle p,y \rangle + \langle q, x-y \rangle; (y,q) \in G(T)\} \tag{5.5}$$

1. If $\mathrm{Dom}(T) \neq \emptyset$, then L_T is a convex and lower semicontinuous function on $X \times X^*$ such that, for every $x \in \mathrm{Dom}(T)$, we have $Tx \subset \overline{\partial}L(x) \cap \delta L(x)$. Moreover, we have

$$L_T^*(p,x) \geq L_T(x,p) \text{ for every } (x,p) \in X \times X^*. \tag{5.6}$$

2. If T is maximal monotone, then $T = \overline{\partial}L = \delta L$ and L_T is Fenchelean, that is

$$L_T(x,p) \geq \langle x,p \rangle \text{ for all } (x,p) \in X \times X^*. \tag{5.7}$$

Proof. (1) If $x \in \mathrm{Dom}(T)$ and $p \in Tx$, then the monotonicity of T yields for any $(y,q) \in G(T)$

$$\langle x,p \rangle \geq \langle y,p \rangle + \langle x-y,q \rangle$$

in such a way that $L_T(x,p) \leq \langle x,p \rangle$. On the other hand, we have

$$L_T(x,p) \geq \langle x,p \rangle + \langle p, x-x \rangle = \langle x,p \rangle,$$

and therefore $p \in \delta L(x)$. Write now for any $(y,q) \in X \times X^*$,

$$\begin{aligned}
L_T(x+y,p+q) - L_T(x,p) &= \sup\{\langle p+q,z \rangle + \langle r, x+y \rangle - \langle z,r \rangle; (z,r) \in G(T)\} \\
&\quad - L_T(x,p) \\
&\geq \langle p+q,x \rangle + \langle p, x+y \rangle - \langle p,x \rangle - \langle p,x \rangle \\
&= \langle q,x \rangle + \langle p,y \rangle,
\end{aligned}$$

which means that $(p,x) \in \partial L(x,p)$ and therefore $p \in \overline{\partial}L(x)$.

Note that $L_T(x,p) = H_T^*(p,x)$, where H_T is the Lagrangian on $X \times X^*$ defined by

$$H_T(x,p) = \begin{cases} \langle x,p \rangle & \text{if } (x,p) \in G(T) \\ +\infty & \text{otherwise.} \end{cases} \tag{5.8}$$

Since $L_T(x,p) = \langle x,p \rangle = H_T(x,p)$ whenever $(x,p) \in G(T)$, it follows that $L_T \leq H_T$ on $X \times X^*$ and so $L_T^*(p,x) \geq H_T^*(p,x) = L_T(x,p)$ everywhere.

For (2), if now T is maximal, then necessarily $Tx = \delta L_T(x) \cap \overline{\partial} L_T(x) = \overline{\partial} L_T(x)$ since $x \to \overline{\partial} L_T(x)$ is a monotone extension of T. In order to show (5.7), assume to the contrary that $L_T(x,p) < \langle x,p \rangle$ for some $(x,p) \in X \times X^*$. Then,

$$\langle p,y \rangle + \langle q,x-y \rangle < \langle p,x \rangle \text{ for all } (y,q) \in G(T),$$

and therefore

$$\langle p-q,x-y \rangle > 0 \text{ for all } (y,q) \in G(T).$$

But since T is maximal monotone, this means that $p \in Tx$. But then $p \in \delta L_T(x)$ by the first part, leading to $L_T(x,p) = \langle x,p \rangle$, which is a contradiction.

Finally, note that since L_T is now subself-dual, we have from Proposition 5.2 that $Tx = \overline{\partial} L_T(x) = \delta L_T(x)$.

Proposition 5.7. *Let X be a reflexive Banach space, and let L be a convex lower semicontinuous Lagrangian on $X \times X^*$ that satisfies*

$$L^*(p,x) \geq L(x,p) \geq \langle x,p \rangle \text{ for every } (x,p) \in X \times X^*. \tag{5.9}$$

Then, there exists a self-dual Lagrangian N on $X \times X^*$ such that $\overline{\partial} L = \overline{\partial} N$ and

$$L^*(p,x) \geq N(x,p) \geq L(x,p) \text{ for every } (x,p) \in X \times X^*. \tag{5.10}$$

Proof. The Lagrangian N is simply the *proximal average* – as defined in Proposition 2.6 – of L and $\tilde{L}$, where $\tilde{L}(x,p) = L^*(p,x)$. It is defined as

$$N(x,p) := \inf \left\{ \frac{1}{2} L(x_1,p_1) + \frac{1}{2} L^*(p_2,x_2) \right.$$
$$\left. + \frac{1}{8} \|x_1 - x_2\|^2 + \frac{1}{8} \|p_1 - p_2\|^2; \ (x,p) = \frac{1}{2}(x_1,p_1) + \frac{1}{2}(x_2,p_2) \right\}.$$

It is easy to see that $L(x,p) \leq N(x,p) \leq L^*(p,x)$, and in view of (11) in Proposition 2.6, N is clearly a self-dual Lagrangian on $X \times X^*$. It remains to show that $\overline{\partial} L(x) = \overline{\partial} N(x)$. Indeed, first it is clear that $\overline{\partial} N(x) \subset \overline{\partial} L(x)$. On the other hand, since L is sub-self-dual, we have from Lemma 5.2 that $\delta L(x) = \partial L(x)$, which means that if $p \in \overline{\partial} L(x)$, then $(p,x) \in \partial L(x,p)$ and therefore $L(x,p) + L^*(p,x) = 2\langle x,p \rangle$. Again, since $p \in \overline{\partial} L(x)$, this implies that $L^*(p,x) = \langle x,p \rangle$ and therefore $N(x,p) = \langle x,p \rangle$ and $p \in \overline{\partial} N(x)$.

Proof of Theorem 5.1: First, assume T is a maximal monotone operator, and associate to it the sub-self-dual Lagrangian L_T via Lemma 5.1,

$$T = \bar{\partial} L_T \text{ and } L_T^*(p,x) \geq L_T(x,p) \geq \langle x, p \rangle.$$

Now apply the preceding proposition to L_T to find a self-dual Lagrangian N_T such that $L_T(x,p) \leq N_T(x,p) \leq L_T^*(p,x)$ for every $(x,p) \in X \times X^*$ and $Tx = \bar{\partial} N_T(x)$ for any $x \in D(T)$.

For the converse, we shall use a result that we establish in the next chapter (Proposition 6.1). Consider a self-dual Lagrangian L and denote by $D : X \to 2^{X^*}$ the duality map between X and X^*,

$$D(x) = \{p \in X^*; \langle p, x \rangle = \|x\|^2\}.$$

It suffices to show that the vector field $\bar{\partial} L + D$ is onto [130]. In other words, we need to find for any $p \in X^*$ an $x \in X$ such that $p = (\bar{\partial} L + D)(x)$. For that, we consider the following Lagrangian on $X \times X^*$:

$$M(x,p) = \inf \left\{ L(x, p-r) + \frac{1}{2}\|x\|_X^2 + \frac{1}{2}\|r\|_{X^*}^2; r \in X^* \right\}.$$

It is a self-dual Lagrangian according to Proposition 3.4. Moreover, assuming without loss of generality that the point $(0,0)$ is in the domain of L, we get the estimate $M(0,p) \leq L(0,0) + \frac{1}{2}\|p\|_{X^*}^2$, and therefore Proposition 6.1 applies and we obtain $\bar{x} \in X$ so that $p \in \bar{\partial} M(\bar{x})$. This means that

$$M(\bar{x}, p) - \langle \bar{x}, p \rangle = \inf \left\{ L(\bar{x}, p-r) - \langle \bar{x}, p-r \rangle + \frac{1}{2}\|\bar{x}\|_X^2 + \frac{1}{2}\|r\|_{X^*}^2 - \langle \bar{x}, r \rangle; r \in X^* \right\}$$
$$= 0.$$

In other words, there exists $\bar{r} \in D(\bar{x})$ such that $p - \bar{r} \in \bar{\partial} L(\bar{x})$ and we are done.

Exercises 5.B. Operations on maximal monotone operators via the corresponding Lagrangian calculus

Let $T : D(T) \subset X \to 2^{X^*}$ be a monotone operator, and consider on $X \times X^*$ the "Fitzpatrick Lagrangian" F_T,

$$F_T(x,p) = \sup\{\langle p, y \rangle + \langle q, x-y \rangle; (y,p) \in G(T)\},$$

and let L_T be the self-dual Lagrangian obtained by the proximal average of F_T and F_T^*. We shall then say that L_T is a self-dual potential for T.

1. Show that $\bar{\partial} L_T$ is a maximal monotone extension of T.
2. Assuming that T satisfies for all $x \in X$

$$\|Tx\| \leq C(1 + \|x\|) \text{ and } \langle Tx, x \rangle \geq \alpha\|x\|^2 - \beta, \tag{5.11}$$

show that we then have for all $(x,p) \in X \times X^*$,

$$F_T(x,p) \leq \frac{(C\|x\| + \|p\|)^2}{2\alpha} + C\|x\| - \beta.$$

3. For a given maximal monotone operator $T : D(T) \subset X \to X^*$, show the following

   - If $\lambda > 0$, then the vector field $\lambda \cdot T$ defined by $(\lambda \cdot T)(x) = \lambda T(\frac{x}{\lambda})$ is maximal monotone with self-dual potential given by $(\lambda \cdot L_T)(x,p) := \lambda^2 L_T(\frac{x}{\lambda}, \frac{p}{\lambda})$.
   - For $y \in X$ and $q \in X^*$, the vector field $T^{1,y}$ (resp., $T^{2,q}$) given by $T^{1,y}(x) = T(x+y)$ (resp., $T^{2,q}(x) = T(x) - q$) is maximal monotone with self-dual potential given by $M_y(x,p) = L_T(x+y,p) - \langle y,p \rangle$ (resp., $N_q(x,p) = L_T(x,p+q) - \langle x,q \rangle$).
   - If X is a Hilbert space and U is a unitary operator ($UU^* = U^*U = I$) on X, then the vector field T_U given by $T_U(x) = U^*T(Ux)$ is maximal monotone with self-dual potential given by $M(x,p) := L_T(Ux,Up)$.
   - If $\Lambda : X \to X^*$ is any bounded skew-adjoint operator, then the vector field $T + \Lambda$ is a maximal monotone operator with self-dual potential given by $M(x,p) = L_T(x, -\Lambda x + p)$.
   - If $\Lambda : X \to X^*$ is an invertible skew-adjoint operator, then the vector field $\Lambda T^{-1}\Lambda - \Lambda$ is maximal monotone with self-dual potential given by $M(x,p) = L_T(x + \Lambda^{-1}p, \Lambda x)$.

4. If φ is a convex lower semicontinuous function on $X \times Y$, where X,Y are reflexive Banach spaces, if $A : X \to Y^*$ is any bounded linear operator, and if J is the symplectic operator on $X \times Y$ defined by $J(x,y) = (-y,x)$, then the vector field $\partial \varphi + (A^*,A) \circ J$ is maximal monotone on $X \times Y$ with self-dual potential given by $L((x,y),(p,q)) = \varphi(x,y) + \varphi^*(A^*y+p, -Ax+q)$.

5. If T_i is maximal monotone on a reflexive Banach space X_i for each $i \in I$, then the vector field $\Pi_{i \in I} T_i$ on $\Pi_{i \in I} X_i$ given by $(\Pi_{i \in I} T_i)((x_i)_i) = \Pi_{i \in I} T_i(x_i)$ is maximal monotone with self-dual potential $M((x_i)_i,(p_i)_i) = \Sigma_{i \in I} L_{T_i}(x_i,p_i)$.

6. If T_1 (resp., T_2) is a maximal operator on X (resp., Y), then for any bounded linear operator $A : X \to Y^*$, the vector field defined on $X \times Y$ by $T = (T_1,T_2) + (A^*,A) \circ J$ is maximal monotone with self-dual potential given by $L((x,y),(p.q)) := L_{T_1}(x,A^*y+p) + L_{T_2}(y, -Ax+q)$.

7. If T and S are two maximal monotone operators on X such that $D(T^{-1}) - D(S^{-1})$ contains a neighborhood of the origin in X^*, then the vector field $T + S$ is maximal monotone with potential given by $L_T \oplus L_S$.

8. If T and S are two maximal monotone operators on X such that $D(T) - D(S)$ contains a neighborhood of the origin in X, then the vector field $T \star S$ whose potential is given by $L_T \star L_S$ is maximal monotone.

Exercises 5.C. Maximal monotone operators on path spaces via the corresponding Lagrangian calculus

Let I be any finite time interval, which we shall take here without loss of generality, to be $[0,1]$, and let X be a reflexive Banach space. A *time-dependent – possibly set-valued – monotone map on* $[0,1] \times X$ (*resp., a time-dependent convex Lagrangian on* $[0,1] \times X \times X^*$) is a map $T : [0,1] \times X \to 2^{X^*}$ (resp., a function $L : [0,1] \times X \times X^* \to \mathbf{R} \cup \{+\infty\}$) such that:

- T (resp., L) is measurable with respect to the σ-field generated by the products of Lebesgue sets in $[0,1]$ and Borel sets in X (resp., in $X \times X^*$).
- For each $t \in [0,1]$, the map $T_t := T(t,\cdot)$ is monotone on X (resp., the Lagrangian $L(t,\cdot,\cdot)$ is convex and lower semicontinuous on $X \times X^*$).

1. If T is a time-dependent maximal monotone operator on $[0,1] \times X$, then the function $L_T(t,x,p) = L_{T_t}(x,p)$ is a time-dependent self-dual Lagrangian L_T on $[0,1] \times X \times X^*$.

2. If T satisfies

$$\|T(t,x)\| \leq C(t)(1+\|x\|) \text{ and } \langle T(t,x),x \rangle \geq \alpha(t)\|x\|^2 - \beta(t) \tag{5.12}$$

for $C(t)$, $\alpha^{-1}(t)$ in $L^\infty([0,1])$, and $\beta(t) \in L^1([0,1])$, then for some $K > 0$ we have for any u in $L^2_X[0,1]$ and p in $L^2_{X^*}[0,1]$

$$\mathcal{L}_T(u,p) = \int_0^1 L_T(t,u(t),p(t))dx \leq K(1+\|u\|_2^2+\|p\|_2^2), \tag{5.13}$$

and the operator $\bar{T}$ defined on L^2_X by $\bar{T}(u(t)_t) = (T(t,u(t)))_t$ is maximal monotone with potential given by $\mathcal{L}(u,p) = \int_0^1 L_{T_t}(u(t),p(t))dt$.
3. Let T be a time-dependent maximal monotone operator on $[0,1] \times H$, where H is a Hilbert space, and let S be a maximal monotone operator on H. Then, the operator

$$\mathcal{T}u = \left(\dot{u} + T_t u, S(u(0)-u(1)) + \frac{u(0)+u(1)}{2}\right)$$

is maximal monotone on A^2_H, with its potential given by the self-dual Lagrangian defined on $A^2_H \times (A^2_H)^* = A^2_H \times (H \times L^2_H)$ by

$$\mathcal{L}(u,p) = \int_0^1 L_{T_t}\left(u(t)+p_0(t),-\dot{u}(t)\right)dt + L_S\left(u(1)-u(0)+p_1,\frac{u(0)+u(1)}{2}\right).$$

Here, the dual of the space $A^2_H := \{u : [0,1] \rightarrow H; \dot{u} \in L^2_H\}$ is identified with $H \times L^2_H$ via the duality formula

$$\langle u,(p_1,p_0)\rangle_{A^2_H,H \times L^2_H} = \left\langle \frac{u(0)+u(1)}{2},p_1 \right\rangle + \int_0^1 \langle \dot{u}(t),p_0(t)\rangle\,dt,$$

where $u \in A^2_H$ and $(p_1,p_0(t)) \in H \times L^2_H$.

Further comments

It is natural to investigate the relationship between maximal monotone operators and self-dual vector fields since both could be seen as extensions of the superposition of subgradients of convex functions with skew-symmetric operators. An early indication was the observation we made in [55] that self-dual vector fields are necessarily maximal monotone. We suggested calling them then *"integrable maximal monotone fields"*, not suspecting that one could eventually prove that all maximal monotone operators are integrable in the sense that they all derive from self-dual Lagrangians [57]. This surprising development actually occurred when we realized through the book of Phelps [130] that Krauss [82] and Fitzpatrick [50] had done some work in this direction in the 1980's, and had managed to associate to a maximal monotone operator T a "convex potential" on phase space. In our terminology, their potential is Fenchelian and sub-self-dual.

The question of whether one can establish the existence of a truly self-dual Lagrangian associated to T was actually one of the original questions of Fitzpatrick [50]. We eventually announced a proof of the equivalence in [57], where we used

Asplund's averaging technique between the sub-self-dual Lagrangian L given by Fitzpatrick and its Legendre dual L^*. This turned out to require a boundedness assumption that could be handled by an additional approximation argument. Later, and upon seeing our argument, Bauschke and Wang [22] gave an explicit formula for the Lagrangian by using the proximal interpolation between L and L^*. It is this formula that we use here. Almost one year later, we eventually learned that the sufficient condition in the equivalence had been established earlier with completely different methods by R.S. Burachik and B. F. Svaiter in [35], while the necessary condition was shown by B. F. Svaiter in [154]. Most of the material in this chapter is taken from Ghoussoub [60].

Part II
COMPLETELY SELF-DUAL SYSTEMS
AND THEIR LAGRANGIANS

Many boundary value problems and evolution equations can be written in the form

$$0 \in \bar{\partial} L(x),$$

where L is a self-dual Lagrangian on phase space $X \times X^*$ and X is a reflexive Banach space. These are the *completely self-dual systems*. They can be solved by minimizing functionals of the form

$$I(x) = L(x, 0).$$

These are the *completely self-dual functionals*. They are convex lower semicontinuous and nonnegative , and under appropriate coercivity conditions they attain their infimum. Their main relevance, though, stems from the fact that this infimum is equal to 0. This property allows for variational formulations and resolutions of several basic differential equations and systems, which – often because of lack of self-adjointness – cannot be expressed as Euler-Lagrange equations, but can be written as a completely self-dual system.

This framework contains all known equations involving maximal monotone operators since they are shown to be self-dual vector fields. Issues of existence and uniqueness of solutions of such equations are easily deduced from the considerations above, with the added benefit of associating to these problems new, completely self-dual energy functionals.

The functional calculus on phase space – developed in Part I – allows for natural constructions of such functionals in many situations where the theory of maximal monotone operators proved delicate. The remarkable permanence properties of self-dual Lagrangians, lead to a fairly large class of completely self-dual equations to which the proposed variational theory applies.

Chapter 6
Variational Principles for Completely Self-dual Functionals

Our basic premise is that many boundary value problems and evolution equations can be solved by minimizing completely self-dual functionals of the form

$$I_p(x) = L(x, p) - \langle Bx, p \rangle,$$

where L is a B-self-dual Lagrangian. Since such functionals I_p are always nonnegative, their main relevance to our study stems from the fact that their infimum is actually equal to 0. This property allows for variational formulations and resolutions of several basic differential systems, which – often because of lack of self-adjointness or linearity – cannot be expressed as Euler-Lagrange equations but can be written in the form

$$p \in \overline{\partial}_B L(x).$$

The fact that the infimum of a completely self-dual functional I is zero follows from the basic duality theory in convex analysis, which in our particular – yet so natural and so encompassing – self-dual setting leads to a situation where the value of the dual problem is exactly the negative of the value of the primal problem, hence leading to zero as soon as there is no duality gap.

Several immediate applications follow from this observation coupled with the remarkable fact that the Lagrangian $L_\Gamma(x, p) = L(x, \Gamma x + p)$ remains self-dual on $X \times X^*$, provided L is and as long as Γ is a skew-symmetric operator. One then quickly obtains variational formulations and resolutions of the Lax-Milgram theorem, variational inequalities, nonself-adjoint semilinear Dirichlet problems, certain differential systems, and other nonpotential operator equations.

6.1 The basic variational principle for completely self-dual functionals

We consider the problem of minimizing self-dual functionals. It is covered by the following very simple – yet far-reaching – proposition. Actually, we shall only need the following relaxed version of self-duality.

N. Ghoussoub, *Self-dual Partial Differential Systems and Their Variational Principles*,
Springer Monographs in Mathematics, DOI 10.1007/978-0-387-84897-6_6,
© Springer Science+Business Media, LLC 2009

Definition 6.1. Given a bounded linear operator $B : X \to X$ and a convex Lagrangian L in $\mathscr{L}(X)$, we shall say that:

1. L is *partially B-self-dual* if

$$L^*(0, Bx) = L(x, 0) \quad \text{for all } x \in X. \tag{6.1}$$

2. L is *B-self-dual on the graph of a map* $\Gamma : D(\Gamma) \subset X$ *into* X^* if

$$L^*(B^*\Gamma x, Bx) = L(x, \Gamma x) \quad \text{for all } x \in D(\Gamma). \tag{6.2}$$

3. More generally, if $Y \times Z$ is a subset of $X \times X^*$, we shall say that L is *self-dual on* $Y \times Z$ if

$$L^*(B^*p, Bx) = L(x, p) \text{ for all } (x, p) \in Y \times Z. \tag{6.3}$$

We start by noting that if L is partially B-self-dual, then again

$$I(x) = L(x, 0) \geq 0 \quad \text{for every } x \in X \tag{6.4}$$

and

$$I(\bar{x}) = \inf_{x \in X} I(x) = 0 \quad \text{if and only if} \quad 0 \in \overline{\partial_B L}(\bar{x}). \tag{6.5}$$

This leads to the following definition.

Definition 6.2. A function $I : X \to \mathbf{R} \cup \{+\infty\}$ is said to be a *completely self-dual functional* on the Banach space X if there exists a bounded linear operator B on X and a partially B-self-dual Lagrangian L on $X \times X^*$ such that $I(x) = L(x, 0)$ for $x \in X$.

We now give sufficient conditions to ensure that the infimum of completely self-dual functionals is zero and is attained.

Proposition 6.1. *Let L be a convex lower semi continuous functional on a reflexive Banach space $X \times X^*$ and let B be a bounded linear operator on X. Assume that L is a partially B-self-dual Lagrangian and that for some $x_0 \in X$ the function $p \to L(x_0, p)$ is bounded above on a neighborhood of the origin in X^*.*

1. *If B is onto, then there exists $\bar{x} \in X$ such that*

$$\begin{cases} L(\bar{x}, 0) = \inf_{x \in X} L(x, 0) = 0, \\ 0 \in \overline{\partial_B L}(\bar{x}). \end{cases} \tag{6.6}$$

2. *If B has dense range and L^* is continuous in the second variable, then there exists $\bar{y} \in X$ such that*

$$L^*(0, \bar{y}) = \inf_{x \in X} L^*(0, x) = \inf_{x \in X} L(x, 0) = 0. \tag{6.7}$$

Moreover, if $\lim_{\|x\| \to \infty} L(x, 0) = +\infty$, then (6.6) also holds with $\bar{y} = B\bar{x}$.

Proof. This follows from the basic duality theory in convex optimization described in Section 2.7. Indeed, if $(\mathscr{P}_p)$ is the primal minimization problem $h(p) = \inf_{x \in X} L(x,p)$ in such a way that $(\mathscr{P}_0)$ is the initial problem $h(0) = \inf_{x \in X} L(x,0)$, then the dual problem $(\mathscr{P}^*)$ is $\sup_{y \in X} -L^*(0,y)$, and we have the weak duality formula

$$\inf \mathscr{P}_0 := \inf_{x \in X} L(x,0) \geq \sup_{y \in X} -L^*(0,y) := \sup \mathscr{P}^*.$$

From the "partial B-self-duality" of L, we get

$$\inf_{x \in X} L(x,0) \geq \sup_{y \in X} -L^*(0,y) \geq \sup_{z \in X} -L^*(0,Bz) = \sup_{z \in X} -L(z,0). \tag{6.8}$$

Note that h is convex on X^* and that its Legendre conjugate satisfies $h^*(By) = L^*(0,By) = L(y,0)$ on X.

If now for some $x_0 \in X$ the function $p \to L(x_0,p)$ is bounded above on a neighborhood of the origin in X^*, then $h(p) = \inf_{x \in X} L(x,p) \leq L(x_0,p)$ and therefore h is subdifferentiable at 0 (i.e., the problem $(\mathscr{P}_0)$ is then stable). Any point $\bar{y} \in \partial h(0)$ is then a minimizer for $x \to L^*(0,x)$ on X, and we have two cases.

(1) If B is onto, then $\bar{y} = B\bar{x} \in \partial h(0)$ and $h(0) + h^*(B\bar{x}) = 0$, which means that

$$-\inf_{x \in X} L(x,0) = -h(0) = h^*(B\bar{x}) = L^*(0,B\bar{x}) = L(\bar{x},0) \geq \inf_{x \in X} L(x,0).$$

It follows that $\inf_{x \in X} L(x,0) = -L(\bar{x},0) \leq 0$, and in view of (6.5), we get that the infimum of $(\mathscr{P})$ is zero and attained at $\bar{x}$, while the supremum of $(\mathscr{P}^*)$ is attained at $B\bar{x}$. In this case, we have

$$L(\bar{x},0) + L^*(0,B\bar{x}) = 0,$$

which yields in view of the limiting case of Legendre duality that $(0,B\bar{x}) \in \partial L(\bar{x},0)$ or $0 \in \bar{\partial}_B L(\bar{x})$.

(2) If B has dense range and L^* is continuous in the second variable, we then get for any sequence $(Bx_n)_n$ going to $\bar{y}$

$$-\inf_{x \in X} L(x,0) = -h(0) = h^*(\bar{y}) = L^*(0,\bar{y}) = \lim_n L^*(0,Bx_n) = \lim_n L(x_n,0)$$
$$\geq \inf_{x \in X} L(x,0).$$

It follows that again $\inf_{x \in X} L(x,0) = L^*(0,\bar{y}) = \inf_{x \in X} L^*(0,x) = 0$.

Finally, if $\lim_{\|x\| \to \infty} L(x,0) = +\infty$, then $(x_n)_n$ is necessarily bounded and a subsequence converges weakly to $\bar{x} \in X$. The rest follows from the lower semicontinuity of I.

Remark 6.1. (i) If B is onto or if B has a dense range and L^* is continuous in the second variable, then (6.6) holds under the condition that $x \to L(x,0)$ is coercive in the following sense:

$$\lim_{\|x\|\to\infty} \frac{L(x,0)}{\|x\|} = +\infty. \tag{6.9}$$

Indeed, since $h^*(By) = L^*(0,By) = L(y,0)$ on X, we get that h^* is coercive on X, which means that h is bounded above on neighborhoods of zero in X^*.

Here is an immediate application of Proposition 6.1.

Corollary 6.1. *Let L be a B-self-dual Lagrangian on a reflexive Banach space $X \times X^*$, where B is a surjective operator on X. Suppose that, for some $x_0 \in X$, the function $p \to L(x_0,p)$ is bounded on the balls of X^*. Then, for each $p \in X^*$, there exists $\bar{x} \in X$ such that*

$$\begin{cases} L(\bar{x},p) - \langle B\bar{x},p\rangle = \inf_{x\in X}\{L(x,p) - \langle Bx,p\rangle\} = 0, \\ p \in \overline{\partial}_B L(\bar{x}). \end{cases} \tag{6.10}$$

Proof. We simply apply Proposition 6.1 to the translated Lagrangian $M(x,q) = L(x,p+q) - \langle Bx,p\rangle$, which is also self-dual on $X \times X^*$.

The following corollary already leads to many applications.

Corollary 6.2. *Let $B : X \to X$ be bounded linear on a reflexive Banach space X and let $\Gamma : X \to X^*$ be an operator such that $\Gamma^* \circ B$ is skew-adjoint. Let L be a Lagrangian on $X \times X^*$ that is B-self-dual on the graph of Γ. Then, $I(x) = L(x,\Gamma x)$ is a completely self-dual functional on X. Moreover, assuming one of the conditions*

(A) $\lim_{\|x\|\to\infty} \frac{I(x)}{\|x\|} = +\infty$ *and B is onto (or B has dense range and L^* is continuous in the second variable) or*
(B) *Γ is invertible, the map $x \to L(x,0)$ is bounded above on the balls of X, and B is onto (or B has dense range and L is continuous in the first variable),*

then there exists $\bar{x} \in X$ such that

$$\begin{cases} I(\bar{x}) = \inf_{x\in X} I(x) = 0, \\ \Gamma\bar{x} \in \overline{\partial}_B L(\bar{x}). \end{cases} \tag{6.11}$$

Proof. We first note that the Lagrangian defined as $M(x,p) = L(x,\Gamma x+p)$ is partially B-self-dual. Indeed, fix $(q,y) \in X^* \times X$, set $r = \Gamma x + p$, and write

$$\begin{aligned} M^*(B^*q,By) &= \sup\{\langle B^*q,x\rangle + \langle By,p\rangle - L(x,\Gamma x+p); (x,p) \in X \times X^*\} \\ &= \sup\{\langle B^*q,x\rangle + \langle By,r-\Gamma x\rangle - L(x,r); (x,r) \in X \times X^*\} \\ &= \sup\{\langle B^*q+\Gamma^*By,x\rangle + \langle By,r\rangle - L(x,r); (x,r) \in X \times X^*\} \\ &= L^*(B^*q+B^*\Gamma y,By). \end{aligned}$$

If $q = 0$, then $M^*(0,By) = L^*(B^*\Gamma y,By) = L(y,\Gamma y) = M(y,0)$, and M is therefore partially B-self-dual.

Similarly, one can show that the Lagrangian $N(x,p) = L(x + \Gamma^{-1}p; \Gamma x)$ is partially B-self-dual, provided Γ is an invertible operator.

In other words, I has two possible representations as a completely self-dual functional:

$$I(x) = M(x,0) = N(x,0).$$

Under assumption (A), we apply Proposition 6.1 to M to obtain $\bar{x} \in X$ such that:

$$L(\bar{x}, \Gamma\bar{x}) = M(\bar{x},0) = \inf_{x \in X} M(x,0) = \inf_{x \in X} L(x, \Gamma x) = 0.$$

Now, note that

$$L(\bar{x}, \Gamma\bar{x}) = L^*(B^*\Gamma\bar{x}, B\bar{x}) = L^*(-\Gamma^*B\bar{x}, B\bar{x}),$$

hence,

$$L(\bar{x}, \Gamma\bar{x}) + L^*(-\Gamma^*B\bar{x}, B\bar{x}) = 0 = \langle (\bar{x}, \Gamma\bar{x}), (-\Gamma^*B\bar{x}, B\bar{x}) \rangle.$$

It follows from the limiting case of Legendre duality that

$$(B^*\Gamma\bar{x}, B\bar{x}) = (-\Gamma^*B\bar{x}, B\bar{x}) \in \partial L(\bar{x}, \Gamma\bar{x}).$$

Under assumption (B), we apply Proposition 6.1 to the partially self-dual Lagrangian N to conclude in a similar fashion. Note that the boundedness condition in this case yields that $p \to N(0,p) = L(\Gamma^{-1}p, 0)$ is bounded on a neighborhood of zero in X^*.

Exercises 6.A. Uniqueness

1. Show that Proposition 6.1 still holds if L is only supposed to be a subself-dual Lagrangian.
2. Let L be a self-dual Lagrangian L on a reflexive Banach space $X \times X^*$. If L is strictly convex in the second variable, show that $x \to \overline{\partial}L(x)$ is then single valued on its domain.
3. Show that the vector field $x \to \overline{\partial}L(x)$ is maximal monotone by proving that for every $\lambda > 0$, the map $(I + \lambda\overline{\partial}L)^{-1}$ is defined everywhere and is single-valued.
4. If L is uniformly convex in the second variable (i.e., if $L(x,p) - \varepsilon\frac{\|p\|^2}{2}$ is convex in p for some $\varepsilon > 0$), show that $x \to \overline{\partial}L(x)$ is then a Lipschitz maximal monotone operator on its domain.
5. Relate various properties of maximal monotone operators (e.g., strict and strong monotonicity, boundedness, linearity, etc.,... see Brézis [25] or Evans [48]) to their counterparts for the corresponding Fitzpatrick subself-dual Lagrangians.

6.2 Complete self-duality in non-selfadjoint Dirichlet problems

Consider a coercive bilinear continuous functional a on a Banach space X, i.e., for some $\lambda > 0$, we have that $a(v,v) \geq \lambda \|v\|^2$ for every $v \in X$. It is well known that if a is symmetric, then, for any $f \in X^*$, one can use a variational approach to find

$u \in X$ such that, for every $v \in X$, we have $a(u,v) = \langle v, f \rangle$. The procedure amounts to minimizing on H the convex functional $\psi(u) = \frac{1}{2}a(u,u) - \langle u, f \rangle$.

The theorem of Lax and Milgram deals with the case where a is not symmetric, for which the variational argument above does not work. Corollary 6.2, however, will allow us to formulate a variational proof of the original Lax-Milgram theorem (Corollary 6.4 below) by means of a completely self-dual functional. The semilinear version of that theorem (Corollary 6.3), as well as the one dealing with nonhomogeneous boundary conditions (Corollary 8.2), can also be resolved via the minimization of a completely self-dual functional.

The rest of this chapter will consist of showing how Corollary 6.2 – applied to the most basic self-dual Lagrangians – already yields variational formulations and resolutions to several nonself-adjoint homogeneous semilinear equations.

A variational resolution for nonsymmetric semilinear homogeneous equations

Corollary 6.3. *Let $\varphi : X \to \mathbf{R} \cup \{+\infty\}$ be a proper convex lower semicontinuous function on a reflexive Banach space X, and let $\Gamma : X \to X^*$ be a bounded positive linear operator. Assume one of the following two conditions:*

(A) $\lim\limits_{\|x\| \to \infty} \|x\|^{-1}(\varphi(x) + \frac{1}{2}\langle \Gamma x, x \rangle) = +\infty$ *or*

(B) *The operator $\Gamma^a = \frac{1}{2}(\Gamma - \Gamma^*) : X \to X^*$ is onto and φ is bounded above on the ball of X.*

Then, there exists for any $f \in X^*$ a solution $\bar{x} \in X$ to the equation

$$-\Gamma x + f \in \partial\varphi(x) \tag{6.12}$$

that can be obtained as a minimizer of the completely self-dual functional

$$I(x) = \psi(x) + \psi^*(-\Gamma^a x), \tag{6.13}$$

where ψ is the functional $\psi(x) = \frac{1}{2}\langle \Gamma x, x \rangle + \varphi(x) - \langle f, x \rangle$.

Proof. Apply Corollary 6.2 to the skew-adjoint operator $\Gamma := \Gamma^a$ and the self-dual Lagrangian $L(x, p) = \psi(x) + \psi^*(p)$. Note that all that is needed in the proposition above is that the function $\varphi(x) + \frac{1}{2}\langle \Gamma x, x \rangle$ (and not necessarily φ itself) be convex and lower semicontinuous.

Example 6.1. A variational formulation for the original Lax-Milgram theorem

Corollary 6.4. *Let a be a coercive continuous bilinear form on $X \times X$, and consider $\Gamma : X \to X^*$ to be the skew-adjoint operator defined by $\langle \Gamma v, w \rangle = \frac{1}{2}(a(v,w) -$*

$a(w,v))$. *For any $f \in X^*$, the completely self-dual functional $I(v) = \psi(v) + \psi^*(\Gamma v)$, where $\psi(v) = \frac{1}{2}a(v,v) - \langle v, f \rangle$, attains its minimum at $\bar{u} \in X$ in such a way that*

$$I(\bar{u}) = \inf_{v \in H} I(v) = 0 \quad and \quad a(v, \bar{u}) = \langle v, f \rangle \text{ for every } v \in X.$$

Proof. Consider the self-dual Lagrangian $L(x,p) = \psi(x) + \psi^*(p)$, and apply Corollary 6.3 to Γ. Note that $\Gamma \bar{u} \in \partial \psi(\bar{u})$ means that for every $v \in X$

$$\frac{1}{2}(a(\bar{u}, v) - a(v, \bar{u})) = \frac{1}{2}(a(\bar{u}, v) + a(v, \bar{u})) - \langle v, f \rangle,$$

which yields our claim.

Example 6.2. Inverting nonself-adjoint matrices by minimizing completely self-dual functionals

An immediate finite-dimensional application of the corollary above is the following variational solution for the linear equation $Ax = y$, where A is a nonsymmetric $n \times n$-matrix and $y \in \mathbf{R}^n$. It then suffices to minimize the completely self-dual functional

$$I(x) = \frac{1}{2}\langle Ax, x \rangle + \frac{1}{2}\langle A_s^{-1}(y - A_a x), y - A_a x \rangle - \langle y, x \rangle$$

on $\mathbf{R}^n$, where A_a is the antisymmetric part of A and A_s^{-1} is the inverse of the symmetric part. If A is coercive (i.e., $\langle Ax, x \rangle \geq c|x|^2$ for all $x \in \mathbf{R}^n$), then there is a solution $\bar{x} \in \mathbf{R}^n$ to the equation obtained as $I(\bar{x}) = \inf_{x \in \mathbf{R}^n} I(x) = 0$. The numerical implementation of this approach to the problem of inverting nonsymmetric matrices was developed in Ghoussoub and Moradifam [66], where it is shown that it has certain advantages on existing numerical schemes.

Example 6.3. A variational solution for variational inequalities

Given again a bilinear continuous functional a on $X \times X$ and $\varphi : X \to \mathbf{R}$ convex and lower semicontinuous, then solving the corresponding variational inequality amounts to constructing for any $f \in X^*$, an element $y \in X$ such that for all $z \in X$

$$a(y, y - z) + \varphi(y) - \varphi(z) \leq \langle y - z, f \rangle. \tag{6.14}$$

It is easy to see that this problem can be rewritten as

$$f \in Ay + \partial \varphi(y),$$

where A is the bounded linear operator from X into X^* defined by $a(u,v) = \langle Au, v \rangle$. This means that the variational inequality (6.14) can be rewritten and solved using the variational principle for completely self-dual functionals. For example, we can formulate and solve variationally the following "obstacle" problem.

Corollary 6.5. *Let a be a bilinear continuous functional on a reflexive Banach space $X \times X$ so that $a(v,v) \geq \lambda \|v\|^2$, and let K be a convex closed subset of X. Then, for any $f \in X^*$, there is $\bar{x} \in K$ such that*

$$a(\bar{x}, \bar{x} - z) \leq \langle \bar{x} - z, f \rangle \quad \text{for all } z \in K. \tag{6.15}$$

The point $\bar{x}$ can be obtained as a minimizer on X of the completely self-dual functional

$$I(x) = \varphi(x) + (\varphi + \psi_K)^*(-Ax),$$

where $\varphi(u) = \frac{1}{2}a(u,u) - \langle f, x \rangle$, $A : X \to X^*$ is the skew-adjoint operator defined by $\langle Au, v \rangle = \frac{1}{2}(a(u,v) - a(v,u))$, and $\psi_K(x) = 0$ on K and $+\infty$ elsewhere.

Example 6.4. A variational principle for a nonsymmetric Dirichlet problem

Let $\mathbf{a} : \Omega \to \mathbf{R}^n$ be a smooth function on a bounded domain Ω of $\mathbf{R}^n$, and consider the first-order linear operator $Av = \mathbf{a} \cdot \nabla v = \sum_{i=1}^n a_i \frac{\partial v}{\partial x_i}$. Assume that the vector field $\sum_{i=1}^n a_i \frac{\partial v}{\partial x_i}$ is actually the restriction of a smooth vector field $\sum_{i=1}^n \bar{a}_i \frac{\partial v}{\partial x_i}$ defined on an open neighborhood X of $\bar{\Omega}$ and that each $\bar{a}_i$ is a $C^{1,1}$ function on X. Consider the Dirichlet problem:

$$\begin{cases} \Delta u + \mathbf{a} \cdot \nabla u = |u|^{p-2}u + f & \text{on } \Omega \\ u = 0 & \text{on } \partial\Omega. \end{cases} \tag{6.16}$$

If $\mathbf{a} = 0$, then to find a solution it is sufficient to minimize the functional

$$\Phi(u) = \frac{1}{2}\int_\Omega |\nabla u|^2 dx + \frac{1}{p}\int_\Omega |u|^p dx + \int_\Omega fu\,dx$$

and find a critical point $\partial\Phi(u) = 0$. However, if the nonself-adjoint term $\mathbf{a}$ is not zero, we can use the approach above to get the following existence theorem.

Theorem 6.1. *Assume $\text{div}(\mathbf{a}) \geq 0$ on Ω, and $1 < p \leq \frac{2n}{n-2}$, then the functional*

$$I(u) := \Psi(u) + \Psi^*\left(\mathbf{a}.\nabla u + \frac{1}{2}\text{div}(\mathbf{a})u\right),$$

where

$$\Psi(u) = \frac{1}{2}\int_\Omega |\nabla u|^2 dx + \frac{1}{p}\int_\Omega |u|^p dx + \int_\Omega fu\,dx + \frac{1}{4}\int_\Omega \text{div}(\mathbf{a})|u|^2 dx$$

is completely self-dual on $L^2(\Omega)$, and there exists $\bar{u} \in H_0^1(\Omega)$ such that $I(\bar{u}) = \inf\{I(u); u \in H_0^1(\Omega)\} = 0$ and $\bar{u}$ is a solution of equation (6.16).

Proof. Indeed, Ψ is clearly convex and lower semicontinuous on $L^2(\Omega)$, while the operator $\Gamma u = \mathbf{a}.\nabla u + \frac{1}{2}\mathrm{div}(\mathbf{a})u$ is skew-adjoint by Green's formula. Again the functional $I(u) = \Psi(u) + \Psi^*(\mathbf{a}.\nabla u + \frac{1}{2}\mathrm{div}(\mathbf{a})u)$ is of the form $L(u,\Gamma u)$, where $L(u,v) = \Psi(u) + \Psi^*(v)$ is a self-dual Lagrangian on $L^2(\Omega) \times L^2(\Omega)$. The existence follows from Corollary 6.3 since Ψ is clearly coercive. Note that $\bar{u}$ then satisfies

$$\mathbf{a}.\nabla\bar{u} + \frac{1}{2}\mathrm{div}(\mathbf{a})\bar{u} = \partial\Psi(\bar{u}) = -\Delta\bar{u} + \bar{u}^{p-1} + f + \frac{1}{2}\mathrm{div}(\mathbf{a})\bar{u},$$

and therefore $\bar{u}$ is a solution for (6.16).

6.3 Complete self-duality and non-potential PDEs in divergence form

Equations of the form

$$\begin{cases} -\mathrm{div}(\partial\varphi(\nabla f(x))) = g(x) & \text{on } \Omega \subset \mathbf{R}^n, \\ f(x) = 0 & \text{on } \partial\Omega, \end{cases} \tag{6.17}$$

where $g \in L^2(\Omega)$ and φ is a convex functional on $\mathbf{R}^n$, are variational since they are the Euler-Lagrange equations associated to the energy functional

$$J(u) = \int_\Omega \{\varphi(\nabla u(x)) - u(x)g(x)\}dx. \tag{6.18}$$

However, those of the form

$$\begin{cases} \mathrm{div}(T(\nabla f(x))) = g(x) & \text{on } \Omega \subset \mathbf{R}^n \\ f(x) = 0 & \text{on } \partial\Omega, \end{cases} \tag{6.19}$$

where T is a nonpotential vector field on $\mathbf{R}^n$, are not variational in the classical sense. We shall now show how solutions can be derived through a self-dual variational principle, at least in the case where T is a general monotone operator.

Proposition 6.2. *Let L be a self-dual Lagrangian on a Hilbert space E, and let $E = Y \oplus Y^\perp$ be a decomposition of E into two orthogonal subspaces, with $\pi : E \to Y$ denoting the orthogonal projection onto Y and $\pi^\perp = I - \pi$.*
If $p \to L(\pi^\perp(p), \pi(p))$ is bounded on the bounded sets of E, then for any $p_0 \in E$, the completely self-dual functional

$$I(u) = L(\pi(u), \pi(p_0)) + \pi^\perp(u)) - \langle u, \pi(p_0)\rangle$$

attains its minimum at some $\bar{u} \in E$ in such a way that

$$\pi(p_0) + \pi^\perp(\bar{u}) \in \bar{\partial}L(\pi(\bar{u})) \tag{6.20}$$

and therefore

$$\pi(p_0) \in \pi \overline{\partial} L(\pi(\bar{u})). \tag{6.21}$$

Proof. First, use Proposition 3.5 to deduce that

$$M(u, p) := L(\pi(u) + \pi^{\perp}(p), \pi(p) + \pi^{\perp}(u))$$

is a self-dual Lagrangian on $E \times E$. Then, apply Proposition 3.4 (2) to infer that the Lagrangian

$$N(u, p) = L(\pi(u) + \pi^{\perp}(p + \pi(p_0)), \pi(p + \pi(p_0)) + \pi^{\perp}(u)) - \langle u, \pi(p_0) \rangle$$

is also a self-dual Lagrangian on $E \times E$ since it is M "translated" by $\pi(p_0)$. The conclusion follows from Proposition 6.1 applied to $I(u) = N(u, 0)$. Indeed, the fact that $I(\bar{u}) = N(\bar{u}, 0) = 0$ yields that $\pi(p_0) + \pi^{\perp}(\bar{u}) \in \overline{\partial} L(\pi(\bar{u}))$. Now apply π to both sides to conclude.

Proposition 6.3. *Let L be a self-dual Lagrangian on a Hilbert space $E \times E$ such that*

$$L(a, b) \leq C(1 + \|a\|^2 + \|b\|^2) \text{ for all } (a, b) \in E \times E. \tag{6.22}$$

Consider $A : X \to E$ to be a bounded linear operator from a reflexive Banach space X into E such that the operator A^*A is an isomorphism from X onto X^*. Then, for any $p_0 \in X^*$, there exists $x_0 \in X$ such that

$$p_0 \in A^* \overline{\partial} L(Ax_0). \tag{6.23}$$

It is obtained as $x_0 = (A^*A)^{-1} A^*(u_0)$, where u_0 is the minimum of the functional

$$I(u) = L(\pi(u), \pi(q_0) + \pi^{\perp}(u)) - \langle u, \pi(q_0) \rangle$$

on the Hilbert space E, q_0 being any element in E such that $A^*q_0 = p_0$ and π denoting the projection $A(A^*A)^{-1}A^*$.

Proof. Let $\Lambda := A^*A$ be the isomorphism from X onto X^*, and note that $\pi := A\Lambda^{-1}A^*$ is a projection from E onto its subspace $Y = A(X)$. Since A^* is onto, there exists $q_0 \in E$ such that $A^*q_0 = p_0$. By the preceding proposition, there exists $u_0 \in E$ such that

$$\pi(q_0) \in \pi \overline{\partial} L(\pi(u_0)). \tag{6.24}$$

By applying A^* to both sides and by using that $A^*\pi = A^*$, it follows that

$$p_0 = A^*q_0 = A^*\pi(q_0) \in A^* \overline{\partial} L(\pi(u_0)) = A^* \overline{\partial} L(A\Lambda^{-1}A^*(u_0)). \tag{6.25}$$

It now suffices to set $x_0 = \Lambda^{-1}A^*(u_0)$ to get (6.23).

Example 6.5. A variational principle for nonpotential quasilinear PDEs in divergence form

In order to resolve (6.19), we assume that T is a maximal monotone operator, and we use Theorem 5.1 to associate to T a self-dual Lagrangian L on $\mathbf{R}^n \times \mathbf{R}^n$ such that $\bar{\partial} L = T$. We then consider the self-dual Lagrangian $\mathscr{L}_T$ on $L^2(\Omega; \mathbf{R}^n) \times L^2(\Omega; \mathbf{R}^n)$ via the formula

$$\mathscr{L}_T(u, p) = \int_\Omega L_T(u(x), p(x)) dx.$$

Assuming that T satisfies for all $x \in \mathbf{R}^n$

$$|Tx| \leq C(1 + |x|) \text{ and } \langle Tx, x \rangle \geq \alpha |x|^2 - \beta, \tag{6.26}$$

this implies that, for all $u, p \in \mathbf{R}^n$,

$$L_T(u, p) \leq \frac{(C|u| + |p|)^2}{2\alpha} + C|u| - \beta,$$

which yields that for some $K > 0$ we have

$$\mathscr{L}_T(u, p) = \int_\Omega L(u(x), p(x)) dx \leq K(1 + \|u\|_2^2 + \|p\|_2^2) \tag{6.27}$$

for any u and p in $L^2(\Omega, \mathbf{R}^n)$. We now apply Proposition 6.3 with the spaces $E = L^2(\Omega; \mathbf{R}^n)$, $X = H_0^1(\Omega)$, $X^* = H^{-1}(\Omega)$, and the operator $A : X \to E$ defined by $Af = \nabla f$. Note that $A^*A = \nabla^*\nabla = -\Delta$, which is an isomorphism from $X = H_0^1(\Omega)$ onto $X^* = H^{-1}(\Omega)$. Note that the projection $\pi(u) = \nabla(-\Delta)^{-1}\nabla^*u$ and its orthogonal $\pi^\perp(u) = u - \nabla(-\Delta)^{-1}\nabla^*u$ are nothing but the Hodge decomposition of any $u \in L^2(\Omega, \mathbf{R}^n)$, into a "pure potential" and a divergence-free vector field.

Let now $p_0 = \nabla(-\Delta)^{-1}g$ in such a way that $\text{div } p_0 = g$. In view of Proposition 6.3, the infimum of the functional

$$J(u) := \int_\Omega \left\{ L_T \left(\nabla(-\Delta)^{-1}\nabla^*u, \, u - \nabla(-\Delta)^{-1}\nabla^*u + p_0 \right) - \langle u, p_0 \rangle \right\} dx \tag{6.28}$$

on $L^2(\Omega, \mathbf{R}^n)$ is equal to zero and is attained at $\bar{u} \in L^2(\Omega, \mathbf{R}^n)$ in such a way that

$$\bar{u} - \nabla(-\Delta)^{-1}\nabla^*\bar{u} + p_0 \in \bar{\partial} L_T(\nabla(-\Delta)^{-1}\nabla^*\bar{u}).$$

By taking their respective divergences, we obtain $\text{div } p_0 \in \text{div}\left(\bar{\partial} L_T(\nabla(-\Delta)^{-1}\nabla^*\bar{u}) \right)$. In other words, by setting $f := (-\Delta)^{-1}\nabla^*\bar{u}$ and recalling that $\bar{\partial} L_T = T$ and $\text{div } p_0 = g$, we finally obtain that $g \in \text{div}(T(\nabla f))$, and f is then a solution of (6.19).

Remark 6.2. Note that an equivalent way to express the variational formulation above consists of minimizing the functional

$$J(f, w) = \int_{\Omega} \left\{ L_T \left(\nabla f(x), w(x) + p_0(x) \right) - f(x) g(x) \right\} dx \qquad (6.29)$$

over all possible $f \in H_0^1(\Omega)$ and all $w \in L^2(\Omega; \mathbf{R}^n)$ with $\operatorname{div} w = 0$.

Exercises 6.B. Orthogonal projections and self-dual Lagrangians

1. Given maximal monotone mappings $T_1, T_2, ..., T_n$ on a Hilbert space H, and positive reals $\alpha_1, \alpha_2, ..., \alpha_n$, consider the problem of finding x, and $y_1, y_2, ..., y_n$ in H such that

$$y_i \in T_i(x) \ \text{ for } i = 1, 2, ..., n \quad \text{and} \quad \sum_{i=1}^{n} \alpha_i y_i = 0. \qquad (6.30)$$

Show that (6.30) can be reduced to the problem of minimizing the completely self-dual functional $\sum_{i=1}^{n} \alpha_i L_{T_i}(x, y_i)$ over all $x \in H$ and $y_1, y_2, ..., y_n$ in H such that $\sum_{i=1}^{n} \alpha_i y_i = 0$, where each L_{T_i} is the self-dual Lagrangian on H associated with T_i.
Hint: Consider the space $E := H_1 \times H_2 \times ... \times H_n$ where each H_i is the Hilbert space H re-equipped with the scalar product $\langle x, y \rangle_{H_i} = \alpha_i \langle x, y \rangle_H$, and the orthogonal projection π from E onto the diagonal subspace $Y = \{(x_1, x_2, ..., x_n); x_1 = x_2 = ... = x_n\}$. Then, use Proposition 3.5 to deduce that $M(u, p) := L(\pi(u) + \pi^{\perp}(p), \pi(p) + \pi^{\perp}(u))$ is a self-dual Lagrangian on $E \times E$, where L is the self-dual Lagrangian on $E \times E$ defined by

$$L((u_i)_{i=1}^{n}, (p_i)_{i=1}^{n}) = \sum_{i=1}^{n} \alpha_i L_{T_i}(u_i, p_i).$$

2. Use the above to develop a self-dual variational approach for locating a point in the intersection of a finite number of closed convex subsets of a Hilbert space.

6.4 Completely self-dual functionals for certain differential systems

The next proposition shows that the theory of self-dual Lagrangians is well suited for "antiHamiltonian" systems of the form

$$(-A^* y, Ax) \in \partial \varphi(x, y), \qquad (6.31)$$

where φ is a convex lower semicontinuous functional on a reflexive Banach space $X \times Y$ and A is any bounded linear operator from X to Y^*.

Proposition 6.4. *Let φ be a proper and coercive convex lower semicontinuous function on $X \times Y$ with $(0,0) \in \operatorname{dom}(\varphi)$, and let $A : X \to Y^*$ be any bounded linear operator. Assume $B_1 : X \to X^*$ (resp., $B_2 : Y \to Y^*$) to be skew-adjoint operators. Then, there exists $(\bar{x}, \bar{y}) \in X \times Y$ such that*

$$(-A^* \bar{y} + B_1 \bar{x}, A\bar{x} + B_2 \bar{y}) \in \partial \varphi(\bar{x}, \bar{y}). \qquad (6.32)$$

The solution is obtained as a minimizer on $X \times Y$ of the completely self-dual functional

$$I(x,y) = \varphi(x,y) + \varphi^*(-A^*y + B_1x, Ax + B_2y).$$

Proof. It is enough to apply Proposition 6.2 to the self-dual Lagrangian

$$L((x,y),(p,q)) = \varphi(x,y) + \varphi^*(-A^*y + B_1x - p, Ax + B_2y - q)$$

obtained by shifting to the right the self-dual Lagrangian $\varphi \oplus_{as} A$ by the skew-adjoint operator (B_1, B_2) (see Proposition 3.4 (9)). This yields that $I(x,y) = L((x,y),(0,0))$ attains its minimum at some $(\bar{x}, \bar{y}) \in X \times Y$ and that the minimum is actually 0. In other words,

$$
\begin{aligned}
0 = I(\bar{x}, \bar{y}) &= \varphi(\bar{x}, \bar{y}) + \varphi^*(-A^*\bar{y} + B_1\bar{x}, A\bar{x} + B_2\bar{y}) \\
&= \varphi(\bar{x}, \bar{y}) + \varphi^*(-A^*\bar{y} + B_1\bar{x}, A\bar{x} + B_2\bar{y}) - \langle(\bar{x}, \bar{y}), (-A^*\bar{y} + B_1\bar{x}, A\bar{x} + B_2\bar{y})\rangle,
\end{aligned}
$$

from which the equation follows.

Corollary 6.6. *Given positive operators $B_1 : X \to X^*$, $B_2 : Y \to Y^*$ and convex functions φ_1 in $\mathscr{C}(X)$ and φ_2 in $\mathscr{C}(Y)$ having 0 in their respective domains, we consider the convex functionals $\psi_1(x) = \frac{1}{2}\langle B_1x, x\rangle + \varphi_1(x)$ and $\psi_2(x) = \frac{1}{2}\langle B_2x, x\rangle + \varphi_2(x)$. Assume*

$$\lim_{\|x\| + \|y\| \to \infty} \frac{\psi_1(x) + \psi_2(y)}{\|x\| + \|y\|} = +\infty.$$

Then, for any $(f,g) \in X^* \times Y^*$, any $c \in \mathbf{R}$, and any bounded linear operator $A : X \to Y^*$, there exists a solution $(\bar{x}, \bar{y}) \in X \times Y$ to the system of equations

$$
\begin{cases}
-A^*y - B_1x + f \in \partial \varphi_1(x) \\
c^2Ax - B_2y + g \in \partial \varphi_2(y).
\end{cases}
\tag{6.33}
$$

It can be obtained as a minimizer of the completely self-dual functional on $X \times Y$

$$I(x,y) = \chi_1(x) + \chi_1^*(-B_1^a x - A^*y) + \chi_2(y) + \chi_2^*(-c^{-2}B_2^a y + Ax), \tag{6.34}$$

where B_1^a (resp., B_2^a) are the skew-symmetric parts of B_1 and B_2 and where $\chi_1(x) = \psi_1(x) - \langle f, x\rangle$ and $\chi_2(x) = c^{-2}(\psi_2(x) - \langle g, x\rangle)$.

Proof. This follows by applying the above proposition to the convex function $\varphi(x,y) = \chi_1(x) + c^2\chi_2(y)$ and the skew-symmetric operators $-B_1^a$ and $-B_2^a$. Note that the operator $\tilde{A} : X \times Y \to X^* \times Y^*$ defined by $\tilde{A}(x,y) = (A^*y, -c^2Ax)$ is skew-adjoint once we equip $X \times Y$ with the scalar product

$$\langle(x,y),(p,q)\rangle = \langle x, p\rangle + c^{-2}\langle y, q\rangle.$$

We then get

$$
\begin{cases}
-A^*y - B_1^a x + f \in \partial \varphi_1(x) + B_1^s(x) \\
c^2Ax - B_2^a y + g \in \partial \varphi_2(y) + B_2^s(y),
\end{cases}
\tag{6.35}
$$

which gives the result.

Example 6.6. A variational principle for coupled equations

Let $\mathbf{b_1} : \Omega \to \mathbf{R^n}$ and $\mathbf{b_2} : \Omega \to \mathbf{R^n}$ be two smooth vector fields on a neighborhood of a bounded domain Ω of $\mathbf{R^n}$ verifying the conditions in Example 6.4. Consider the system:

$$\begin{cases} \Delta(v+u) + \mathbf{b_1} \cdot \nabla u = |u|^{p-2}u + f & \text{on } \Omega, \\ \Delta(v - c^2 u) + \mathbf{b_2} \cdot \nabla v = |v|^{q-2}v + g & \text{on } \Omega, \\ \qquad\qquad u = v = 0 & \text{on } \partial\Omega. \end{cases} \tag{6.36}$$

We can use the above to get the following result.

Theorem 6.2. *Assume* $\operatorname{div}(\mathbf{b_1}) \geq 0$ *and* $\operatorname{div}(\mathbf{b_2}) \geq 0$ *on* Ω, $1 < p, q \leq \frac{n+2}{n-2}$, *and consider on* $L^2(\Omega) \times L^2(\Omega)$ *the completely self-dual functional*

$$I(u,v) = \Psi(u) + \Psi^* \left(\mathbf{b_1}.\nabla u + \frac{1}{2}\operatorname{div}(\mathbf{b_1})u + \Delta v \right)$$
$$+ \Phi(v) + \Phi^* \left(\frac{1}{c^2}\mathbf{b_2}.\nabla v + \frac{1}{2c^2}\operatorname{div}(\mathbf{b_2})v - \Delta u \right),$$

where

$$\Psi(u) = \frac{1}{2}\int_\Omega |\nabla u|^2 dx + \frac{1}{p}\int_\Omega |u|^p dx + \int_\Omega fu\,dx + \frac{1}{4}\int_\Omega \operatorname{div}(\mathbf{b_1})\,|u|^2 dx,$$

$$\Phi(v) = \frac{1}{2c^2}\int_\Omega |\nabla v|^2 dx + \frac{1}{qc^2}\int_\Omega |v|^q dx + \frac{1}{c^2}\int_\Omega gv\,dx + \frac{1}{4c^2}\int_\Omega \operatorname{div}(\mathbf{b_2})\,|v|^2 dx$$

and Ψ^* *and* Φ^* *are their Legendre transforms. Then, there exists* $(\bar{u}, \bar{v}) \in H_0^1(\Omega) \times H_0^1(\Omega)$ *such that*

$$I(\bar{u}, \bar{v}) = \inf\{I(u,v); (u,v) \in H_0^1(\Omega) \times H_0^1(\Omega)\} = 0$$

and $(\bar{u}, \bar{v})$ *is a solution of equation* (6.36).

We can also reduce general minimization problems of nonself-dual functionals of the form $I(x) = \varphi(x) + \psi(Ax)$ to the much easier problem of minimizing self-dual functionals in two variables.

Proposition 6.5. *Let* φ *(resp.,* ψ*) be a convex lower semicontinuous function on a reflexive Banach space* X *(resp.* Y^**) and let* $A : X \to Y^*$ *be a bounded linear operator. Consider on* $X \times Y$ *the completely self-dual functional*

$$I(x,y) = \varphi(x) + \psi^*(y) + \varphi^*(-A^*y) + \psi(Ax).$$

Assuming $\lim_{\|x\|+\|y\|\to\infty} I(x,y) = +\infty$, *then the infimum of* I *is zero and is attained at a point* $(\bar{x}, \bar{y})$ *that determines the extremals of the min-max problem:*

$$\sup\{-\psi^*(y) - \varphi^*(-A^*y); y \in Y\} = \inf\{\varphi(x) + \psi(Ax); x \in X\}.$$

The pair $(\bar{x}, \bar{y})$ also satisfies the system

$$\begin{cases} -A^*y \in \partial \varphi(x) \\ Ax \in \partial \psi^*(y). \end{cases} \tag{6.37}$$

Proof. It is sufficient to note that $I(x,y) = L((x,y),(0,0))$, where L is the self-dual Lagrangian defined on $X \times Y$ by

$$L((x,y),(p,q)) = \varphi(x) + \psi^*(y) + \varphi^*(-A^*y + p) + \psi(Ax + q).$$

By considering more general twisted sum Lagrangians, we obtain the following application.

Corollary 6.7. *Let X and Y be two reflexive Banach spaces and let $A : X \to Y^*$ be any bounded linear operator. Assume L and M are self-dual Lagrangians on X and Y, respectively, such that*

$$\lim_{\|x\|+\|y\| \to \infty} \frac{L(x, A^*y) + M(y, -Ax)}{\|x\| + \|y\|} = +\infty.$$

Then, $I(x,y) := L(\bar{x}, A^*\bar{y}) + M(\bar{y}, -A\bar{x})$ is a completely self-dual functional that attains its infimum at $(\bar{x}, \bar{y}) \in X \times Y$ in such a way that $I(\bar{x}, \bar{y}) = 0$ and

$$\begin{cases} -A^*\bar{y} \in \bar{\partial}L(\bar{x}) \\ A\bar{x} \in \bar{\partial}M(\bar{y}). \end{cases} \tag{6.38}$$

Proof. It suffices to apply Corollary 6.2 to the self-dual Lagrangian $L \oplus_A M$.

6.5 Complete self-duality and semilinear transport equations

Theorem 6.3. *Let L be a B-self-dual Lagrangian on a reflexive space X, where $B : X \to X$ is a bounded linear operator on X. Consider a linear operator $\Gamma : D(\Gamma) \subset X \to X^*$ such that $B^*\Gamma$ is antisymmetric , and assume one of the following two conditions:*

1. L is standard and $\mathrm{Dom}_1(L) \subset D(\Gamma)$.
2. $B^*\Gamma$ is skew-adjoint and $x \to L(x,0)$ is bounded on the unit ball of X.

Then,

$$I(x) = \begin{cases} L(x, \Gamma x) & \text{if } x \in D(\Gamma) \\ +\infty & \text{if } x \notin D(\Gamma) \end{cases}$$

is a completely self-dual functional on X. Moreover, if $\lim\limits_{\|x\|\to+\infty} \frac{L(x,\Gamma x)}{\|x\|} = +\infty$, *then there exists* $\bar{x} \in D(\Gamma)$ *such that*

$$\begin{cases} I(\bar{x}) = \inf_{x \in X} I(x) = 0, \\ \Gamma\bar{x} \in \bar{\partial}_B L(\bar{x}). \end{cases} \tag{6.39}$$

Proof. In both cases, the Lagrangian L_Γ is B-self-dual by Proposition 4.1. We can therefore apply Corollary 6.2 to get $\bar{x} \in \text{Dom}_1(L_\Gamma) = D(\Gamma) \cap \text{Dom}_1(L)$, which satisfies the claim.

Corollary 6.8. *Let* φ *be a bounded below proper convex lower semicontinuous function on a reflexive Banach space X such that* $\lim\limits_{\|x\|\to+\infty} \frac{\varphi(x)}{\|x\|} = +\infty$. *Suppose* $\Gamma : D(\Gamma) \subset X \to X^*$ *is an antisymmetric operator such that one of the following conditions holds:*

1. $\text{Dom}(\varphi) \subset D(\Gamma)$.
2. Γ *is skew-adjoint, and* φ *is bounded on the unit ball of X.*

Then, for every $f \in X^*$, *there exists a solution* $\bar{x} \in \text{Dom}(\varphi) \cap D(\Gamma)$ *to the equation*

$$f + \Gamma x \in \partial\varphi(x). \tag{6.40}$$

It is obtained as a minimizer of the functional $I(x) = \varphi(x) - \langle f, x\rangle + \varphi^*(\Gamma x + f)$.

Proof. It is an immediate consequence of Theorem 6.3 applied to the self-dual Lagrangian $L(x, p) = \psi(x) + \psi^*(p)$, where $\psi(x) = \varphi(x) + \langle f, x\rangle$.

Example 6.7. Transport equations with the p-Laplacian

Consider the equation

$$\begin{cases} -\Delta_p u + \frac{1}{2}a_0 u + u|u|^{m-2} + f = \mathbf{a} \cdot \nabla u & \text{on} \quad \Omega \subset \mathbf{R}^n, \\ u = 0 & \text{on} \quad \partial\Omega. \end{cases} \tag{6.41}$$

We then have the following application of Corollary 6.8 in the case of an unbounded antisymmetric operator.

Theorem 6.4. *Let* $\mathbf{a} \in C^\infty(\bar{\Omega})$ *be a smooth vector field on* $\Omega \subset \mathbf{R}^n$, *and let* $a_0 \in L^\infty(\Omega)$. *Let* $p \geq 2$, $1 \leq m \leq \frac{np}{n-p}$, *and assume the following condition:*

$$\text{div}(\mathbf{a}) + a_0 \geq 0 \text{ on } \Omega. \tag{6.42}$$

Consider the following convex and lower semicontinuous functional on $L^2(\Omega)$:

$$\varphi(u) = \frac{1}{p}\int_\Omega |\nabla u|^p\, dx + \frac{1}{4}\int_\Omega (\text{div}(\mathbf{a}) + a_0)|u|^2\, dx + \frac{1}{m}\int_\Omega |u|^m dx + \int_\Omega uf\, dx$$

if $u \in W_0^{1,p}(\Omega)$, and $+\infty$ if $u \notin W_0^{1,p}(\Omega)$. The functional

$$I(u) = \begin{cases} \varphi(u) + \varphi^*(\mathbf{a} \cdot \nabla u + \frac{1}{2}\mathrm{div}(\mathbf{a})u) & \text{if } u \in W_0^{1,p}(\Omega) \\ +\infty & \text{otherwise} \end{cases} \qquad (6.43)$$

is completely self-dual on $L^2(\Omega)$, and it attains its minimum at $\bar{u} \in W_0^{1,p}(\Omega)$ in such a way that $I(\bar{u}) = \inf\{I(u); u \in L^2(\Omega)\} = 0$, while solving equation (6.41).

Proof. Let $X = L^2(\Omega)$, and note that the operator $\Gamma : D(\Gamma) \to X^*$ defined by $u \mapsto \mathbf{a} \cdot \nabla u + \frac{1}{2}\mathrm{div}(\mathbf{a})u$ with domain $D(\Gamma) = \{u \in L^2(\Omega); \mathbf{a} \cdot \nabla u + \frac{1}{2}\mathrm{div}(\mathbf{a})u \in L^2(\Omega)\}$ is antisymmetric on X. The functional φ has a symmetric domain, namely $W_0^{1,p}(\Omega)$, that is contained in the domain of Γ. Moreover, φ is obviously coercive on $L^2(\Omega)$, and Corollary 6.8(1) therefore applies to yield $\bar{u} \in L^2(\Omega)$ such that $0 = I(\bar{u}) = \inf\{I(u); u \in L^2(\Omega)\}$. Clearly, $\bar{u} \in W_0^{1,p}(\Omega)$, and the rest follows as in the preceding examples.

Example 6.8. Transport equation with no diffusion term

Let now $\mathbf{a} : \Omega \to \mathbf{R}^\mathbf{n}$ be a vector field in $C_0^\infty(\Omega)$, where Ω is a bounded domain Ω of $\mathbf{R}^\mathbf{n}$. The first-order linear operator $\Gamma u = \mathbf{a} \cdot \nabla u + \frac{1}{2}\mathrm{div}(\mathbf{a})u$ is then skew-adjoint from L^p into L^q whenever $p > 1$ and $\frac{1}{p} + \frac{1}{q} = 1$. Its domain is the space $H_\Gamma^{p,q}(\Omega) = \{u \in L^p; \mathbf{a} \cdot \nabla u \in L^q\}$. Let $a_0 \in L^\infty(\Omega)$, and consider the following problem:

$$\begin{cases} \mathbf{a} \cdot \nabla u = |u|^{p-2}u + a_0 u + f & \text{on} \quad \Omega, \\ u = 0 & \text{on} \quad \partial\Omega. \end{cases} \qquad (6.44)$$

Theorem 6.5. *Assume either one of the following two conditions:*

$$2 \le p < \infty \text{ and } a_0 + \frac{1}{2}\mathrm{div}(\mathbf{a}) \ge 0 \text{ on } \Omega, \qquad (6.45)$$

$$1 < p \le 2 \text{ and } a_0 + \frac{1}{2}\mathrm{div}(\mathbf{a}) \ge \delta > 0 \text{ on } \Omega. \qquad (6.46)$$

For $f \in L^q(\Omega)$ where $\frac{1}{p} + \frac{1}{q} = 1$, consider on $L^p(\Omega)$ the functional

$$\Psi(u) = \frac{1}{p}\int_\Omega |u|^p dx + \frac{1}{2}\int_\Omega \left(a_0 + \frac{1}{2}\mathrm{div}(\mathbf{a})\right)|u|^2 dx + \int_\Omega f u \, dx$$

and its Legendre transform Ψ^*. The completely self-dual functional on $L^p(\Omega)$

$$I(u) = \begin{cases} \Psi(u) + \Psi^*(\mathbf{a} \cdot \nabla u + \frac{1}{2}\mathrm{div}(\mathbf{a})u) & \text{if } u \in H_\Gamma^{p,q}(\Omega) \\ +\infty & \text{otherwise} \end{cases} \qquad (6.47)$$

then attains its minimum at $\bar{u} \in H_\Gamma^{p,q}(\Omega)$ in such a way that $I(\bar{u}) = \inf\{I(u); u \in L^p(\Omega)\} = 0$ and $\bar{u}$ is a solution of equation (6.44).

Proof. This is an immediate application of Corollary 6.8 in the case where the operator Γ is skew-adjoint.

Exercises 6.C. More on the composition of a self-dual Lagrangian with an unbounded skew-adjoint operator

Let L be a self-dual Lagrangian on $X \times X^*$ and let $\Gamma : D(\Gamma) \subset X \to X^*$ be a skew-adjoint operator.

1. Show that, for every $\lambda > 0$, the Lagrangian

$$M_\lambda^1(x,p) = \begin{cases} L_\lambda^1(x, \Gamma x + p) & \text{if } x \in D(\Gamma) \\ +\infty & \text{if } x \notin D(\Gamma) \end{cases}$$

   is self-dual on $X \times X^*$.

2. Assume that for some $C > 0$

$$C(\|x\|^2 - 1) \le L(x,0) \quad \text{for every } x \in X, \tag{6.48}$$

   and show that for every $\lambda > 0$ there exists $x_\lambda \in X$ such that

$$\Gamma x_\lambda \in \bar\partial L(J_\lambda(x_\lambda, \Gamma x_\lambda)), \tag{6.49}$$

   where $J_\lambda(x, p)$ is the unique point where

$$L_\lambda^1(x,p) = L(J_\lambda(x,p), p) + \frac{1}{2\lambda}\|x - J_\lambda(x,p)\|^2 + \frac{\lambda}{2}\|p\|^2.$$

3. Show that if $(\Gamma x_\lambda)_\lambda$ is bounded in X^*, then there exists $\bar x \in X$ such that $\Gamma \bar x \in \bar\partial L(\bar x)$.
4. Show that if X is a Hilbert space H, then $(\Gamma x_\lambda)_\lambda$ is bounded, provided that for every $\lambda > 0$ the following condition holds:

$$\langle \Gamma x, \bar\partial L(x)\rangle \ge -C\big(1 + (1 + \|x\|)\|\bar\partial L_\lambda(x)\|\big) \quad \text{for every } x \in D(\Gamma).$$

5. Consider for $\lambda > 0$ the bounded operators $\Gamma_\lambda = \frac{I - (I + \lambda\Gamma)^{-1}}{\lambda} = \Gamma(I + \lambda\Gamma)$. Show that

$$(I - \lambda\Gamma)^{-1} = \big((I + \lambda\Gamma)^{-1}\big)^* \quad \text{and} \quad (\Gamma_\lambda)^* = (-\Gamma)_\lambda$$

   and that Γ_λ is a normal operator meaning that $\Gamma_\lambda\Gamma_\lambda^* = \Gamma_\lambda^*\Gamma_\lambda$.
6. By considering the self-dual Lagrangian $M_\lambda^2(x,p) = L(x, \Gamma_\lambda^a x + p)$, where $\Gamma_\lambda^a y_\lambda := \frac{\Gamma_\lambda y_\lambda - \Gamma_\lambda^* y_\lambda}{2}$ is the antisymmetric part of γ_λ, show that if L satisfies (6.48), then for each $\lambda > 0$ there exists $y_\lambda \in H$ such that

$$\Gamma_\lambda^a y_\lambda \in \bar\partial L(y_\lambda).$$

7. Show that if $(\Gamma^a y_\lambda)_\lambda$ is bounded in H, then there exists $\bar y \in H$ such that $\Gamma \bar y \in \bar\partial L(\bar y)$.
8. If φ is a proper convex lower semicontinuous function on H such that $\lim_{\|x\|\to+\infty} \frac{\varphi(x)}{\|x\|} = +\infty$ and if for some $h \in H$ and $C > 0$ we have

$$\varphi\big((I + \lambda\Gamma)^{-1}(x + \lambda h)\big) \le \varphi(x) + C\lambda \quad \text{for all } x \in H,$$

   then there exists $y \in H$ such that $\Gamma y \in \partial\varphi(y)$.

Further comments

The basic variational principle for completely self-dual functionals and its first applications to Lax-Milgram type results were given in Ghoussoub [55]. Particular cases of this approach in the case of Lagrangians of the form $L(x, p) = \varphi(x) + \varphi^*(p)$, where ψ is a convex lower semicontinuous function were formulated by many authors (Aubin [6], Auchmuty [10]–[13], Barbu-Kunish [20], Lemaire [84], Mabrouk [91], Nayroles [118], Rios [131] – [134], Roubicek [138], Telega [155], and Visintin [159]). Auchmuty also noted in [11] that variational inequalities can be formulated in terms of self-dual variational equalities. Proofs of resolution – as opposed to formulation – were eventually given in [13] for the case of cyclically monotone operators. Completely self-dual functionals for quasilinear PDEs in divergence form were constructed in [60]. The applications to systems – by exploiting the fact that any operator can be made skew-adjoint on phase space – were given in Ghoussoub [55]. The numerical application to the problem of inverting nonsymmetric matrices was developed in Ghoussoub-Moradifam [66]. The two ways of regularizing a completely self-dual functional described in Exercises 6.C were motivated by Barbu [18] and were kindly communicated to us by L. Tzou.

Chapter 7
Semigroups of Contractions Associated to Self-dual Lagrangians

We develop here a variational theory for dissipative initial-value problems via the theory of self-dual Lagrangians. We consider semilinear parabolic equations, with homogeneous state-boundary conditions, of the form

$$\begin{cases} -\dot{u}(t) + \Gamma u(t) + \omega u(t) \in \partial \varphi(t, u(t)) & \text{on} \quad [0, T] \\ u(0) = u_0, \end{cases} \tag{7.1}$$

where Γ is an antisymmetric – possibly unbounded – operator on a Hilbert space H, φ is a convex lower semicontinuous function on H, $\omega \in \mathbf{R}$, and $u_0 \in H$. Assuming for now that $\omega = 0$, the framework proposed in the last chapter for the stationary case already leads to a formulation of (7.1) as a time-dependent self-dual equation on state space of the form

$$\begin{cases} -\dot{u}(t) \in \overline{\partial} L(t, u(t)) \\ u(0) = u_0, \end{cases} \tag{7.2}$$

where the self-dual Lagrangians $L(t, \cdot, \cdot)$ on $H \times H$ are associated to the convex functional φ and the operator Γ in the following way:

$$L(t, u, p) = \varphi(t, u) + \varphi^*(t, \Gamma u + p).$$

We shall see that a (time-dependent) self dual Lagrangian $L : [0, T] \times H \times H \to \mathbf{R}$ on a Hilbert space H, "lifts", to a partially self-dual Lagrangian $\mathscr{L}$ on path space $A_H^2 = \{u : [0, T] \to H; \dot{u} \in L_H^2\}$ via the formula

$$\mathscr{L}(u, p) = \int_0^T L(t, u(t) - p(t), -\dot{u}(t)) dt + \ell_{u_0}(u(0) - a, u(T)),$$

where ℓ_{u_0} is an appropriate time-boundary Lagrangian and where $(p(t), a) \in L_H^2 \times H$, which happens to be a convenient representation for the dual of A_H^2. Equation (7.2) can then be formulated as a stationary equation on path space of the form

$$0 \in \overline{\partial} \mathscr{L}(u), \tag{7.3}$$

N. Ghoussoub, *Self-dual Partial Differential Systems and Their Variational Principles*, Springer Monographs in Mathematics, DOI 10.1007/978-0-387-84897-6_7, © Springer Science+Business Media, LLC 2009

hence reducing the dynamic problem to the stationary case already considered in the previous chapter. A solution $\bar{u}(t)$ of (7.3) can then be obtained by simply minimizing the completely self-dual functional

$$I(u) = \int_0^T L(t, u(t), -\dot{u}(t))dt + \ell_{u_0}(u(0), u(T)).$$

As such, one can naturally associate to the Lagrangian L a semigroup of contractive maps $(S_t)_t$ on H via the formula $S_t u_0 := \bar{u}(t)$. This chapter is focused on the implementation of this approach with a minimal set of hypotheses, and on the application of this variational approach to various standard parabolic equations.

7.1 Initial-value problems for time-dependent Lagrangians

Self-dual Lagrangians on path space

Definition 7.1. Let B be a bounded linear operator on a reflexive Banach space X and let $[0, T]$ be a fixed time interval. We shall say that a time-dependent convex Lagrangian L on $[0, T] \times X \times X^*$ is a *B-self-dual Lagrangian* if for any $t \in [0, T]$ the map $L_t : (x, p) \to L(t, x, p)$ is in $\mathscr{L}_B^{sd}(X)$, that is, if for all $t \in [0, T]$

$$L^*(t, B^* p, Bx) = L(t, x, p) \quad \text{for all } (x, p) \in X \times X^*,$$

where here L^* is the Legendre transform of L in the last two variables.

The most basic time-dependent B-self-dual Lagrangians are again of the form

$$L(t, x, p) = \varphi(t, Bx) + \varphi^*(t, p),$$

where for each t the function $x \to \varphi(t, x)$ is convex and lower semicontinuous on X and B is either onto or has dense range, while $x \to \varphi(t, x)$ is continuous. Theorem 2.4 shows that self-duality naturally "lifts" to – at least – certain subsets of path space. Indeed, an immediate application of that result is the following proposition.

Proposition 7.1. *Suppose L is a time-dependent self-dual Lagrangian on $[0, T] \times H \times H$. Then,*

1. *for each $\omega \in \mathbf{R}$, the Lagrangian $\mathscr{M}(u, p) := \int_0^T e^{2wt} L(t, e^{-wt} u(t), e^{-wt} p(t))dt$ is self-dual on $L_H^2 \times L_H^2$.*
2. *if ℓ is a convex lower semicontinuous function on $H \times H$ that satisfies*

$$\ell^*(x, p) = \ell(-x, p) \text{ for } (x, p) \in H \times H, \tag{7.4}$$

then the Lagrangian defined for $(u, p) = (u, (p_0(t), p_1)) \in A_H^2 \times (A_H^2)^* = A_H^2 \times (H \times L_H^2)$ by

$$\mathscr{L}(u,p) = \int_0^T L(t,u(t)-p_0(t),-\dot{u}(t))dt + \ell(u(0)-p_1,u(T)) \qquad (7.5)$$

satisfies $\mathscr{L}^*(p,u) = \mathscr{L}(u,p)$ for any $(u,p) \in A_H^2 \times (L_H^2 \times \{0\})$.

In particular, $\mathscr{L}$ is partially self-dual on the space $A_H^2 \times (A_H^2)^*$, where the duality is given by the identification of $(A_H^2)^*$ with $L_H^2 \times H$.

The proposition above combined with Proposition 6.1 has the following immediate application.

Theorem 7.1. Let L be a time-dependent self-dual Lagrangian on $[0,T] \times H \times H$, and consider ℓ to be a boundary Lagrangian on $H \times H$ satisfying $\ell^*(x,p) = \ell(-x,p)$ for all $(x,p) \in H \times H$. Suppose there exist $C_1 > 0, C_2 > 0$ such that

$$\int_0^T L(t,x(t),0)dt \leq C_1(1+\|x\|_{L_H^2}^2) \text{ for all } x \in L_H^2, \qquad (7.6)$$

$$\ell(a,0) \leq C_2(1+\|a\|_H^2) \text{ for all } a \in H. \qquad (7.7)$$

(1) The functional $I_{L,\ell}(u) := \int_0^T L(t,u(t),-\dot{u}(t))dt + \ell(u(0),u(T))$ is then completely self-dual on A_H^2.
(2) There exists $v \in A_H^2$ such that $\big(v(t),\dot{v}(t)\big) \in \text{Dom}(L)$ for almost all $t \in [0,T]$ and

$$I_{L,\ell}(v) = \inf_{u \in A_H^2} I_{\ell,L}(u) = 0, \qquad (7.8)$$

$$\frac{d}{dt}\partial_p L(t,v(t),\dot{v}(t)) = \partial_x L(t,v(t),-\dot{v}(t)), \qquad (7.9)$$

$$-\dot{v}(t) \in \overline{\partial}L(t,v(t)), \qquad (7.10)$$

$$\big(-v(0),v(T)\big) \in \partial\ell(v(0),v(T)). \qquad (7.11)$$

(3) In particular, for every $v_0 \in H$, the completely self-dual functional

$$I_{L,v_0}(u) = \int_0^T L(t,u(t),-\dot{u}(t))dt + \frac{1}{2}\|u(0)\|^2 - 2\langle v_0,u(0)\rangle + \|v_0\|^2 + \frac{1}{2}\|u(T)\|^2$$

has minimum zero on A_H^2. It is attained at a path v that solves (7.9), (7.10), while satisfying

$$v(0) = v_0, \qquad (7.12)$$

and

$$\|v(t)\|_H^2 = \|v_0\|^2 - 2\int_0^t L(s,v(s),-\dot{v}(s))ds \quad \text{for } t \in [0,T]. \qquad (7.13)$$

(4) If L is strictly convex, then v is unique, and if L is autonomous and uniformly convex, then v belongs to $C^1([0,T],H)$ and we have

$$\|\dot{v}(t)\| \leq \|\dot{v}(0)\| \text{ for all } t \in [0,T]. \qquad (7.14)$$

Proof. (1) Apply Proposition 7.1 to get that

$$\mathcal{L}(u,p) = \int_0^T L(t,u(t) - p_0(t), -\dot{u}(t))dt + \ell(u(0) - p_1, u(T))$$

is a partially self-dual Lagrangian on A_H^2.

(2) Apply Proposition 6.1 to $I_{L,\ell}(u) = \mathcal{L}(u,0)$ since in this case

$$\mathcal{L}(0,p) = \int_0^T L(t, -p_0(t), 0)dt + \ell(-p_1, 0) \leq C(1 + \|p_0\|_{L_H^2}^2) + \|p_1\|_H^2,$$

which means that $p \to \mathcal{L}(0,p)$ is bounded on the bounded sets of $(A_H^2)^*$. There exists therefore $v \in A_H^2$ such that (7.8) holds. Note that (7.9) is nothing but the corresponding Euler-Lagrange equation. For the rest, note that (7.8) yields

$$I_{L,\ell}(v) = \int_0^T \{L(t, v(t), -\dot{v}(t)) + \langle v(t), \dot{v}(t)\rangle\} \, dt$$

$$+ \ell(v(0), v(T)) - \frac{1}{2}(\|v(T)\|^2 - \|v(0)\|^2)$$

$$= 0.$$

Since $L(t,x,p) \geq \langle x,p \rangle$ for $(t,x,p) \in [0,T] \times H \times H$, and $\ell(a,b) \geq \frac{1}{2}(\|b\|^2 - \|a\|^2)$ for all $a,b \in H$, we get

$$L(t, v(t), -\dot{v}(t)) + \langle v(t), \dot{v}(t)\rangle = 0 \text{ for a.e } t \in [0,T]$$

and

$$\ell(v(0), v(T)) - \frac{1}{2}(\|v(T)\|^2 - \|v(0)\|^2) = 0,$$

which translate into (7.10), (7.11) respectively.

(3) Given now $v_0 \in H$, use the boundary Lagrangian

$$\ell_{v_0}(r, s) = \frac{1}{2}\|r\|^2 - 2\langle v_0, r\rangle + \|v_0\|^2 + \frac{1}{2}\|s\|^2,$$

which clearly satisfies $\ell_{v_0}^*(x, p) = \ell_{v_0}(-x, p)$, to obtain

$$I_{L,v_0}(u) = \int_0^T [L(t, u(t), -\dot{u}(t)) + \langle u(t), \dot{u}(t)\rangle] \, dt + \|u(0) - v_0\|^2,$$

and consequently

$$v(0) = v_0 \text{ and } L(s, v(s), -\dot{v}(s)) + \langle v(s), \dot{v}(s)\rangle = 0 \text{ for a.e } s \in [0,T]. \qquad (7.15)$$

This yields equation (7.13) since then $\frac{d(|v(s)|^2)}{ds} = -2L(s, v(s), -\dot{v}(s))$.

(4) If now L is strictly convex, then I_{L,v_0} is strictly convex and the minimum is attained uniquely. If L is uniformly convex, then (7.10) combined with Lemma 3.1 yields that $v \in C^1([0,T], H)$, and (7.14) then follows from the following observation.

Lemma 7.1. *Suppose $L(t,\,,\,)$ is convex on $H \times H$ for each $t \in [0,T]$, and that $x(t)$ and $v(t)$ are two paths in $C^1([0,T],H)$ satisfying $x(0) = x_0$, $v(0) = v_0$, $-\dot{x}(t) \in \overline{\partial}L(t,x(t))$ and $-\dot{v}(t) \in \overline{\partial}L(t,v(t))$. Then, $\|x(t) - v(t)\| \le \|x(0) - v(0)\|$ for all $t \in [0,T]$.*

Proof. In view of the monotonicity of $\overline{\partial}L$, we can estimate $\alpha(t) = \frac{d}{dt}\frac{\|x(t)-v(t)\|^2}{2}$ as follows:

$$\alpha(t) = \langle v(t) - x(t), \dot{v}(t) - \dot{x}(t)\rangle = -\left\langle v(t) - x(t), \overline{\partial}L(t,v(t)) - \overline{\partial}L(t,x(t))\right\rangle \le 0.$$

It then follows that $\|x(t) - v(t)\| \le \|x(0) - v(0)\|$ for all $t > 0$. Now if L is autonomous, $v(t)$ and $x(t) = v(t+h)$ are solutions for any $h > 0$, so that (7.14) follows from the above.

We now give a couple of immediate applications to parabolic equations involving time-dependent convex or semiconvex dissipative terms as well as – possibly unbounded – skew-adjoint operators. We shall, however, see in the next sections that the boundedness hypothesis can be relaxed considerably when dealing with autonomous Lagrangians.

Parabolic equations with time-dependent semilinearities and no diffusive term

Theorem 7.2. *Let $\Gamma : D(\Gamma) \subset H \to H$ be a skew-adjoint operator on a Hilbert space H, and let $\Phi : [0,T] \times H \to \mathbf{R} \cup \{+\infty\}$ be a time-dependent convex Gâteaux-differentiable function on H such that for some $m,n > 1$ and $C_1, C_2 > 0$ we have*

$$C_1\left(\|x\|_{L_H^2}^m - 1\right) \le \int_0^T \Phi(t,x(t))dt \le C_2\left(1 + \|x\|_{L_H^2}^n\right) \text{ for every } x \in L_H^2. \quad (7.16)$$

Let $(S_t)_{t \in \mathbf{R}}$ be the C_0-unitary group of operators on H associated to Γ, and consider for any given $\omega \in \mathbf{R}$ the following functional on A_H^2:

$$I(x) = \int_0^T e^{-2\omega t}\left\{\Phi(e^{\omega t}S_t x(t)) + \Phi^*(-e^{\omega t}S_t \dot{x}(t))\right\}dt$$
$$+ \frac{1}{2}(|x(0)|^2 + |x(T)|^2) - 2\langle x(0), v_0\rangle + |v_0|^2.$$

Then, I is a completely self-dual functional on A_H^2, and there exists a path $\hat{x} \in A_H^2$ such that:

1. $I(\hat{x}) = \inf\limits_{x \in A_H^2} I(x) = 0$.

2. The path $v(t) := S_t e^{\omega t}\hat{x}(t)$ is a mild solution of the equation

$$\dot{v}(t) + \Gamma v(t) - \omega v(t) \in -\partial\Phi(t,v(t)) \quad \text{for a.e. } t \in [0,T] \quad (7.17)$$
$$v(0) = v_0, \quad (7.18)$$

meaning that it satisfies the following integral equation:

$$v(t) = S_t v_0 - \int_0^t \left\{ S_{t-s} \partial \Phi(s, v(s)) - \omega S_t v(s) \right\} ds \text{ for } t \in [0, T].   (7.19)$$

Proof. Apply Theorem 7.1 with the Lagrangian

$$L(t, x, p) = e^{-2t\omega} \left\{ \Phi(t, e^{t\omega} S_t x) + \Phi^*(t, e^{t\omega} S_t x + e^{t\omega} S_t p) \right\},$$

which is self-dual thanks to properties (1) and (4) of Proposition 3.4. We then obtain $\hat{x}(t) \in A_H^2$ such that

$$I(\hat{x}) = \int_0^T e^{-2\omega t} \Phi\big(t, S_t e^{\omega t} \hat{x}(t)\big) + \Phi^*\big(- S_t e^{\omega t} \dot{\hat{x}}(t)\big) dt$$

$$+ \frac{1}{2}\big(|\hat{x}(0)|^2 + |\hat{x}(T)|^2\big) - 2\langle \hat{x}(0), v_0 \rangle + |v_0|^2$$

$$= 0,$$

which gives

$$0 = \int_0^T e^{-2\omega t} \left[\Phi\big(t, S_t e^{\omega t} \hat{x}(t)\big) + \Phi^*\big(- S_t e^{\omega t} \dot{\hat{x}}(t)\big) + \langle S_t e^{\omega t} \hat{x}(t), S_t e^{\omega t} \dot{\hat{x}}(t) \rangle \right] dt$$

$$- \int_0^T \langle S_t \hat{x}(t), S_t \dot{\hat{x}}(t) \rangle dt + \frac{1}{2}\big(|\hat{x}(0)|^2 + |\hat{x}(T)|^2\big) - 2\langle \hat{x}(0), v_0 \rangle + |v_0|^2$$

$$= \int_0^T e^{-2\omega t} \left[\Phi\big(t, S_t e^{\omega t} \hat{x}(t)\big) + \Phi^*\big(- S_t e^{\omega t} \dot{\hat{x}}(t)\big) + \langle S_t e^{\omega t} \hat{x}(t), S_t e^{\omega t} \dot{\hat{x}}(t) \rangle \right] dt$$

$$+ \|\hat{x}(0) - v_0\|^2.$$

Since $\Phi\big(t, S_t e^{\omega t} \hat{x}(t)\big) + \Phi^*\big(- S_t e^{\omega t} \dot{\hat{x}}(t)\big) + \langle S_t e^{\omega t} \hat{x}(t), S_t e^{\omega t} \dot{\hat{x}}(t) \rangle \geq 0$ for every $t \in [0, T]$, we get equality, from which we can conclude that

$$-S_t e^{\omega t} \dot{\hat{x}}(t) = \partial \Phi(t, S_t e^{\omega t} \hat{x}(t)) \text{ for almost all } t \in [0, T] \text{ and } \hat{x}(0) = v_0.   (7.20)$$

In order to show that $v(t) := S_t e^{\omega t} \hat{x}(t)$ is a mild solution for (7.17), we set $u(t) = e^{\omega t} \hat{x}(t)$ and write

$$-S_t(\dot{u}(t) - \omega u(t)) = \partial \Phi(t, S_t u(t)),$$

and hence $\dot{u}(t) - \omega u(t) = -S_{-t} \partial \Phi(t, v(t))$. By integrating between 0 and t, we get

$$u(t) = u(0) - \int_0^t \left\{ S_{-s} \partial \Phi(s, v(s)) - \omega u(s) \right\} ds.$$

Substituting $v(t) = S_t u(t)$ in the above equation gives

$$S_{-t} v(t) = v(0) - \int_0^t \left\{ S_{-s} \partial \Phi(s, v(s)) - \omega v(s) \right\} ds,$$

and consequently

$$v(t) = S_t v(0) - S_t \int_0^t \{S_{-s} \partial \Phi(s, v(s)) - \omega v(s)\} \, ds$$

$$= S_t v_0 - \int_0^t \{S_{t-s} \partial \Phi(s, v(s)) - \omega S_t v(s)\} \, ds,$$

which means that $v(t)$ is a mild solution for (7.17).

Remark 7.1. One can actually drop the coercivity condition (the lower bound) on $\Phi(t, u(t))$ in (7.16). Indeed, by applying the result to the coercive convex functional $\bar{\Phi}(t, u(t)) := \Phi(t, u(t)) + \frac{\varepsilon}{2} \|u(t)\|_H^2$ and $\bar{\omega} = \omega + \varepsilon$, we then obtain a solution of (7.17).

Example 7.1. The complex Ginzburg-Landau initial-value problem on $\mathbf{R}^N$

As an illustration, we consider the complex Ginzburg-Landau equations on $\mathbf{R}^N$

$$\dot{u}(t) - i\Delta u + \partial \varphi(t, u(t)) - \omega u(t) = 0. \tag{7.21}$$

Theorem 7.2 yields under the condition

$$-C < \int_0^T \varphi(t, u(t)) \, dt \leq C \left(\int_0^T \|u(t)\|_{L^2(\mathbf{R}^N)}^2 \, dt + 1 \right), \tag{7.22}$$

where $C > 0$, that there exists a mild solution of

$$\begin{cases} \dot{u}(t) - i\Delta u + \partial \varphi(t, u(t)) - \omega u(t) = 0, \\ \qquad\qquad\qquad\qquad\qquad\qquad u(0, x) = u_0. \end{cases}$$

7.2 Initial-value parabolic equations with a diffusive term

Given $0 < T < \infty$, $1 < p < \infty$, $1 < q < \infty$ with $\frac{1}{p} + \frac{1}{q} = 1$, and a Hilbert space H such that $X \subseteq H \subseteq X^*$ is an evolution triple, we recall the definition of the space

$$\mathscr{X}_{p,q} := W^{1,p}(0, T; X, H) = \{u : u \in L^p(0, T : X), \dot{u} \in L^q(0, T : X^*)\}$$

equipped with the norm $\|u\|_{\mathscr{X}_{p,q}} = \|u\|_{L^p(0,T:X)} + \|\dot{u}\|_{L^q(0,T:X^*)}$.

Theorem 7.3. *Let $X \subset H \subset X^*$ be an evolution triple, and consider a time-dependent self-dual Lagrangian $L(t, x, p)$ on $[0, T] \times X \times X^*$ and a Lagrangian ℓ on $H \times H$ verifying $\ell^*(x, p) = \ell(-x, p)$ on $H \times H$. Assume the following conditions are satisfied:*

(A_1) *For some $p \geq 2$, $m, n > 1$, and $C_1, C_2 > 0$, we have*

$$C_1 \left(\|x\|_{L_X^p}^m - 1 \right) < \int_0^T L(t, x(t), 0) \, dt \leq C_2 \left(1 + \|x\|_{L_X^p}^n \right) \text{ for every } x \in L_X^p.$$

(A_2) *For some $C_3 > 0$, we have*

$$\ell(a,b) \leq C_3(1 + \|a\|_H^2 + \|b\|_H^2) \text{ for all } a,b \in H.$$

The functional $I(x) = \int_0^T L\big(t,x(t),-\dot{x}(t)\big)\,dt + \ell(x(0),x(T))$ is then completely self-dual on $\mathscr{X}_{p,q}$ and attains its minimum at a path v such that

$$I(v) = \inf\{I(x); x \in \mathscr{X}_{p,q}\} = 0, \tag{7.23}$$

$$-\dot{v}(t) \in \overline{\partial}L\big(t,v(t)\big) \text{ a.e., } t \in [0,T], \tag{7.24}$$

$$(-v(0),v(T)) \in \partial\ell(v(0),v(T)). \tag{7.25}$$

Proof. Use Lemma 3.4 and condition (A_1) to lift L to a self-dual Lagrangian on $H \times H$. Then, consider for $\lambda > 0$, the λ-regularization of L, namely

$$L_\lambda^1(t,x,p) := \inf\left\{L(t,z,p) + \frac{\|x-z\|^2}{2\lambda} + \frac{\lambda}{2}\|p\|^2; z \in H\right\}. \tag{7.26}$$

It is clear that L_λ^1 and ℓ satisfy conditions (7.6) and (7.7) of Theorem 7.1. It follows that there exists a path $v_\lambda(t) \in A_H^2$ such that

$$\int_0^T L_\lambda^1\big(t,v_\lambda(t),-\dot{v}_\lambda(t)\big)\,dt + \ell(v_\lambda(0),v_\lambda(T)) = 0. \tag{7.27}$$

Since L is convex and lower semicontinuous, there exists $i_\lambda(v_\lambda)$ such that

$$L_\lambda^1(t,v_\lambda,-\dot{v}_\lambda) = L(t,i_\lambda(v_\lambda),-\dot{v}_\lambda) + \frac{\|v_\lambda(t) - i_\lambda(v_\lambda)\|^2}{2\lambda} + \frac{\lambda}{2}\|\dot{v}_\lambda\|^2. \tag{7.28}$$

Combine the last two identities to get

$$0 = \int_0^T \left(L(t,i_\lambda(v_\lambda),-\dot{v}_\lambda) + \frac{\|v_\lambda - i_\lambda(v_\lambda)\|^2}{2\lambda} + \frac{\lambda}{2}\|\dot{v}_\lambda\|^2\right)dt$$
$$+ \ell(v_\lambda(0),v_\lambda(T)). \tag{7.29}$$

By the coercivity assumptions in (A_1), we obtain that $(i_\lambda(v_\lambda))_\lambda$ is bounded in $L^p(0,T;X)$ and $(v_\lambda)_\lambda$ is bounded in $L^2(0,T;H)$. According to Lemma 3.5, condition (A_1) yields that $\int_0^T L(t,x(t),p(t))\,dt$ is coercive in $p(t)$ on $L^q(0,T;X^*)$, and therefore it follows from (7.29) that $(\dot{v}_\lambda)_\lambda$ is bounded in $L^q(0,T;X^*)$. Also, since all the other terms in (7.29) are bounded below, it follows that $\int_0^T \|v_\lambda(t) - i_\lambda(v_\lambda)\|^2\,dt \leq 2\lambda C$ for a constant $C > 0$.

Condition (A_2) combined with the fact that $\ell^*(x,p) = \ell(-x,p)$ yields that ℓ is coercive on H, from which we can deduce that $v_\lambda(0)$ and $v_\lambda(T)$ are also bounded in H. Therefore there exists $v \in L_H^2$ with $\dot{v} \in L^q(0,T;X^*)$ and $v(0), v(T) \in X^*$ such that

$$i_\lambda(v_\lambda) \rightharpoonup v \quad \text{in} \quad L^p(0,T;X),$$
$$\dot{v}_\lambda \rightharpoonup \dot{v} \quad \text{in} \quad L^q(0,T;X^*),$$
$$v_\lambda \rightharpoonup v \quad \text{in} \quad L^2(0,T;H),$$
$$v_\lambda(0) \rightharpoonup v(0), \quad v_\lambda(T) \rightharpoonup v(T) \quad \text{in} \quad H.$$

By letting λ go to zero in (7.29), we obtain from the above and the lower semicontinuity of L and ℓ that

$$\ell\big(v(0), v(T)\big) + \int_0^T L\big(t, v(t), \dot{v}(t)\big)\, dt \le 0. \tag{7.30}$$

The reverse inequality follows from self-duality and we therefore have equality. The rest follows as in Theorem 7.1.

Example 7.2. The heat equation

Let Ω be a smooth bounded domain in $\mathbf{R}^n$, and consider the evolution triple $H_0^1(\Omega) \subset H := L^2(\Omega) \subset H^{-1}$. Consider on $L^2(\Omega)$ the Dirichlet functional

$$\varphi(u) = \begin{cases} \frac{1}{2}\int_\Omega |\nabla u|^2 & \text{on} \quad H_0^1(\Omega) \\ +\infty & \text{elsewhere.} \end{cases} \tag{7.31}$$

Its Legendre conjugate is then given by $\varphi^*(u) = \frac{1}{2}\int_\Omega |\nabla(-\Delta)^{-1}u|^2\, dx$.

Theorem 7.3 allows us to conclude that, for any $u_0 \in L^2(\Omega)$ and any $f \in L^2\big([0,T];H^{-1}(\Omega)\big)$, the infimum of the completely self-dual functional

$$I(u) = \frac{1}{2}\int_0^T \int_\Omega \big(|\nabla u(t,x)|^2 - 2f(t,x)u(x,t)\big)\, dxdt$$
$$+ \int_0^T \frac{1}{2}\int_\Omega \left|\nabla(-\Delta)^{-1}\left(f(t,x) - \frac{\partial u}{\partial t}(t,x)\right)\right|^2 dxdt - 2\int_\Omega u(0,x)u_0(x)\, dx$$
$$+ \int_\Omega |u_0(x)|^2\, dx + \frac{1}{2}\int_\Omega (|u(0,x)|^2 + |u(T,x)|^2)\, dx \tag{7.32}$$

on the space A_H^2 is equal to zero and is attained uniquely at an $H_0^1(\Omega)$-valued path u such that $\int_0^T \|\dot{u}(t)\|_2^2 dt < +\infty$ and is a solution of the equation:

$$\begin{cases} \frac{\partial u}{\partial t} = \Delta u + f & \text{on} \quad [0,T] \times \Omega, \\ u(0,x) = u_0(x) & \text{on} \quad \Omega. \end{cases} \tag{7.33}$$

Parabolic equations with time-dependent dissipative and conservative terms

In the case of parabolic equations involving bounded skew-adjoint operators, we can already deduce the following general result.

Theorem 7.4. *Let $X \subset H \subset X^*$ be an evolution triple, $A_t : X \to X^*$ bounded positive operators on X, and $\varphi : [0,T] \times X \to \mathbf{R} \cup \{+\infty\}$ a time-dependent convex, lower semicontinuous and proper function. Consider the convex function $\Phi(t,x) = \varphi(t,x) + \frac{1}{2}\langle A_t x, x \rangle$ as well as the antisymmetric part $A_t^a := \frac{1}{2}(A_t - A_t^*)$ of A_t. Assume that for some $p \geq 2$, $m,n > 1$, and $C_1, C_2 > 0$, we have for every $x \in L_X^p$*

$$C_1 \left(\|x\|_{L_X^p}^m - 1 \right) \leq \int_0^T \{ \Phi(t,x(t)) + \Phi^*(t, -A_t^a x(t)) \} \, dt \leq C_2 \left(1 + \|x\|_{L_X^p}^n \right).$$

For any $T > 0$, $\omega \in \mathbf{R}$, and $v_0 \in H$, consider the following functional on $\mathscr{X}_{p,q}$:

$$I(x) = \int_0^T e^{-2\omega t} \left\{ \Phi(t, e^{\omega t} x(t)) + \Phi^*(t, -e^{\omega t}(A_t^a x(t) + \dot{x}(t))) \right\} dt$$

$$+ \frac{1}{2}(|x(0)|^2 + |x(T)|^2) - 2\langle x(0), v_0 \rangle + |v_0|^2.$$

Then, I is a completely self-dual functional on $\mathscr{X}_{p,q}$, and there exists $\hat{x} \in L^p(0,T:X)$ with $\dot{\hat{x}} \in L^q(0,T:X^*)$ such that:

1. $I(\hat{x}) = \inf\limits_{x \in \mathscr{X}_{p,q}} I(x) = 0$.

2. If $v(t)$ is defined by $v(t) := e^{\omega t} \hat{x}(t)$, then it satisfies

$$-\dot{v}(t) - A_t v(t) + \omega v(t) \in \partial \varphi(t,v(t)) \quad \textit{for a.e. } t \in [0,T], \qquad (7.34)$$

$$v(0) = v_0. \qquad (7.35)$$

This is a direct corollary of Theorem 7.3 applied to the self-dual Lagrangian

$$L(t,x,p) = e^{-2\omega t} \left\{ \Phi(t, e^{\omega t} x) + \Phi^*(t, -e^{\omega t} A^a x + e^{\omega t} p) \right\}$$

associated to a convex lower semicontinuous function Φ, a skew-adjoint operator A^a, and a scalar ω, and $\ell(r,s) = \frac{1}{2}\|r\|^2 - 2\langle v_0, r \rangle + \|v_0\|^2 + \frac{1}{2}\|s\|^2$.

Example 7.3. Initial-value Ginzburg-Landau evolution with diffusion

Consider complex Ginzburg-Landau equations of the type,

$$\begin{cases} \dfrac{\partial u}{\partial t} - (\kappa + i)\Delta u + \partial \varphi(t,u) - \omega u = 0 & (t,x) \in (0,T) \times \Omega, \\ u(t,x) = 0 & x \in \partial \Omega, \qquad (7.36) \\ u(0,x) = u_0(x), \end{cases}$$

where $\kappa > 0$, $\omega \geq 0$, Ω is a bounded domain in $\mathbf{R}^N$, $u_0 \in L^2(\Omega)$, and Ψ is a time-dependent convex lower semicontinuous function. An immediate corollary of Theorem 7.4 is the following.

Corollary 7.1. *Let $X := H_0^1(\Omega)$, $H := L^2(\Omega)$, and $X^* = H^{-1}(\Omega)$. If for some $C > 0$ we have*

$$-C \leq \int_0^T \Psi(t, u(t)) \, dt \leq C(\int_0^T \|u(t)\|_X^2 \, dt + 1) \text{ for every } u \in L_X^2[0, T],$$

then there exists a solution $u \in \mathscr{X}_{2,2}$ for equation (7.36).

Proof. Set $A = -(\kappa + i)\Delta u$, so that $\Phi(t, u) := \frac{k}{2} \int |\nabla u|^2 dx + \varphi(t, u(t))$ and $A^a = -i\Delta$. Note that since

$$c_1(\|u\|_{L_X^2}^2 - 1) \leq \int_0^T \Phi(t, u) \, dt \leq c_2(\|u\|_{L_X^2}^2 + 1) \tag{7.37}$$

for some $c_1, c_2 > 0$, we therefore have

$$c_1'(\|v\|_{L_{X^*}^2}^2 - 1) \leq \int_0^T \Phi^*(t, v) \, dt \leq c_2'(\|v\|_{L_{X^*}^2}^2 + 1)$$

for some $c_1', c_2' > 0$ and hence,

$$c_1'\left(\int_0^T \int_\Omega |\nabla(-\Delta)^{-1}v|^2 \, dx \, dt - 1\right) \leq \int_0^T \Phi^*(t, v) \, dt$$
$$\leq c_2'\left(\int_0^T \int_\Omega |\nabla(-\Delta)^{-1}v|^2 \, dx \, dt + 1\right),$$

from which we obtain

$$c_1'\left(\int_0^T \int_\Omega |\nabla u|^2 \, dx \, dt - 1\right) \leq \int_0^T \Phi^*(t, i\Delta u) \, dt \leq c_2'\left(\int_0^T \int_\Omega |\nabla u|^2 \, dx \, dt + 1\right),$$

which, once coupled with (7.37), yields the required boundedness in Theorem 7.4.

7.3 Semigroups of contractions associated to self-dual Lagrangians

In this section, we shall associate to a self-dual Lagrangian L on a Hilbert space $H \times H$ a semigroup of maps $(T_t)_{t \in \mathbf{R}^+}$ on H, which is defined for any $x_0 \in \text{Dom}(\bar{\partial}L)$ as $T_t x_0 = x(t)$, where $x(t)$ is the unique solution of the following:

$$\begin{cases} -\dot{x}(t) \in \bar{\partial}L(x(t)) & t \in [0, T] \\ x(0) = x_0. \end{cases} \tag{7.38}$$

As noted above, such a solution can be obtained by minimizing the completely self-dual functional

$$I(u) = \int_0^T L(u(t), -\dot{u}(t))dt + \frac{1}{2}\|u(0)\|^2 - 2\langle x_0, u(0)\rangle + \|x_0\|^2 + \frac{1}{2}\|u(T)\|^2$$

on A_H^2 and showing that $I(x) = \inf_{u \in A_H^2} I(u) = 0$. Now, according to Theorem 7.1, this can be done whenever the Lagrangian L satisfies the condition

$$L(x, 0) \leq C(\|x\|^2 + 1) \quad \text{for all } x \in H, \tag{7.39}$$

which is too stringent for most applications. We shall, however, see that this condition can be relaxed considerably via the λ-regularization procedure of self-dual Lagrangians. The following is the main result of this section.

Theorem 7.5. *Let L be a self-dual Lagrangian on a Hilbert space $H \times H$ that is uniformly convex in the first variable. Assuming $\mathrm{Dom}(\bar{\partial}L)$ is nonempty, then for any $\omega \in \mathbf{R}$ there exists a semigroup of maps $(T_t)_{t \in \mathbf{R}^+}$ defined on $\mathrm{Dom}(\bar{\partial}L)$ such that:*

1. $T_0 x = x$ and $\|T_t x - T_t y\| \leq e^{-\omega t}\|x - y\|$ for any $x, y \in \mathrm{Dom}(\bar{\partial}L)$.
2. *For any $x_0 \in \mathrm{Dom}(\bar{\partial}L)$, we have $T_t x_0 = x(t)$, where $x(t)$ is the unique path that minimizes on A_H^2 the completely self-dual functional*

$$I(u) = \int_0^T e^{2\omega t} L(u(t), -\omega u(t) - \dot{u}(t))dt$$
$$+ \frac{1}{2}\|u(0)\|^2 - 2\langle x_0, u(0)\rangle + \|x_0\|^2 + \frac{1}{2}\|e^{\omega T}u(T)\|^2.$$

3. *For any $x_0 \in \mathrm{Dom}(\bar{\partial}L)$, the path $x(t) = T_t x_0$ satisfies*

$$-\dot{x}(t) - \omega x(t) \in \bar{\partial}L(x(t)) \quad t \in [0, T] \tag{7.40}$$
$$x(0) = x_0.$$

First, we shall prove the following improvement of Theorem 7.1, provided L is autonomous. The boundedness condition is still there, but we first show regularity in the semiconvex case.

Proposition 7.2. *Assume $L : H \times H \to \mathbf{R} \cup \{+\infty\}$ is an autonomous self-dual Lagrangian that is uniformly convex and suppose*

$$L(x, 0) \leq C(\|x\|^2 + 1) \quad \text{for } x \in H. \tag{7.41}$$

Then, for any $\omega \in \mathbf{R}$, $x_0 \in H$, there exists $u \in C^1([0, T] : H)$ such that $u(0) = x_0$ and

$$0 = \int_0^T e^{2\omega t} L\left(e^{-\omega t}u(t), -e^{-\omega t}\dot{u}(t)\right)dt$$
$$+ \frac{1}{2}\|u(0)\|^2 - 2\langle x_0, u(0)\rangle + \|x_0\|^2 + \frac{1}{2}\|u(T)\|^2, \tag{7.42}$$

$$-e^{-\omega t}\big(\dot{u}(t),u(t)\big) \in \partial L\big(e^{-\omega t}u(t),e^{-\omega t}\dot{u}(t)\big) \qquad t \in [0,T], \qquad (7.43)$$

$$\|\dot{u}(t)\| \leq C(\omega,T)\|\dot{u}(0)\| \qquad t \in [0,T], \qquad (7.44)$$

where $C(\omega,T)$ is a positive constant.

Proof of Proposition 7.2: Apply Theorem 7.1 to the Lagrangian $M(t,x,p) = e^{2\omega t}L(e^{-\omega t}x,e^{-\omega t}p)$, which is also self-dual by Proposition 3.4. There exists then $u \in A_H^2$ such that $\big(u(t),\dot{u}(t)\big) \in \mathrm{Dom}(M)$ for a.e. $t \in [0,T]$ and $I(u) = \inf_{x \in A_H^2} I(x) = 0$,

where

$$I(x) = \int_0^T M(t,x(t),-\dot{x}(t))dt + \frac{1}{2}\|x(0)\|^2 - 2\langle x_0,x(0)\rangle + \|x_0\|^2 + \frac{1}{2}\|x(T)\|^2.$$

The path u then satisfies $u(0) = x_0$, and for a.e. $t \in [0,T]$,

$$-\big(\dot{u}(t),u(t)\big) \in \partial M\big(t,u(t),\dot{u}(t)\big)$$

and via the chain rule

$$\partial M(t,x,p) = e^{\omega t}\partial L\big(e^{-\omega t}x,e^{-\omega t}p\big),$$

and get that for almost all $t \in [0,T]$

$$-e^{-\omega t}\big(\dot{u}(t),u(t)\big) \in \partial L\big(e^{-\omega t}u(t),e^{-\omega t}\dot{u}(t)\big).$$

Apply Lemma 3.1 to $x(t) = e^{-\omega t}u(t)$ and $v(t) = e^{-\omega t}\dot{u}(t)$ to conclude that $\dot{u} \in C\big([0,T]:H\big)$, and therefore $u \in C^1\big([0,T]:H\big)$. Since L is self-dual and uniformly convex, we get from Lemma 3.1 that $(x,p) \mapsto \partial L(x,p)$ is Lipschitz, and so by continuity we have now for all $t \in [0,T]$

$$-e^{-\omega t}\big(\dot{u}(t),u(t)\big) \in \partial L\big(e^{-\omega t}u(t),e^{-\omega t}\dot{u}(t)\big),$$

and (7.43) is verified.

To establish (7.44), we first differentiate to obtain

$$e^{-2\omega t}\frac{d}{dt}\|u(t)-e^{-\omega h}u(t+h)\|^2 = 2e^{-2\omega t}\langle u(t)-e^{-\omega h}u(t+h),\dot{u}(t)-e^{-\omega h}\dot{u}(t+h)\rangle.$$

Setting now $v_1(t) = \partial_1 L\big(e^{-\omega t}u(t),e^{-\omega t}\dot{u}(t)\big)$ and $v_2(t) = \partial_2 L\big(e^{-\omega t}u(t),e^{-\omega t}\dot{u}(t)\big)$, we obtain from (7.43) and monotonicity that

$$e^{-2\omega t}\frac{d}{dt}\|u(t)-e^{-\omega h}u(t+h)\|^2 = \langle e^{-\omega t}u(t)-e^{-\omega(t+h)}u(t+h),-v_1(t)+v_1(t+h)\rangle +$$

$$\langle e^{-\omega t}\dot{u}(t)-e^{-\omega(t+h)}\dot{u}(t+h),-v_2(t)+v_2(t+h)\rangle$$

$$\leq 0.$$

We conclude from this that

$$\frac{\|u(t) - e^{-h\omega}u(t+h)\|}{h} \leq \frac{\|u(0) - e^{-h\omega}u(h)\|}{h},$$

and as we take $h \to 0$, we get $\|\omega u(t) + \dot{u}(t)\| \leq \|\omega x_0 + \dot{u}(0)\|$. Therefore

$$\|\dot{u}(t)\| \leq \|\dot{u}(0)\| + |\omega|\|u(t)\|$$

and

$$\|u(t)\| \leq \int_0^t \|\dot{u}(s)\| \, ds \leq \|\dot{u}(0)\| T + |\omega| \int_0^t \|u(s)\| \, ds.$$

It follows from Grönwall's inequality that $\|u(t)\| \leq \|\dot{u}(0)\| \left(C + |\omega| e^{|\omega| T} \right)$ for all $t \in [0, T]$ and finally that

$$\|\dot{u}(t)\| \leq \|\dot{u}(0)\| \left(C + |\omega| + |\omega|^2 e^{|\omega| T} \right).$$

We now proceed with the proof of Theorem 7.5. For that we first consider the λ-regularization of the Lagrangian so that the boundedness condition (7.41) in Proposition 7.2 is satisfied, and then make sure that all goes well when we take the limit as λ goes to 0.

End of proof of Theorem 7.5. Let $M_\lambda(t, x, p) = e^{2\omega t} L_\lambda^1 \left(e^{-\omega t} x, e^{-\omega t} p \right)$, which is also self-dual and uniformly convex by Lemma 3.3.

We have $L_\lambda^1(t, x, 0) \leq L(0, 0) + \frac{\|x\|^2}{2\lambda}$, and hence Proposition 7.2 applies and we get for all $\lambda > 0$ a solution $x_\lambda \in C^1 \left([0, T] : H \right)$ such that $x_\lambda(0) = x_0$,

$$\int_0^T M_\lambda \left(t, x_\lambda(t), -\dot{x}_\lambda(t) \right) dt + \ell \left(x_\lambda(0), x_\lambda(T) \right) = 0, \tag{7.45}$$

$$-e^{-\omega t} \left(\dot{x}_\lambda(t), x_\lambda(t) \right) \in \partial L_\lambda^1 \left(e^{-\omega t} x_\lambda(t), e^{-\omega t} \dot{x}_\lambda(t) \right) \text{ for } t \in [0, T], \tag{7.46}$$

$$\|\dot{x}_\lambda(t)\| \leq C(\omega, T)\|\dot{x}_\lambda(0)\|. \tag{7.47}$$

Here $\ell \left(x_\lambda(0), x_\lambda(T) \right) = \frac{1}{2}\|x_\lambda(0)\|^2 - 2\langle x_0, x_\lambda(0) \rangle + \|x_0\|^2 + \frac{1}{2}\|u_\lambda(T)\|^2$. By the definition of $M_\lambda(t, x, p)$, identity (7.45) can be written as

$$\int_0^T e^{2\omega t} L_\lambda^1 \left(e^{-\omega t} x_\lambda(t), -e^{-\omega t} \dot{x}_\lambda(t) \right) dt + \ell \left(x_\lambda(0), x_\lambda(T) \right) = 0, \tag{7.48}$$

and since

$$L_\lambda^1(x, p) = L \left(J_\lambda(x, p), p \right) + \frac{\|x - J_\lambda(x, p)\|^2}{2\lambda} + \frac{\lambda \|p\|^2}{2},$$

equation (7.45) can be written as

$$\int_0^T e^{2\omega t} \left(L\big(v_\lambda(t), -e^{-\omega t}\dot{x}_\lambda(t)\big) + \frac{\|e^{-\omega t}x_\lambda(t) - v_\lambda(t)\|^2}{2\lambda} + \frac{\lambda \|e^{-\omega t}\dot{x}_\lambda(t)\|^2}{2} \right) dt$$
$$+ \ell\big(x_\lambda(0), x_\lambda(T)\big) = 0,$$

where $v_\lambda(t) = J_\lambda \big(e^{-\omega t}x_\lambda(t), -e^{-\omega t}\dot{x}_\lambda(t)\big)$. Using Lemma 3.3, we get from (7.46) that for all t

$$-e^{-\omega t}\dot{x}_\lambda(t) = \partial_1 L_\lambda^1 \big(e^{-\omega t}x_\lambda(t), e^{-\omega t}\dot{x}_\lambda(t)\big) = \frac{e^{-\omega t}x_\lambda(t) - v_\lambda(t)}{\lambda}. \qquad (7.49)$$

Setting $t = 0$ in (7.46) and noting that $x_\lambda(0) = x_0$, we get that $-\big(\dot{x}_\lambda(0), x_0\big) \in \partial L_\lambda^1\big(x_0, \dot{x}_\lambda(0)\big)$, and since $x_0 \in \text{Dom}(\overline{\partial L})$, we can apply (7.14) in Theorem 7.1 to get that $\|\dot{x}_\lambda(0)\| \le C$ for all $\lambda > 0$. Now we plug this inequality into (7.47) and obtain

$$\|\dot{x}_\lambda(t)\| \le D(\omega, T) \quad \forall \lambda > 0 \quad \forall t \in [0, T].$$

This yields by (7.49) that

$$\|e^{-\omega t}x_\lambda(t) - v_\lambda(t)\| \le e^{|\omega|T} D(\omega, T)\lambda \quad \forall t \in [0, T],$$

and hence

$$\frac{\|e^{-\omega t}x_\lambda(t) - v_\lambda(t)\|^2}{\lambda} \to 0 \text{ uniformly in } t. \qquad (7.50)$$

Moreover, since $\|\dot{x}_\lambda\|_{A_H^2} \le D(\omega, T)$ for all $\lambda > 0$, there exists $\hat{x} \in A_H^2$ such that, up to a subsequence,

$$x_\lambda \rightharpoonup \hat{x} \text{ in } A_H^2, \qquad (7.51)$$

and again by (7.49) we have

$$\int_0^T \|v_\lambda(t) - e^{-\omega t}\hat{x}(t)\|_H^2 dt \to 0, \qquad (7.52)$$

while clearly implies that

$$\lambda \frac{\|e^{-\omega t}\dot{x}_\lambda(t)\|^2}{2} \to 0 \text{ uniformly.} \qquad (7.53)$$

Now use (7.50)–(7.53) and the lower semicontinuity of L, to deduce from (7.48) that as $\lambda \to 0$ we have

$$I(\hat{x}) = \int_0^T e^{2\omega t} L\big(e^{-\omega t}\hat{x}(t), -e^{-\omega t}\hat{x}(t)\big) dt + \ell\big(\hat{x}(0), \hat{x}(T)\big) \le 0.$$

Since we already know that $I(x) \ge 0$ for all $x \in A_H^2$, we finally get our claim that $0 = I(\hat{x}) = \inf_{x \in A_H^2} I(x)$.

Now define $T_t x_0 := e^{-\omega t}\hat{x}(t)$. It is easy to see that $x(t) := T_t x_0$ satisfies equation (7.40) and that $T_0 x_0 = x_0$. We need to check that $\{T_t\}_{t \in \mathbf{R}^+}$ is a semigroup. By the uniqueness of minimizers, this is equivalent to showing that for all $s < T$, the function $v(t) := x(t+s)$ satisfies

$$0 = \int_0^{T-s} e^{2\omega t} L\left(v(t), -e^{-\omega t}\left(\frac{d}{dt}e^{\omega t}v(t)\right)\right) dt$$
$$+ \frac{1}{2}\|v(0)\|^2 - 2\langle T_s x_0, v(0)\rangle + \|T_s x_0\|^2 + \frac{1}{2}\|e^{\omega(T-s)}v(T)\|^2.$$

By the definition of $x(t)$ and the fact that $I(\hat{x}) = 0$, we have

$$0 = \int_0^T e^{2\omega t} L\left(x(t), -e^{-\omega t}\left(\frac{d}{dt}e^{\omega t}x(t)\right)\right) dt$$
$$+ \frac{1}{2}\|x(0)\|^2 - 2\langle x_0, x(0)\rangle + \|x_0\|^2 + \frac{1}{2}\|e^{\omega T}x(T)\|^2.$$

Since $x(t)$ satisfies equation (7.40), we have

$$0 = \int_0^s e^{2\omega t} L\left(x(t), -e^{-\omega t}\left(\frac{d}{dt}e^{\omega t}x(t)\right)\right) dt$$
$$+ \frac{1}{2}\|x(0)\|^2 - 2\langle x_0, x(0)\rangle + \|x_0\|^2 + \frac{1}{2}\|e^{\omega s}x(s)\|^2.$$

By subtracting the two equations we get

$$0 = \int_s^T e^{2\omega t} L\left(x(t), -e^{-\omega t}\left(\frac{d}{dt}e^{\omega t}x(t)\right)\right) dt + \frac{1}{2}\|e^{\omega T}x(T)\|^2 - \frac{1}{2}\|e^{\omega s}x(s)\|^2.$$

Make a substitution $r = t - s$ and we obtain

$$0 = e^{2\omega s}\left\{\int_0^{T-s} e^{2\omega r} L\left(v(r), -e^{-\omega r}\left(\frac{d}{dr}e^{\omega r}v(r)\right)\right) dr + \frac{1}{2}\|e^{\omega(T-s)}x(T)\|^2 - \frac{1}{2}\|x(s)\|^2\right\}$$

and finally

$$0 = \int_0^{T-s} e^{2\omega r} L\left(v(r), -e^{-\omega r}\left(\frac{d}{dr}e^{\omega r}v(r)\right)\right) dr$$
$$+ \frac{1}{2}\|v(0)\|^2 - 2\langle T_s x_0, v(0)\rangle + \|T_s x_0\|^2 + \frac{1}{2}\|e^{\omega(T-s)}v(T)\|^2.$$

It follows that $T_s(T_t x_0) = T_{s+t} x_0$.

To check the Lipschitz constant of the semigroup, we differentiate $\|T_t x_0 - T_t x_1\|^2$ and use equation (7.40) in conjunction with monotonicity to see that

$$\frac{d}{dt}\|T_t x_0 - T_t x_1\|^2 \leq -\omega\|T_t x_0 - T_t x_1\|^2$$

whenever $x_0, x_1 \in \text{Dom}(\overline{\partial L})$. A simple application of Grönwall's inequality gives the desired conclusions.

Exercise 7. A.

1. Show that the semi-group associated to a self-dual Lagrangian L by Theorem 7.5 maps $\mathrm{Dom}(\bar{\partial}L)$ into itself, and that it can be extended to a continuous semi-group of contractions $(S_t)_{t \in \mathbf{R}^+}$ on the closure of $\mathrm{Dom}(\bar{\partial}L)$, in such a way that $\lim\limits_{t \to 0} \|S_t u - u\| = 0$ for every u in the closure of $\mathrm{Dom}(\bar{\partial}L)$.
2. Conversely, assuming that $(S_t)_t$ is a continuous semi-group of contractions on a closed convex subset D of a Hilbert space H, show that there exists a self-dual Lagrangian L such that for every $u_0 \in D$, we have $0 \in \frac{d}{dt} S_t u_0 + \bar{\partial}L(S_t u_0)$.

7.4 Variational resolution for gradient flows of semiconvex functions

Corollary 7.2. *Let $\varphi : H \to \mathbf{R} \cup \{+\infty\}$ be a bounded below proper convex and lower semicontinuous on a Hilbert space H. For any $u_0 \in \mathrm{Dom}(\partial\varphi)$ and any $\omega \in \mathbf{R}$, consider the completely self-dual functional*

$$I(u) = \int_0^T e^{2\omega' t} \left[\psi(e^{-\omega' t} u(t)) + \psi^*(-e^{-\omega' t} \dot{u}(t)) \right] dt$$
$$+ \frac{1}{2}(|u(0)|^2 + |u(T)|^2) - 2\langle u(0), u_0 \rangle + |u_0|^2, \qquad (7.54)$$

where $\psi(x) = \varphi(x) + \frac{1}{2}\|x\|^2$ and $\omega' = \omega - 1$. Then, I has a unique minimizer $\bar{u}$ in A_H^2 such that $I(\bar{u}) = \inf\limits_{u \in A_H^2} I(u) = 0$, and the path $v(t) = e^{-\omega' t} \bar{u}(t)$ is the unique solution of the equation

$$\begin{cases} \dot{v}(t) - \omega v(t) \in -\partial\varphi(v(t)) & \text{a.e.} \quad \text{on} \quad [0,T] \\ \qquad\qquad v(0) = u_0. \end{cases} \qquad (7.55)$$

Proof. This is a direct application of the above with $L(x,p) = \psi(x) + \psi^*(p)$, which is clearly uniformly convex in the first variable.

Example 7.4. Quasilinear parabolic equations

Let Ω be a smooth bounded domain in $\mathbf{R}^n$. For $p \geq \frac{n-2}{n+2}$, the Sobolev space $W_0^{1,p+1}(\Omega) \subset H := L^2(\Omega)$, and so we define on $L^2(\Omega)$ the functional

$$\varphi(u) = \begin{cases} \frac{1}{p+1} \int_\Omega |\nabla u|^{p+1} & \text{on} \quad W_0^{1,p+1}(\Omega) \\ +\infty & \text{elsewhere} \end{cases} \qquad (7.56)$$

and let φ^* be its Legendre conjugate. For any $\omega \in \mathbf{R}$, any $u_0 \in L^2(\Omega)$, and any $f \in L^2([0,T]; W^{-1, \frac{p+1}{p}}(\Omega))$, the infimum of the completely self-dual functional

$$I(u) = \frac{1}{p+1} \int_0^T e^{2\omega t} \int_\Omega \left(|\nabla u(t,x)|^{p+1} - (p+1)f(t,x)u(x,t) \right) dxdt$$

$$+ \int_0^T e^{2\omega t} \varphi^* \left(f(t) - \omega u(t) - \frac{\partial u}{\partial t} \right) dt$$

$$-2 \int_\Omega u(0,x)u_0(x)\,dx + \int_\Omega |u_0(x)|^2\,dx + \frac{1}{2} \int_\Omega (|u(0,x)|^2 + e^{2T}|u(T,x)|^2)dx$$

on the space A_H^2 is equal to zero and is attained uniquely at a $W_0^{1,p+1}(\Omega)$-valued path u such that $\int_0^T \|\dot{u}(t)\|_2^2 dt < +\infty$ and is a solution of the equation

$$\begin{cases} \frac{\partial u}{\partial t} = \Delta_p u + \omega u + f & \text{on } \Omega \times [0,T], \\ u(0,x) = u_0(x) & \text{on } \Omega, \end{cases} \tag{7.57}$$

Similarly, we can deal with the equation

$$\begin{cases} \frac{\partial u}{\partial t} = \Delta_p u - Au + \omega u + f & \text{on } \Omega \times [0,T], \\ u(0,x) = u_0(x) & \text{on } \Omega, \end{cases} \tag{7.58}$$

whenever A is a positive operator on $L^2(\Omega)$.

Example 7.5. Porous media equations

Let $H = H^{-1}(\Omega)$ equipped with the scalar product $\langle u,v \rangle_{H^{-1}(\Omega)} = \int_\Omega u(-\Delta)^{-1}v dx$. For $m \geq \frac{n-2}{n+2}$, we have $L^{m+1}(\Omega) \subset H^{-1}$, and so we can consider the functional

$$\varphi(u) = \begin{cases} \frac{1}{m+1} \int_\Omega |u|^{m+1} & \text{on } L^{m+1}(\Omega) \\ +\infty & \text{elsewhere} \end{cases} \tag{7.59}$$

and its conjugate

$$\varphi^*(v) = \frac{m}{m+1} \int_\Omega |(-\Delta)^{-1}v|^{\frac{m+1}{m}}\,dx. \tag{7.60}$$

Then, for any $\omega \in \mathbf{R}$, $u_0 \in H^{-1}(\Omega)$, and $f \in L^2([0,T];H^{-1}(\Omega))$, the infimum of the completely self-dual functional

$$I(u) = \frac{1}{m+1} \int_0^T e^{2\omega t} \int_\Omega |u(t,x)|^{m+1}\,dxdt$$

$$+ \frac{m}{m+1} \int_0^T e^{2\omega t} \int_\Omega \left| (-\Delta)^{-1} \left(f(t,x) - \omega u(t,x) - \frac{\partial u}{\partial t}(t,x) \right) \right|^{\frac{m+1}{m}} dxdt$$

$$- \int_0^T e^{2\omega t} \int_\Omega u(x,t)(-\Delta)^{-1}f(t,x)dxdt + \int_\Omega |\nabla(-\Delta)^{-1}u_0(x)|^2\,dx$$

$$-2 \int_\Omega u_0(x)(-\Delta)^{-1}u(0,x)\,dx + \frac{1}{2} \left(\|u(0)\|_{H^{-1}}^2 + e^{2\omega T}\|u(T)\|_{H^{-1}}^2 \right)$$

on the space A_H^2 is equal to zero and is attained uniquely at an $L^{m+1}(\Omega)$-valued path u such that $\int_0^T \|\dot{u}(t)\|_H^2 dt < +\infty$ and is a solution of the equation

$$\begin{cases} \frac{\partial u}{\partial t}(t,x) = \Delta u^m(t,x) + \omega u(t,x) + f(t,x) & \text{on } \Omega \times [0,T], \\ u(0,x) = u_0(x) & \text{on } \Omega. \end{cases} \tag{7.61}$$

7.5 Parabolic equations with homogeneous state-boundary conditions

Corollary 7.3. Let $\varphi : H \to \mathbf{R} \cup \{+\infty\}$ be a convex, lower semicontinuous, and proper function on a Hilbert space H, and let Γ be an antisymmetric linear operator into H whose domain $D(\Gamma)$ contains $D(\varphi)$. Assume that

$$x_0 \in D(\Gamma) \cap D(\partial \varphi). \tag{7.62}$$

Then, for all $\omega \in \mathbf{R}$ and all $T > 0$, there exists $v \in A_H^2([0,T])$ such that

$$\begin{cases} -\dot{v}(t) + \Gamma v(t) - \omega v(t) \in \partial \varphi(v(t)) & \text{for a.e. } t \in [0,T] \\ v(0) = x_0. \end{cases} \tag{7.63}$$

Proof. Consider the uniformly convex and lower semicontinuous function

$$\psi(x) := \varphi(x+x_0) + \frac{\|x\|^2}{2} - \langle x, \Gamma x_0 \rangle + \langle x, \omega x_0 \rangle.$$

Setting $\psi_t(x) := e^{2(\omega-1)t} \psi(e^{-(\omega-1)t}x)$, the assumptions ensure that

$$L(t,x,p) := \psi_t(x) + \psi_t^*(\Gamma x + p) \text{ if } x \in D(\varphi)$$

and $+\infty$ elsewhere, is a self-dual Lagrangian by Proposition 4.1. The fact that $x_0 \in D(\Gamma) \cap D(\partial \varphi)$ implies that $0 \in \text{Dom}(\overline{\partial L})$. Consider now the following completely self-dual functional on $A_H^2([0,T])$:

$$I(u) = \int_0^T e^{2(\omega-1)t} \psi(e^{-(\omega-1)t}x(t)) + e^{2(\omega-1)t} \psi^*(e^{-(\omega-1)t}(\Gamma x(t) - \dot{x}(t))) dt$$

$$+ \frac{1}{2}(\|x(0)\|^2 + \|x(T)\|^2).$$

All the hypotheses of Theorem 7.5 are then satisfied, and we can deduce that there exists a path $\bar{x}$ in $A_H^2([0,T])$ such that $I(\bar{x}) = \inf_{x \in A_H^2([0,T])} I(x) = 0$. Therefore

$$0 = \int_0^T \psi_t(\bar{x}(t)) + \psi_t^*(\Gamma \bar{x}(t) - \dot{\bar{x}}(t)) dt + \frac{1}{2}(\|\bar{x}(0)\|^2 + \|\bar{x}(T)\|^2)$$

$$\geq \int_0^T \langle \bar{x}(t), \Gamma \bar{x}(t) - \dot{\bar{x}}(t) \rangle dt + \frac{1}{2}(\|\bar{x}(0)\|^2 + \|\bar{x}(T)\|^2)$$
$$= \|\bar{x}(0)\|^2 \geq 0.$$

It follows that

$$-\dot{\bar{x}}(t) + \Gamma \bar{x}(t) \in e^{(\omega-1)t} \partial \psi(e^{-(\omega-1)t} \bar{x}(t))$$
$$\bar{x}(0) = 0,$$

and by a simple application of the product-rule we see that $\bar{v}(t)$ defined by $\bar{v}(t) := e^{-(\omega-1)t} \bar{x}(t)$ satisfies

$$-\dot{\bar{v}}(t) + \Gamma \bar{v}(t) - (\omega - 1)\bar{v}(t) \in \partial \psi(\bar{v}(t)) \qquad \text{for a.e. } t \in [0,T] \qquad (7.64)$$
$$\bar{v}(0) = 0.$$

Since $\partial \psi(x) = \partial \varphi(x + x_0) + x - \Gamma x_0 + \omega x_0$, we get that $v(t) := \bar{v}(t) + x_0$ satisfies equation (7.63).

Example 7.6. Evolution driven by the transport operator and the p-Laplacian

Consider the following evolution equation on a smooth bounded domain of $\mathbf{R}^n$:

$$\begin{cases} -\frac{\partial u}{\partial t} + \mathbf{a}(x) \cdot \nabla u = -\Delta_p u + \frac{1}{2} a_0(x) u + \omega u & \text{on } [0,T] \times \Omega \\ u(0,x) = \qquad u_0(x) & \text{on } \Omega \\ u(t,x) = \qquad 0 & \text{on } [0,T] \times \partial \Omega. \end{cases} \qquad (7.65)$$

We can now establish variationally the following existence result.

Corollary 7.4. *Let* $\mathbf{a} : \mathbf{R}^n \to \mathbf{R}^n$ *be a smooth vector field and* $a_0 \in L^\infty(\Omega)$. *For* $p \geq 2$, $\omega \in \mathbf{R}$, *and* u_0 *in* $W_0^{1,p}(\Omega) \cap \{u; \Delta_p u \in L^2(\Omega)\}$, *there exists* $\bar{u} \in A_{L^2(\Omega)}^2([0,T])$, *which solves equation (7.65). Furthermore,* $\Delta_p \bar{u}(x,t) \in L^2(\Omega)$ *for almost all* $t \in [0,T]$.

Proof. The operator $\Gamma u = \mathbf{a} \cdot \nabla u + \frac{1}{2}(\nabla \cdot \mathbf{a})u$ with domain $D(\Gamma) = H_0^1(\Omega)$ is antisymmetric . In order to apply Corollary 7.3 with $H = L^2(\Omega)$, we need to ensure the convexity of the potential, and for that we pick $K > 0$ such that $\nabla \cdot \mathbf{a}(x) + a_0(x) + K \geq 1$ for all $x \in \Omega$.

Now define $\varphi : H \to \mathbf{R} \cup \{+\infty\}$ by

$$\varphi(u) = \frac{1}{p} \int_\Omega |\nabla u(x)|^p dx + \frac{1}{4} \int_\Omega (\nabla \cdot \mathbf{a}(x) + a_0(x) + K)|u(x)|^2 dx$$

if $u \in W_0^{1,p}(\Omega)$ and $+\infty$ otherwise.

By observing that φ is a convex lower semicontinuous function with symmetric domain $D(\varphi) \subset D(\Gamma)$, we can apply Corollary 7.3 with the linear factor $(\omega - \frac{K}{2})$ to obtain the existence of a path $\bar{u} \in A_H^2([0,T])$ such that $\bar{u}(0) = x_0$, and

$$-\dot{\bar{u}}(t) + \Gamma\bar{u}(t) \in \partial\varphi(\bar{u}(t)) + \left(\omega - \frac{K}{2}\right)\bar{u}(t) \text{ for a.e. } t \in [0,T],$$

which is precisely equation (7.65). Since now $\partial\varphi(\bar{u}(t))$ is a nonempty set in H for almost all $t \in [0,T]$, we have $\Delta_p\bar{u}(x,t) \in L^2(\Omega)$ for almost all $t \in [0,T]$.

The following result capitalizes on the fact that if Γ is skew-adjoint, then its domain need not be large.

Corollary 7.5. *Let $X \subset H \subset X^*$ be an evolution triple and let $\Gamma : D(\Gamma) \subseteq H \to X^*$ be a skew-adjoint operator. Let $\Phi : X \to \bar{\mathbf{R}}$ be a uniformly convex, lower semicontinuous, and proper function on X that is bounded on the bounded sets of X and is also coercive on X. Assume that $x_0 \in D(\Gamma)$ and that $\partial\Phi(x_0) \cap H$ is not empty. Then, for all $\omega \in \mathbf{R}$ and all $T > 0$, the self-dual functional*

$$I(x) = \int_0^T e^{2\omega t}\{\Phi(e^{-\omega t}x(t)) + \Phi^*(-\Gamma e^{-\omega t}x(t) - e^{-\omega t}\dot{x}(t))\}\,dt$$
$$+ \frac{1}{2}(|x(0)|^2 + |x(T)|^2) - 2\langle x(0), v_0\rangle + |v_0|^2$$

attains its minimum on A_H^2 at a path $\tilde{v} \in A_H^2$ such that $I(\tilde{v}) = \inf_{x \in A_H^2} I(x) = 0$. Moreover, the path $v(t) = e^{-\omega t}\tilde{v}(t)$ solves the following equation:

$$-\dot{v}(t) - \Gamma v(t) - \omega v(t) \in \partial\Phi(v(t)) \quad \text{for a.e. } t \in [0,T]$$
$$v(0) = v_0.$$

Example 7.7. Ginzburg-Landau evolution without diffusion

Consider the following evolution equation in $\mathbf{R}^N$:

$$\begin{cases} \dot{u}(t) - i\alpha\Delta u + \gamma|u|^{q-1}u + i\beta u - \omega u = 0 & \text{on } \mathbf{R}^N, \\ u(x,0) = u_0. \end{cases} \tag{7.66}$$

In order to find a solution for equation (7.66), it suffices to apply Corollary 7.5 with $H := L^2(\mathbf{R}^N; \mathbf{C})$, $X = H \cap L^q(\mathbf{R}^N; \mathbf{C})$ equipped with their real structure and to consider the skew-adjoint operator defined by $\Gamma u := -i\alpha\Delta u + 2i\beta u$ with domain $D(\Gamma) = \{u \in L^2(\mathbf{R}^n); \Delta u \in L^2(\mathbf{R}^n)\}$ and the convex function

$$\Phi(u) = \frac{\gamma}{q+1}\int_{\mathbf{R}^N} |u|^{q+1}\,dx.$$

We then obtain the following corollary.

Corollary 7.6. *For every $q > 1$ and $u_0 \in L^2(\mathbf{R}^N) \cap L^{2q}(\mathbf{R}^N)$ with $\Delta u_0 \in L^2(\mathbf{R}^N)$, the equation*

$$\begin{cases} \dot{u}(t) - i\Delta u + |u|^{q-1}u = 0 & \text{on } \mathbf{R}^N, \\ u(x,0) = u_0, \end{cases} \tag{7.67}$$

has a solution $u \in A_H^2$ that can be obtained by minimizing the completely self-dual functional

$$I(u) = \int_0^T \left\{ \frac{1}{p+1} \int_{\mathbf{R}^N} |u(t,x)|^{p+1} dx + \frac{p+1}{p} \int_{\mathbf{R}^N} |i\Delta u(t,x) - \frac{\partial u}{\partial t}(t,x)|^{\frac{p}{p+1}} dx \right\} dt$$

$$- 2 \int_{\mathbf{R}^N} u(0,x)u_0(x) dx + \int_{\mathbf{R}^N} |u_0(x)|^2 dx$$

$$+ \frac{1}{2} \int_{\mathbf{R}^N} (|u(0,x)|^2 + |u(T,x)|^2) dx. \tag{7.68}$$

7.6 Variational resolution for coupled flows and wave-type equations

Self-dual Lagrangians are also suited to treating variationally certain coupled evolution equations.

Proposition 7.3. *Let φ be a proper, convex, lower semicontinuous function on $X \times Y$, and let $A : X \to Y^*$ be any bounded linear operator. Assume $\Gamma_1 : X \to X$ (resp., $\Gamma_2 : Y \to Y$) are positive operators. Then, for any $(x_0, y_0) \in \text{Dom}(\partial \varphi)$ and any $(f,g) \in X^* \times Y^*$, there exists a path $(x(t), y(t)) \in A_X^2 \times A_Y^2$ such that*

$$\begin{cases} -\dot{x}(t) - A^*y(t) - \Gamma_1 x(t) + f \in \partial_1 \varphi(x(t), y(t)), \\ -\dot{y}(t) + Ax(t) - \Gamma_2 y(t) + g \in \partial_2 \varphi(x(t), y(t)), \\ \quad\quad\quad\quad\quad x(0) = x_0, \\ \quad\quad\quad\quad\quad y(0) = y_0. \end{cases} \tag{7.69}$$

The solution is obtained as a minimizer on $A_X^2 \times A_Y^2$ of the completely self-dual functional

$$I(x,y) = \int_0^T \left\{ \psi(x(t),y(t)) + \psi^*(-A^*y(t) - \Gamma_1^a x(t) - \dot{x}(t), Ax(t) - \Gamma_2^a y(t) - \dot{y}(t)) \right\} dt$$

$$+ \frac{1}{2} \|x(0)\|^2 - 2\langle x_0, x(0) \rangle + \|x_0\|^2 + \frac{1}{2} \|x(T)\|^2$$

$$+ \frac{1}{2} \|y(0)\|^2 - 2\langle y_0, y(0) \rangle + \|y_0\|^2 + \frac{1}{2} \|y(T)\|^2,$$

whose infimum is zero. Here Γ_1^a and Γ_2^a are the skew-symmetric parts of Γ_1 and Γ_2, respectively, and

$$\psi(x,y) = \varphi(x,y) + \frac{1}{2} \langle \Gamma_1 x, x \rangle - \langle f, x \rangle + \frac{1}{2} \langle \Gamma_2 y, y \rangle - \langle g, x \rangle.$$

Proof. It suffices to apply Theorem 7.5 to the self-dual Lagrangian

$$L((x,y),(p,q)) = \psi(x,y) + \psi^*(-A^*y - \Gamma_1^a x + p, Ax - \Gamma_2^a y + q).$$

If $(\bar{x}(t),\bar{y}(t))$ is where the infimum is attained, then we get

$$
\begin{aligned}
0 = I(\bar{x},\bar{y}) \\
= \int_0^T \big\{ \psi(\bar{x}(t),\bar{y}(t)) + \psi^*(-A^*\bar{y}(t) - \Gamma_1^a\bar{x}(t) - \dot{\bar{x}}(t), A\bar{x}(t) - \Gamma_2^a\bar{y}(t) - \dot{\bar{y}}(t)) \\
-\langle(\bar{x}(t),\bar{y}(t)),(-A^*\bar{y}(t) - \Gamma_1^a\bar{x}(t) - \dot{\bar{x}}(t), A\bar{x}(t) - \Gamma_2^a\bar{y}(t) - \dot{\bar{y}}(t))\rangle \big\} dt \\
+\|x(0) - x_0\|^2 + \|y(0) - y_0\|^2.
\end{aligned}
$$

It follows that $\bar{x}(0) = x_0$, $\bar{y}(0) = 0$, and the integrand is zero for a.e. t, which yields

$$
\begin{aligned}
-\dot{x}(t) - A^*y(t) - \Gamma_1^a x(t) &\in \partial_1 \psi(x(t),y(t)) = \partial_1 \varphi(x(t),y(t)) + \Gamma_1^s x(t) - f, \\
-\dot{y}(t) + Ax(t) - \Gamma_2^a y(t) &\in \partial_2 \psi(x(t),y(t)) = \partial_2 \varphi(x(t),y(t)) + \Gamma_2^s y(t) - g, \\
x(0) &= x_0, \\
y(0) &= y_0.
\end{aligned}
$$

By applying Theorem 7.5, we also obtain the following proposition.

Proposition 7.4. *Consider two convex lower semicontinuous functions φ_1 and φ_2 on Hilbert spaces X and Y, respectively, as well as two positive operators Γ_1 on X and Γ_2 on Y. For any pair $(f,g) \in X \times Y$ and $c > 0$, we consider the convex functionals*

$$
\psi_1(x) = \tfrac{1}{2}\langle \Gamma_1 x, x\rangle + \varphi_1(x) - \langle f, x\rangle \quad and \quad \psi_2(y) = c^{-2}\big(\tfrac{1}{2}\langle \Gamma_2 y, y\rangle + \varphi_2(y) - \langle g, y\rangle\big).
$$

Then, for any bounded linear operator $A : X \to Y$, $x_0 \in \mathrm{Dom}(\partial \varphi_1)$, $y_0 \in \mathrm{Dom}(\partial \varphi_2)$, and $\omega, \omega' \in \mathbf{R}$, the following functional is completely self-dual on $A_X^2 \times A_Y^2$:

$$
\begin{aligned}
I(u,v) = \int_0^T e^{-2\omega t} \big\{ \psi_1(e^{\omega t}u(t)) + \psi_1^*(e^{\omega t}(-A^*v(t) - \Gamma_1^a u(t) - \dot{u}(t))) \big\} dt \\
+ \int_0^T e^{-2\omega' t} \big\{ \psi_2(e^{\omega' t}v(t)) + \psi_2^*(e^{\omega' t}(Au(t) - c^{-2}\Gamma_2^a v(t) - c^{-2}\dot{v}(t))) \big\} dt \\
+ \frac{1}{2}\|u(0)\|^2 - 2\langle x_0, u(0)\rangle + \|x_0\|^2 + \frac{1}{2}\|u(T)\|^2 \\
+ \frac{1}{2c^2}\|v(0)\|^2 - \frac{2}{c^2}\langle y_0, v(0)\rangle + \frac{1}{c^2}\|y_0\|^2 + \frac{1}{2c^2}\|v(T)\|^2.
\end{aligned}
$$

Moreover, its infimum is zero and is attained at a path $(\bar{x}(t),\bar{y}(t))$, in such a way that $x(t) = e^{\omega t}\bar{x}(t)$ and $y(t) = e^{\omega' t}\bar{y}(t)$ form a solution of the system of equations

$$
\begin{cases}
-\dot{x}(t) + \omega x(t) - A^*y(t) - \Gamma_1 x(t) + f & \in \partial \varphi_1(x(t)) \\
-\dot{y}(t) + \omega' y(t) + c^2 Ax(t) - \Gamma_2 y(t) + g & \in \partial \varphi_2(y(t)) \\
\qquad\qquad x(0) = x_0 \\
\qquad\qquad y(0) = y_0.
\end{cases} \tag{7.70}
$$

Example 7.8. A variational principle for coupled equations

Let $\mathbf{b}_1 : \Omega \to \mathbf{R}^n$ and $\mathbf{b}_2 : \Omega \to \mathbf{R}^n$ be two smooth vector fields on a neighborhood of a bounded domain Ω of $\mathbf{R}^n$, and consider the system of evolution equations

$$
\begin{cases}
-\dfrac{\partial u}{\partial t} - \Delta(v-u) + \mathbf{b}_1 \cdot \nabla u = |u|^{p-2}u + f & \text{on } (0,T] \times \Omega, \\
-\dfrac{\partial v}{\partial t} + \Delta(v+c^2 u) + \mathbf{b}_2 \cdot \nabla v = |v|^{q-2}v + g & \text{on } (0,T] \times \Omega \\
\quad u(t,x) = v(t,x) = 0 & \text{on } (0,T] \times \partial\Omega, \\
\quad u(0,x) = u_0(x) & \text{for } x \in \Omega, \\
\quad v(0,x) = v_0(x) & \text{for } x \in \Omega.
\end{cases}
\tag{7.71}
$$

Theorem 7.6. *Assume* $\operatorname{div}(\mathbf{b}_1) \geq 0$ *and* $\operatorname{div}(\mathbf{b}_2) \geq 0$ *on* Ω, $1 < p,q \leq \frac{n+2}{n-2}$, *and consider the functional on the space* $A^2_{L^2(\Omega)} \times A^2_{L^2(\Omega)}$

$$
\begin{aligned}
I(u,v) = & \int_0^T \left\{ \Psi(u(t)) + \Psi^*\left(\mathbf{b}_1.\nabla u(t) + \frac{1}{2}\operatorname{div}(\mathbf{b}_1)u(t) - \Delta v(t) - \dot{u}(t) \right) \right\} dt \\
& + \int_0^T \left\{ \Phi(v(t)) + \Phi^*\left(\frac{1}{c^2}\mathbf{b}_2.\nabla v(t) + \frac{1}{2c^2}\operatorname{div}(\mathbf{b}_2)v(t) + \Delta u(t) - \frac{1}{c^2}\dot{v}(t) \right) \right\} dt \\
& + \int_\Omega \left\{ \frac{1}{2}(|u(0,x)|^2 + |u(T,x)|^2) - 2u(0,x)u_0(x) + |u_0(x)|^2 \right\} dx \\
& + \frac{1}{c^2} \int_\Omega \left\{ \frac{1}{2}(|v(0,x)|^2 + |v(T,x)|^2) - 2v(0,x)v_0(x) + |v_0(x)|^2 \right\} dx,
\end{aligned}
$$

where

$$
\Psi(u) = \tfrac{1}{2}\int_\Omega |\nabla u|^2 dx + \tfrac{1}{p}\int_\Omega |u|^p dx + \int_\Omega fu\,dx + \tfrac{1}{4}\int_\Omega \operatorname{div}(\mathbf{b}_1)|u|^2 dx
$$

and

$$
\Phi(v) = \frac{1}{2c^2}\int_\Omega |\nabla v|^2 dx + \frac{1}{qc^2}\int_\Omega |v|^q dx + \frac{1}{c^2}\int_\Omega gv\,dx + \frac{1}{4c^2}\int_\Omega \operatorname{div}(\mathbf{b}_2)|v|^2 dx
$$

and Ψ^* *and* Φ^* *are their Legendre transforms. I is then a completely self-dual functional on* $A^2_{H_0^1(\Omega))} \times A^2_{H_0^1(\Omega)}$, *and there exists then* $(\bar{u}, \bar{v}) \in A^2_{H_0^1(\Omega))} \times A^2_{H_0^1(\Omega)}$ *such that* $I(\bar{u}, \bar{v}) = \inf\left\{ I(u,v); (u,v) \in A^2_{H_0^1(\Omega))} \times A^2_{H_0^1(\Omega)} \right\} = 0$, *and* $(\bar{u}, \bar{v})$ *is a solution of* (7.71).

Example 7.9. Pressureless gas of sticky particles

Motivated by the recent work of Brenier [24], we consider equations of the form

$$
\partial_{tt} X = c^2 \partial_{yy} X - \partial_t \partial_a \mu, \quad \partial_a X \geq 0, \quad \mu \geq 0,
\tag{7.72}
$$

where here $X(t) := X(t,a,y)$ is a function on $K = [0,1] \times \mathbf{R}/\mathbf{Z}$ and $\mu(t,a,y)$ is a nonnegative measure that plays the role of a Lagrange multiplier for the constraint $\partial_a X \geq 0$. Following Brenier, we reformulate the above problem as the following system governing the variables $(X,U) : [0,T] \to A_{\mathbf{R}}^2[K] \times A_{\mathbf{R}}^2[K]$

$$
\begin{cases}
-\dot{X}(t) - \frac{\partial U}{\partial y}(t) \in \partial \varphi_1(X(t)) \\
\dot{U}(t) + \frac{\partial X}{\partial y}(t) = 0, \\
X(0) = X_0, \\
U(0) = U_0,
\end{cases}
\tag{7.73}
$$

where φ_1 is the convex function defined on $L^2(K)$ by

$$
\varphi_1(X) = \begin{cases} 0 & \text{if } \partial_a X \geq 0 \\ +\infty & \text{elsewhere.} \end{cases}
\tag{7.74}
$$

Actually, we may consider the Hilbert space $\mathscr{X}$ to be the subspace of $A_{\mathbf{R}}^2[K]$ consisting of those functions on K that are periodic in y. Define the operator $\Gamma(X,U) = (-\frac{\partial U}{\partial y}, \frac{\partial X}{\partial y})$ from $\mathscr{X}$ to $\mathscr{X}^*$. We can solve this system with the above method by setting $\varphi_2(U) = 0$ for every $U \in L^2(K)$ and considering the completely self-dual functional on the space $A_{\mathscr{X}}^2[0,T]$:

$$
\begin{aligned}
I(X,U) = &\int_0^T \left\{ \varphi_1(X(t)) + \varphi_1^* \left(-\frac{\partial U}{\partial y}(t) - \dot{X}(t) \right) \right\} dt \\
&+ \int_0^T \left\{ \varphi_2^* \left(-\frac{\partial X}{\partial y}(t) - \dot{U}(t) \right) \right\} dt \\
&+ \frac{1}{2}\|X(0)\|^2 - 2\langle X_0, X(0) \rangle + \|X_0\|^2 + \frac{1}{2}\|X(T)\|^2 \\
&+ \frac{1}{2}\|U(0)\|^2 - 2\langle Y_0, U(0) \rangle + \|Y_0\|^2 + \frac{1}{2}\|U(T)\|^2.
\end{aligned}
$$

It follows from Theorem 7.5 that if (X_0, U_0) are such that $\partial_a X_0 \geq 0$, then the minimum of I is then zero and is attained at a path $(\bar{X}(t), \bar{U}(t))$ that solves the system of equations (7.73).

7.7 Variational resolution for parabolic-elliptic variational inequalities

Consider for each time t, a bilinear continuous functional a_t on a Hilbert space $H \times H$ and a time-dependent convex lower semicontinuous function $\varphi(t,\cdot) : H \to \mathbf{R} \cup \{+\infty\}$. Solving the corresponding parabolic variational inequality amounts to constructing for a given $f \in L^2([0,T];H)$ and $x_0 \in H$, a path $x(t) \in A_H^2([0,T])$ such that for all $z \in H$

$$\langle \dot{x}(t), x(t) - z \rangle + a_t(x(t), x(t) - z) + \varphi(t, x(t)) - \varphi(t, z) \le \langle x(t) - z, f(t) \rangle \quad (7.75)$$

for almost all $t \in [0,T]$. This problem can be rewritten as $f(t) \in \dot{y}(t) + A_t y(t) + \partial \varphi(t, y)$, where A_t is the bounded linear operator on H defined by $a_t(u, v) = \langle A_t u, v \rangle$. This means that the variational inequality (7.75) can be solved using the variational principle in Theorem 7.4. For example, one can then formulate and solve variationally the following "obstacle" problem.

Corollary 7.7. *Let $(a_t)_t$ be bilinear continuous functionals on $H \times H$ satisfying:*

- *For some $\lambda > 0$, $a_t(v, v) \ge \lambda \|v\|^2$ on H for every $t \in [0,T]$.*
- *The map $u \to \int_0^T a_t(u(t), u(t)) dt$ is continuous on L_H^2.*

If K is a convex closed subset of H, then for any $f \in L^2([0,T]; H)$ and any $x_0 \in K$, there exists a path $x \in A_H^2([0,T])$ such that $x(0) = x_0$, $x(t) \in K$ for almost all $t \in [0,T]$ and

$$\langle \dot{x}(t), x(t) - z \rangle + a_t(x(t), x(t) - z) \le \langle x(t) - z, f \rangle \quad \text{for all } z \in K.$$

The path $x(t)$ is obtained as a minimizer on $A_H^2([0,T])$ of the completely self-dual functional

$$I(y) = \int_0^T \left\{ \varphi(t, y(t)) + (\varphi(t, \cdot) + \psi_K)^*(-\dot{y}(t) - \Gamma_t y(t)) \right\} dt$$
$$+ \frac{1}{2}(|y(0)|^2 + |y(T)|^2) - 2\langle y(0), x_0 \rangle + |x_0|^2.$$

Here ψ_K is the convex function defined as $\psi_K(y) = 0$ on K and $+\infty$ elsewhere, $\varphi(t, y) = \frac{1}{2}a_t(y, y) - \langle f(t), y \rangle$, while $\Gamma_t : H \to H$ is the skew-adjoint operator defined by $\langle \Gamma_t u, v \rangle = \frac{1}{2}(a_t(u, v) - a_t(v, u))$.

Note that Theorem 7.4 is not directly applicable to this situation since the Lagrangian does not satisfy the necessary boundedness condition, nor is it autonomous. Moreover, one cannot use Theorem 7.5 because the Lagrangian is not autonomous. However, one can still replace ψ_K by its λ-regularization ψ_K^λ, apply Theorem 7.4 to the function $\varphi_\lambda(t, \cdot) = \varphi(t, \cdot) + \psi_K^\lambda$, and then let $\lambda \to 0$ to conclude. The details are left to the interested reader.

Exercises 7. B.

1. Verify all the examples of this chapter.
2. Establish Corollary 7.7.

Further comments

As mentioned in the introduction, the initial impetus for the theory came from a conjecture of Brézis and Ekeland [29], [30], who formulated a (somewhat) self-dual variational principle for gradient flows. Their formulation was extended to more general parabolic equations by Auchmuty [10], who eventually gave a proof in [12] for the case of gradient flows under certain boundedness conditions. A proof – based on self-duality – was given by Ghoussoub and Tzou [67] for gradient flows of convex energies under less restrictive conditions. The case of semiconvex energies was handled by Ghoussoub and McCann [61]. The variational construction of a semigroup of contractions associated to a self-dual Lagrangian was given by Ghoussoub and Tzou in [68]. It gives an alternative proof for associating a continuous semigroup of contractions to a maximal monotone operator [25]. Parabolic equations involving unbounded antisymmetric operators were considered in [68].

Chapter 8
Iteration of Self-dual Lagrangians and Multiparameter Evolutions

Nonhomogeneous boundary conditions translate into a lack of antisymmetry in the differential system. The iteration of a self-dual Lagrangian on phase space $X \times X^*$ with an operator that is skew-adjoint modulo a boundary triplet $(H, \mathscr{B}, R)$ needs to be combined with an R-self-dual function ℓ on the boundary H, in order to restore self-duality to the whole system. This is done via the Lagrangian

$$L_{\Gamma, \ell}(x, p) = \begin{cases} L(x, \Gamma x + p) + \ell(\mathscr{B}x) & \text{if } x \in D(\Gamma) \cap D(\mathscr{B}) \\ +\infty & \text{if } x \notin D(\Gamma) \cap D(\mathscr{B}), \end{cases} \qquad (8.1)$$

which is then self-dual, and as a consequence one obtains solutions for the boundary value problem

$$\begin{cases} \Gamma x \in \bar{\partial} L(x) \\ R\mathscr{B}x \in \partial \ell(\mathscr{B}x) \end{cases} \qquad (8.2)$$

by inferring that the infimum on X of the completely self-dual functional $I(x) := L_{\Gamma, \ell}(x, 0) = L(x, \Gamma x) + \ell(\mathscr{B}x)$ is attained and is equal to zero. Moreover, the addition of the R-self-dual boundary Lagrangian required to restore self-duality often leads to the natural boundary conditions.

The latter Lagrangian can then be lifted to path space, provided one adds a suitable self-dual time-boundary Lagrangian. This iteration is used to solve initial-value parabolic problems whose state-boundary values are evolving in time such as

$$\begin{cases} -\dot{x}(t) + \Gamma_t x(t) \in \bar{\partial} L(t, x(t)) & \text{for } t \in [0, T] \\ R_t \mathscr{B}_t(x(t)) \in \partial \ell_t(\mathscr{B}_t x(t)) & \text{for } t \in [0, T] \\ x(0) = x_0, \end{cases} \qquad (8.3)$$

where L is a time-dependent self-dual Lagrangian on a Banach space X anchored on a Hilbert space H (i.e., $X \subset H \subset X^*$), x_0 is a prescribed initial state in X, $\Gamma_t : D(\Gamma_t) \subset X \to X^*$ is antisymmetric modulo a boundary pair $(H_t, R_t, \mathscr{B}_t)$ with $\mathscr{B}_t : D(\mathscr{B}_t) \subset X \to H_t$ as a boundary operator, R_t is a self-adjoint automorphism on H_t, and ℓ_t is an R_t-self-dual function on the boundary space H_t. The corresponding self-dual Lagrangian on $L_X^2[0, T] \times L_{X^*}^2[0, T]$ is then

N. Ghoussoub, *Self-dual Partial Differential Systems and Their Variational Principles*, Springer Monographs in Mathematics, DOI 10.1007/978-0-387-84897-6_8, © Springer Science+Business Media, LLC 2009

$$\mathscr{L}(u,p) = \int_0^T L_{\Gamma,\ell_t}(t,u(t),p(t)-\dot{u}(t))dt$$
$$+ \frac{1}{2}|u(0)|_H^2 + 2\langle x_0,u(0)\rangle + |x_0|_H^2 + \frac{1}{2}|u(T)|_H^2 \qquad (8.4)$$

if $\dot{u} \in L_{X^*}^2[0,T]$ and $+\infty$ otherwise.

This process can be iterated again by considering the path space $L_X^2[0,T]$ as a new state space for the newly obtained self-dual Lagrangian, leading to the construction of multiparameter flows such as

$$\begin{cases} -\frac{\partial x}{\partial t}(s,t) - \frac{\partial x}{\partial s}(s,t) \in \overline{\partial}M((s,t),x(s,t),\frac{\partial x}{\partial t}(s,t)+\frac{\partial x}{\partial s}(s,t)) \\ x(0,t) = x_0 \text{ a.e. } t \in [0,T] \\ x(s,0) = x_0 \text{ a.e. } s \in [0,S]. \end{cases} \qquad (8.5)$$

This method is quite general and far-reaching, but may be limited by the set of conditions needed to accomplish the above mentioned iterations. This chapter is focussed on cases where this can be done.

8.1 Self-duality and nonhomogeneous boundary value problems

Propositions 4.2 and 6.1 combine to yield the following variational principle for nonhomogeneous boundary value problems.

Theorem 8.1. *Let $B : X \to X$ be a bounded linear operator on a reflexive Banach space X, and let $\Gamma : D(\Gamma) \subset X \to X^*$ be a linear operator such that the pair (B,Γ) is antisymmetric modulo a boundary triplet $(H,R,\mathscr{B})$. Let $L : X \times X^* \to \mathbf{R}$ be a convex Lagrangian that is continuous in the first variable, and let $\ell : H \to \mathbf{R} \cup \{+\infty\}$ be a convex continuous function on H such that one of the following two conditions holds:*

1. *L is standard, $\mathrm{Dom}_1(L) \subset D(\Gamma) \cap D(\mathscr{B})$, and $\mathscr{B}(\mathrm{Dom}_1(L)) \cap \mathrm{Dom}(\ell) \neq \emptyset$.*
2. *The pair (B,Γ) is skew-adjoint modulo $(H,R,\mathscr{B})$, and for some $p_0 \in X^*$, the map $x \to L(x,p_0)$ is bounded on the ball of X, while $\ell(s) \leq C(1+\|s\|^2)$ for $s \in H$.*

Assume that

$$\lim_{\|x\| \to +\infty} \frac{L(x,\Gamma x)+\ell(\mathscr{B}x)}{\|x\|} = +\infty. \qquad (8.6)$$

If B has dense range, then the functional

$$I(x) := L(x,\Gamma x)+\ell(\mathscr{B}x)$$

is completely self-dual, and it attains its minimum at a point $\bar{x} \in D(\mathscr{B}) \cap D(\Gamma)$ in such a way that $I(\bar{x}) = \inf_{x \in X} I(x) = 0$ and

$$\Gamma\bar{x} \in \overline{\partial}_B L(\bar{x}) \qquad (8.7)$$
$$R\mathscr{B}\bar{x} \in \partial\ell(\mathscr{B}\bar{x}).$$

Proof. By Proposition 4.2, the Lagrangian $L_{(\Gamma,\ell)}$ defined in (8.1) above is B-self-dual. The hypotheses then allows us to apply Proposition 6.1 to the functional $I(x) = L_{(\Gamma,\ell)}(x,0)$ and find $\bar{x} \in X$ such that

$$0 = I(\bar{x}) = L(\bar{x}, \Gamma\bar{x}) - \langle B\bar{x}, \Gamma\bar{x} \rangle - \frac{1}{2}\langle \mathscr{B}\bar{x}, R\mathscr{B}\bar{x} \rangle + \ell(\mathscr{B}\bar{x}).$$

Since $L(x,p) \geq \langle Bx, p \rangle$ and $\ell(s) \geq \frac{1}{2}\langle s, Rs \rangle$, we get

$$\begin{cases} L(\bar{x}, \Gamma\bar{x}) = \langle B\bar{x}, \Gamma\bar{x} \rangle \\ \ell(\mathscr{B}\bar{x}) = \frac{1}{2}\langle \mathscr{B}\bar{x}, R\mathscr{B}\bar{x} \rangle, \end{cases} \tag{8.8}$$

and we are done.

Self-dual formulation of classical boundary conditions

Before we apply Theorem 8.1, we shall show how various classical boundary conditions can be expressed as $R\mathscr{B}x \in \partial\ell(\mathscr{B}x)$, where $\mathscr{B}$ is a boundary operator.

Self-dual formulation of Dirichlet boundary conditions. Suppose, as is often the case, that we have a Green's formula of the form

$$\langle x, \Gamma x \rangle = \frac{1}{2}(\|\mathscr{B}_1 x\|^2 - \|\mathscr{B}_2 x\|^2) \text{ for all } x \in D(\Gamma) \cap D(\mathscr{B}), \tag{8.9}$$

where the boundary operator $\mathscr{B}x = (\mathscr{B}_1 x, \mathscr{B}_2 x)$ is from X into some Hilbert space $H = H_1 \times H_2$. This can also be written as

$$\langle x, \Gamma x \rangle + \frac{1}{2}\langle \mathscr{B}x, R\mathscr{B}x \rangle = 0 \text{ for all } x \in D(\Gamma) \cap D(\mathscr{B}), \tag{8.10}$$

where R is the automorphism on $H = H_1 \times H_2$ given by $R(s,r) = (-s,r)$.

Now the simplest R-self-dual function on $H = H_1 \times H_2$ is clearly of the form $\ell(s,r) = \psi_1(s) + \psi_2(r)$ with $\psi_1^*(s) = \psi(-s)$ and $\psi_2^*(r) = \psi_2(r)$. This means that we must have $\psi_2(r) = \frac{1}{2}\|r\|_{H_2}^2$. On the other hand, to any given $a \in H_1$ we can associate the antiself-dual function $\psi_1(s) = \frac{1}{2}\|s\|^2 - 2\langle a,s \rangle + \|a\|^2$. In other words, the function

$$\ell_a(s,r) = \frac{1}{2}\|s\|_{H_1}^2 - 2\langle a,s \rangle_{H_1} + \|a\|_{H_1}^2 + \frac{1}{2}\|r\|_{H_2}^2 \tag{8.11}$$

is the most natural R-self-dual function on $H_1 \times H_2$. But then the boundary equation $R\mathscr{B}x \in \partial\ell_a(\mathscr{B}x)$ means that $-\mathscr{B}_1 x = \mathscr{B}_1 x - 2a$ and $\mathscr{B}_2 x = \mathscr{B}_2 x$, which is nothing but the Dirichlet boundary condition $\mathscr{B}_1 x = a$. In other words, the Dirichlet boundary condition could be formulated as a self-dual boundary condition since

$$R\mathscr{B}x \in \partial\ell_a(\mathscr{B}x) \text{ is equivalent to } \mathscr{B}_1 x = a. \tag{8.12}$$

Since this situation will often occur in what follows, we shall formalize it in the following definition.

Definition 8.1. (1) Say that the operator $\Gamma : D(\Gamma) \subset X \to X^*$ is *antisymmetric modulo a "trace boundary operator"* $\mathscr{B} = (\mathscr{B}_1, \mathscr{B}_2)$ into a space $H_1 \times H_2$, if it satisfies (1), (2) of Definition 4.3 as well as

$$\langle \Gamma x, x \rangle = \tfrac{1}{2}(\|\mathscr{B}_2 x\|^2 - \|\mathscr{B}_1 x\|^2) \text{ for every } x \in D(\Gamma) \cap D(\mathscr{B}). \tag{8.13}$$

(2) Say that Γ is *skew-symmetric modulo a "trace boundary operator"* $\mathscr{B} = (\mathscr{B}_1, \mathscr{B}_2)$ if it is antisymmetric modulo $\mathscr{B}$, while verifying (4) of Definition 4.3.

Self-dual formulation for boundary conditions of periodic type. Suppose now that we have a Green's formula of the form

$$\langle x, \Gamma x \rangle = \langle \mathscr{B}_1 x, \mathscr{B}_2 x \rangle \text{ for all } x \in D(\Gamma) \cap D(\mathscr{B}), \tag{8.14}$$

where the boundary operator $\mathscr{B}x = (\mathscr{B}_1 x, \mathscr{B}_2 x)$ is from X into a Hilbert space $H = E \times E^*$. This can also be written as

$$\langle x, \Gamma x \rangle + \tfrac{1}{2}\langle \mathscr{B}x, R\mathscr{B}x \rangle = 0 \text{ for all } x \in D(\Gamma) \cap D(\mathscr{B}), \tag{8.15}$$

where R is the automorphism on $H = E \times E^*$ given by $R(r,s) = (-s,-r)$. In this case, any self-dual Lagrangian ℓ on $E \times E^*$ is an R-self-dual function. In particular, if we take the self-dual Lagrangian ℓ associated to any convex lower semicontinuous function ψ and any skew-adjoint operator T on E (that is, $\ell(x,p) = \psi(x) + \psi^*(-Tx - p)$), then we have

$$R\mathscr{B}x \in \partial\ell(\mathscr{B}x) \text{ is equivalent to } \mathscr{B}_2 x + T\mathscr{B}_1 x \in -\partial\psi(\mathscr{B}_1 x). \tag{8.16}$$

Self-dual formulation of boundary conditions of linking type. More generally, we may have a Green's formula of the form

$$\langle x, \Gamma x \rangle = \langle \mathscr{B}_1^1 x, \mathscr{B}_1^2 x \rangle - \langle \mathscr{B}_2^1 x, \mathscr{B}_2^2 x \rangle \text{ for all } x \in D(\Gamma) \cap D(\mathscr{B}), \tag{8.17}$$

where the boundary operator $\mathscr{B}x = (\mathscr{B}_1 x, \mathscr{B}_2 x)$ is from X into a Hilbert space $H = (E_1 \times E_1^*) \times (E_2 \times E_2^*)$. This can also be written as

$$\langle x, \Gamma x \rangle + \tfrac{1}{2}\langle \mathscr{B}x, R\mathscr{B}x \rangle = 0 \text{ for all } x \in D(\Gamma) \cap D(\mathscr{B}), \tag{8.18}$$

where R is the automorphism on $H = (E_1 \times E_1^*) \times (E_2 \times E_2^*)$ given by $R := (R_1, R_2)$, where $R_1(r,s) = (-s,-r)$ and $R_2(r,s) = (s,r)$. In this case, any function on H of the form

$$\ell(r_1, s_1, r_2, s_2) = \ell_1(r_1, s_1) + \ell_2(r_2, s_2),$$

where ℓ_1 is an antiself-dual function on $E_1 \times E_1^*$ and ℓ_2 is a self-dual Lagrangian on $E_2 \times E_2^*$ is an R-self-dual function on H. Typical examples of those are of the form

$$\ell_1(s,r) = \psi_1(s) + \psi_1^*(-T_1 s - r) \text{ and } \ell_2(s,r) = \psi_2(s) + \psi_2^*(T_2 s + r),$$

defined on $E_1 \times E_1^*$ and $E_2 \times E_2^*$, respectively, where ψ_i is convex lower semicontinuous on E_i and T_i is a skew-adjoint operator on E_i $(i = 1, 2)$. In this case, we have that

$$R \mathscr{B} x \in \partial \ell(\mathscr{B} x) \text{ is equivalent to } \begin{cases} \mathscr{B}_1^2 x + T_1(\mathscr{B}_1^1 x) \in -\partial \psi_1(\mathscr{B}_1^1 x) \\ \mathscr{B}_2^2 x + T_2(\mathscr{B}_2^1 x) \in \partial \psi_2(\mathscr{B}_2^1 x). \end{cases} \quad (8.19)$$

Self-dual formulations of more general boundary conditions. A richer class of automorphisms may appear in other situations. Indeed, assume the boundary space H is the product of k Hilbert spaces, say $H = E^k$. Besides the operators $\pm \mathrm{Id}$, we can consider R_σ-self-dual boundary Lagrangians on E^k, where R_σ is the self-adjoint automorphism on E^k associated to a permutation σ of $\{1, ..., k\}$, that is,

$$R_\sigma((x_i)_{i=1}^k) = (\pm x_{\sigma(i)})_{i=1}^k \text{ for any } (x_i)_{i=1}^k \in H^k. \quad (8.20)$$

The following corollary of Theorem 8.1 covers a wide range of applications.

Theorem 8.2. *Let Φ be a convex and lower semicontinuous function on a reflexive Banach space X such that, for some constant $C > 0$ and $p_1, p_2 > 1$, we have*

$$\frac{1}{C} \left(\|x\|_X^{p_1} - 1 \right) \leq \Phi(x) \leq C \left(\|x\|_X^{p_2} + 1 \right) \text{ for every } x \in X. \quad (8.21)$$

Let ψ_1 (resp., ψ_2) be a bounded below proper convex lower semicontinuous function on a Hilbert space E_1 (resp., E_2), and consider boundary Hilbert spaces $H_1 = E_1^k$ (resp., $H_2 = E_2^l$) with the automorphisms $R_1 = -R_\sigma$ on H_1 and $R_2 = R_\tau$ on H_2, where σ and τ are the permutations $\sigma(i) = k + 1 - i$ on $\{1, ..., k\}$ and $\tau(i) = l + 1 - i$ on $\{1, ..., l\}$. Consider the following framework.

- *$B : X \to X$ is a linear operator with dense range, and $\Gamma : D(\Gamma) \subset X \to X^*$ is a linear operator such that $B^*\Gamma$ is antisymmetric modulo the boundary triplet $(H, \mathscr{B}, R)$, where $\mathscr{B} := (\mathscr{B}_1, \mathscr{B}_2) : D(\mathscr{B}) \subset X \to H := E_1^k \times E_2^l$ and R is the automorphism (R_1, R_2) on $H := E_1^k \times E_2^l$. Assume one of the following two conditions holds:*

  1. $\mathrm{Dom}(\Phi) \subset D(\Gamma) \cap D(\mathscr{B})$ and $\mathscr{B}(\mathrm{Dom}(\Phi)) \cap \mathrm{Dom}(\psi_1) \cap \mathrm{Dom}(\psi_2) \neq \emptyset$.
  2. $B^*\Gamma$ is skew-adjoint modulo the triplet $(H, \mathscr{B}, R)$, and

  $$C_i^{-1}(\|s\|^2 - 1) \leq \psi_i(s) \leq C_i(1 + \|s\|^2) \text{ for } s \in E_i, \ i = 1, 2. \quad (8.22)$$

Then, the following results hold:

1. If $k = l = 1$, then for any $a \in H_1$ there exists a solution $\bar{x} \in X$ to the boundary value problem

$$\begin{cases} \Gamma x + f \in \partial \Phi(Bx) \\ \mathscr{B}_1(x) = a \end{cases} \quad (8.23)$$

that is obtained as a minimizer on E of the completely self-dual functional

$$I(x) = \Phi(Bx) - \langle f, Bx \rangle + \Phi^*(\Gamma x + f)$$

$$+ \frac{1}{2} \|\mathscr{B}_1(x)\|^2 - 2\langle a, \mathscr{B}_1(x)\rangle + \|a\|^2 + \frac{1}{2} \|\mathscr{B}_2(x)\|^2.$$

2. If $k = 2$ and $l = 0$, then for any bounded skew-adjoint operator T on H_1 there exists a solution $\bar{x} \in X$ to the boundary value problem

$$\begin{cases} \Gamma x + f \in \partial \Phi(Bx) \\ \mathscr{B}_1^2 x + T(\mathscr{B}_1^1 x) \in -\partial \psi_1(\mathscr{B}_1^1 x) \end{cases} \tag{8.24}$$

that is obtained as a minimizer on E of the completely self-dual functional

$$I(x) = \Phi(Bx) - \langle f, Bx\rangle + \Phi^*(\Gamma x - f) + \psi_1(\mathscr{B}_1^1 x) + \psi_1^*(-T\mathscr{B}_1^1 x - \mathscr{B}_1^2 x).$$

3. If $k = l = 2$, then for any bounded skew-adjoint operator T_1 on H_1 (resp., T_2 on H_2), there exists a solution $\bar{x} \in X$ to the boundary value problem

$$\begin{cases} \Gamma x + f \in \partial \Phi(Bx) \\ \mathscr{B}_1^2 x + T_1(\mathscr{B}_1^1 x) \in -\partial \psi_1(\mathscr{B}_1^1 x) \\ \mathscr{B}_2^2 x + T_2(\mathscr{B}_2^1 x) \in \partial \psi_2(\mathscr{B}_2^1 x) \end{cases} \tag{8.25}$$

that is obtained as a minimizer on E of the completely self-dual functional

$$\begin{aligned} I(x) = \ & \Phi(Bx) - \langle f, Bx\rangle + \Phi^*(\Gamma x + f) \\ & + \psi_1(\mathscr{B}_1^1 x) + \psi_1^*(-T_1\mathscr{B}_1^1 x - \mathscr{B}_2^1 x) + \psi_2(\mathscr{B}_2^1 x) + \psi_2^*(T_2\mathscr{B}_2^1 x + \mathscr{B}_2^2 x). \end{aligned}$$

All functionals above are defined to be equal to $+\infty$ when x is not in $D(\Gamma) \cap D(\mathscr{B})$.

Proof. Let $\Psi(x) = \Phi(x) + \langle f, x\rangle$, and consider the self-dual Lagrangian $L(x, p) := \Psi(x) + \Psi^*(p)$ and $B^*\Gamma$, which is skew-adjoint modulo the triplet $(H, \mathscr{B}, R) = \big(E_1^k \times E_2^l, (\mathscr{B}_1, \mathscr{B}_2), (R_1, R_2)\big)$. The boundary Lagrangian will differ according to k, l.

(1) If $k = l = 1$, then we take $\ell(r, s) = \ell_1(r) + \ell_2(s)$, where ℓ_1 is the antiself-dual function on H_1 defined by $\ell_1(r) = \frac{1}{2}\|r\|^2 - 2\langle a, r\rangle + \|a\|^2$, while ℓ_2 is the self-dual function $\ell_2(s) = \frac{1}{2}\|s\|^2$ on H_2. The completely self-dual functional I can then be rewritten as the sum of two nonnegative terms:

$$I(x) = \Psi(Bx) + \Psi^*(\Gamma x) - \langle Bx, \Gamma x\rangle + \|\mathscr{B}_1 x - a\|^2 \geq 0.$$

(2) If $k = 2$ and $l = 0$, then we take $\ell(r, s) = \ell_1(r_1, r_2)$, where ℓ_1 is the antiself-dual Lagrangian on $H_1 \times H_1$ defined by $\ell_1(r_1, r_2) = \psi(r_1) + \psi^*(-Tr_1 - r_2)$. Since we have

$$\langle Bx, \Gamma x\rangle + \frac{1}{2}\big\langle (\mathscr{B}_1^1 x, \mathscr{B}_1^2 x), (-\mathscr{B}_1^2 x, -\mathscr{B}_1^1 x)\big\rangle = 0,$$

the completely self-dual functional I can then be rewritten as the sum of two non-negative terms:

$$\begin{aligned} I(x) = \ & \Psi(Bx) + \Psi^*(\Gamma x) - \langle Bx, \Gamma x\rangle \\ & + \psi(\mathscr{B}_1^1 x) + \psi^*(T\mathscr{B}_1^1 x - \mathscr{B}_1^2 x) + \langle \mathscr{B}_1^1 x, \mathscr{B}_1^2 x\rangle. \end{aligned}$$

(3) If $k = l = 2$, then we take $\ell(r,s) = \ell_1(r_1,r_2) + \ell_2(s_1,s_2)$, where ℓ_1 is the antiself-dual Lagrangian on H_1^2 defined by $\ell_1(r_1,r_2) = \psi_1(r_1) + \psi_1^*(-T_1 r_1 - r_2)$, while ℓ_2 is the self-dual Lagrangian $\ell_2(s_1,s_2) = \psi_2(s_1) + \psi_2^*(T_2 s_1 + s_2)$ on H_2^2. Since we have

$$\langle Bx, \Gamma x\rangle + \frac{1}{2}\langle(\mathscr{B}_1^1 x, \mathscr{B}_1^2 x), (-\mathscr{B}_1^2 x, -\mathscr{B}_1^1 x)\rangle + \frac{1}{2}\langle(\mathscr{B}_2^1 x, \mathscr{B}_2^2 x), (\mathscr{B}_2^2 x, \mathscr{B}_2^1 x)\rangle = 0,$$

the completely self-dual functional I can then be rewritten as the sum of three non-negative terms:

$$\begin{aligned}
I(x) = &\; \Psi(Bx) + \Psi^*(\Gamma x) - \langle Bx, \Gamma x\rangle \\
&+ \psi_1(\mathscr{B}_1^1 x) + \psi_1^*(-T_1\mathscr{B}_1^1 x - \mathscr{B}_1^2 x) + \langle\mathscr{B}_1^1 x, \mathscr{B}_1^2 x\rangle \\
&+ \psi_2(\mathscr{B}_2^1 x) + \psi_2^*(T_2\mathscr{B}_2^1 x + \mathscr{B}_2^2 x) - \langle\mathscr{B}_2^1 x, \mathscr{B}_2^2 x\rangle.
\end{aligned}$$

In all cases, we can apply Theorem 4.2 to conclude.

We now give examples reflecting the various situations.

8.2 Applications to PDEs involving the transport operator

We now deal with the following transport equations without diffusion terms:

$$\begin{cases} \mathbf{a}(x) \cdot \nabla u + a_0(x)u + u|u|^{p-2} = f(x) & \text{for } x \in \Omega \\ \qquad\qquad\qquad\qquad\qquad\quad u(x) = u_0 & \text{for } x \in \Sigma_+. \end{cases} \tag{8.26}$$

We shall assume that the domain Ω and the vector field $\mathbf{a}$ satisfy all the assumptions used in Lemma 4.1 above to guarantee that the transport operator $\Gamma : D(\Gamma) \subset L^p(\Omega) \to L^q(\Omega)$ defined as

$$\Gamma u = \mathbf{a} \cdot \nabla u + \tfrac{1}{2}(\nabla \cdot \mathbf{a})u \text{ with domain } D(\Gamma) = \{u \in L^p(\Omega); \mathbf{a} \cdot \nabla u + \tfrac{\nabla \mathbf{a}}{2}u \in L^q(\Omega)\}$$

is skew-adjoint modulo the boundary operators $\mathscr{B}u = (\mathscr{B}_1 u, \mathscr{B}_2 u) = (u|_{\Sigma_+}, u|_{\Sigma_-})$ whose domain is

$$D(\mathscr{B}) = \{u \in L^p(\Omega); (u|_{\Sigma_+}, u|_{\Sigma_-}) \in L^2(\Sigma_+; |\mathbf{a} \cdot \hat{n}|d\sigma) \times L^2(\Sigma_-; |\mathbf{a} \cdot \hat{n}|d\sigma)\}.$$

Example 8.1. Nonhomogeneous transport equation ($k = l = 1$)

We distinguish the cases when $p \geq 2$ and when $1 < p < 2$.

Theorem 8.3. *Assume $p \geq 2$, and let $f \in L^q$, where $\frac{1}{p} + \frac{1}{q} = 1$, and $a_0 \in L^\infty(\Omega)$. Suppose there exists $\tau \in C^1(\bar{\Omega})$ such that*

$$\mathbf{a} \cdot \nabla \tau + \tfrac{1}{2}(\nabla \cdot \mathbf{a} + a_0) \geq 0 \text{ on } \Omega. \tag{8.27}$$

Define on $X = L^p(\Omega)$ the convex function

$$\varphi(u) := \frac{1}{p}\int_\Omega e^\tau |e^{-\tau}u|^p dx + \frac{1}{2}\int_\Omega \mathbf{a}\cdot\nabla\tau |u|^2 dx + \frac{1}{4}\int_\Omega (\nabla\cdot\mathbf{a}+a_0)|u|^2 dx + \int_\Omega ue^\tau f dx.$$

For any $v_0 \in L^2(\Sigma_+; |\mathbf{a}\cdot\hat{n}|d\sigma)$, consider the functional defined by

$$I(u) := \varphi(u) + \varphi^*\left(\mathbf{a}\cdot\nabla u + \frac{(\nabla\cdot\mathbf{a})}{2}u\right)$$
$$+ \int_{\Sigma_+}\left\{\frac{1}{2}|u|^2 - 2e^\tau uv_0 + e^{2\tau}|v_0|^2\right\}|\mathbf{a}\cdot\hat{n}|d\sigma + \int_{\Sigma_-}\frac{1}{2}|u|^2|\mathbf{a}\cdot\hat{n}|d\sigma$$

on $X_1 := D(\Gamma)\cap D(\mathscr{B})$ and $+\infty$ elsewhere on $L^p(\Omega)$. Then:

1. I is completely self-dual on X, and there exists $\bar{u} \in X_1$ such that

$$I(\bar{u}) = \inf\{I(u); u \in L^p(\Omega)\} = 0.$$

2. The function $\bar{v} := e^{-\tau}\bar{u}$ satisfies the nonlinear transport equation

$$\begin{cases} \mathbf{a}\cdot\nabla\bar{v} - \frac{a_0}{2}\bar{v} = \bar{v}|\bar{v}|^{p-2} + f & \text{on } \Omega, \\ \bar{v} = v_0 & \text{on } \Sigma_+. \end{cases} \tag{8.28}$$

Proof. Let $X = L^p(\Omega)$, and use Lemma 4.1 to deduce that the operator $\Gamma : D(\Gamma) \to X^*$ is skew-adjoint modulo the boundary operators $\mathscr{B}u = (u|_{\Sigma_+}, u|_{\Sigma_-})$. By Proposition 4.2, the Lagrangian defined by

$$M(u,p) = \varphi(u) + \varphi^*(\Gamma u + p) + \ell(\mathscr{B}_1(u), \mathscr{B}_2(u)) \tag{8.29}$$

if $u \in X_1$ and $+\infty$ if $u \notin X_1$ is self-dual on $L^p(\Omega) \times L^q(\Omega)$, where the Lagrangian ℓ is defined on $L^2(\Sigma_+; |\mathbf{a}\cdot\hat{n}|d\sigma) \times L^2(\Sigma_-; |\mathbf{a}\cdot\hat{n}|d\sigma) \to \mathbf{R}$ by

$$\ell(h,k) := \int_{\Sigma_+}\left\{\frac{1}{2}|h|^2 - 2e^\tau hv_0 + e^{2\tau}|v_0|^2\right\}|\mathbf{a}\cdot\hat{n}|d\sigma + \frac{1}{2}\int_{\Sigma_-}|k|^2|\mathbf{a}\cdot\hat{n}|d\sigma.$$

Note that the boundary Lagrangian here corresponds to the case where $k = l = 1$ in the preceding theorem. The hypotheses of Theorem 8.1 are satisfied, and therefore there exists then $\bar{u} \in L^p(\Omega)$ such that $0 = M(\bar{u},0) = I(\bar{u}) = \inf\{I(u); u \in X_1\}$ and assertion (1) is verified.

To get (2), we observe again that by Green's formula we have

$$0 = I(\bar{u}) = \varphi(\bar{u}) + \varphi^*(\Gamma\bar{u}) - \langle\bar{u}, \Gamma\bar{u}\rangle + \frac{1}{2}\int_{\Sigma_+}|\bar{u} - e^\tau v_0|^2|\mathbf{a}\cdot\hat{n}|d\sigma.$$

In particular, $\varphi(\bar{u}) + \varphi^*(\Gamma\bar{u}) = \langle u, \Gamma u\rangle$ and $\int_{\Sigma_+}|\bar{u} - e^\tau v_0|^2|\mathbf{a}\cdot\hat{n}|d\sigma = 0$ in such a way that $\Gamma\bar{u} \in \partial\varphi(\bar{u})$ and $\bar{u}|_{\Sigma_+} = e^\tau v_0$. In other words,

$$\mathbf{a}\cdot\nabla\bar{u}+\frac{1}{2}(\nabla\cdot\mathbf{a})\bar{u}=\bar{u}|e^{-\tau}\bar{u}|^{p-2}+(\mathbf{a}\cdot\nabla\tau)\bar{u}+\frac{1}{2}(\nabla\cdot\mathbf{a}+a_0)\bar{u}+e^{\tau}f$$

and $\bar{u}|_{\Sigma_+}=e^{\tau}v_0$. Multiply both equations by $e^{-\tau}$ and use the product rule for differentiation to get $\mathbf{a}\cdot\nabla\bar{v}-\frac{a_0}{2}\bar{v}=\bar{v}|\bar{v}|^{p-2}+f$ and $\bar{v}|_{\Sigma_+}\equiv v_0$, where $\bar{v}:=e^{-\tau}\bar{u}$. In the case where $1<p\le 2$, we have the following result.

Theorem 8.4. *Assume $1<p\le 2$, and let $f\in L^2(\Omega)$ and $a_0\in L^{\infty}(\Omega)$. Suppose there exists $\tau\in C^1(\bar{\Omega})$ such that for some $\varepsilon>0$ we have*

$$\mathbf{a}\cdot\nabla\tau+\tfrac{1}{2}(\nabla\cdot\mathbf{a}+a_0)\ge\varepsilon>0 \text{ on }\Omega. \tag{8.30}$$

Define on $L^2(\Omega)$ the convex functional

$$\varphi(u):=\frac{1}{p}\int_{\Omega}e^{\tau}|e^{-\tau}u|^p dx+\frac{1}{2}\int_{\Omega}\mathbf{a}\cdot\nabla\tau|u|^2 dx+\frac{1}{4}\int_{\Omega}(\nabla\cdot\mathbf{a}+a_0)|u|^2 dx+\int_{\Omega}ue^{\tau}f dx.$$

For $v_0\in L^2(\Sigma_+;|\mathbf{a}\cdot\hat{n}|d\sigma)$, consider the completely self-dual functional defined by

$$I(u):=\varphi(u)+\varphi^*\left(\mathbf{a}\cdot\nabla u+\frac{(\nabla\cdot\mathbf{a})}{2}u\right)$$

$$+\int_{\Sigma_+}\left\{\frac{1}{2}|u|^2-2e^{\tau}uv_0+e^{2\tau}|v_0|^2\right\}|\mathbf{a}\cdot\hat{n}|d\sigma+\int_{\Sigma_-}\frac{1}{2}|u|^2|\mathbf{a}\cdot\hat{n}|d\sigma$$

on $X_1:=D(\Gamma)\cap D(\mathcal{B})$ and $+\infty$ elsewhere on $L^2(\Omega)$

1. There exists then $\bar{u}\in X_1$ such that $I(\bar{u})=\inf\{I(u); u\in L^2(\Omega)\}=0$.
2. The function $\bar{v}:=e^{-\tau}\bar{u}$ satisfies the nonlinear transport equation (8.28).

Proof. In this case, the right space is $X=L^2(\Omega)$, and $\Gamma:D(\Gamma)\to X^*$ is defined as in the first case but with domain $D(\Gamma)=\{u\in L^2(\Omega):\mathbf{a}\cdot\nabla u\in L^2(\Omega)\}$. It is again skew-adjoint modulo the boundary operator $\mathcal{B}=(u|_{\Sigma_+},u|_{\Sigma_-})$ whose domain is

$$D(\mathcal{B})=\{u\in L^2(\Omega); (u|_{\Sigma_+},u|_{\Sigma_-})\in L^2(\Sigma_+;|\mathbf{a}\cdot\hat{n}|d\sigma)\times L^2(\Sigma_-;|\mathbf{a}\cdot\hat{n}|d\sigma)\}.$$

Defining M again as in (8.29), and since now φ is bounded on the bounded sets of L^2, we can now invoke Proposition 4.2 to conclude that $M(u,p)$ is a self-dual Lagrangian on the space $L^2(\Omega)\times L^2(\Omega)$. But in this case φ is coercive because of condition (8.30), and therefore φ^* is bounded on bounded sets. All the hypotheses of Theorem 8.1 are now satisfied, so there exists $\bar{u}\in L^2(\Omega)$ such that $0=M(\bar{u},0)=\inf\{M(u,0); u\in L^2\}$. The rest follows as in the case where $p\ge 2$.

8.3 Initial-value problems driven by a maximal monotone operator

We apply Theorem 8.1 to solve variationally the initial-value problem

$$\begin{cases} -\dot{u}(t) \in T(t, u(t)) \text{ for } t \in [0, T] \\ u(0) = u_0 \end{cases} \tag{8.31}$$

and more generally

$$\begin{cases} -\dot{x}(t) \in A^* T(t, Ax(t)) \text{ for } t \in [0, T] \\ x(0) = x_0, \end{cases} \tag{8.32}$$

where $T(t, \cdot)$ is a time-dependent maximal monotone operator on a Hilbert space E and A is a bounded linear operator from a Banach space X into E. By Theorem 5.1, we can associate to T a time-dependent self-dual Lagrangian L_T on $[0, 1] \times X \times X^*$ in such a way that $L_T(t, x, p) = L_{T_t}(x, p)$, where L_{T_t} is a self-dual Lagrangian such that $\overline{\partial} L_{T_t} = T_t$. If now T satisfies

$$\|T(t, x)\| \leq C(t)(1 + \|x\|) \text{ and } \langle T(t, x), x \rangle \geq \alpha(t) \|x\|^2 - \beta(t) \tag{8.33}$$

for $C(t)$, $\alpha^{-1}(t)$ in $L^\infty([0, 1])$, and $\beta(t) \in L^1([0, 1])$, then for some $K > 0$ we have for any u in $L_X^2[0, 1]$ and p in $L_{X^*}^2[0, 1]$,

$$\mathscr{L}_T(u, p) = \int_0^1 L_T(t, u(t), p(t)) dx \leq K(1 + \|u\|_2^2 + \|p\|_2^2). \tag{8.34}$$

Variational resolutions for (8.31) and (8.32) can therefore be derived from the following two theorems.

Theorem 8.5. *Consider an evolution triple $X \subset H \subset X^*$, and let B be an automorphism of H whose restriction to X is also an automorphism of X. Suppose ℓ is a $(-B, B)$-self-dual function on $H \times H$ (i.e., $\ell^*(Bx, Bp) = \ell(-Bx, Bp)$) such that*

$$-C \leq \ell(a, b) \leq C(1 + \|a\|_H^2 + \|b\|_H^2) \text{ for all } (a, b) \in H \times H. \tag{8.35}$$

Let L be a time-dependent B-self-dual Lagrangian on $X \times X^*$ that satisfies (8.34). Consider the functional

$$I(u) = \begin{cases} \int_0^T L(t, u(t), -\dot{u}(t)) dt + \ell(u(0), u(T)) & \text{if } u \in \mathscr{X}_{2,2} \\ +\infty & \text{otherwise.} \end{cases}$$

(1) The infimum of I on L_X^2 is then equal to zero and is attained at some v in $\mathscr{X}_{2,2}$. Moreover, if

$$\int_0^T \langle Bv(t), \dot{v}(t) \rangle dt = \frac{1}{2}(\langle Bv(T), v(T) \rangle - \langle Bv(0), v(0) \rangle), \tag{8.36}$$

then

$$-\dot{v}(t) \in \overline{\partial}_B L(t, v(t)) \text{ for almost all } t \in [0, T], \tag{8.37}$$

$$(-Bv(0), Bv(T)) \in \partial \ell(v(0), v(T)). \tag{8.38}$$

(2) In particular, if B is positive and self-adjoint, then for any $v_0 \in H$ there is $v \in \mathscr{X}_{2,2}$ such that $v(0) = v_0$ and which satisfies (8.37) as well as

$$\langle Bv(t), v(t) \rangle = \langle Bv_0, v_0 \rangle - \int_0^t L(s, v(s), \dot{v}(s))ds \quad \text{for all } t \in [0, T]. \qquad (8.39)$$

Proof. (1) This follows immediately from Theorem 8.1 applied to the B-self-dual Lagrangian $\mathscr{L}(u, p) := \int_0^T L(t, u(t), p(t))dt$ on $L_X^2 \times L_{X^*}^2$, the linear unbounded operator $u \to -\dot{u}$, which is skew-adjoint modulo the boundary operator $\mathscr{B} : D(\mathscr{B}) : L_X^2 \to H \times H$ defined by $\mathscr{B}u = (u(0), u(T))$, and where the automorphism is $R(a, b) = (-a, b)$ since for u and v in $\mathscr{X}_{2,2}$ we have $\int_0^T \langle v, \dot{u} \rangle = -\int_0^T \langle \dot{v}, u \rangle + \langle v(T), u(T) \rangle - \langle v(0), u(0) \rangle$. Note also that the subspace $\mathscr{X}_{2,0} = \{u \in \mathscr{X}_{2,2}; u(0) = u(T) = 0\}$ is dense in L_X^2. Moreover, for each $(a, b) \in X \times X$, there is $w \in \mathscr{X}_{2,2}$ such that $w(0) = a$ and $w(T) = b$, namely the linear path $w(t) = \frac{(T-t)}{T}a + \frac{t}{T}b$. Since also X is dense in H and ℓ is continuous on H, all the required hypotheses of Theorem 8.1 are satisfied.

It follows that there exists $v \in L_X^2$ such that $I(v) = 0$. Necessarily, $v \in \mathscr{X}_{2,2}$, and using (8.36), we may write

$$\begin{aligned} 0 = I(v) &= \int_0^T L(t, v(t), -\dot{v}(t))dt + \ell(v(0), v(T)) \\ &= \int_0^T \{L(t, v(t), -\dot{v}(t)) + \langle Bv(t), \dot{v}(t) \rangle\} dt \\ &\quad - \frac{1}{2}\big(\langle Bv(T), v(T) \rangle + \langle Bv(0), v(0) \rangle\big) + \ell(v(0), v(T)). \end{aligned}$$

Since L is B-self-dual, we have $L(t, x, p) \geq \langle Bx, p \rangle$, and since ℓ is a $(-B, B)$-self-dual function on $H \times H$, then $\ell(x, p) \geq \frac{1}{2}\big(\langle Bp, p \rangle - \langle Bx, x \rangle\big)$. It follows that

$$L(t, v(t), -\dot{v}(t)) + \langle Bv(t), \dot{v}(t) \rangle = 0 \text{ for a.e. } t \in [0, T]$$

and $\ell(v(0), v(T)) = \frac{1}{2}\big(\langle Bv(T), v(T) \rangle - \langle Bv(0), v(0) \rangle\big)$. This translates into $-\dot{v}(t) \in \bar{\partial}_B L(t, v(t))$ for almost all $t \in [0, T]$ and $(-Bv(0), Bv(T)) \in \partial \ell(v(0), v(T))$.

(2) This follows by applying the above to the $(-B, B)$-self-dual boundary function

$$\ell(x, p) = \frac{1}{2}\langle Bx, x \rangle - 2\langle v_0, Bx \rangle + \langle Bv_0, v_0 \rangle + \frac{1}{2}\langle Bp, p \rangle.$$

Theorem 8.6. *Let E be a Hilbert space and let L be a time-dependent self-dual Lagrangian on $E \times E$ such that for some $C \in L^\infty([0, T], \mathbf{R})$*

$$L(t, a, b) \leq C(t)(1 + \|a\|^2 + \|b\|^2) \text{ for all } (a, b) \in E \times E. \qquad (8.40)$$

Consider an evolution triple $X \subset H \subset X^*$, and let $A : X \to E$ be a bounded linear operator from X into E such that the operator A^*A is an isomorphism from X onto X^*. Then, for any $p_0(t) \in L_{X^*}^2[0, T]$ and $x_0 \in X$, there exists $x(t) \in \mathscr{X}_{2,2}$ such that

$$\begin{cases} -\dot{x}(t) + p_0(t) \in A^* \overline{\partial} L(t, Ax(t)) \\ \quad\quad\quad x(0) = x_0. \end{cases} \tag{8.41}$$

Proof. Denote by $\Lambda := A^*A$ the isomorphism from X onto X^*, and note that $\pi := A\Lambda^{-1}A^*$ is a projection from E onto its subspace $Y := A(X)$. Since A^* is onto, there exists $q_0(t) \in L_E^2$ such that $A^* q_0(t) = p_0(t)$ for every $t \in [0, 1]$. By Proposition 3.5, the Lagrangian

$$N(t, u, p) = L(t, \pi(u) + \pi^{\perp}(p), \pi(p) + \pi(q_0(t)) + \pi^{\perp}(u)) - \langle u, \pi(q_0(t)) \rangle$$

is also self-dual on $E \times E$ for every $t \in [0, 1]$, and therefore the Lagrangian

$$\mathcal{N}(x, p) := \int_0^T N(t, x(t), p(t)) \, dt$$

is self-dual on $L_E^2 \times L_E^2$. Let now $B := \Lambda^{-1}A^*$ be the bounded linear operator from E onto X, and denote by J the injection of X into X^*. Define on L_E^2 the operator

$$u(t) \to \Gamma u(t) := B^* JB\dot{u}(t) \text{ with domain } D(\Gamma) = A_E^2[0, T]$$

and the boundary operator

$$u(t) \to \big(b_1(u(t)), b_2(u(t))\big) := \big(Bu(0), Bu(T)\big) \text{ from its domain in } L_E^2 \text{ into } H \times H.$$

Note that

$$\begin{aligned} \langle \Gamma u, u \rangle_{L_E^2} &= \int_0^T \langle B^* JB\dot{u}(t), u(t) \rangle_E \, dt = \int_0^T \langle B\dot{u}(t), Bu(t) \rangle_H \, dt \\ &= \frac{1}{2} \big(|Bu(T)|_H^2 - |Bu(0)|_H^2 \big) \end{aligned} \tag{8.42}$$

and that the operator Γ is actually *skew-adjoint modulo the boundary operators* (b_1, b_2). Proposition 8.1 applied to the space $\mathcal{E} = L_E^2$, the operator Γ, the boundary operators $u \to \big(Bu(0), Bu(T)\big)$, and the boundary space $H_1 \times H_2 := H \times H$ yields that the Lagrangian defined by

$$\mathcal{L}(u, p) = \mathcal{N}(u, p - \Gamma u) + \frac{1}{2}|Bu(0)|_H^2 - 2\langle x_0, Bu(0) \rangle_H + |x_0|_H^2 + \frac{1}{2}|Bu(T)|^2$$

if $u \in A_E^2[0, T]$ and $+\infty$ otherwise, is then self-dual on $L_E^2 \times L_E^2$. In view of (8.40), we can use Proposition 6.1 to deduce that the minimum of the completely self-dual functional $I(u) = \mathcal{L}(u, 0)$ on L_E^2 is zero and is attained at some v in L_E^2 such that $I(v) = 0$. Using (8.42), we may write

$$\begin{aligned} 0 &= I(v) \\ &= \int_0^1 \{N(t, v(t), -\Gamma v(t)) + \langle \Gamma v(t), v(t) \rangle_E\} \, dt - \frac{1}{2}\big(|Bv(T)|_H^2 - |Bv(0)|_H^2\big) \end{aligned}$$

$$+\frac{1}{2}|Bv(0)|_H^2 - 2\langle x_0, Bv(0)\rangle_H + |x_0|_H^2 + \frac{1}{2}|Bv(T)|_H^2$$
$$= \int_0^1 \{N(t,v(t),-\Gamma v(t)) + \langle \Gamma v(t), v(t)\rangle_E\}\, dt + |Bu(0) - x_0|_H^2.$$

Now, since $N(t,x,p) \geq \langle x,p\rangle$ for each $(x,p) \in E \times E$, it follows that

$$N(t,v(t),-\Gamma v(t)) + \langle \Gamma v(t), v(t)\rangle = 0 \text{ for a.e. } t \in [0,1]$$

and $Bv(0) = x_0$. This translates into

$$\pi^\perp v(t) - B^* JB\dot{v}(t) + \pi q_0(t) \in \overline{\partial} L(t, \pi(v(t)))$$

for almost all $t \in [0,1]$. By projecting both sides with π and then applying A^* and using that $A^*\pi = A^*$ and $\pi\Gamma = \Gamma$, we get that

$$-A^* B^* JB\dot{v}(t) + A^* q_0(t) \in A^* \overline{\partial} L(t, A\Lambda^{-1} A^* v(t)).$$

Since $A^* B^* JB = JB$, we obtain

$$-JB\dot{v}(t) + A^* q_0(t) \in A^* \overline{\partial} L(t, ABv(t)).$$

Setting $x(t) = Bv(t)$, we get that $-\dot{x}(t) + p_0(t) \in A^* \overline{\partial} L(t, Ax(t))$ and $x(0) = x_0$.

Example 8.2. Variational resolution for a nonpotential evolution equation

Consider the initial-value problem

$$\begin{cases} f_t(t,x) + \operatorname{div}(T(\nabla_x f(t,x))) = g(t,x) & \text{on } [0,1] \times \Omega, \\ f(t,x) = 0 & \text{on } [0,1] \times \partial\Omega, \\ f(0,x) = f_0(x) & \text{on } \partial\Omega, \end{cases} \qquad (8.43)$$

where $g \in L^2([0,1] \times \Omega)$, $f_0 \in H_0^1(\Omega)$ and T is maximal monotone on $\mathbf{R}^n$.

In order to resolve (8.43), we use Theorem 5.1 to associate to the maximal monotone operator T a self-dual Lagrangian L on $\mathbf{R}^n \times \mathbf{R}^n$ such that $\overline{\partial} L = T$. We then consider the self-dual Lagrangian $\mathscr{L}_T$ on $L^2(\Omega; \mathbf{R}^n) \times L^2(\Omega; \mathbf{R}^n)$ via the formula

$$\mathscr{L}_T(u,p) = \int_\Omega L_T(u(x), p(x))\, dx.$$

We shall now be able to solve the evolution equations in (8.43) on the time interval $[0,1]$, by minimizing

$$\mathscr{I}(u) = \int_0^1 \int_\Omega L_T \left(\nabla(-\Delta)^{-1} \nabla^* u, u - \nabla(-\Delta)^{-1} \nabla^* u - \nabla(-\Delta)^{-2} \nabla^* \dot{u} + p_0 \right) dx dt$$
$$- \int_0^1 \int_\Omega u p_0 \, dx dt + \frac{1}{2} \|(-\Delta)^{-1} \nabla^* u(0)\|_2^2 - 2\langle f_0, (-\Delta)^{-1} \nabla^* u(0)\rangle$$
$$+ \|f_0\|_2^2 + \frac{1}{2} \|(-\Delta)^{-1} \nabla^* u(T)\|_2^2$$

on $L^2([0,1] \times \Omega; \mathbf{R}^n)$, and where $p_0 \in L^2([0,1] \times \Omega, \mathbf{R}^n)$ is chosen such that $\nabla^* p_0(t) = g(t)$ for every $t \in [0,1]$. This follows directly from Theorem 8.6 applied again with the spaces $E = L^2(\Omega; \mathbf{R}^n)$, $X = H_0^1(\Omega)$, $X^* = H^{-1}(\Omega)$, and the operator $A : X \rightarrow E$ defined by $Af = \nabla f$. Note that $H_0^1(\Omega) \subset L^2(\Omega) \subset X^* = H^{-1}(\Omega)$ is an evolution triple. The operator B from E into X is then $B := (-\Delta)^{-1} \nabla^*$, and $\Gamma u(t) = B^* J B \dot{u}(t) = \nabla(-\Delta)^{-2} \nabla^* \dot{u}(t)$, the space $\mathscr{E}$ being $L_E^2 = L^2([0,1] \times \Omega, \mathbf{R}^n)$. Note that

$$\mathscr{I}(u) := \int_0^1 \int_\Omega L_T \left(\nabla(-\Delta)^{-1} \nabla^* u, u - \nabla(-\Delta)^{-1} \nabla^* u - \nabla(-\Delta)^{-2} \nabla^* \dot{u} + p_0 \right)$$
$$+ \int_0^1 \int_\Omega \{ u \cdot \nabla(-\Delta)^{-2} \nabla^* \dot{u} - u p_0 \} \, dx dt$$
$$+ \|(-\Delta)^{-1} \nabla^* u(0) - f_0\|_2^2.$$

If now $\mathscr{I}(u) = 0$, then

$$\begin{cases} u - \nabla(-\Delta)^{-1} \nabla^* u - \nabla(-\Delta)^{-2} \nabla^* \dot{u} + p_0 \in \bar{\partial} L_T (\nabla(-\Delta)^{-1} \nabla^* u) \\ (-\Delta)^{-1} \nabla^* u(0) = f_0. \end{cases} \quad (8.44)$$

By taking the divergence on both sides, we get that

$$\begin{cases} (-\Delta)^{-1} \nabla^* \dot{u}(t) + g(x) \in -\mathrm{div} \bar{\partial} L_T (\nabla(-\Delta)^{-1} \nabla^* u(t)) \\ (-\Delta)^{-1} \nabla^* u(0) = f_0. \end{cases} \quad (8.45)$$

It is now clear that $f(t) = (-\Delta)^{-1} \nabla^* u(t)$ is a solution for (8.43).

Remark 8.1. Alternatively, we can replace the latter by a minimization of the functional

$$\mathscr{I}(f, w) = \int_0^1 \int_\Omega L_T \left(\nabla f(t,x), w(t,x) - \nabla(-\Delta)^{-1} \frac{\partial f}{\partial t}(t,x) \right) dx dt$$
$$- \int_0^1 \int_\Omega f(t,x) g(t,x) dx dt$$
$$+ \int_\Omega \left\{ \frac{1}{2} f(0,x)^2 - 2 f_0(x) f(0,x) + f_0(x)^2 + \frac{1}{2} f(T,x)^2 \right\} dx$$

over all possible $f \in L^2([0,1]; H_0^1(\Omega))$ with $\dot{f} \in L^2([0,1]; H^{-1}(\Omega))$ and all $w \in L^2([0,1] \times \Omega; \mathbf{R}^n)$ with $\mathrm{div}\, w(t,x) = g(t,x)$ for a.e. $t \in [0,1]$.

8.4 Lagrangian intersections of convex-concave Hamiltonian systems

Example 8.3. Convex-concave Hamiltonian systems ($k = l = 2$)

The following is an application for the existence of a Hamiltonian path that connects two Lagrangian submanifolds.

Theorem 8.7. *Let E be a Hilbert space and let $\mathscr{H} : [0,T] \times E \times E \to \mathbf{R}$ be a Hamiltonian of the form $\mathscr{H}(t,x_1,x_2) = \varphi_1(t,x_1) - \varphi_2(t,x_2)$, where for each $t \in [0,T]$ the functions $\varphi_1(t,\cdot)$ and $\varphi_2(t,\cdot)$ are convex lower semicontinuous on E satisfying for some $C(t) \in L_\infty^+([0,T])$,*

$$-C(t) \leq \varphi_1(t,x_1) + \varphi_2(t,x_2) \leq C(t)\left(1 + \|x_1\|_E^2 + \|x_2\|_E^\beta\right).$$

Let ψ_1, ψ_2 be convex lower semicontinuous functions on E such that

$$C_i^{-1}(\|s\|^2 - 1) \leq \psi_i(s) \leq C_i(1 + \|s\|^2) \text{ for } s \in E, \ i = 1,2.$$

Then, there exists $(x_1,x_2) \in A_{E \times E}^2([0,T])$ such that for almost all $t \in [0,T]$,

$$-\dot{x}_2(t) \in \partial_1 \mathscr{H}(t,x_1(t),x_2(t)),$$

$$\dot{x}_1(t) \in \partial_2 \mathscr{H}(t,x_1(t),x_2(t)),$$

and satisfying the boundary conditions

$$-A_1 x_1(0) - x_2(0) \in \partial \psi_1(x_1(0))$$

$$-A_2 x_1(T) + x_2(T) \in \partial \psi_2(x_1(T)).$$

The solution can be obtained by minimizing the completely self-dual functional

$$
\begin{aligned}
I(x,y) = \int_0^T & \mathscr{K}(t,x(t),y(t)) + \mathscr{K}^*(t,-\dot{y}(t),-\dot{x}(t))dt \\
& + \psi_1(x(0)) + \psi_1^*(-y(0) - A_1 x(0)) + \psi_2(x(T)) + \psi_2^*(y(T) - A_2 x(T))
\end{aligned}
$$

on the space $A^2([0,T]; E \times E)$, where $\mathscr{K}$ is the convex function

$$\mathscr{K}(t,x,y) = \varphi_1(t,x) + \varphi_2(t,y)$$

on $\mathbf{R}^{2n}$ and $\mathscr{K}^*$ is its Legendre transform for each $t \in [0,T]$.

Proof. Consider on $L^2([0,T]; E \times E)$ the function $\Phi(x,y) = \int_0^T \mathscr{K}(t,x(t),y(t))dt$, which is convex and lower semicontinuous, and note that the operator $\Gamma(p,q) = (\dot{q},\dot{p})$ satisfies:

$$\langle \Gamma(p,q),(p,q)\rangle = 2 \int_0^T (\dot{q}p + \dot{p}q)dt = 2\langle p(T),q(T)\rangle - 2\langle p(0),q(0)\rangle.$$

In other words, Γ is skew-adjoint modulo the boundary pair $(\mathscr{B}, R)$, where $\mathscr{B} : L^2_{E \times E}[0, T] \to E^2 \times E^2$ is defined via the formula

$$\mathscr{B}(p, q) = (\mathscr{B}^1(p, q), \mathscr{B}^2(p, q)) = (p(T), q(T)), (p(0), q(0))$$

and R is the automorphism on $E^2 \times E^2$ defined by

$$R((a, b), (c, d)) = ((-b, -a), (d, c)).$$

All the hypotheses of Theorem 8.2 are satisfied, and a solution can then be obtained by minimizing the completely self-dual functional I, whose infimum is attained and is equal to zero.

8.5 Parabolic equations with evolving state-boundary conditions

When dealing with general parabolic equations of type (8.3), one can impose suitable conditions on L, Γ_t, and the boundary triplets $(H_t, R_t, \mathscr{B}_t)$ to ensure that the function L_{Γ_t, ℓ_t} given by Theorem 4.2 is a time-dependent self-dual Lagrangian. One can then use Proposition 7.1 to lift it to a partially self-dual Lagrangian $\mathscr{L}$ on the path space $A^2_H[0, T]$. However, one cannot apply the variational principle of Proposition 7.1 because $\mathscr{L}_{\Gamma_t, \ell_t}$ is not bounded in the first variable.

    Now the only case where such a boundedness condition can be relaxed is when the Lagrangian L is autonomous, or if it is of the form $L(t, x, p) = e^{2wt}L(e^{-wt}x, e^{-wt}p)$ (See Theorem 7.5). This seriously restricts the applicability of our approach to problems involving time-dependent state-boundary conditions. However, one can still show the following result.

Theorem 8.8. *Let L be an autonomous self-dual Lagrangian on a Hilbert space $H \times H$ that is continuous and uniformly convex in the first variable. Let $\Gamma : D(\Gamma) \subset H \to H$ be a skew-adjoint operator modulo a boundary triplet $(E, R, \mathscr{B})$ and let ℓ be an R-self-dual function on E such that, for some $C > 0$, $\ell(s) \leq C(1 + \|s\|^2)$ for all $s \in E$.*
Then, for any $x_0 \in \mathrm{Dom}(\bar{\partial}L) \cap D(\Gamma)$ such that $R\mathscr{B}x_0 \in \partial\ell(\mathscr{B}x_0)$ and for any $\omega \in \mathbf{R}$, the functional

$$I(u) = \int_0^T e^{2\omega t} \{L(u(t), \Gamma u(t) - \omega u(t) - \dot{u}(t)) + \ell(\mathscr{B}u(t))\} dt$$

$$+ \frac{1}{2}\|u(0)\|^2 - 2\langle x_0, u(0)\rangle + \|x_0\|^2 + \frac{1}{2}\|e^{\omega T}u(T)\|^2$$

attains its minimum at a path $x \in A^2_H$ such that $I(x) = \inf_{u \in A^2_H} I(u) = 0$ and therefore $x(t)$ solves the equation

$$\begin{cases} -\dot{x}(t) + \Gamma x(t) - \omega x(t) \in \bar{\partial} L(x(t)) & \text{for all } t \in [0,T], \\ x(0) = x_0, \\ R\mathcal{B}(x(t)) \in \partial \ell(\mathcal{B}x(t)) & \text{for all } t \in [0,T]. \end{cases} \tag{8.46}$$

Proof. By Theorem 4.2, the Lagrangian

$$L_{\Gamma,\ell}(x,p) = \begin{cases} L(x, \Gamma x + p) + \ell(\mathcal{B}x) & \text{if } x \in D(\Gamma) \cap D(\mathcal{B}) \\ +\infty & \text{if } x \notin D(\Gamma) \cap D(\mathcal{B}) \end{cases} \tag{8.47}$$

is also self-dual on $H \times H$ and satisfies all the hypotheses of Theorem 7.5. Since $x_0 \in \text{Dom}(\bar{\partial}L) \cap D(\Gamma)$ and $\mathcal{B}x_0 \in \mathcal{C}_R\ell$, it follows that $x_0 \in \text{Dom}(\bar{\partial}L_{\Gamma,\ell})$ and Theorem 7.5 applies to yield the existence of $x \in A_H^2$ such that $I(x) = \inf_{u \in A_H^2} I(u) = 0$. Now write that

$$\begin{aligned} 0 &= I(x) \\ &= \int_0^T e^{2\omega t} \{ L(x(t), \Gamma x(t) - \omega x(t) - \dot{x}(t)) - \langle x(t), \Gamma x(t) - \omega x(t) - \dot{x}(t) \rangle \} dt \\ &\quad + \int_0^T e^{2\omega t} \left\{ \ell(\mathcal{B}x(t)) - \frac{1}{2} \langle \mathcal{B}x(t), R\mathcal{B}x(t) \rangle \right\} dt \\ &\quad + \| x(0) - x_0 \|^2. \end{aligned}$$

The conclusion follows since each one of the three terms above is nonnegative .

Parabolic equations driven by first-order operators

We now apply the results of the last section to the particular class of self-dual Lagrangians of the form $L(x,p) = \varphi(x) + \varphi^*(\Gamma x + p)$ to obtain variational formulations and proofs of existence for parabolic equations of the form

$$\begin{cases} -\dot{x}(t) + \Gamma x(t) - \omega x(t) \in \partial \varphi(x(t)), \\ x(0) = x_0, \\ \mathcal{B}_1(x(t)) = \mathcal{B}_1(x_0), \end{cases} \tag{8.48}$$

where Γ is a skew-adjoint operator modulo a "trace boundary operator" $\mathcal{B} = (\mathcal{B}_1, \mathcal{B}_2)$ into a space $H_1 \times H_2$. In other words, we are assuming that

$$\langle \Gamma x, x \rangle = \frac{1}{2}(\|\mathcal{B}_2 x\|^2 - \|\mathcal{B}_1 x\|^2).$$

Corollary 8.1. *Let $X \subset H \subset X^*$ be an evolution triple and let $\Gamma : D(\Gamma) \subset X \to X^*$ be a skew-adjoint operator modulo a trace boundary operator $\mathcal{B} = (\mathcal{B}_1, \mathcal{B}_2)$: $D(\mathcal{B}) \subset X \to H_1 \times H_2$. Let $\varphi : X \to \mathbf{R}$ be a convex lower semicontinuous function on X that is bounded on the bounded sets of X and also coercive on X. Assume that*

$$x_0 \in D(\Gamma) \cap D(\mathscr{B}) \text{ and } \partial\varphi(x_0) \cap H \text{ is nonempty.} \qquad (8.49)$$

Then, for all $\omega \in \mathbf{R}$ and all $T > 0$, there exists $x(t) \in A_H^2([0,T])$, which solves (8.48). It is obtained by minimizing over A_H^2 the completely self-dual functional

$$I(u) = \int_0^T e^{2\omega't} \left\{ \psi(e^{-\omega't}u(t)) + \psi^*(e^{-\omega't}(\Gamma u(t) - \dot{u}(t))) \right\} dt$$

$$+ \int_0^T \left\{ \frac{1}{2} \|\mathscr{B}_1(u(t))\|_{H_1}^2 - 2\langle e^{\omega't}\mathscr{B}_1(u(t)), \mathscr{B}_1(x_0)\rangle_{H_1} \right.$$

$$\left. + e^{2\omega't} \|\mathscr{B}_1(x_0)\|_{H_1}^2 + \frac{1}{2} \|\mathscr{B}_2(u(t))\|_{H_2}^2 \right\} dt$$

$$+ \frac{1}{2} \|u(0)\|_H^2 - 2\langle u(0), x_0\rangle_H + \|x_0\|_H^2 + \frac{1}{2} \|u(T)\|_H^2,$$

where $\psi(x) = \varphi(x) + \frac{\|x\|^2}{2}$ and $\omega' = \omega - 1$. The minimum of I is then zero and is attained at a path $y(t)$ such that $x(t) = e^{-\omega't}y(t)$ is a solution of (8.48).

Proof. The function $\psi : X \to \mathbf{R}$ by $\psi(x) = \varphi(x) + \frac{\|x\|^2}{2}$ is clearly uniformly convex. Set $X_1 = D(\Gamma) \cap D(\mathscr{B})$, and use Proposition 4.2 to deduce that the Lagrangian

$$M(x,p) := \psi(x) + \psi^*(\Gamma x + p)$$

$$+ \frac{1}{2} \|\mathscr{B}_1(x)\|_{H_1}^2 - 2\langle \mathscr{B}_1(x), \mathscr{B}_1(x_0)\rangle_{H_1} + \|\mathscr{B}_1(x_0)\|_{H_1}^2 + \frac{1}{2} \|\mathscr{B}_2(x)\|_{H_2}^2$$

if $x \in X_1$ and $+\infty$ elsewhere is a self-dual Lagrangian on $X \times X^*$. Indeed, in this case, $R(h_1, h_2) = (-h_1, h_2)$ on $H_1 \times H_2$ and

$$\ell(h_1, h_2) = \frac{1}{2} \|h_1\|_{H_1}^2 - 2\langle h_1, \mathscr{B}_1(x_0)\rangle_{H_1} + \|\mathscr{B}_1(x_0)\|_{H_1}^2 + \frac{1}{2} \|h_2\|_{H_2}^2.$$

The coercivity condition on ψ ensures – via Lemma 3.4 – that $M(x,p)$ lifts to a self-dual Lagrangian on $H \times H$ that is uniformly convex in the first variable. It is easy to check that all the conditions of Theorem 8.8 are satisfied by $M(x,p)$. Note now that the conditions on x_0 ensure that $x_0 \in \text{Dom}(\overline{\partial}M)$. Indeed, condition (8.49) and the definition of ℓ yield that $x_0 \in D(\Gamma) \cap D(\mathscr{B})$ and that $(-\mathscr{B}_1(x_0), \mathscr{B}_2(x_0)) \in \partial\ell(\mathscr{B}_1(x_0), \mathscr{B}_2(x_0))$. It follows that there exists $y \in A_H^2([0,T])$ such that $I(y) = 0$. We then have

$$0 = \int_0^T e^{2\omega't} \left\{ \psi(e^{-\omega't}y(t)) + \psi^*(e^{-\omega't}(\Gamma y(t) - \dot{y}(t))) - \langle y(t), \Gamma y(t) - \dot{y}(t)\rangle \right\} dt$$

$$+ \int_0^T \|\mathscr{B}_1(y(t)) - e^{\omega't}\mathscr{B}_1(x_0)\|_{H_1}^2 dt$$

$$+ \|y(0) - x_0\|_H^2.$$

It follows that $x(t) = e^{-\omega't}y(t)$ satisfies

$$-\dot{x}(t) + \Gamma x(t) \in \partial \psi(x(t)) + (\omega - 1)x(t) \text{ for } t \in [0,T]$$
$$\mathcal{B}_1(x(t)) = \mathcal{B}_1(x_0) \text{ for } t \in [0,T]$$
$$x(0) = x_0.$$

Since $\partial \psi(x) = \partial \varphi(x) + x$, we have that $x(t)$ solves (8.48).

Example 8.4. Evolutions driven by transport operators

Consider the following evolution equation on $[0,T] \times \Omega$:

$$\begin{cases} -\frac{\partial u}{\partial t} + \mathbf{a}(x) \cdot \nabla u = \frac{1}{2}a_0(x)u + u|u|^{p-2} + \omega u, \\ \quad u(0,x) = u_0(x) \quad \text{on } \Omega, \\ \quad u(t,x) = u_0(x) \quad \text{on } [0,T] \times \Sigma_+. \end{cases} \tag{8.50}$$

We assume again that the domain Ω and the vector field $\mathbf{a}(\cdot)$ satisfy all the assumptions in Lemma 4.1.

Corollary 8.2. *Let $p > 1$, $f \in L^2(\Omega)$, and $a_0 \in L^\infty(\Omega)$. Then, for any $\omega \in \mathbf{R}$ and $u_0 \in L^\infty(\Omega) \cap H^1(\Omega)$, there exists $\bar{u} \in A^2_{L^2(\Omega)}([0,T])$ that solves equation (8.50).*

Proof. We distinguish two cases.

Case 1: $p \geq 2$. We then take $X = L^p(\Omega)$, $H = L^2(\Omega)$ since again the operator $\Gamma u = \mathbf{a} \cdot \nabla u + \frac{1}{2}(\nabla \cdot \mathbf{a})u$ is skew-adjoint modulo the trace boundary, $\mathcal{B}u = (u|_{\Sigma_+}, u|_{\Sigma_-})$.

Case 2: $1 < p < 2$. The space is then $X = H = L^2(\Omega)$.

In both cases, pick $K > 0$ such that $\nabla \cdot \mathbf{a}(x) + a_0(x) + K \geq 1$ for all $x \in \Omega$, and define the function $\varphi : X \to \mathbf{R}$ by

$$\varphi(u) := \frac{1}{p} \int_\Omega |u(x)|^p dx + \frac{1}{4} \int_\Omega (\nabla \cdot \mathbf{a}(x) + a_0(x) + K)|u(x)|^2 dx.$$

Then, φ is a convex lower semicontinuous function that is bounded on bounded sets of X and also is coercive on X. Since $u_0 \in L^\infty(\Omega) \cap H^1(\Omega)$, $\partial \varphi(u_0)$ is therefore nonempty and $u_0 \in D(\Gamma) \cap D(\mathcal{B})$.

So by Corollary 8.1, there exists $\bar{u} \in A^2_H([0,T])$ such that

$$-\ddot{\bar{u}}(t) + \Gamma \bar{u}(t) \in \partial \varphi(\bar{u}(t)) + \left(\omega - \frac{K}{2}\right)\bar{u}(t) \quad \text{for} \quad t \in [0,T]$$

$$\mathcal{B}_1(\bar{u}(t)) = \mathcal{B}_1(u_0) \quad \text{for} \quad t \in [0,T]$$

$$\bar{u}(0) = u_0,$$

and this is precisely equation (8.50).

8.6 Multiparameter evolutions

Let H be a Hilbert space and let $L : [0,T] \times H \times H \to \mathbf{R}$ be a time-dependent self-dual Lagrangian such that for every $p \in H$ and $t \in [0,T]$ the map

$$x \mapsto L(t,x,p) \quad \text{is bounded on the bounded sets of } H. \tag{8.51}$$

For every $x_0 \in H$, the Lagrangian defined on $L_H^2[0,T] \times L_H^2[0,T]$ by

$$\mathcal{L}(x,p) = \int_0^T L(t,x(t),-\dot{x}(t)+p(t))dt$$
$$+ \frac{1}{2}\|x(0)\|_H^2 + 2\langle x_0, x(0)\rangle + \|x_0\|_H^2 + \frac{1}{2}\|x(T)\|_H^2$$

if $x \in A_{H'}^2$ and $+\infty$ otherwise is then itself a self-dual Lagrangian on $L_H^2[0,T] \times L_H^2[0,T]$. Setting now $H' := L_H^2[0,T]$ as a state space, we can then lift the Lagrangian $\mathcal{L}$ to a new path space $L_{H'}^2[0,S]$ and obtain a new completely self-dual functional

$$\mathcal{I}(x) := \int_0^S \mathcal{L}\left(x(s), -\frac{dx}{ds}(s)\right)ds + \ell'(x(0),x(S)),$$

that we can minimize on $A_{H'}^2[0,S]$. Here is the main result of this section.

Theorem 8.9. *Let H be a Hilbert space and let $L : H \times H \to \mathbf{R} \cup \{+\infty\}$ be a self-dual Lagrangian on $H \times H$ that is uniformly convex in the first variable. For $x_0 \in H$, we consider on $L^2([0,S];L_H^2[0,T])$ the functional defined as*

$$I(x) = \int_0^S \int_0^T L\left(x(s,t), -\frac{\partial x}{\partial t}(s,t) - \frac{\partial x}{\partial s}(s,t)\right)dtds$$
$$+ \int_0^S \left(\frac{1}{2}\|x(s,0)\|_H^2 - 2\langle x(s,0),x_0\rangle + \|x_0\|_H^2 + \frac{1}{2}\|x(s,T)\|_H^2\right)ds$$
$$+ \int_0^T \left(\frac{1}{2}\|x(0,t)\|_H^2 - 2\langle x(0,t),x_0\rangle + \|x_0\|_H^2 + \frac{1}{2}\|x(S,t)\|_H^2\right)dt \tag{8.52}$$

on the space $A^2([0,S];A_H^2[0,T])$ and $+\infty$ elsewhere.

If $x_0 \in \mathrm{Dom}(\bar{\partial}L)$, then I is a completely self-dual functional and there exists $\hat{x} \in A^2([0,S];L_H^2[0,T])$ such that $\hat{x}(s,\cdot) \in A_H^2[0,T]$ for almost all $s \in [0,S]$ and

$$I(\hat{x}) = \inf\{I(x); x \in L^2([0,S];L_H^2[0,T])\} = 0. \tag{8.53}$$

Furthermore, for almost all $(s,t) \in [0,S] \times [0,T]$, we have

$$-\frac{\partial \hat{x}}{\partial t}(s,t) - \frac{\partial \hat{x}}{\partial s}(s,t) \in \bar{\partial}L(\hat{x}(s,t)) \quad a.e. [0,S] \times [0,T], \tag{8.54}$$
$$\hat{x}(0,t) = x_0 \quad a.e. t \in [0,T], \tag{8.55}$$
$$\hat{x}(s,0) = x_0 \quad a.e. s \in [0,S]. \tag{8.56}$$

We first note in the following proposition that if L is finitely valued, then the conclusions of the theorem are easy to establish, even if L is not autonomous. The main difficulty of the proof is to get rid of this boundedness condition.

Proposition 8.1. *Let H be a Hilbert space and let $L : [0,T] \times H \times H \to \mathbf{R}$ be a finite time-dependent self-dual Lagrangian on $H \times H$ that is uniformly convex in the first variable. For $x_0 \in H$, we consider on $L^2([0,S]; L_H^2[0,T])$ the functional defined as*

$$
I(x) = \int_0^S \int_0^T L\left(t, x(s,t), -\frac{\partial x}{\partial t}(s,t) - \frac{\partial x}{\partial s}(s,t)\right) dt\,ds
$$
$$
+ \int_0^S \left(\frac{1}{2}\|x(s,0)\|_H^2 - 2\langle x(s,0), x_0\rangle + \|x_0\|_H^2 + \frac{1}{2}\|x(s,T)\|_H^2\right) ds
$$
$$
+ \int_0^T \left(\frac{1}{2}\|x(0,t)\|_H^2 - 2\langle x(0,t), x_0\rangle + \|x_0\|_H^2 + \frac{1}{2}\|x(S,t)\|_H^2\right) dt \quad (8.57)
$$

on the space $A^2([0,S]; A_H^2[0,T])$ and $+\infty$ elsewhere.

If $x_0 \in \mathrm{Dom}(\bar\partial L)$, then I is a completely self-dual functional and there exists $\hat{x} \in A^2([0,S]; L_H^2[0,T])$ such that $\hat{x}(s, \cdot) \in A_H^2[0,T]$ for almost all $s \in [0,S]$ and

$$
I(\hat{x}) = \inf\{I(x);\, x \in L^2([0,S]; L_H^2[0,T])\} = 0. \tag{8.58}
$$

Furthermore, for almost all $(s,t) \in [0,S] \times [0,T]$, $\hat{x}$ satisfies (8.54), (8.55) and (8.56).

Proof. Indeed, by Theorem 4.3, $\mathscr{L}$ is a self-dual Lagrangian on the space $H' \times H' := L_H^2[0,T] \times L_H^2[0,T]$, and is uniformly convex in the first variable. Since $x_0 \in \mathrm{Dom}(\bar\partial L)$, it is easy to see that $0 \in \mathrm{Dom}(\bar\partial \mathscr{L})$ in path space. Therefore, by Theorem 7.5, we can find an $\hat{x} \in A_{H'}^2([0,S]) = A^2([0,S]; L_H^2[0,T])$ such that

$$
0 = \int_0^S \mathscr{L}(\hat{x}(s), -\dot{\hat{x}}(s))ds + \frac{1}{2}\|\hat{x}(0)\|_{H'}^2 - 2\langle\hat{x}(0), x_0\rangle_{H'} + \|x_0\|_H^2 + \frac{1}{2}\|\hat{x}(S)\|_{H'}^2.
$$

From the definition of $\mathscr{L}$, we get that $\hat{x}(s, \cdot) \in A_H^2([0,T])$ for a.e., $s \in [0,S]$, while satisfying (8.53). We therefore get the following chain of inequalities:

$$
0 = \int_0^S \int_0^T L\left(t, \hat{x}(s,t), -\frac{\partial \hat{x}}{\partial t}(s,t) - \frac{\partial \hat{x}}{\partial s}(s,t)\right) dt\,ds
$$
$$
+ \int_0^S \left(\frac{1}{2}\|\hat{x}(s,0)\|_H^2 - 2\langle\hat{x}(s,0), x_0\rangle_H + \|x_0\|_H^2 + \frac{1}{2}\|\hat{x}(s,T)\|_H^2\right) ds
$$
$$
+ \int_0^T \left(\frac{1}{2}\|\hat{x}(0,t)\|_H^2 - 2\langle\hat{x}(0,t), x_0\rangle_H + \|x_0\|_H^2 + \frac{1}{2}\|\hat{x}(S,t)\|_H^2\right) dt
$$
$$
\geq \int_0^S \int_0^T -\left\langle x(s,t), \frac{\partial \hat{x}}{\partial t}(s,t) + \frac{\partial \hat{x}}{\partial s}(s,t)\right\rangle dt\,ds
$$
$$
+ \int_0^S \left(\frac{1}{2}\|\hat{x}(s,0)\|_H^2 - 2\langle\hat{x}(s,0), x_0\rangle + \|x_0\|_H^2 + \frac{1}{2}\|\hat{x}(s,T)\|_H^2\right) ds
$$
$$
+ \int_0^T \left(\frac{1}{2}\|\hat{x}(0,t)\|_H^2 - 2\langle\hat{x}(0,t), x_0\rangle + \|x_0\|_H^2 + \frac{1}{2}\|\hat{x}(S,t)\|_H^2\right) dt
$$

$$\geq \int_0^T \|\hat{x}(0,t) - x_0\|_H^2 dt + \int_0^S \|\hat{x}(s,0) - x_0\|_H^2 ds \geq 0.$$

This clearly yield equations (8.54), (8.55), and (8.56).

In the next proposition, we do away with the assumption of boundedness of the self-dual Lagrangian L that was used in Proposition 8.1. We first λ-regularize the Lagrangian L, and then derive some uniform bounds to ensure convergence in the proper topology when λ goes to 0. To do this, we need to first state some precise estimates on approximate solutions obtained using inf-convolution. Recall first from Proposition 3.3 that the Lagrangian

$$L_\lambda^1(x,p) := \inf_{z \in H} \left\{ L(z,p) + \frac{1}{2\lambda} \|x - z\|_H^2 \right\} + \frac{\lambda}{2} \|p\|_H^2$$

is self-dual for each $\lambda > 0$.

Lemma 8.1. *Let $L : H \times H \to \mathbf{R}$ be a self-dual Lagrangian that is uniformly convex in the first variable. If $p_0 \in \partial L(x_0)$ and if $\hat{x} \in A_H^2[0,T]$ satisfies*

$$\int_0^T L(\hat{x}(t), -\dot{\hat{x}}(t)) dt + \frac{1}{2} \|\hat{x}(0)\|_H^2 + \|x_0\|_H^2 + \langle \hat{x}(0), x_0 \rangle + \frac{1}{2} \|\hat{x}(T)\|_H^2 = 0,$$

then we have the estimate

$$\int_0^T \|\dot{\hat{x}}(t)\|_H^2 dt \leq T \|p_0\|_H^2. \tag{8.59}$$

Proof. By the uniqueness of the minimizer, $\hat{x}$ is the weak limit in $A_H^2([0,T])$ of the net $(x_\lambda)_\lambda$ in $C^{1,1}([0,T])$, where $-\dot{x}_\lambda(t) \in \partial L_\lambda^1(x_\lambda(t))$ and $x_\lambda(0) = x_0$. By Proposition 7.1, we have that $\|\dot{x}_\lambda(t)\|_H \leq \|\dot{x}_\lambda(0)\|_H$ for all $t \in [0,T]$. Since $(-\dot{x}_\lambda(0), x_0) \in \partial L_\lambda^1(x_0, \dot{x}_\lambda(0))$, we get from Lemma 3.3 that $\|x_\lambda(0)\|_H \leq \|p_0\|_H$ for all $\lambda > 0$. Therefore, letting $\lambda \to 0$ and using the weak lower semicontinuity of the norm, we get that $\int_0^T \|\dot{\hat{x}}(t)\|_H^2 dt \leq T \|p_0\|_H^2$.

Proof of Theorem 8.9. By applying Proposition 8.1 to the regularized Lagrangian L_λ^1, we obtain $\hat{x}_\lambda \in A^2([0,S]; L_H^2[0,T])$ satisfying for all $(s,t) \in [0,S] \times [0,T]$

$$-\frac{d\hat{x}_\lambda}{dt}(s,t) - \frac{d\hat{x}_\lambda}{ds}(s,t) \in \overline{\partial} L_\lambda^1(\hat{x}_\lambda(s,t)), \tag{8.60}$$

$$\hat{x}_\lambda(0,t) = x_0 \ \forall t \in [0,T], \tag{8.61}$$

$$\hat{x}_\lambda(s,0) = x_0 \ \forall s \in [0,S], \tag{8.62}$$

and

$$0 = \int_0^S \int_0^T L_\lambda^1 \left(\hat{x}_\lambda(s,t), -\frac{d\hat{x}_\lambda}{dt}(s,t) + \frac{d\hat{x}_\lambda}{ds}(s,t) \right) dt ds$$
$$+ \int_0^S \left(\frac{1}{2} \|\hat{x}_\lambda(s,0)\|_H^2 - 2\langle x_\lambda(s,0), x_0 \rangle + \|x_0\|_H^2 + \frac{1}{2} \|\hat{x}_\lambda(s,T)\|_H^2 \right) ds$$

$$+ \int_0^T \left(\frac{1}{2} \|\hat{x}_\lambda(0,t)\|_H^2 - 2\langle x_\lambda(0,t), x_0 \rangle + \|x_0\|_H^2 + \frac{1}{2} \|\hat{x}_\lambda(S,t)\|_H^2 \right) dt. \quad (8.63)$$

Now consider the self-dual Lagrangian $\mathscr{L}_\lambda^1$ on $L_H^2([0,T])$ defined by

$$\mathscr{L}_\lambda(x,p) := \int_0^T L_\lambda^1 \left(x(t), \frac{dx}{dt}(t) + p(t) \right) dt$$
$$+ \frac{1}{2} \|x(0)\|_H^2 + \frac{1}{2} \|x(T)\|_H^2 - 2\langle x_0, x(0) \rangle + \|x_0\|_H^2$$

if $x \in A_H^2([0,T])$ and $+\infty$ elsewhere. Let $\hat{\mathscr{X}}_\lambda : [0,S] \to L_H^2[0,T]$ be the map $s \mapsto \hat{x}_\lambda(s,\cdot) \in L_H^2[0,T]$, and denote by $\mathscr{X}_0 \in L_H^2[0,T]$ the constant map $t \mapsto x_0$. Then, we infer from (8.63) that $\hat{\mathscr{X}}_\lambda$ is an arc in $A^2([0,S];L_H^2[0,T])$ that satisfies

$$0 = \int_0^S \mathscr{L}_\lambda(\hat{\mathscr{X}}_\lambda(s), -\frac{d\hat{\mathscr{X}}_\lambda}{ds}(s))ds + \frac{1}{2}\left(\|\hat{\mathscr{X}}_\lambda(0)\|_{L_H^2[0,T]}^2 + \|\hat{\mathscr{X}}_\lambda(S)\|_{L_H^2[0,T]}^2 \right)$$
$$-2\langle \mathscr{X}_0, \hat{\mathscr{X}}_\lambda(S) \rangle_{L_H^2[0,T]} + \|\mathscr{X}_0\|_{L_H^2[0,T]}$$

with $\mathscr{X}_0 \in \text{Dom}(\bar{\partial}\mathscr{L}_\lambda)$. Applying Lemma 8.1 to the self-dual Lagrangian $\mathscr{L}_\lambda$ and the Hilbert space $L_H^2[0,T]$ we get that

$$\int_0^S \int_0^T \left\| \frac{d\hat{x}_\lambda(s,t)}{ds} \right\|_H^2 dt ds \leq S \int_0^T \|\mathscr{P}_\lambda(t)\|_H^2 dt,$$

where $\mathscr{P}_\lambda \in L_H^2[0,T]$ is any arc that satisfies $(-\mathscr{P}_\lambda, \mathscr{X}_0) \in \partial\mathscr{L}_\lambda(\mathscr{X}_0, -\mathscr{P}_\lambda)$. Observe that if the point $p_\lambda \in H$ satisfies the equation $(-p_\lambda, x_0) \in \partial L_\lambda(x_0, -p_\lambda)$, then we can just take $\mathscr{P}_\lambda$ to be the constant arc $t \mapsto p_\lambda$. Combining this fact with Lemma 8.1, we obtain that, for all $s \in [0,S]$ and all $\lambda > 0$,

$$\int_0^S \int_0^T \|\frac{d\hat{x}_\lambda(s,t)}{ds}\|_H^2 dt ds \leq ST \|p_0\|_H^2.$$

In deriving the above estimates, we have interpreted $\hat{x}_\lambda(s,t)$ as a map $\hat{\mathscr{X}}_\lambda : [0,S] \to L_H^2[0,T]$. However, we can also view it as a map from $[0,T] \to L_H^2[0,S]$ and run the above argument in this new setting. By doing this, we obtain that for all $\lambda > 0$

$$\int_0^S \int_0^T \left\| \frac{d\hat{x}_\lambda(s,t)}{ds} \right\|_H^2 dt ds + \int_0^S \int_0^T \left\| \frac{d\hat{x}_\lambda(s,t)}{dt} \right\|_H^2 dt ds \leq 2TS \|p_0\|_H^2. \quad (8.64)$$

Now, for any $(v_1(s,t), v_2(s,t))$ satisfying equation (8.60), we can use monotonicity to derive the bound:

$$\frac{d}{dt} \|v_1(s,t) - v_2(s,t)\|_H^2 + \frac{d}{ds} \|v_1(s,t) - v_2(s,t)\|_H^2 \leq 0.$$

So we obtain

$$\int_0^S \|v_1(s,t) - v_2(s,t)\|_H^2 ds + \int_0^T \|v_1(s,t) - v_2(s,t)\|_H^2 dt$$

$$\leq \int_0^S \|v_1(s,0) - v_2(s,0)\|_H^2 ds + \int_0^T \|v_1(0,t) - v_2(0,t)\|_H^2 dt.$$

Now, picking $v_1(s,t) = \hat{x}_\lambda(s,t)$ and $v_2(s,t) = \hat{x}_\lambda(s+h,t)$, we get that

$$\int_0^S \left\|\frac{d\hat{x}_\lambda(s,t)}{ds}\right\|^2 ds + \int_0^T \left\|\frac{d\hat{x}_\lambda(s,t)}{ds}\right\|^2 dt$$

$$\leq \int_0^S \left\|\frac{d\hat{x}_\lambda(s,0)}{ds}\right\|^2 ds + \int_0^T \left\|\frac{d\hat{x}_\lambda(0,t)}{ds}\right\|^2 dt. \quad (8.65)$$

Setting $s = 0$ in equation (8.60), we get that for all $t \in [0,T]$

$$-\left(\frac{d\hat{x}_\lambda}{dt}(0,t) + \frac{d\hat{x}_\lambda}{ds}(0,t)\right) \in \bar{\partial}L_\lambda(x_0).$$

Therefore, by Proposition 3.3, we have that for all $t \in [0,T]$ and $\lambda > 0$,

$$\left\|\frac{d\hat{x}_\lambda}{dt}(0,t) + \frac{d\hat{x}_\lambda}{ds}(0,t)\right\|_H \leq \|p_0\|_H.$$

Observe that if we take $v_2(s,t) = \hat{x}_\lambda(s,t+h)$, we can use the same argument as above to get that for all $s \in [0,S]$,

$$\left\|\frac{d\hat{x}_\lambda}{dt}(s,0) + \frac{d\hat{x}_\lambda}{ds}(s,0)\right\|_H \leq \|p_0\|_H.$$

Therefore, for all $s \in [0,S], t \in [0,T]$, and $\lambda > 0$:

$$\left\|\frac{d\hat{x}_\lambda}{dt}(0,t) + \frac{d\hat{x}_\lambda}{ds}(0,t)\right\|_H + \left\|\frac{d\hat{x}_\lambda}{dt}(s,0) + \frac{d\hat{x}_\lambda}{ds}(s,0)\right\|_H \leq 2\|p_0\|_H. \quad (8.66)$$

Combining (8.66), (8.65), and (8.64), we get that

$$\int_0^S \int_0^T \left\|\frac{d\hat{x}_\lambda}{ds}(s,t)\right\|_H^2 + \left\|\frac{d\hat{x}_\lambda}{dt}(s,t)\right\|_H^2 dt ds \leq C \quad (8.67)$$

for some constant independent of λ.

If $J_\lambda(x,p)$ is such that $L_\lambda(x,p) = L(J_\lambda(x,p),p) + \frac{\lambda}{2}\|p\|_H^2$, then setting $v_\lambda(s,t) := J_\lambda\left(\hat{x}_\lambda(s,t), \frac{d\hat{x}_\lambda}{dt}(s,t) + \frac{d\hat{x}_\lambda}{ds}(s,t)\right)$, we can deduce from equation (8.60) that

$$-\frac{d\hat{x}_\lambda}{dt}(s,t) - \frac{d\hat{x}_\lambda}{ds}(s,t) = \frac{\hat{x}_\lambda(s,t) - v_\lambda(s,t)}{\lambda}.$$

The estimate given by equation (8.67) then implies

$$\lim_{\lambda \to 0} \int_0^T \int_0^S \|\hat{x}_\lambda(s,t) - v_\lambda(s,t)\|_H^2 ds dt = 0.$$

Therefore, combining this with (8.67) we obtain – modulo passing to a subsequence –

$$\hat{x}_\lambda \rightharpoonup \hat{x} \text{ in } A^2([0,S]; L^2_H[0,T]), \tag{8.68}$$
$$\hat{x}_\lambda \rightharpoonup \hat{x} \text{ in } A^2([0,T]; L^2_H[0,S]), \tag{8.69}$$
$$v_\lambda \rightharpoonup \hat{x} \text{ in } L^2_H([0,S] \times [0,T]). \tag{8.70}$$

Write (8.63) in the form

$$0 = \int_0^S \int_0^T L\left(v_\lambda(s,t), -\frac{d\hat{x}_\lambda}{dt}(s,t) - \frac{d\hat{x}_\lambda}{ds}(s,t)\right) + \frac{\lambda}{2}\left\|\frac{d\hat{x}_\lambda}{dt}(s,t) + \frac{d\hat{x}_\lambda}{ds}(s,t)\right\|^2_H dt ds$$

$$+ \int_0^S \frac{1}{2}\|\hat{x}_\lambda(s,0)\|^2_H - 2\langle x_\lambda(s,0), x_0\rangle + \|x_0\|^2_H + \frac{1}{2}\|\hat{x}_\lambda(s,T)\|^2_H ds$$

$$+ \int_0^T \frac{1}{2}\|\hat{x}_\lambda(0,t)\|^2_H - 2\langle x_\lambda(0,t), x_0\rangle + \|x_0\|^2_H + \frac{1}{2}\|\hat{x}_\lambda(S,t)\|^2_H dt$$

and taking $\lambda \to 0$, using the convergence results in (8.68) in conjunction with lower semicontinuity, we get

$$0 \geq \int_0^S \int_0^T L\left(\hat{x}(s,t), -\frac{\partial\hat{x}}{\partial t}(s,t) - \frac{\partial\hat{x}}{\partial s}(s,t)\right) dt ds$$

$$+ \int_0^S \frac{1}{2}\|\hat{x}(s,0)\|^2_H - 2\langle\hat{x}(s,0), x_0\rangle + \|x_0\|^2_H + \frac{1}{2}\|\hat{x}(s,T)\|^2_H ds$$

$$+ \int_0^T \frac{1}{2}\|\hat{x}(0,t)\|^2_H - 2\langle\hat{x}(0,t), x_0\rangle + \|x_0\|^2_H + \frac{1}{2}\|\hat{x}(S,t)\|^2_H dt \geq 0.$$

The rest is now straightforward.

Clearly, this argument can be extended to obtain N-parameter gradient flow. We state the result without proof.

Corollary 8.3. *Let $L : H \times H \to \mathbf{R} \cup \{+\infty\}$ be a self-dual Lagrangian that is uniformly convex in the first variable and let $u_0 \in \mathrm{Dom}(\overline{\partial}L)$. Then, for all $T_1 \geq T_2.. \geq T_N > 0$, there exists $u \in L^2_H\left(\prod_{j=0}^N [0,T_j]\right)$ such that $\frac{\partial u}{\partial t_j} \in L^2_H\left(\prod_{j=0}^N [0,T_j]\right)$ for all $j = 1,...,N$, and satisfies the differential equation*

$$-\sum_{j=1}^N \frac{\partial u}{\partial t_j}(t_1,...,t_N) \in \overline{\partial}L(u(t_1,..,t_N))$$

with boundary data $u(t_1,...,t_N) = u_0$ if one of the $t_j = 0$.

We conclude this chapter with some remarks.

Remark 8.2. Let $u : [0,T] \to H$ be the one-parameter gradient flow associated to the self-dual Lagrangian L, namely

$$-\frac{du}{dt}(t) \in \overline{\partial}L(u(t)) \text{ and } u(0) = u_0.$$

If we make the change of variables $v(s',t') = u(\frac{s'+t'}{2})$, then $v(\cdot,\cdot)$ obviously solves (8.54), with the boundary condition $v(s',t') = u_0$ on the hyperplane $s' = -t'$. In comparison, Theorem 8.9 above yields a solution u for (8.54) with a boundary condition that is prescribed on two hyperplanes, namely $u(0,t) = u(s,0) = u_0$ for all $(s,t) \in [0,S] \times [0,T]$.

Remark 8.3. Let now $u : [0,\infty) \times [0,\infty) \times [0,\infty) \to H$ be a solution for the three-parameter self-dual flow.

$$-\left(\frac{\partial u}{\partial r} + \frac{\partial u}{\partial s} + \frac{\partial u}{\partial t}\right)(r,s,t) \in \bar{\partial}L(u(r,s,t))$$

$$u(0,s,t) = u(r,0,t) = u(r,s,0) = u_0.$$

With the change of variable $v(r',s',t') = u(\frac{s'+r'}{2}, \frac{t'+r'}{2}, \frac{s'+t'}{2})$, $v(r',s',t')$ again solves the differential equation

$$-\frac{\partial v}{\partial r'} - \frac{\partial v}{\partial s'} - \frac{\partial v}{\partial t'} \in \bar{\partial}L(u)$$

on the domain

$$D = \{(r',s',t') \mid s' \geq -r',\ r' \geq -t',\ s' \geq -t'\}$$

with boundary conditions

$$v(r',s',t') = u_0 \text{ if } s' = -r' \text{ or } r' = -t' \text{ or } s' = -t'.$$

Looking now at (r',s') as "state variables" and t' as the time variable, we see that at any given time t', $v(r',s',t')$ solves the equation on $\{(r',s') \mid s' \geq -r',\ r' \geq -t',\ s' \geq -t'\}$ with $v = u_0$ on the boundary of this domain. This essentially describes a simple PDE with a time-evolving boundary.

Exercises 8. B.

1. Establish the N-parameter version of Theorem (i.e., Corollary 8.3).
2. Show that in the case of a gradient flow of a convex energy – as opposed to a more general Lagrangian – one can use the method of characteristics to obtain the result in Remark 8.2 even for nonconstant data:

$$x_0(r) := \hat{x}(r,0) \text{ if } r \geq 0 \text{ and } x_0(r) := \hat{x}(0,-r) \text{ if } r \leq 0.$$

Verify that the solution can be obtained by

$$\hat{x}(t,s) := T\left(\frac{1}{2}(t+s) - \frac{1}{2}|s-t|x_0\left(\frac{1}{2}(s-t)\right)\right),$$

where the semigroup $T(t,x_0) := u(t)$ is the one associated to the gradient flow of φ starting at x_0.

Further comments

The material for this chapter is taken from Ghoussoub [58]. The particular case where the operator is skew-adjoint modulo a Dirichlet-type boundary was given in [68]. Lagrangian intersections of Hamiltonian systems driven by convex-concave potentials were studied by Ghoussoub and Tzou [69]. The iterative properties of self-dual Lagrangians, and their applications to multiparameter evolutions were also studied in [69].

Chapter 9
Direct Sum of Completely Self-dual Functionals

If $\Gamma : X \to X^*$ is an invertible skew-adjoint operator on a reflexive Banach space X and L is a self-dual Lagrangian on $X \times X^*$, then we have seen in Chapter 3 that $M(x,p) = L(x + \Gamma^{-1}p, \Gamma x)$ is also a self-dual Lagrangian. By minimizing over X the completely self-dual functional $I(x) = M(x,0) = L(x, \Gamma x)$ with $L(x,p) = \varphi(x) + \varphi^*(p)$, one can then find solutions of $\Gamma x \in \partial\varphi(x)$ as long as φ is convex lower semicontinuous and bounded above on the unit ball of X. In other words, the theory of self-dual Lagrangians readily implies that if the linear system $\Gamma x = p$ is uniquely solvable, then the semilinear system $\Gamma x \in \partial\varphi(x)$ is also solvable for slowly growing convex nonlinearities φ.

This chapter deals with an extension of the above observation to systems of equations. Namely, if $\Gamma_i : Z \to X_i^*$ are linear operators from a Banach space Z into reflexive spaces X_i^* for $i = 1, ..., n$, in such a way that the system of linear equations

$$\Gamma_i x = p_i \text{ for } i = 1, ..., n$$

can be uniquely solved, then one can also solve uniquely the semilinear system

$$\Gamma_i x \in \partial\varphi_i(A_i x) \quad \text{for } i = 1, ..., n,$$

provided $A_i : Z \to X_i$ are bounded linear operators that satisfy the identity

$$\sum_{i=1}^{n} \langle A_i x, \Gamma_i x \rangle = 0 \text{ for all } x \in Z,$$

and the φ_i's are slowly growing convex nonlinearities. This is done by minimizing the functional

$$I(z) = \sum_{i=1}^{n} \varphi_i(A_i z) + \varphi_i^*(\Gamma_i z),$$

which is then completely self-dual on Z. This result is applied to derive variational resolutions to various evolution equations.

N. Ghoussoub, *Self-dual Partial Differential Systems and Their Variational Principles*, Springer Monographs in Mathematics, DOI 10.1007/978-0-387-84897-6_9, © Springer Science+Business Media, LLC 2009

9.1 Self-dual systems of equations

Consider $(n+1)$ reflexive Banach spaces $Z, X_1, X_2, ..., X_n$, and bounded linear operators $(A_i, \Gamma_i) : Z \to X_i \times X_i^*$ for $i = 1, ..., n$. We shall consider a framework where one can associate a self-dual Lagrangian on Z to a given family of self-dual Lagrangians L_i on $X_i \times X_i^*$.

For that, we assume that the linear operator $\Gamma := (\Gamma_1, \Gamma_2, ..., \Gamma_n) : Z \to X_1^* \times X_2^* \times ... \times X_n^*$ is an isomorphism of Z onto $X_1^* \times X_2^* \times ... \times X_n^*$. We can therefore put Z in duality with the space $X_1 \oplus X_2 \oplus ... \oplus X_n$ via the formula

$$\langle z, (p_1, p_2, ..., p_n) \rangle = \sum_{i=1}^{n} \langle \Gamma_i z, p_i \rangle, \tag{9.1}$$

where $z \in Z$ and $(p_1, p_2, ..., p_n) \in X_1 \oplus X_2 \oplus ... \oplus X_n$.

If B_i is a bounded linear operator on X_i for $i = 1, ..., n$, and $B = (B_i)_{i=1}^{n}$ is the product operator on $X_1 \oplus X_2 \oplus ... \oplus X_n$, then we can define an operator $\bar{B} : Z \to Z$ via the formula

$$\langle \bar{B} z, (p_1, p_2, ..., p_n) \rangle = \langle z, (B_1 p_1, B_2 p_2, ..., B_n p_n) \rangle.$$

In other words, $\bar{B} = \Gamma^{-1} B^* \Gamma$, with its coordinates $(\bar{B}_i)_{i=1,...,n}$ being representations of B_i^* on the space Z.

Theorem 9.1. *In the above framework, consider L_i to be a convex lower semicontinuous Lagrangian on $X_i \times X_i^*$ for each $i = 1, ..., n$. Suppose that the following formula holds:*

$$\sum_{i=1}^{n} \langle A_i y, \Gamma_i z \rangle + \langle A_i z, \Gamma_i y \rangle = 0 \text{ for all } y, z \in Z. \tag{9.2}$$

1. *The Lagrangian defined on $Z \times Z^*$ by $M(z, p) = \sum_{i=1}^{n} L_i(A_i z + p_i, \Gamma_i z)$ then has a Legendre transform equal to $M^*(p, z) = \sum_{i=1}^{n} L_i^*(\Gamma_i z, A_i z + p_i)$ on $Z^* \times Z$.*
2. *Assume each L_i is B_i-self-dual for some bounded linear operator B_i on X_i that satisfies the following commutation relations*

$$A_i \bar{B}_i^* = B_i^* A_i \quad \text{for } i = 1, ..., n. \tag{9.3}$$

   Then L is $\bar{B}$-self-dual on $Z \times Z^*$, and the functional $I(z) = \sum_{i=1}^{n} L_i(A_i z, \Gamma_i z)$ is completely self-dual over Z.
3. *If each B_i is onto and if $p \to L_i(p, x_i^*)$ is bounded on the unit ball of X_i for some $x_i^* \in X_i^*$, then the infimum of I over Z is zero and is attained at some $\bar{z} \in Z$. Moreover, if $\bar{z}$ satisfies*

$$\sum_{i=1}^{n} \langle B_i A_i \bar{z}, \Gamma_i \bar{z} \rangle = 0, \tag{9.4}$$

   then it solves the system of equations

$$\Gamma_i \bar{z} \in \overline{\partial}_{B_i} L_i(A_i \bar{z}), \quad i = 1, \dots, n. \tag{9.5}$$

Proof. (1) Fix $((q_1, q_2, \dots, q_n), y) \in (X_1 \oplus X_2 \oplus \dots \oplus X_n) \times Z$ and calculate

$$M^*(q, y) = \sup \Big\{ \sum_{i=1}^{n} \langle \Gamma_i z, q_i \rangle + \langle \Gamma_i y, p_i \rangle - \sum_{i=1}^{n} L_i(A_i z + p_i, \Gamma_i z); z \in Z, p_i \in X_i \Big\}.$$

Setting $x_i = A_i z + p_i \in X_i$, we obtain that

$$M^*(q, y) = \sup \Big\{ \sum_{i=1}^{n} \langle \Gamma_i z, q_i \rangle + \langle \Gamma_i y, x_i - A_i z \rangle - \sum_{i=1}^{n} L_i(x_i, \Gamma_i z) : z \in Z, x_i \in X_i \Big\}$$

$$= \sup \Big\{ \sum_{i=1}^{n} \langle \Gamma_i z, q_i \rangle + \langle \Gamma_i y, x_i \rangle + \langle A_i y, \Gamma_i z \rangle - \sum_{i=1}^{n} L_i(x_i, \Gamma_i z); z \in Z, x_i \in X_i \Big\}$$

$$= \sup \Big\{ \sum_{i=1}^{n} \langle \Gamma_i z, q_i + A_i y \rangle + \langle \Gamma_i y, x_i \rangle - \sum_{i=1}^{n} L_i(x_i, \Gamma_i z); z \in Z, x_i \in X_i \Big\}.$$

Since Z can be identified with $X_1^* \oplus X_2^* \oplus \dots \oplus X_n^*$ via the correspondence $z \to (\Gamma_1 z, \Gamma_2 z, \dots, \Gamma_n z)$, we obtain

$$M^*(q, y) = \sup \Big\{ \sum_{i=1}^{n} \langle z_i, q_i + A_i y \rangle + \langle \Gamma_i y, x_i \rangle - \sum_{i=1}^{n} L_i(x_i, z_i); z_i \in X_i^*, x_i \in X_i \Big\}$$

$$= \sum_{i=1}^{n} \sup \{ \langle z_i, q_i + A_i y \rangle + \langle \Gamma_i y, x_i \rangle - L_i(x_i, z_i); z_i \in X_i^*, x_i \in X_i \}$$

$$= \sum_{i=1}^{n} L_i^*(\Gamma_i y, q_i + A_i y).$$

The proof of (2) is straightforward, while to prove (3) we first use Proposition 6.1 to deduce that the infimum of $I(z) = L(z, 0)$ is zero and that it is attained at some $\bar{z} \in Z$. It follows that

$$0 = I(\bar{z}) = \sum_{i=1}^{n} L_i(A_i \bar{z}, \Gamma_i \bar{z}) = \sum_{i=1}^{n} L_i(A_i \bar{z}, \Gamma_i \bar{z}) - \langle B_i A_i \bar{z}, \Gamma_i \bar{z} \rangle.$$

Since each term is nonnegative , we get that $L_i(A_i \bar{z}, \Gamma_i \bar{z}) - \langle B_i A_i \bar{z}, \Gamma_i \bar{z} \rangle = 0$ for each $i = 1, \dots, n$, and therefore

$$(B_i^* \Gamma_i \bar{z}, B_i A_i \bar{z}) \in \partial L_i(A_i \bar{z}, \Gamma_i \bar{z}), \quad i = 1, \dots, n.$$

The following corollary is now immediate.

Corollary 9.1. *Let $\Gamma_i : Z \to X_i^*$ be bounded linear operators such that the linear system*

$$\Gamma_i z = p_i, \ i = 1, \dots, n \tag{9.6}$$

can be solved uniquely for any $(p_1, p_2, ..., p_n) \in X_1^* \times X_2^* ... \times X_n^*$. *Then, for any convex lower semicontinuous functions* φ_i *on* X_i *verifying for some* $C_i > 0$ *and* $\alpha_i > 0$,

$$-C_i \leq \varphi_i(x) \leq C_i(1 + \|x\|^{\alpha_i}) \text{ for all } x \in X_i \qquad (9.7)$$

and for any bounded linear operators $A_i : Z \to X_i$ *such that (9.2) holds, there exists a solution to the system*

$$\Gamma_i z \in \partial \varphi_i(A_i z), \ i = 1, ..., n. \qquad (9.8)$$

It is obtained as the minimum over Z of the completely self-dual functional

$$I(z) = \sum_{i=1}^{n} \varphi_i(A_i z) + \varphi_i^*(\Gamma_i z).$$

9.2 Lifting self-dual Lagrangians to $A_H^2[0, T]$

We consider again the space $A_H^2 := \{u : [0, T] \to H; \dot{u} \in L_H^2\}$ consisting of all absolutely continuous arcs $u : [0, T] \to H$ equipped with the norm

$$\|u\|_{A_H^2} = \left\{ \left\| \frac{u(0) + u(T)}{2} \right\|_H^2 + \int_0^T \|\dot{u}\|_H^2 \, dt \right\}^{\frac{1}{2}}.$$

The space A_H^2 can be identified with the product space $H \times L_H^2$, in such a way that its dual $(A_H^2)^*$ can also be identified with $H \times L_H^2$ via the formula

$$\left\langle u, (p_1, p_0) \right\rangle_{A_H^2, H \times L_H^2} = \left\langle \frac{u(0) + u(T)}{2}, p_1 \right\rangle + \int_0^T \langle \dot{u}(t), p_0(t) \rangle \, dt,$$

where $u \in A_H^2$ and $(p_1, p_0(t)) \in H \times L_H^2$.

Theorem 9.2. *Suppose L is a time-dependent Lagrangian on* $[0, T] \times H \times H$ *and that* ℓ *is a Lagrangian on* $H \times H$, *and consider the Lagrangian defined on* $A_H^2 \times (A_H^2)^* = A_H^2 \times (H \times L_H^2)$ *by*

$$\mathcal{L}(u, p) = \int_0^T L(t, u(t) + p_0(t), -\dot{u}(t)) \, dt + \ell \left(u(T) - u(0) + p_1, \frac{u(0) + u(T)}{2} \right).$$

1. The Legendre transform of $\mathcal{L}$ *on* $A_H^2 \times (L_H^2 \times H)$ *is given by*

$$\mathcal{L}^*(p, u) = \int_0^T L^* \left(t, -\dot{u}(t), u(t) + p_0(t)\right) dt + \ell^* \left(\frac{u(0) + u(T)}{2}, u(T) - u(0) + p_1 \right).$$

2. If $L(t, \cdot, \cdot)$ *and* ℓ *are B-self-dual Lagrangians on* $H \times H$, *where B is a bounded linear operator on H, then* $\mathcal{L}$ *is a* $\bar{B}$-*self-dual Lagrangian on* $A_H^2 \times (A_H^2)^*$, *where*

$\bar{B}$ is defined on A_H^2 by $(\bar{B}u)(t) = B(u(t))$ and

$$I(u) = \mathscr{L}(u,0) = \int_0^T L\big(t,u(t),-\dot{u}(t)\big)\,dt + \ell\Big(u(T)-u(0),\frac{u(0)+u(T)}{2}\Big)$$

is then a completely self-dual functional on A_H^2.

3. If in addition B is onto and we have the boundedness conditions

$$\int_0^T L(t,x(t),0)\,dt \le C\big(1+\|x\|_{L_H^2}^2\big) \text{ for all } x \in L_H^2, \tag{9.9}$$

ℓ is bounded from below and $\ell(a,0) \le C(\|a\|_H^2+1)$ for all $a \in H$,   (9.10)

then the infimum of the functional I over A_H^2 is zero and is attained at some $v \in A_H^2$. Moreover, if

$$\int_0^T \langle Bv(t),\dot{v}(t)\rangle\,dt + \frac{1}{2}\langle Bv(0)-Bv(T),v(0)+v(T)\rangle = 0, \tag{9.11}$$

then v solves the boundary value problem

$$-\dot{v}(t) \in \bar{\partial}_B L\big(t,v(t)\big) \quad \text{for a.e. } t \in [0,T], \tag{9.12}$$

$$\frac{v(0)+v(T)}{2} \in \bar{\partial}_B \ell\big(v(T)-v(0)\big). \tag{9.13}$$

Proof. (1) follows from the previous proposition. Indeed, in the terminology of Proposition 9.1, we have identified $Z = A_H^2$ with the product $X_1^* \times X_2^*$, where $X_1 = X_1^* = L_H^2$ and $X_2 = X_2^* = H$, via the map

$$u \in Z \longmapsto (\Gamma_1 u, \Gamma_2 u) := \Big(-\dot{u}(t),\frac{u(0)+u(T)}{2}\Big) \in L_H^2 \times H,$$

which is clearly an isomorphism, the inverse map being

$$(x,f(t)) \in H \times L_H^2 \longmapsto x + \frac{1}{2}\Big(\int_0^t f(s)\,ds - \int_t^T f(s)\,ds\Big) \in Z.$$

Define now the maps $A_1 : Z \to X_1$ by $A_1 u = u$ and $A_2 Z \to X_2$ by $A_2(u) = u(T)-u(0)$ in such a way that

$$\langle A_1 u, \Gamma_1 u\rangle + \langle A_2 u, \Gamma_2 u\rangle = -\int_0^T u(t)\dot{u}(t)\,dt + \Big\langle u(T)-u(0),\frac{u(0)+u(T)}{2}\Big\rangle = 0.$$

The proof of (2) is obvious in view of (1), as for (3), we apply Proposition 6.1 to the Lagrangian

$$\mathscr{L}(u,p) = \int_0^T L\big(t,u(t)+p_0(t),-\dot{u}(t)\big)\,dt + \ell\Big(u(T)-u(0)+p_1,\frac{u(0)+u(T)}{2}\Big),$$

which is B-self-dual on A_H^2. Since $I(x) = \mathscr{L}(x,0)$, we obtain $v(t) \in A_H^2$ such that

$$\int_0^T L\big(t,v(t),-\dot{v}(t)\big)\,dt + \ell\left(v(T)-v(0),\frac{v(0)+v(T)}{2}\right) = 0,$$

which gives, as long as (9.11) is satisfied

$$0 = \int_0^T \big[L\big(t,Bv(t),-\dot{v}(t)\big) + \langle Bv(t),\dot{v}(t)\rangle\big]\,dt$$
$$- \int_0^T \langle Bv(t),\dot{v}(t)\rangle\,dt + \ell\left(v(T)-v(0),\frac{v(0)+v(T)}{2}\right)$$
$$= \int_0^T \big[L\big(t,v(t),-\dot{v}(t)\big) + \langle Bv(t),\dot{v}(t)\rangle\big]\,dt - \frac{1}{2}\langle Bv(T)-Bv(0),v(0)+v(T)\rangle$$
$$+\ell\left(v(T)-v(0),\frac{v(0)+v(T)}{2}\right).$$

Since L and ℓ are B-self-dual Lagrangians, $L\big(t,v(t),-\dot{v}(t)\big) + \langle Bv(t),\dot{v}(t)\rangle \geq 0$ and

$$\ell\left(v(T)-v(0),\frac{v(0)+v(T)}{2}\right) - \left\langle Bv(T)-Bv(0),\frac{v(0)+v(T)}{2}\right\rangle \geq 0,$$

which means that $L\big(t,v(t),-\dot{v}(t)\big) + \langle Bv(t),\dot{v}(t)\rangle = 0$ for almost all $t \in [0,T]$, and

$$\ell\left(v(T)-v(0),\frac{v(0)+v(T)}{2}\right) - \left\langle Bv(T)-Bv(0),\frac{v(0)+v(T)}{2}\right\rangle = 0.$$

The result now follows from the above identities and the limiting case in Fenchel-Legendre duality. In other words, v solves the following boundary value problem

$$\big(-B^*\dot{v}(t),Bv(t)\big) \in \partial L\big(t,v(t),\dot{v}(t)\big) \quad \text{for} \quad t \in [0,T],$$
$$\left(B^*\frac{v(0)+v(T)}{2},B(v(T)-v(0))\right) \in \partial\ell\left(v(T)-v(0),\frac{v(0)+v(T)}{2}\right).$$

9.3 Lagrangian intersections via self-duality

Let now $Z = A_H^2 \times A_H^2$, with H being a Hilbert space. We shall identify it with the product space $X_1^* \oplus X_2^* \oplus X_3^*$, with $X_1 = X_1^* = L_H^2 \times L_H^2$, $X_2 = X_2^* = H$, and $X_3 = X_3^* = H$, in the following way. To $(u,v) \in Z$, we associate

$$\big(\Gamma_1(u,v),\Gamma_2(u,v),\Gamma_3(u,v)\big) = \big((\dot{u}(t),\dot{v}(t)),u(0),v(T)\big) \in (L_H^2 \times L_H^2) \times H \times H.$$

The inverse map from $(L_H^2 \times L_H^2) \times H \times H$ to $Z = A_H^2 \times A_H^2$ is then given by the map

$$\big((f(t),g(t)),x,y\big) \longmapsto \left(x + \int_0^t f(s)\,ds,\ y - \int_t^T g(s)\,ds\right).$$

The map $A_1 : Z \to X_1 := L_H^2 \times L_H^2$ is defined as $A_1(u,v) = (v,u)$, while $A_2 : Z \to X_2 := H$ is defined as $A_2(u,v) = v(0)$ and $A_3(u,v) = -u(T)$. It is clear that

$$\sum_{i=1}^{3} \langle A_i(u,v), \Gamma_i(u,v) \rangle = \int_0^T (\dot{u}(t)v(t) + \dot{v}(t)u(t)) dt \tag{9.14}$$

$$+ \langle u(0), v(0) \rangle - \langle u(T), v(T) \rangle = 0.$$

Theorem 9.1 now yields the following.

Theorem 9.3. *Suppose L is a time-dependent Lagrangian on $[0,T] \times H^2 \times H^2$ and that ℓ_1, ℓ_2 are two Lagrangians on $H \times H$. Consider the Lagrangian defined for $((u,v),(p_0^1(t), p_0^2(t), p_1, p_2)) \in Z \times Z^* = (A_H^2 \times A_H^2) \times (L_{H^2}^2 \times H \times H)$ by*

$$\mathscr{L}((u,v),p) = \int_0^T L\big(t, (v(t) + p_0^1(t), u(t) + p_0^2(t)), (\dot{u}(t), \dot{v}(t))\big) dt$$

$$+ \ell_1(v(0) + p_1, u(0)) + \ell_2(-u(T) + p_2, v(T)).$$

1. The Legendre transform of $\mathscr{L}$ on $(A_H^2 \times A_H^2) \times (L_{H^2}^2 \times H^2)$ is then given by

$$\mathscr{L}^*((u,v),p) = \int_0^T L^*\big(t, (\dot{u}(t), \dot{v}(t)), (v(t) + p_0^1(t), u(t) + p_0^2(t))\big) dt$$

$$+ \ell_1^*\big(u(0), v(0) + p_1\big) + \ell_2^*\big(v(T), -u(T) + p_2\big).$$

2. Suppose $L(t, \cdot, \cdot)$ (resp., ℓ_1) (resp., ℓ_2) is a $B_1 \times B_2$-self-dual (resp., B_1-self-dual) (resp., B_2-self-dual) Lagrangian on $H^2 \times H^2$ (resp., on $H \times H$) for some bounded linear operators B_1 and B_2 on H. Then, $\mathscr{L}$ is a $\bar{B}$-self-dual Lagrangian on $Z \times Z^*$, where $\bar{B}$ is defined on Z by $\bar{B}(u,v)(t) = (B_1u(t), B_2v(t))$.

3. If in addition B_1 and B_2 are onto, and if we have the following boundedness conditions: for all $(p_1, p_2) \in L_{H^2}^2$, and all $a \in H$,

$$\int_0^T L(t, (p_1(t), p_2(t)), 0) dt \leq C\big(1 + \|p_1\|_{L_H^2}^2 + \|p_2\|_{L_H^2}^2\big), \tag{9.15}$$

$$\ell_i(a, 0) \leq C_i\big(\|a\|_H^2 + 1\big), \tag{9.16}$$

then the infimum over $Z := A_H^2 \times A_H^2$ of the completely self-dual functional

$$I(u,v) = \mathscr{L}((u,v),0) = \int_0^T L\big(t, (v(t), u(t)), (\dot{u}(t), \dot{v}(t))\big) dt$$

$$+ \ell_1\big(v(0), u(0)\big) + \ell_2\big(-u(T), v(T)\big)$$

is zero and is attained at some $(\bar{u}, \bar{v}) \in A_H^2 \times A_H^2$. Moreover, if

$$\int_0^T \{\langle B_2 v(t), \dot{u}(t) \rangle + \langle B_1 u(t), \dot{v}(t) \rangle\} dt + \langle B_1 u(0), v(0) \rangle - \langle u(T), B_2 v(T) \rangle = 0, \tag{9.17}$$

then (u,v) solves the boundary value problem

$$(\dot{v}(t), \dot{u}(t)) \in \overline{\partial}_{B_1 \times B_2} L(t, u(t), v(t)) \quad \text{for} \quad t \in [0, T], \tag{9.18}$$

$$u(0) \in \overline{\partial}_{B_1} \ell_1(v(0)), \tag{9.19}$$

$$-u(T) \in \overline{\partial}_{B_2} \ell_2(v(T)). \tag{9.20}$$

Example 9.1. Connecting Lagrangian manifolds with orbits of convex-concave Hamiltonian systems

The following result is now a direct application of Theorem 9.3. It proves the existence of a path connecting in prescribed time T two given self-dual Lagrangian submanifolds in H^2 through a self-dual Lagrangian submanifold in phase space H^4. More precisely, it is possible to connect in any time $T > 0$ the graph of $\overline{\partial} L_1$, where L_1 is an antiself-dual Lagrangian on H^2, to the graph of $\overline{\partial} L_2$, where L_2 is a self-dual Lagrangian on H^2, through a path in phase space $(x(t), \dot{x}(t))$ that lies on a given S-self-dual submanifold in H^4, where S is the automorphism $S(x, p) = (p, x)$ (in particular, the graph of the vector field $S\partial \varphi$, where φ is a convex function on H^2).

Theorem 9.4. *Let ψ_1 and ψ_2 be two convex and lower semicontinuous functions on a Hilbert space E, let $A_1, A_2 : E \to E$ be skew-adjoint operators, and consider the following antiself-dual (resp., self-dual) Lagrangian submanifold:*

$$\mathcal{M}_1 := \{(x_1, x_2) \in E^2; -x_2 + A_1 x_1 \in \partial \psi_1(x_1)\}$$

(resp.,

$$\mathcal{M}_2 := \{(x_1, x_2) \in E^2; x_1 + A_2 x_2 \in \partial \psi_2(x_1)\}).$$

Let $\Phi : [0, T] \times E \times E \to \mathbf{R}$ be such that $\Phi(t, \cdot, \cdot)$ is convex and lower semicontinuous for each $t \in [0, T]$, and consider the following evolving S-self-dual Lagrangian submanifold of E^4:

$$\mathcal{M}_3(t) := \{((x_1, x_2), (p_1, p_2)) \in E^2 \times E^2; (p_2, p_1) \in (\partial_1 \Phi(t, x_1, x_2), \partial_2 \Phi(t, x_1, x_2))\}.$$

Assume that for some $C_i > 0$, $i = 1, 2$, we have

$$-\infty < \psi_i(a) \leq C_i \left(\|a\|_H^2 + 1 \right) \text{ for all } a \in E \tag{9.21}$$

and that for some $C(t) \in L_\infty^+([0, T])$ we also have

$$\Phi(t, x_1, x_2) \leq C(t)(1 + \|x_1\|_E^2 + \|x_2\|_E^2). \tag{9.22}$$

Then, there exists $x \in A_{E \times E}^2$ such that

$$x(0) \in \mathcal{M}_1, \ x(T) \in \mathcal{M}_2, \text{ and } (x(t), \dot{x}(t)) \in \mathcal{M}_3(t) \text{ for a.e. } t \in [0, T].$$

Proof. Let $H = E \times E$, and consider the self-dual Lagrangians on $H \times H$ defined by

$$L(t, x, p) := \Phi(t, x) + \Phi^*(t, p),$$

$\ell_1(a_1,a_2) := \psi_1(a_1) + \psi_1^*(A_1a_1 - a_2)$ and $\ell_2(b_1,b_2) = \psi_2(b_1) + \psi_2^*(A_2b_1 + b_2)$,

as well as the completely self-dual functional defined on $A_{E\times E}^2[0,T]$ by

$$I(u,v) = \int_0^T \{\Phi(t,(v(t),u(t))) + \Phi^*(t,\dot{u}(t)),\dot{v}(t)\}\,dt$$
$$+\psi_1(v(0)) + \psi_1^*(A_1v(0) - u(0)) + \psi_2(v(T)) + \psi_2^*(A_2u(T) + v(T)).$$

The conditions on Φ, ψ_1, and ψ_2 ensure that all the hypotheses of Theorem 9.3 are satisfied, and hence there exists a path $(u,v) \in A_{E\times E}^2$ such that $I(u,v) = 0$. It is then easy to check that $(x_1,x_2) := (v,u)$ satisfy

$$\big(\dot{x}_2(t),\dot{x}_1(t)\big) \in \partial\Phi(x_1(t),x_2(t)) \text{ a.e. } t \in [0,T],$$

$$A_1x_1(0) - x_2(0) \in \partial\psi_1(x_1(0)),$$

$$A_2x_2(T) + x_1(T) \in \partial\psi_2(x_1(T)).$$

In other words, $x \in A_{E\times E}^2$ is such that $x(0) \in \mathcal{M}_1$, $x(T) \in \mathcal{M}_1$, and $(x(t),\dot{x}(t)) \in \mathcal{M}_3(t)$ for a.e. $t \in [0,T]$.

It is worth noting that the analogous result for the more standard case of convex Hamiltonians $\mathcal{H}$ (as opposed to convex-concave) will be proved in Chapter 13 but only in the finite-dimensional case.

Corollary 9.2. *Let E be a Hilbert space and $\mathcal{H}(\cdot,\cdot) : E \times E \to \mathbf{R}$ be a Hamiltonian of the form $\mathcal{H}(x_1,x_2) = \varphi_1(x_1) - \varphi_2(x_2)$, where φ_1,φ_2 are convex lower semicontinuous functions satisfying*

$$-C \le \varphi_1(x_1) + \varphi_2(x_2) \le C\big(1 + \|x_1\|_E^\beta + \|x_2\|_E^\beta\big).$$

Furthermore, let ψ_1, ψ_2, A_1, and A_2 be as in Theorem 9.4. Then, there exists $(x_1,x_2) \in A_{E\times E}^\alpha([0,T])$ such that for almost all $t \in [0,T]$,

$$-\dot{x}_2(t) \in \partial_1\mathcal{H}(x_1(t),x_2(t)),$$

$$\dot{x}_1(t) \in \partial_2\mathcal{H}(x_1(t),x_2(t)),$$

and satisfying the boundary conditions

$$-A_1x_1(0) - x_2(0) \in \partial\psi_1(x_1(0)),$$

$$-A_2x_1(T) + x_2(T) \in \partial\psi_2(x_1(T)).$$

Proof. This is a restatement of Theorem 9.4 for $\Phi(x_1,x_2) = \varphi_1(x_1) + \varphi_2(x_2)$.

Corollary 9.3. *Let E be a Hilbert space and let φ be a convex lower semicontinuous function on E satisfying $\varphi(x) \le C\big(1 + \|x\|_H^\beta\big)$. Let ψ_1, ψ_2, A_1, and A_2 be as in Theorem 9.4. Then, there exists $x \in A_E^\alpha([0,T])$ such that, for almost all $t \in [0,T]$,*

$$\ddot{x}(t) \in \partial\varphi(x(t)),$$

$$-\dot{x}(0) \in \partial \psi_1(x(0)) + A_1 x(0),$$
$$\dot{x}(T) \in \partial \psi_2(x(T)) + A_2 x(T).$$

Proof. It is enough to apply the above to $\varphi_2 = \varphi$ and $\varphi_1(x_1) = \frac{1}{2}\|x_1\|_H^2$.

An application to infinite-dimensional Hamiltonian systems.

Consider an evolution triple $Y \subset E \subset Y^*$, where Y is a reflexive Banach space that is densely embedded in a Hilbert space E. Then, the product $X := Y \times Y$ is clearly a reflexive Banach space that is densely embedded in the Hilbert space $H = E \times E$ and $X \subset H \subset X^*$ is also an evolution triple.

We consider now a simple but illustrative example. Let φ_1, φ_2 be convex lower semicontinuous functions on E whose domain is Y and which are coercive on Y. Define the convex function $\Phi : H \to \mathbf{R} \cup \{+\infty\}$ by $\Phi(x) = \Phi(y_1, y_2) := \varphi_1(y_1) + \varphi_2(y_2)$, and consider the linear automorphism $S : X^* \to X^*$ defined by

$$Sx^* = S(y_1^*, y_2^*) := (-y_2^*, -y_1^*).$$

Clearly S is an automorphism whose restriction to H and X are also automorphisms.

Consider now the Lagrangians $L : X^* \times X^* \to \mathbf{R} \cup \{+\infty\}$ defined as:

$$L(x,v) = \Phi(x) + (\Phi|_X)^*(Sv). \tag{9.23}$$

Now, for the boundary, consider the convex lower semicontinuous functions ψ_1, $\psi_2 : Y^* \to \mathbf{R} \cup \{\infty\}$ assuming that both are coercive on Y. To these functions, we associate the boundary Lagrangian $\ell : X^* \times X^* \to \mathbf{R} \cup \{+\infty\}$ by

$$\ell((a_1,a_2),(b_1,b_2)) = \psi_1(a_1) + (\psi_1|_X)^*(-a_2) + \psi_2(b_1) + (\psi_2|_X)^*(b_2). \tag{9.24}$$

It is then easy to show that L is S-self-dual on $X^* \times X^*$ since the convex function Φ is coercive on X and that ℓ is S-compatible.

Proposition 9.1. *Suppose that $\varphi_j(y) \leq C(\|y\|_Y^\beta + 1)$ for $j = 1,2$, and that ψ_1 is bounded on the bounded sets of Y, and consider the Hamiltonian $\mathcal{H}(p,q) = \varphi_1(p) - \varphi_2(q)$. Then, for any $T > 0$, there exists a solution $(\bar{p}, \bar{q}) \in A_{H,X^*}^\alpha$ to the following Hamiltonian system:*

$$\dot{p}(t) \in \partial_2 \mathcal{H}(p(t), q(t)),$$
$$-\dot{q}(t) \in \partial_1 \mathcal{H}(p(t), q(t)),$$
$$-p(0) \in \partial \psi_1(q(0))'$$
$$p(T) \in \partial \psi_2(q(T)).$$

It can be obtained by minimizing the completely self-dual functional on the space A_{H,X^*}^α

$$I(p,q) = \int_0^T \varphi(p(t), q(t)) + (\varphi_{|X})^*(-\dot{q}(t), -\dot{p}(t)) dt$$

$$+\ell\big((p(0),q(0)),(p(T),q(T))\big),$$

where φ is the convex function $\varphi(p,q) = \varphi_1(p) + \varphi_2(q)$ and ℓ is as in (9.24).

Exercises 9.B.

1. Suppose ℓ_1 (resp., ℓ_2) is an antiself-dual (resp., self-dual) Lagrangian on the Hilbert space $E \times E$. Show that the convex function $\ell : E^2 \times E^2 \to \mathbf{R}$ defined by

$$\ell((a_1,a_2),(b_1,b_2)) = \ell_1(a_1,a_2) + \ell_2(b_1,b_2)$$

   is T-self-dual, where $T(x_1,x_2,y_1,y_2) = (x_2,x_1,-y_2,-y_1)$.

2. In particular, if ψ_1 and ψ_2 are convex lower semicontinuous on E and if A_1, A_2 are bounded skew-adjoint operators on E, then what can be said about the Lagrangian $\ell : E^2 \times E^2 \to \mathbf{R}$ defined by

$$\ell(a,b) := \psi_1(a_1) + \psi_1^*(-A_1a_1 - a_2) + \psi_2(b_1) + \psi_2^*(-A_2b_1 + b_2).$$

3. Verify Theorem 9.3 and its corollaries.

.

Chapter 10
Semilinear Evolution Equations with Self-dual Boundary Conditions

This chapter is concerned with existence results for evolutions of the form

$$\begin{cases} \dot{u}(t) \in -\overline{\partial}L(t,u(t)) & \forall t \in [0,T] \\ \frac{u(0)+u(T)}{2} \in -\overline{\partial}\ell(u(0) - u(T)), \end{cases}$$

where both L and ℓ are self-dual Lagrangians. We then apply it to equations

$$\begin{cases} \dot{u}(t) + Au(t) + \omega u(t) \in -\partial\varphi(t,u(t)) & \text{for a.e. } t \in [0,T] \\ \frac{u(0)+e^{TA}e^{-\omega T}u(T)}{2} \in -\partial\psi\left(u(0) - e^{TA}e^{-\omega T}u(T)\right), \end{cases}$$

where φ and ψ are convex lower semicontinuous energies, A is a skew-adjoint operator, and $\omega \in \mathbf{R}$. The time-boundary conditions we obtain are quite general, though they include as particular cases the following more traditional ones:

- initial-value problems: $x(0) = x_0$;
- periodic orbits: $x(0) = x(T)$;
- antiperiodic orbits: $x(0) = -x(T)$;
- periodic orbits up to an isometry: $x(T) = e^{-T(\omega I + A)}x(0)$.

10.1 Self-dual variational principles for parabolic equations

Consider a general semilinear evolution equation of the form

$$\dot{x}(t) + Ax(t) + \omega x(t) \in -\partial\varphi(t,x(t)) \quad \text{for a.e. } t \in [0,T], \tag{10.1}$$

where $\omega \in \mathbf{R}$, $\varphi(t,\cdot) : H \to \mathbf{R} \cup \{+\infty\}$ is a proper convex and lower semicontinuous functional on a Hilbert space H and $A : \text{Dom}(A) \subseteq H \to H$ is a linear operator, a typical example being the complex Ginsburg-Landau equation on $\Omega \subseteq \mathbf{R}^N$

$$\frac{\partial u}{\partial t} - (\kappa + i\alpha)\Delta u + \omega u = -\partial\varphi(t,u(t)) \quad \text{for } t \in (0,T]. \tag{10.2}$$

N. Ghoussoub, *Self-dual Partial Differential Systems and Their Variational Principles,* 187
Springer Monographs in Mathematics, DOI 10.1007/978-0-387-84897-6_10,
© Springer Science+Business Media, LLC 2009

In order to solve these parabolic equations via a self-dual variational principle, we first need to write them in a self-dual form such as

$$\dot{u}(t) \in -\overline{\partial}L(t,u(t)) \quad \text{for } t \in [0,T],\tag{10.3}$$

where L is a self-dual Lagrangian. We shall see that there are many ways to associate a self-dual Lagrangian L to the given vector field $Ax(t) + \omega x(t) + \partial\varphi$. The choices will depend on the nature of the operator A, the sign of ω, and the available boundedness conditions on the convex function φ.

Moreover, we shall also see that many standard time-boundary conditions can be formulated in a self-dual form such as

$$\frac{u(0)+u(T)}{2} \in -\overline{\partial}\ell(u(0)-u(T)),\tag{10.4}$$

where ℓ is a self-dual boundary Lagrangian. The choice of ℓ will depend on whether we are dealing with initial-value problems, periodic problems, or other – possibly nonlinear – time-boundary conditions.

Many aspects of the evolution equation (10.1) can therefore be reduced to study self-dual evolutions of the form (10.1). It is important to note that the interior Lagrangians L are expected in the applications to be smooth and hence their subdifferentials will coincide with their differentials and the corresponding inclusions will often be equations. On the other hand, it is crucial that the boundary Lagrangians ℓ be allowed to be degenerate so as to cover the various boundary conditions we need.

We start by describing the various choices we have for writing an evolution equation in a self-dual form.

The selection of self-dual Lagrangians

1. *The diffusive case* corresponds for instance to when $w \geq 0$, A is a positive operator, and the – then convex – function $\Phi(t,x) = \varphi(t,x) + \frac{1}{2}\langle Ax,x\rangle + \frac{w}{2}\|x\|_H^2$ is coercive on the right space. In this case, the self-dual Lagrangian is

$$L(t,x,p) = \Phi(t,x) + \Phi^*(t, -A^a x + p),\tag{10.5}$$

   where A^a is the antisymmetric part of the operator A.
2. *The non-diffusive case* essentially means that one of the above requirements is not satisfied, e.g., $\omega < 0$ or if A is unbounded and purely skew-adjoint ($\kappa = 0$). The self-dual Lagrangian is then

$$L(t,x,p) = e^{-2\omega t}\left\{\varphi(t,e^{\omega t}S_t x) + \varphi^*(t,e^{\omega t}S_t p)\right\},\tag{10.6}$$

   where S_t is the C_0-unitary group associated to the skew-adjoint operator A. This nondiffusive case cannot be formulated on "energy spaces" and therefore requires less stringent coercivity conditions. However, the equation may not in this case have solutions satisfying the standard boundary conditions. Instead, and as we shall see below, one has to settle for solutions that are periodic but only up to the isometry e^{-TA}.

3. *The mixed case* deals with

$$\dot{x}(t) + A_1 x(t) + A_2 x(t) + \omega x(t) \in -\partial\varphi(t, x(t)) \quad \text{for a.e. } t \in [0, T], \quad (10.7)$$

where A_1 is a bounded positive operator and A_2 is an unbounded and purely skew-adjoint operator "playing a different role" in the equation. One example we consider, is the following evolution equation with an advection term.

$$\dot{u}(t) + \mathbf{a}.\nabla u(t) - i\Delta u + \omega u(t) = -\partial\varphi(t, u(t)) \quad \text{for } t \in [0, T]. \quad (10.8)$$

The self-dual Lagrangian is then

$$L(t, x, p) = e^{-2\omega t} \left\{ \varphi(t, e^{\omega t} S_t x) + \varphi^*(t, -e^{\omega t} A_1^a S_t x + e^{\omega t} S_t p) \right\}, \quad (10.9)$$

where S_t is the C_0-unitary group associated to the skew-adjoint operator A_2. Again, one then gets the required boundary condition up to the isometry e^{-TA_2}.

The selection of boundary Lagrangians

The simplest version of the self-dual boundary condition (10.1) is simply

$$\frac{v(0) + e^{-\omega T} S_{-T} v(T)}{2} \in -\partial\psi\left(v(0) - e^{-\omega T} S_{-T} v(T)\right), \quad (10.10)$$

where ψ is a convex function on H and $(S_t)_t$ is the C_0-unitary group associated to the skew-adjoint part of the operator. Here is a sample of the various boundary conditions that one can obtain by choosing ψ accordingly in (10.10):

1. *Initial boundary condition*, say $v(0) = v_0$ for a given $v_0 \in H$. Then it suffices to choose $\psi(u) = \frac{1}{4}\|u\|_H^2 - \langle u, v_0 \rangle$.
2. *Periodic type solutions* of the form $v(0) = S_{-T} e^{-\omega T} v(T)$. Then ψ is chosen as

$$\psi(u) = \begin{cases} 0 & u = 0 \\ +\infty & \text{elsewhere.} \end{cases}$$

3. *Antiperiodic type solutions* $v(0) = -S_{-T} e^{-\omega T} v(T)$. Then $\psi \equiv 0$ on H.

In the latter case, we shall sometimes say that the solutions are periodic and antiperiodic orbits up to an isometry.

We start by giving a general self-dual variational principle for parabolic equations, but under restrictive boundedness conditions. They will be relaxed later when dealing with more specific situations.

Proposition 10.1. *Suppose L is a time-dependent self-dual Lagrangian on $[0,T] \times H \times H$, and let ℓ be a self-dual Lagrangian on $H \times H$. Assume that for some $n > 1$ and $C > 0$, we have:*

(A'_1) $\int_0^T L(t,x(t),0)\,dt \leq C\big(\|x\|_{L_H^2}^n + 1\big)$ *for all $x \in L_H^2$.*

(A'_2) $\int_0^T L(t,x(t),p(t))\,dt \to \infty$ *as $\|x\|_{L_H^2} \to \infty$ for every $p \in L_H^2$.*

(A'_3)   *ℓ is bounded from below and $0 \in \mathrm{Dom}(\ell)$.*

The completely self-dual functional

$$I(x) = \int_0^T L\big(t,x(t),-\dot{x}(t)\big)\,dt + \ell\Big(x(0)-x(T), -\frac{x(0)+x(T)}{2}\Big)$$

then attains its minimum at a path $u \in A_H^2$ satisfying

$$I(u) = \inf_{x \in A_H^2} I(x) = 0, \tag{10.11}$$

$$-\dot{u}(t) \in \overline{\partial}L\big(t,u(t)\big) \quad \forall t \in [0,T], \tag{10.12}$$

$$-\frac{u(0)+u(T)}{2} \in \overline{\partial}\ell(u(0)-u(T)). \tag{10.13}$$

Proof. Define for each $\lambda > 0$ the λ-regularization ℓ_λ^1 of the boundary Lagrangian ℓ. By Lemma 3.3, ℓ_λ^1 is also seldual on $H \times H$, and by Theorem 9.2, the Lagrangian

$$\mathscr{M}_\lambda(u,p) = \int_0^T L\big(t,u+p_0,-\dot{u}(t)\big)\,dt + \ell_\lambda^1\Big(u(0)-u(T)+p_1, -\frac{u(0)+u(T)}{2}\Big)$$

is a self-dual Lagrangian on $A_H^2 \times (L_H^2 \times H)$. It also satisfies the hypotheses of Proposition 6.1. It follows that the infimum of the functional

$$I_\lambda(x) = \int_0^T L\big(t,x(t),-\dot{x}(t)\big)\,dt + \ell_\lambda^1\Big(x(0)-x(T), -\frac{x(0)+x(T)}{2}\Big)$$

on A_H^2 is zero and is attained at some $x_\lambda \in A_H^2$ satisfying

$$\int_0^T L\big(t,x_\lambda,-\dot{x}_\lambda\big)\,dt + \ell_\lambda^1\Big(x_\lambda(0)-x_\lambda(T), -\frac{x_\lambda(0)+x_\lambda(T)}{2}\Big) = 0, \tag{10.14}$$

$$-\dot{x}_\lambda(t) \in \overline{\partial}L\big(t,x_\lambda(t)\big), \tag{10.15}$$

$$-\frac{x_\lambda(0)+x_\lambda(T)}{2} \in \overline{\partial}\ell_\lambda^1\big(x_\lambda(0)-x_\lambda(T)\big). \tag{10.16}$$

We now show that $(x_\lambda)_\lambda$ is bounded in A_H^2. Indeed, since ℓ is bounded from below, so is ℓ_λ, which together with (10.14) implies that $\int_0^T L(t,x_\lambda(t),-\dot{x}_\lambda(t))\,dt$ is bounded. It follows from (A'_1) and Lemma 3.5 that $\{\dot{x}_\lambda(t)\}_\lambda$ is bounded in L_H^2. It also follows from (A'_2) that $\{x_\lambda(t)\}_\lambda$ is bounded in L_H^2, and hence x_λ is bounded in A_H^2 and thus, up to a subsequence $x_\lambda(t) \rightharpoonup u(t)$ in A_H^2, $x_\lambda(0) \rightharpoonup u(0)$ and $x_\lambda(T) \rightharpoonup u(T)$ in H.

From (10.14), we have that $\ell_\lambda^1\left(x_\lambda(0) - x_\lambda(T), -\frac{x_\lambda(0)+x_\lambda(T)}{2}\right)$ is bounded from above. Hence, it follows from Lemma 3.3 that

$$\ell\left(u(0) - u(T), -\frac{u(0)+u(T)}{2}\right) \leq \liminf_{\lambda \to 0} \ell_\lambda^1\left(x_\lambda(0) - x_\lambda(T), -\frac{x_\lambda(0)+x_\lambda(T)}{2}\right).$$

By letting $\lambda \to 0$ in (10.14), we get

$$\int_0^T L\left(t, u(t), -\dot{u}(t)\right) dt + \ell\left(u(0) - u(T), -\frac{u(0)+u(T)}{2}\right) \leq 0.$$

On the other hand, for every $x \in A_H^2$, we have

$$I(x) = \int_0^T L\left(t, x(t), -\dot{x}(t)\right) dt + \ell\left(x(0) - x(T), -\frac{x(0)+x(T)}{2}\right)$$

$$= \int_0^T \left\{ L\left(t, x(t), -\dot{x}(t)\right) + \langle x(t), \dot{x}(t) \rangle \right\} dt$$

$$+ \ell\left(x(0) - x(T), -\frac{x(0)+x(T)}{2}\right) + \left\langle x(0) - x(T), \frac{x(0)+x(T)}{2} \right\rangle$$

$$\geq 0,$$

which means $I(u) = 0$ and therefore $u(t)$ satisfies (10.12) and (10.13) as well.

10.2 Parabolic semilinear equations without a diffusive term

We now consider the case where A is a purely skew-adjoint operator and therefore cannot contribute to the coercivity of the problem.

Theorem 10.1. *Let $(S_t)_{t \in \mathbf{R}}$ be a C_0-unitary group of operators associated to a skew-adjoint operator A on a Hilbert space H, and let $\varphi : [0,T] \times H \to \mathbf{R} \cup \{+\infty\}$ be a time-dependent convex Gâteaux-differentiable function on H. Assume the following conditions:*

(A_1)   *For some $m, n > 1$ and $C_1, C_2 > 0$, we have for every $x \in L_H^2$*

$$C_1\left(\|x\|_{L_H^2}^m - 1\right) \leq \int_0^T \left\{ \varphi(t, x(t)) + \varphi^*(t, 0) \right\} dt \leq C_2\left(1 + \|x\|_{L_H^2}^n\right).$$

(A_2)   *ψ is a bounded below convex lower semicontinuous function on H with $0 \in \mathrm{Dom}(\psi)$.*

For any given $\omega \in \mathbf{R}$ and $T > 0$, consider the following functional on A_H^2:

$$I(x) = \int_0^T e^{-2\omega t} \left\{ \varphi(t, e^{\omega t} S_t x(t)) + \varphi^*(t, -e^{\omega t} S_t \dot{x}(t)) \right\} dt$$

$$+ \psi(x(0) - x(T)) + \psi^*\left(-\frac{x(0)+x(T)}{2}\right).$$

Then, there exists a path $u \in A_H^p$ such that:

1. $I(u) = \inf\limits_{x \in A_H^2} I(x) = 0.$
2. *The path $v(t) := S_t e^{\omega t} u(t)$ is a mild solution of the equation*

$$\dot{v}(t) + Av(t) + \omega v(t) \in -\partial \varphi(t, v(t)) \quad \text{for a.e. } t \in [0, T], \quad (10.17)$$

$$\frac{v(0) + S_{-T} e^{-\omega T} v(T)}{2} \in -\partial \psi \left(v(0) - S_{-T} e^{-\omega T} v(T) \right). \quad (10.18)$$

Equation (10.17) means that v satisfies the following integral equation:

$$v(t) = S_t v(0) - \int_0^t \left(S_{t-s} \partial \varphi(s, v(s)) - \omega S_t v(s) \right) ds \text{ for all } t \in [0, T]. \quad (10.19)$$

Proof. Consider the self-dual Lagrangians $M(t, x, p) = \varphi(t, x) + \varphi^*(t, p)$ and $\ell(x, p) = \psi(x) + \psi^*(p)$, and apply Proposition 10.1 to the Lagrangian

$$L(t, x, p) = e^{-2\omega t} M\left(t, S_t e^{\omega t} x, S_t e^{\omega t} p\right), \quad (10.20)$$

which is self-dual according to Proposition 3.4. We then obtain $u(t) \in A_H^2$ such that

$$0 = \int_0^T e^{-2\omega t} \varphi\left(t, S_t e^{\omega t} u(t)\right) + \varphi^*\left(-S_t e^{\omega t} \dot{u}(t)\right) dt$$

$$+ \psi(u(0) - u(T)) + \psi^*\left(-\frac{u(0) + u(T)}{2}\right),$$

which gives

$$0 = \int_0^T e^{-2\omega t} \left[\varphi\left(t, S_t e^{\omega t} u(t)\right) + \varphi^*\left(-S_t e^{\omega t} \dot{u}(t)\right) + \langle S_t e^{\omega t} u(t), S_t e^{\omega t} \dot{u}(t)\rangle\right] dt$$
$$- \int_0^T \langle S_t u(t), S_t \dot{u}(t)\rangle dt + \psi(u(0) - u(T)) + \psi^*\left(-\frac{u(0) + u(T)}{2}\right)$$
$$= \int_0^T e^{-2\omega t} \left[\varphi\left(t, S_t e^{\omega t} u(t)\right) + \varphi^*\left(-S_t e^{\omega t} \dot{u}(t)\right) + \langle S_t e^{\omega t} u(t), S_t e^{\omega t} \dot{u}(t)\rangle\right] dt$$
$$- \int_0^T \langle u(t), \dot{u}(t)\rangle dt + \psi(u(0) - u(T)) + \psi^*\left(-\frac{u(0) + u(T)}{2}\right)$$
$$= \int_0^T e^{-2\omega t} \left[\varphi\left(t, S_t e^{\omega t} u(t)\right) + \varphi^*\left(-S_t e^{\omega t} \dot{u}(t)\right) + \langle S_t e^{\omega t} u(t), S_t e^{\omega t} \dot{u}(t)\rangle\right] dt$$
$$- \frac{1}{2}\|u(T)\|^2 + \frac{1}{2}\|u(0)\|^2 + \psi(u(0) - u(T)) + \psi^*\left(-\frac{u(0) + u(T)}{2}\right)$$
$$= \int_0^T e^{-2\omega t} \left[\varphi\left(t, S_t e^{\omega t} u(t)\right) + \varphi^*\left(-S_t e^{\omega t} \dot{u}(t)\right) + \langle S_t e^{\omega t} u(t), S_t e^{\omega t} \dot{u}(t)\rangle\right] dt$$
$$+ \left\langle u(0) - u(T), \frac{u(0) + u(T)}{2}\right\rangle + \psi(u(0) - u(T)) + \psi^*\left(-\frac{u(0) + u(T)}{2}\right).$$

Since clearly

$$\varphi\left(t, S_t e^{\omega t} u(t)\right) + \varphi^*\left(-S_t e^{\omega t} \dot{u}(t)\right) + \langle S_t e^{\omega t} u(t), S_t e^{\omega t} \dot{u}(t)\rangle \geq 0$$

for every $t \in [0,T]$ and since

$$\psi(u(0) - u(T)) + \psi^*\left(-\frac{u(0) + u(T)}{2}\right) + \left\langle u(0) - u(T), \frac{u(0) + u(T)}{2}\right\rangle \geq 0,$$

we get equality, from which we can conclude, that for almost all $t \in [0,T]$,

$$-S_t e^{\omega t}\dot{u}(t) = \partial\varphi(t, S_t e^{\omega t}u(t)) \text{ and } \frac{u(0)+u(T)}{2} \in -\partial\psi(u(0) - u(T)). \quad (10.21)$$

In order to show that $v(t) := S_t e^{\omega t}u(t)$ is a mild solution for (10.17), we set $x(t) = e^{\omega t}u(t)$ and write

$$-S_t(\dot{x}(t) - \omega x(t)) = \partial\varphi(t, S_t x(t)),$$

and hence, $-(\dot{x}(t) + \omega x(t)) = S_{-t}\partial\varphi(t, v(t))$. By integrating between 0 and t, we get

$$x(t) = x(0) - \int_0^t \{S_{-s}\partial\varphi(s, v(s)) - \omega u(s)\}\, ds$$

Substituting $v(t) = S_t x(t)$ in the above equation gives

$$v(t) = S_t v(0) - \int_0^t \left(S_{t-s}\partial\varphi(s, v(s)) - S_t \omega v(s)\right) ds,$$

which means that $v(t)$ is a mild solution for (10.17).

On the other hand, the boundary condition $\frac{u(0)+u(T)}{2} \in -\partial\psi(u(0) - u(T))$ translates after the change of variables into

$$\frac{v(0) + e^{-\omega T}S(-T)v(T)}{2} \in -\partial\psi\left(v(0) - e^{-\omega T}S(-T)v(T)\right)$$

and we are done.

Example 10.1. The complex Ginzburg-Landau equations in $\mathbf{R}^N$

Consider the following evolution on $\mathbf{R}^N$

$$\dot{u}(t) + i\Delta u + \partial\varphi(t, u(t)) + \omega u(t) = 0 \quad \text{for } t \in [0,T]. \quad (10.22)$$

(1) Under the condition:

$$C_1\left(\int_0^T \|u(t)\|_2^2\, dt - 1\right) \leq \int_0^T \varphi(t, u(t))\, dt \leq C_2\left(\int_0^T \|u(t)\|_2^2\, dt + 1\right), \quad (10.23)$$

where $C_1, C_2 > 0$, Theorem 10.1 yields a solution of

$$\begin{cases} \dot{u}(t) + i\Delta u + \partial\varphi(t, u(t)) + \omega u(t) = 0 & \text{for } t \in [0,T] \\ e^{-\omega T}e^{-iT\Delta}u(T) = u(0). \end{cases} \quad (10.24)$$

(2) If $\omega \geq 0$, then one can replace φ with the convex function $\Phi(x) = \varphi(x) + \frac{\omega}{2}\|x\|^2$ to obtain solutions such that

$$u(0) = e^{-iT\Delta}u(T) \text{ or } u(0) = -e^{-iT\Delta}u(T). \tag{10.25}$$

(3) One can also drop the coercivity condition (the lower bound) on $\varphi(t, u(t))$ in (10.23) and still get periodic-type solutions. Indeed, by applying our result to the now coercive convex functional $\Psi(t, u(t)) := \varphi(t, u(t)) + \frac{\varepsilon}{2}\|u(t)\|_H^2$ and $\omega - \varepsilon$, we obtain a solution such that

$$e^{(-\omega+\varepsilon)T}e^{-iT\Delta}u(T) = u(0). \tag{10.26}$$

Example 10.2. Almost Periodic solutions for linear Schrödinger equations:

Consider now the linear Schrodinger equation

$$i\frac{\partial u}{\partial t} = \Delta u - V(x)u. \tag{10.27}$$

Assuming that the space $\{u \in H^{2,2}(\mathbf{R}^N) : \int_{\mathbf{R}^N} |V(x)|u^2\,dx < \infty\}$ is dense in $H := L^2(\mathbf{R}^N)$, we get that the operator $Au := i\Delta u - iV(x)u$ is skew-adjoint on H. In order to introduce some coercivity and to avoid the trivial solution, we can consider for any $\varepsilon, \delta > 0$ and $0 \neq f \in H$ the convex function $\varphi_\varepsilon(u) := \frac{\varepsilon}{2}\|u\|_H^2 + \delta\langle f, u\rangle$.

By applying Theorem 10.1 to A, φ_ε, and $\omega = \varepsilon$, we get a nontrivial solution $u \in A_H^2$ for the equation

$$\begin{cases} i\frac{\partial u}{\partial t} = \Delta u - V(x)u + \delta f, \\ u(0) = e^{-\varepsilon T}e^{iT(-\Delta + V(x))}u(T). \end{cases} \tag{10.28}$$

Example 10.3. Coupled flows and wave-type equations

Let $A : D(A) \subseteq H \to H$ be a linear operator with a dense domain in H. Suppose $D(A) = D(A^*)$, and define the operator $\mathscr{A}$ on the product space $H \times H$ as follows:

$$\begin{cases} \mathscr{A} : D(\mathscr{A}) \subseteq H \times H \to H \times H \\ \mathscr{A}(x, y) := (Ay, -A^*x). \end{cases}$$

It is easily seen that $\mathscr{A} : D(\mathscr{A}) \subseteq H \times H \to H \times H$ is a skew-adjoint operator, and hence, by virtue of Stone's Theorem, $\mathscr{A}$ is the generator of a C_0 unitary group $\{S_t\}$ on $H \times H$. Here is another application of Theorem 10.1.

Theorem 10.2. *Let $\varphi(t, \cdot)$ and ψ be proper convex lower semicontinuous functionals on $H \times H$. Assume the following conditions:*

(A''$_1$)   *For some $m, n > 1$ and $C_1, C_2 > 0$, we have*

$$C_1\left(\|x\|^m_{L^2_{H\times H}} - 1\right) \le \int_0^T \varphi(t,x(t))\,dt \le C_2\left(1 + \|x\|^n_{L^2_{H\times H}}\right) \text{ for every } x \in L^2_{H\times H}.$$

(A_2'') $\psi : H \times H \to \mathbf{R} \cup \{+\infty\}$ is bounded below and $0 \in \mathrm{Dom}(\psi)$.

Then, there exists a mild solution $(u(t),v(t)) \in A^2_{H\times H}$ for the system

$$\begin{cases} -\dot{u}(t) + Av(t) - \omega u(t) = \partial_1 \varphi(t,u(t),v(t)), \\ -\dot{v}(t) - A^* u(t) - \omega v(t) = \partial_2 \varphi(t,u(t),v(t)), \end{cases}$$

with a boundary condition of the form (10.18).

10.3 Parabolic semilinear equation with a diffusive term

For a given $0 < T < \infty$, $1 < p < \infty$, and a Hilbert space H such that $X \subseteq H \subseteq X^*$ is an evolution triple, we consider again the space

$$\mathscr{X}_{p,q} = \{u : u \in L^p(0,T : X), \dot{u} \in L^q(0,T : X^*)\}$$

equipped with the norm $\|u\|_{\mathscr{X}_{p,q}} = \|u\|_{L^p(0,T:X)} + \|\dot{u}\|_{L^q(0,T:X^*)}$, which leads to a continuous injection $\mathscr{X}_{p,q} \subseteq C(0,T : H)$.

Theorem 10.3. *Let $X \subset H \subset X^*$ be an evolution triple, and consider a time-dependent self-dual Lagrangian $L(t,x,p)$ on $[0,T] \times X \times X^*$ and a self-dual Lagrangian ℓ on $H \times H$ such that the following conditions are satisfied:*

(B_1') *For some $p \ge 2$, $m,n > 1$, and $C_1, C_2 > 0$, we have*

$$C_1\left(\|x\|^m_{L^p_X} - 1\right) \le \int_0^T L(t,x(t),0)\,dt \le C_2\left(1 + \|x\|^n_{L^p_X}\right) \text{ for every } x \in L^p_X.$$

(B_2') *ℓ is bounded from below.*

The functional $I(x) = \int_0^T L\left(t, -x(t), \dot{x}(t)\right)dt + \ell\left(x(0) - x(T), -\frac{x(0)+x(T)}{2}\right)$ then attains its minimum on $\mathscr{X}_{p,q}$ at a path $u \in \mathscr{X}_{p,q}$ such that

$$I(u) = \inf\{I(x); x \in \mathscr{X}_{p,q}\} = 0, \tag{10.29}$$

$$-\dot{u}(t) \in \overline{\partial}L(t,u(t)) \quad \forall t \in [0,T], \tag{10.30}$$

$$-\frac{u(0)+u(T)}{2} \in \overline{\partial}\ell\left(u(0) - u(T)\right). \tag{10.31}$$

Proof. Use Lemma 3.4 to lift the Lagrangian L to a time-dependent self-dual Lagrangian on $[0,T] \times H \times H$ via the formula

$$M(t,u,p) := \begin{cases} L(t,u,p) & \text{if } u \in X \\ +\infty & \text{if } u \in H \setminus X. \end{cases}$$

We start by assuming that $\ell(a,b) \to \infty$ as $\|b\| \to \infty$. Consider for $\lambda > 0$ the λ-regularization of M, namely

$$M_\lambda^1(t,x,p) := \inf\left\{ M(t,z,p) + \frac{\|x-z\|^2}{2\lambda} + \frac{\lambda}{2}\|p\|^2; z \in H \right\}. \tag{10.32}$$

It is easy to check that M_λ^1 satisfies the conditions (A_1') and (A_2') of Proposition 10.1. It follows that there exists a path $x_\lambda(t) \in A_H^2$ such that

$$\int_0^T M_\lambda^1\left(t,x_\lambda,-\dot{x}_\lambda\right)dt + \ell\left(x_\lambda(0) - x_\lambda(T), -\frac{x_\lambda(0) + x_\lambda(T)}{2}\right) = 0. \tag{10.33}$$

We now show that $(x_\lambda)_\lambda$ is bounded in an appropriate function space. Indeed, since L is convex and lower semicontinuous, there exists $i_\lambda(x_\lambda)$ such that the infimum in (10.32) is attained at $i_\lambda(x_\lambda) \in X$, i.e.,

$$M_\lambda(t,x_\lambda,-\dot{x}_\lambda) = L(t,i_\lambda(x_\lambda),-\dot{x}_\lambda) + \frac{\|x_\lambda - i_\lambda(x_\lambda)\|^2}{2\lambda} + \frac{\lambda}{2}\|\dot{x}_\lambda\|^2. \tag{10.34}$$

Plug (10.34) into equality (10.33) to get

$$0 = \int_0^T \left(L(t,i_\lambda(x_\lambda),-\dot{x}_\lambda(t)) + \frac{\|x_\lambda(t) - i_\lambda(x_\lambda)\|^2}{2\lambda} + \frac{\lambda}{2}\|\dot{x}_\lambda\|^2 \right) dt \tag{10.35}$$

$$+\ell\left(x_\lambda(0) - x_\lambda(T), -\frac{x_\lambda(0) + x_\lambda(T)}{2}\right). \tag{10.36}$$

By the coercivity assumptions in (B_1'), we obtain that $(i_\lambda(x_\lambda))_\lambda$ is bounded in $L^p(0,T;X)$ and $(x_\lambda)_\lambda$ is bounded in $L^2(0,T;H)$. According to Lemma 3.5, condition (B_1') yields that $\int_0^T L(t,x(t),p(t))\,dt$ is coercive in $p(t)$ on $L^q(0,T;X^*)$, and therefore it follows from (10.35) that $(\dot{x}_\lambda)_\lambda$ is bounded in $L^q(0,T;X^*)$. Also, since L and ℓ are bounded from below, it follows again from (10.35) that $\int_0^T \|x_\lambda(t) - i_\lambda(x_\lambda)\|^2\,dt \le 2\lambda C$ for some $C > 0$. Since $x_\lambda(T) - x_\lambda(0) = \int_0^T \dot{x}_\lambda\,dt$, therefore $x_\lambda(T) - x_\lambda(0)$ is bounded in X^*. Also, since $\ell(a,b) \to \infty$ as $\|b\| \to \infty$, it follows that $x_\lambda(0) + x_\lambda(T)$ is also bounded in H and consequently in X^*. Therefore there exists $u \in L_H^2$ with $\dot{u} \in L^q(0,T;X^*)$ and $u(0), u(T) \in X^*$ such that

$$i_\lambda(x_\lambda) \rightharpoonup u \quad \text{in} \quad L^p(0,T;X),$$
$$\dot{x}_\lambda \rightharpoonup \dot{u} \quad \text{in} \quad L^q(0,T;X^*),$$
$$x_\lambda \rightharpoonup u \quad \text{in} \quad L^2(0,T;H),$$
$$x_\lambda(0) \rightharpoonup u(0), \quad x_\lambda(T) \rightharpoonup u(T) \quad \text{in} \quad X^*.$$

By letting λ go to zero in (10.35), we obtain from the above that

$$\ell\left(u(0) - u(T), -\frac{u(0) + u(T)}{2}\right) + \int_0^T L\left(t,u(t),-\dot{u}(t)\right)dt \le 0. \tag{10.37}$$

It follows from (B_1'), Lemma 3.5, and (10.37) that $u \in \mathscr{X}_{p,q}$ and consequently $u(0), u(T) \in H$.

Now we show that one can actually do without the coercivity condition on ℓ. Indeed, by using the λ-regularization ℓ_λ^1 of ℓ, we get the required coercivity condition on the second variable of ℓ_λ and we obtain from the above that there exists $x_\lambda \in \mathscr{X}_{p,q}$ such that

$$\int_0^T L\left(t, x_\lambda, -\dot{x}_\lambda\right) dt + \ell_\lambda \left(x_\lambda(0) - x_\lambda(T), -\frac{x_\lambda(0) + x_\lambda(T)}{2} \right) \leq 0. \quad (10.38)$$

It follows from (B_1') and the boundedness of ℓ_λ^1 from below that $(x_\lambda)_\lambda$ is bounded in $L^p(0,T;X)$ and $(\dot{x}_\lambda)_\lambda$ is bounded in $L^q(0,T;X^*)$ again by virtue of Lemma 3.5. Hence, $(x_\lambda)_\lambda$ is bounded in $\mathscr{X}_{p,q}$ and therefore $(x_\lambda(0))_\lambda$ and $(x_\lambda(T))_\lambda$ are bounded in H. We therefore get, up to a subsequence, that

$$x_\lambda \rightharpoonup u \quad \text{in} \quad L^p(0,T;X),$$
$$\dot{x}_\lambda \rightharpoonup \dot{u} \quad \text{in} \quad L^q(0,T;X^*),$$
$$x_\lambda(0) \rightharpoonup u(0) \quad \text{in} \quad H,$$
$$x_\lambda(T) \rightharpoonup u(T) \quad \text{in} \quad H.$$

By letting λ go to zero in (10.38), it follows from the above that

$$\int_0^T L\left(t, u(t), -\dot{u}(t)\right) dt + \ell\left(u(0) - u(T), -\frac{u(0) + u(T)}{2} \right) \leq 0.$$

So $I(u) = 0$ and u is a solution of (10.30) and (10.31).

Corollary 10.1. *Let $X \subseteq H \subseteq X^*$ be an evolution triple, let $A : X \to X^*$ be a bounded positive operator on X, and let $\varphi : [0,T] \times X \to \mathbf{R} \cup \{+\infty\}$ be a time-dependent convex, lower semicontinuous, and proper function on X. Consider the convex function $\Phi(x) = \varphi(x) + \frac{1}{2}\langle Ax, x \rangle$ as well as the antisymmetric part $A^a := \frac{1}{2}(A - A^*)$ of A. Assume the following conditions hold:*

(B_1) *For some $p \geq 2$, $m, n > 1$, and $C_1, C_2 > 0$, we have for every $x \in L_X^p$*

$$C_1\left(\|x\|_{L_X^p}^m - 1\right) \leq \int_0^T \left\{ \Phi(t, x(t)) + \Phi^*(t, -A^a x(t)) \right\} dt \leq C_2\left(1 + \|x\|_{L_X^p}^n\right).$$

(B_2) *ψ is a bounded below convex lower semicontinuous function on H with $0 \in \mathrm{Dom}(\psi)$.*

For any $T > 0$ and $\omega \geq 0$, consider the following functional on $\mathscr{X}_{p,q}$:

$$I(x) = \int_0^T e^{-2\omega t} \left\{ \Phi(t, e^{\omega t} x(t)) + \Phi^*(t, -e^{\omega t}(A^a x(t) + \dot{x}(t))) \right\} dt$$
$$+ \psi(x(0) - x(T)) + \psi^*\left(-\frac{x(0) + x(T)}{2} \right).$$

Then, there exists a path $u \in L^p(0,T:X)$ with $\dot{u} \in L^q(0,T:X^*)$ such that

1. $I(u) = \inf\limits_{x \in \mathscr{X}_{p,q}} I(x) = 0.$
2. *If $v(t)$ is defined by $v(t) := e^{\omega t} u(t)$, then it satisfies*

$$\dot{v}(t) + Av(t) + \omega v(t) \in -\partial\varphi(t,v(t)) \quad \text{for a.e. } t \in [0,T], \qquad (10.39)$$

$$\frac{v(0) + e^{-wT} v(T)}{2} \in -\partial\psi\left(v(0) - e^{-wT} v(T)\right). \qquad (10.40)$$

Proof. It suffices to apply Theorem 10.3 to the self-dual Lagrangian

$$L(t,x,p) = e^{-2\omega t}\left\{\Phi(t,e^{\omega t}x) + \Phi^*(t, -e^{\omega t}A^a x + e^{\omega t}p)\right\}$$

associated to a convex lower semicontinuous function Φ, a skew-adjoint operator A^a, and a scalar ω.

Example 10.4. Complex Ginzburg-Landau evolution with diffusion

Consider a complex Ginzburg-Landau equation of the type.

$$\begin{cases} \frac{\partial u}{\partial t} - (\kappa + i)\Delta u + \partial\Psi(t,u) + \omega u = 0, & (t,x) \in (0,T) \times \Omega, \\ u(t,x) = 0, & x \in \partial\Omega, \qquad (10.41) \\ e^{-\omega T} u(T) = u(0), \end{cases}$$

where $\kappa > 0$, $\omega \le 0$, Ω is a bounded domain in $\mathbf{R}^N$, and Ψ is a time-dependent convex lower semicontinuous function. An immediate corollary of Theorem 10.3 is the following.

Corollary 10.2. *Let $X := H_0^1(\Omega)$, $H := L^2(\Omega)$, and $X^* = H^{-1}(\Omega)$. If for some $C > 0$, we have*

$$-C \le \int_0^T \Psi(t,u(t))\,dt \le C(\int_0^T \|u(t)\|_{H_0^1}^2\,dt + 1) \text{ for every } u \in L_X^2,$$

then there exists a solution $u \in \mathscr{X}_{2,2}$ for equation (10.41).

Proof. Set $\varphi(t,u) := \frac{k}{2}\int_\Omega |\nabla u|^2 dx + \Psi(t,u(t))$, $A = -(1+i)\Delta$, $A^a = -i\Delta$, and note that since

$$c_1(\|u\|_{L_X^2}^2 - 1) \le \int_0^T \varphi(t,u)\,dt \le c_2(\|u\|_{L_X^2}^2 + 1) \qquad (10.42)$$

for some $c_1, c_2 > 0$, we therefore have

$$c_1'(\|v\|_{L_{X^*}^2}^2 - 1) \le \int_0^T \varphi^*(t,v)\,dt \le c_2'(\|v\|_{L_{X^*}^2}^2 + 1)$$

for some $c_1', c_2' > 0$, and hence,

$$c_1' \left(\int_0^T \int_\Omega |\nabla(-\Delta)^{-1}v|^2 \, dx \, dt - 1 \right) \le \int_0^T \varphi^*(t,v) \, dt$$

$$\le c_2' \left(\int_0^T \int_\Omega |\nabla(-\Delta)^{-1}v|^2 \, dx \, dt + 1 \right),$$

from which we obtain

$$c_1' \left(\int_0^T \int_\Omega |\nabla u|^2 \, dx \, dt - 1 \right) \le \int_0^T \varphi^*(t, i\Delta u) \, dt \le c_2' \left(\int_0^T \int_\Omega |\nabla u|^2 \, dx \, dt + 1 \right),$$

which, once coupled with (10.42), yields the required boundedness in (B_1').

10.4 More on skew-adjoint operators in evolution equations

We now show how one can sometimes combine the two ways to define a self-dual Lagrangian that deals with a superposition of an unbounded skew-adjoint operator with another bounded positive operator. Note the impact on the boundeness condition (B_1) above.

Corollary 10.3. *Let $X \subseteq H \subseteq X^*$ be an evolution triple in such a way that the duality map $D : X \to X^*$ is linear and symmetric. Let $A_1 : X \to X^*$ be a bounded positive operator on X and let $A_2 : D(A) \subseteq X \to X^*$ be a – possibly unbounded – skew-adjoint operator. Let $\varphi : [0,T] \times X \to \mathbf{R} \cup \{+\infty\}$ be a time-dependent convex, lower semicontinuous, and proper function on X, and consider the convex function $\Phi(x) = \varphi(x) + \frac{1}{2}\langle A_1 x, x \rangle$ as well as the antisymmetric part $A_1^a := \frac{1}{2}(A_1 - A_1^*)$ of A_1. Let $\bar{S}_t : X^* \to X^*$ be the unitary group generated by A_2 and $S_t = D\bar{S}_t D^{-1}$. Assume the following conditions:*

(D_1) *For some $p \ge 2$, $m, n > 1$, and $C_1, C_2 > 0$, we have for every $x \in L_X^p$*

$$C_1 \left(\|x\|_{L_X^p}^m - 1 \right) \le \int_0^T \left\{ \Phi(t, S_t x(t)) + \Phi^*(t, -A_1^a S_t x(t)) \right\} dt \le C_2 \left(1 + \|x\|_{L_X^p}^n \right).$$

(D_2) *ψ is a bounded below convex lower semicontinuous function on H with $0 \in \text{Dom}(\psi)$.*

For any $T > 0$ and $\omega \in \mathbf{R}$, consider the following functional on $\mathscr{X}_{p,q}$

$$I(x) = \int_0^T e^{-2\omega t} \left\{ \Phi(t, e^{\omega t} S_t x(t)) + \Phi^*(t, -e^{\omega t}(A^a S_t x(t) + \bar{S}_t \dot{x}(t))) \right\} dt$$

$$+ \psi(x(0) - x(T)) + \psi^* \left(-\frac{x(0) + x(T)}{2} \right).$$

Then, there exists a path $u \in L^p(0,T : X)$ with $\dot{u} \in L^q(0,T : X^*)$ such that

1. $I(u) = \inf\limits_{x \in \mathscr{X}_{p,q}} I(x) = 0.$

2. *Moreover, if $\bar{S}_t = S_t$ on X, then $v(t) := e^{\omega t}\bar{S}_t u(t)$ satisfies for a.e. $t \in [0, T]$*

$$\dot{v}(t) + A_1 v(t) + A_2 v(t) + \omega v(t) \in -\partial \varphi(t, v(t)), \tag{10.43}$$

$$\frac{v(0) + S_{(-T)} e^{-\omega T} v(T)}{2} \in -\partial \psi \left(v(0) - S_{(-T)} e^{-\omega T} v(T) \right). \tag{10.44}$$

Proof. It suffices to apply Theorem 10.3 to the self-dual Lagrangian

$$L_S(t, x, p) = e^{-2\omega t} \left\{ \Phi(t, e^{\omega t} S_t x) + \Phi^*(t, -e^{\omega t} A^a S_t x + e^{\omega t} \bar{S}_t p) \right\},$$

which is self-dual in view of Proposition 3.4.

Example 10.5. The complex Ginzburg-Landau equations with advection in a bounded domain

We consider the evolution equation

$$\dot{u}(t) - i\Delta u + a\nabla u(t) + \partial \varphi(t, u(t)) + \omega u(t) = 0 \quad \text{for } t \in [0, T], \tag{10.45}$$

on a bounded domain Ω in $\mathbf{R}^n$. Under the condition that a is a constant vector and

$$C_1\left(\int_0^T \|u(t)\|_2^2 \, dt - 1\right) \leq \int_0^T \varphi(t, u(t)) \, dt \leq C_2\left(\int_0^T \|u(t)\|_2^2 \, dt + 1\right), \tag{10.46}$$

where $C_1, C_2 > 0$, Corollary 10.3 yields a solution of (10.45) u such that

$$e^{-\omega T} e^{-iT\Delta} u(T) = u(0). \tag{10.47}$$

Proof. Set $A_1 u = a\nabla u$, $A_2 = -i\Delta$, and $H = L^2(\Omega)$ in Corollary 10.3. Define the Banach space $X_1 = \{u \in H; A_1 u \in H\}$ equipped with the norm $\|u\|_X = (\|u\|_H^2 + \|A_1 u\|_H^2)^{\frac{1}{2}}$. Therefore $X^* = \{(I + A_1^* A_1)u; u \in X\}$ and the norm in X^* is $\|f\|_{X^*} = \|(I + A_1^* A_1)^{-1} f\|_X$. Note that $D = I + A_1^* A_1$ is the duality map between X and X^* since $\langle u, Du \rangle = \langle u, (I + A_1^* A_1)u \rangle = \|u\|_X^2$ and $\|Du\|_{X^*} = \|D^{-1} Du\|_X = \|u\|_X$.

Exercises 10.A.

1. Verify that Examples 10.1–10.5 satisfy the conditions of the general theorems from which they follow.
2. Show that, if in the above examples $\omega \geq 0$, then one can obtain truly periodic or anti-periodic solutions (possibly up to an isometry).

Further comments

There is a large literature about the existence of periodic solutions for evolutions driven by monotone operators. See for example Brezis [25], Browder [33], Vrabie

[160], [161], and Showalter [144]. Actually, the existence of periodic solutions follows from a very general result of J. L. Lions (See [144] Proposition III.5.1]) that deals with equations of the form

$$\begin{cases} -\dot{u}(t) = \Lambda(t, u(t)) \\ u(0) = u(T), \end{cases}$$

where $\Lambda(t, .)$ is a family of "hemi-continuous" and monotone operators from a Banach space X to X^* that satisfy for some $p \geq 2$, $k \in L^q[0, T]$ ($\frac{1}{p} + \frac{1}{q} = 1$) and constants $c_1, c_2 > 0$, the conditions:

$$\|\Lambda(t, x)\|_{X^*} \leq c_1(\|x\|_X^{p-1} + k(t))$$

and

$$\langle \Lambda(t, x), x \rangle \geq c_2 \|u\|_X^p - k(t)$$

for all $(t, x) \in [0, T] \times X$.

The self-dual approach allows for more general – even nonlinear – time-boundary conditions. It does give true periodic solutions, when the strong coercivity conditions of Lions are satisfied. However, it also applies when such conditions are not satisfied, yielding periodic solutions up to an isometry given by the infinitesimal generator of the underlying skew-adjoint operator. All the results of this chapter come from Ghoussoub and Moameni [64].

Part III
SELF-DUAL SYSTEMS AND THEIR ANTISYMMETRIC HAMILTONIANS

Stationary Navier-Stokes equations, finite-dimensional Hamiltonian systems, and equations driven by nonlocal operators such as the Choquard-Pekar equation are not completely self-dual systems but can be written in the form

$$0 \in \Lambda u + \bar{\partial} L(u),$$

where L is a self-dual Lagrangian on $X \times X^*$ and $\Lambda : D(\Lambda) \subset X \to X^*$ is a, not necessarily linear, operator. They can be solved by minimizing the functionals

$$I(u) = L(u, -\Lambda u) + \langle \Lambda u, u \rangle$$

on X by showing that their infimum is zero and that it is attained. These functionals are typical examples of a class of *self-dual functionals* that we introduce and study in this part of the book. They are defined as functionals of the form

$$I(u) = \sup_{v \in X} M(u, v),$$

where M is an *antisymmetric Hamiltonian* on $X \times X$, which contain and extend in a nonconvex way, the Hamiltonians associated to self-dual Lagrangians by standard Legendre duality (i.e., in one of the variables).

The class of antisymmetric Hamiltonians is quite large and easier to handle than the class of self-dual Lagrangians. It contains "Maxwellian" Hamiltonians of the form $M(x, y) = \varphi(y) - \varphi(x)$, with φ being convex and lower semicontinuous, but also those of the form $M(x, y) = \langle \Lambda x, x - y \rangle$, where Λ is a suitable – possibly nonlinear – operator, as well as their sum.

Self-dual functionals turn out to have many of the variational properties of the completely self-dual functionals, yet they are much more encompassing since they allow for the variational resolution of a larger class of linear and nonlinear partial differential equations.

Chapter 11
The Class of Antisymmetric Hamiltonians

Completely self-dual functionals can be written as

$$I(x) = L(x,0) = \sup_{y \in X} H_L(y, Bx) \quad \text{for all } x \in X,$$

where L is a B-self-dual Lagrangian for some operator B on X and H_L is the Hamiltonian associated to L. These Hamiltonians have some remarkable properties and are typical members of the class $\mathscr{H}_B^{skew}(X)$ of those concave-convex functions on state space $X \times X$ that are B-skew-symmetric, i.e., they satisfy $H(x, By) = -H(y, Bx)$. Besides being stable under addition, this class naturally leads to the much larger class $\mathscr{H}^{asym}(X)$ of *antisymmetric Hamiltonians* consisting of those functions on $X \times X$ that are weakly lower semicontinuous in the first variable, concave in the second, and zero on the diagonal. Functionals of the form

$$I(x) = \sup_{y \in X} M(x, y)$$

with M antisymmetric will be called *self-dual functionals* as they turn out to have many of the variational properties of the completely self-dual functionals yet are much more encompassing since they allow for the variational resolution of many more linear and nonlinear partial differential equations. The class of antisymmetric Hamiltonians is a convex cone that contains Hamiltonians of the form

$$M(x, y) = \varphi(y) - \varphi(x),$$

with φ being convex and lower semicontinuous, as well as their sum with functions of the form

$$M(x, y) = \langle \Lambda x, x - y \rangle,$$

provided $\Lambda : X \to X^*$ is a not necessarily linear *regular operator*, that is, Λ is weak-to-weak continuous, while satisfying that $u \to \langle u, \Lambda u \rangle$ is weakly lowerr semicontinuous. Examples of such operators are of course the linear positive operators but also include some linear but non-necessarily positive operators such as $\Lambda u = J\dot{u}$, which is regular on the Sobolev space $H^1[0, T]$ of $\mathbf{R}^{2N}$-valued functions on $[0, T]$, where J is the symplectic matrix. They also include important nonlinear operators such as

N. Ghoussoub, *Self-dual Partial Differential Systems and Their Variational Principles*, Springer Monographs in Mathematics, DOI 10.1007/978-0-387-84897-6_11, © Springer Science+Business Media, LLC 2009

the Stokes operator $u \to u \cdot \nabla u$ acting on the subspace of $H_0^1(\Omega, \mathbf{R}^n)$ consisting of divergence-free vector fields (up to dimension 4).

11.1 The Hamiltonian and co-Hamiltonians of self-dual Lagrangians

Let X be a reflexive Banach space. For each Lagrangian L on $X \times X^*$, we can define its corresponding Hamiltonian $H_L : X \times X \to \bar{\mathbf{R}}$ (resp., co-Hamiltonian $\tilde{H}_L : X^* \times X^* \to \bar{\mathbf{R}}$) by

$$H_L(x,y) = \sup\{\langle y, p\rangle - L(x,p); p \in X^*\},$$

resp.,

$$\tilde{H}_L(p,q) = \sup\{\langle y, p\rangle - L(y,q); y \in X\},$$

which is the Legendre transform in the second variable (resp., first variable). Their effective domains are

$$\mathrm{Dom}_1(H_L) := \{x \in X; H_L(x,y) > -\infty \text{ for all } y \in X\}$$
$$= \{x \in X; H_L(x,y) > -\infty \text{ for some } y \in X\}$$

and

$$\mathrm{Dom}_2(\tilde{H}_L) := \{q \in X^*; \tilde{H}_L(p,q) > -\infty \text{ for all } p \in X^*\}$$
$$= \{q \in X^*; \tilde{H}_L(p,q) > -\infty \text{ for some } p \in X^*\}$$

It is clear that $\mathrm{Dom}_1(L) = \mathrm{Dom}_1(H_L)$ and $\mathrm{Dom}_2(L) = \mathrm{Dom}_2(\tilde{H}_L)$, where the (partial) domains of a Lagrangian L have already been defined as

$$\mathrm{Dom}_1(L) = \{x \in X; L(x,p) < +\infty \text{ for some } p \in X^*\}$$

and

$$\mathrm{Dom}_2(L) = \{p \in X^*; L(x,p) < +\infty \text{ for some } x \in X\}.$$

We now procced to identify the class of Hamiltonians associated to self-dual Lagrangians. We denote by K_2^* (resp., K_1^*) the Legendre dual of a functional $K(x,y)$ with respect to the second variable (resp., the first variable).

Proposition 11.1. *If L is a B-self-dual Lagrangian on $X \times X^*$, then its Hamiltonian H_L on $X \times X$ (resp., its co-Hamiltonian $\tilde{H}_L$ on $X^* \times X^*$) satisfies the following properties:*

1. *For $y \in X$, the function $x \to H_L(x,y)$ $\big(\text{resp., for } p \in X^*, \text{ the function } q \to \tilde{H}_L(p,q)\big)$ is concave.*
2. *For $x \in X$, the function $y \to H_L(x,y)$ $\big(\text{resp., for } q \in X^*, \text{ the function } p \to \tilde{H}_L(p,q)\big)$ is convex and lower semicontinuous.*

3. For $x, y \in X$, $p, q \in X^*$, we have

$$H_L(y, Bx) \leq -H_L(x, By) \quad (resp., \ \tilde{H}_L(B^*q, p) \leq -\tilde{H}_L(B^*p, q)). \tag{11.1}$$

Proof. (1) Since L is convex in both variables, its corresponding Hamiltonian H_L is then concave in the first variable. Indeed, it suffices to prove the convexity of the function

$$x \to T(x, y) := -H_L(x, y) = \inf\{L(x, p) - \langle y, p \rangle; p \in X^*\}.$$

For that, consider $\lambda \in (0, 1)$ and elements $x_1, x_2 \in X$ such that $H_L(x_1, y)$ and $H_L(x_2, y)$ are finite. For every $a > T(x_1, y)$ (resp., $b > T(x_1, y)$), find $p_1, p_2 \in X^*$ such that

$$T(x_1, y) \leq L(x_1, p_1) - \langle y, p_1 \rangle \leq a \quad \text{and} \quad T(x_2, y) \leq L(x_2, p_2) - \langle y, p_2 \rangle \leq b.$$

Now use the convexity of L in both variables to write

$$\begin{aligned}
T(\lambda x_1 + (1 - \lambda)x_2, y) &= \inf\{L(\lambda x_1 + (1 - \lambda)x_2, p) - \langle y, p \rangle; p \in X^*\} \\
&\leq L(\lambda x_1 + (1 - \lambda)x_2, \lambda p_1 + (1 - \lambda)p_2) \\
&\quad - \langle y, \lambda p_1 + (1 - \lambda)p_2 \rangle \\
&\leq \lambda (L(x_1, p_1) - \langle y, p_1 \rangle) + (1 - \lambda)(L(x_2, p_2) - \langle y, p_2 \rangle) \\
&\leq \lambda a + (1 - \lambda)b.
\end{aligned}$$

This completes the proof of the concavity of $x \to H_L(x, y)$.

The proof of (2) is straightfroward since $y \to H_L(x, y)$ is the supremum of continuous and affine functionals, and hence, is convex and lower semicontinuous in the second variable.

(3) First note that the Legendre transform of $-H_L(\cdot, y)$ with respect to the first variable is related to the Legendre transform in both variables of its Lagrangian in the following way:

$$\begin{aligned}
(-H_L)_1^*(p, y) &= \sup\{\langle p, x \rangle + H_L(x, y); x \in X\} \\
&= \sup\{\langle p, x \rangle + \sup\{\langle y, q \rangle - L(x, q); q \in X^*\}; x \in X\} \\
&= L^*(p, y).
\end{aligned}$$

If now L is a B-self-dual Lagrangian, then the convex lower semicontinuous envelope of the function $\ell_y : x \to -H_L(x, By)$ (i.e., the largest convex lower semicontinuous function below ℓ_y) is

$$\begin{aligned}
\ell_y^{**}(x) &= \sup\{\langle p, x \rangle - (-H_L)_1^*(p, By); p \in X^*\} \\
&= \sup\{\langle p, x \rangle - L^*(p, By); p \in X^*\} \\
&\geq \sup\{\langle B^*q, x \rangle - L^*(B^*q, By); q \in X^*\} \\
&= \sup\{\langle q, Bx \rangle - L(y, q); q \in X^*\} \\
&= H_L(y, Bx).
\end{aligned}$$

It then follows that $H_L(y,Bx) \leq \ell_y^{**}(x) \leq \ell_y(x) = -H_L(x,By)$.

Remark 11.1. If B^* is onto or if it has dense range, while L^* is continuous in the first variable, then the function $x \to H_L(y,Bx)$ is equal to the convex lower semiconvex envelope of the function $x \to -H_L(x,By)$. Note also that, for any $z \in \mathrm{Dom}_1(H) = \{x \in X; H(x,y) > -\infty \text{ for all } y \in X\}$, we have that the function $H_{Bz}: x \to -H(x,Bz)$ is convex and valued in $\mathbf{R} \cup \{+\infty\}$. Moreover, $\mathrm{Dom}_1(H) \subset \mathrm{Dom}(H_{Bz})$ hence, for any $z, y \in \mathrm{Dom}_1(H)$,

$$H(z,By) = -H(y,Bz) \text{ if and only if } x \to H(x,Bz) \text{ is upper semicontinuous at } y.$$

As mentioned above, since a Lagrangian $L \in \mathscr{L}(X)$ is convex in both variables, its corresponding Hamiltoninan H_L is always concave in the first variable. However, H_L is not necessarily upper semicontinuous in the first variable, even if L is a self-dual Lagrangian. This leads to the following notion.

Definition 11.1. A Lagrangian $L \in \mathscr{L}(X)$ will be called *tempered* if for each $y \in \mathrm{Dom}_1(L)$, the map $x \to H_L(x,-y)$ from X to $\mathbf{R} \cup \{-\infty\}$ is upper semicontinuous.

The Hamiltonians of tempered self-dual Lagrangians satisfy the following remarkable antisymmetry relation:

$$H_L(y,Bx) = -H_L(x,By) \text{ for all } (x,y) \in X \times \mathrm{Dom}_1(L). \tag{11.2}$$

For such Hamiltonians, we obviously have

$$H_L(x,Bx) = 0 \quad \text{for all } x \in \mathrm{Dom}_1(L). \tag{11.3}$$

The Hamiltonians of basic self-dual Lagrangians

1. To any pair of proper convex lower semicontinuous functions φ and ψ on a Banach space X, one can associate a Lagrangian on phase space $X \times X^*$, via the formula $L(x,p) = \varphi(x) + \psi^*(p)$. Its Hamiltonian is $H_L(x,y) = \psi(y) - \varphi(x)$ if $x \in \mathrm{Dom}(\varphi)$ and $-\infty$ otherwise, while its co-Hamiltonian is $\tilde{H}_L(p,q) = \varphi^*(p) - \psi^*(q)$ if $q \in \mathrm{Dom}(\psi^*)$ and $-\infty$ otherwise. The domains are then $\mathrm{Dom}_1 H_L := \mathrm{Dom}(\varphi)$ and $\mathrm{Dom}_2(\tilde{H}_L) := \mathrm{Dom}(\psi^*)$. These Lagrangians are readily tempered.
2. More generally, if B is a bounded operator on X, $\Gamma: X \to X^*$ a linear operator such that $B^*\Gamma$ is skew-adjoint, $f \in X^*$, and $\varphi: X \to \mathbf{R} \cup \{+\infty\}$ a proper convex and lower semicontinuous, then the B-self-dual Lagrangian on $X \times X^*$

$$L(x,p) = \varphi(Bx) + \langle f,x \rangle + \varphi^*(\Gamma x + p - f)$$

has a Hamiltonian equal to

$$H_L(x,y) = \begin{cases} \varphi(y) - \varphi(Bx) + \langle \Gamma x, y \rangle + \langle f, -Bx + y \rangle & \text{if } Bx \in \text{Dom}(\varphi) \\ -\infty & \text{if } Bx \notin \text{Dom}(\varphi). \end{cases} \quad (11.4)$$

It is again a tempered Lagrangian.

We shall see later that not all self-dual Lagrangians are automatically tempered. The following lemma shows that it is the case under certain coercivity conditions.

Proposition 11.2. *Let L be a self-dual Lagrangian on a reflexive Banach space X. If for some $p_0 \in X$ and $\alpha > 1$ we have that $L(x, p_0) \le C(1 + \|x\|^\alpha)$ for all $x \in X$, then L is tempered.*

Proof. Note that in this case we readily have that $\text{Dom}_1(L) = \text{Dom}_1(H_L) = X$. Assume first that $\lim\limits_{\|x\| + \|p\| \to +\infty} \frac{L(x,p)}{\|x\| + \|p\|} = \infty$, and write

$$\begin{aligned} H_L(x,y) &= \sup\{\langle p, y \rangle - L(x,p); p \in X^*\} \\ &= \sup\{\langle p, y \rangle - L^*(p,x); p \in X^*\} \\ &= \sup\{\langle p, y \rangle - \sup\{\langle p, z \rangle + \langle x, q \rangle - L(z,q); z \in X, q \in X^*\}; p \in X^*\} \\ &= \sup\{\langle p, y \rangle + \inf\{\langle -p, z \rangle - \langle x, q \rangle + L(z,q); z \in X, q \in X^*\}; p \in X^*\} \\ &= \sup_{p \in X^*} \inf_{(z,q) \in X \times X^*} \{\langle p, y - z \rangle - \langle x, q \rangle + L(z,q)\}. \end{aligned}$$

The function S defined on the product space $(X \times X^*) \times X^*$ as

$$S((z,q), p) = \langle p, y - z \rangle - \langle x, q \rangle + L(z,q)$$

is convex and lower semicontinuous in the first variable (z,q) and concave and upper semicontinuous in the second variable p. Hence, in view of the coercivity condition, von Neuman's min-max theorem [8] applies and we get

$$\begin{aligned} H(x,y) &= \sup_{p \in X^*} \inf_{(z,q) \in X \times X^*} \{\langle p, y - z \rangle - \langle x, q \rangle + L(z,q)\} \\ &= \inf_{(z,q) \in X \times X^*} \sup_{p \in X^*} \{\langle p, y - z \rangle - \langle x, q \rangle + L(z,q)\} \\ &= \inf_{q \in X^*} \{\langle x, q \rangle + L(y,q)\} \\ &= -H_L(y, -x). \end{aligned}$$

It follows that L is tempered under the coercivity assumption.

Suppose now that $L(x, p_0) \le C(1 + \|x\|^\alpha)$, and consider the λ-regularization of its conjugate L^*, that is, $M_\lambda = L^* \star T_\lambda^*$, where $T_\lambda^*(p,x) = \frac{\|p\|^2}{2\lambda^2} + \frac{\lambda^2 \|x\|^2}{2}$. Since L^* is self-dual on $X^* \times X$, we get from Lemma 3.3 that M_λ is self-dual on $X^* \times X$. Moreover,

$$M_\lambda(p,x) = \inf\left\{ L^*(q,x) + \frac{\|p - q\|^2}{2\lambda^2} + \frac{\lambda^2 \|x\|^2}{2}; q \in X^* \right\}$$

$$\leq L^*(p_0, x) + \frac{1}{2\lambda^2}\|p - p_0\|^2 + \frac{\lambda^2\|x\|^2}{2}$$

$$\leq L(x, p_0) + \frac{1}{2\lambda^2}\|p - p_0\|^2 + \frac{\lambda^2\|x\|^2}{2}$$

$$\leq C_1 + C_2\|x\|^2 + C_3\|p\|^2,$$

which means that its dual M_λ^* is a self-dual Lagrangian on $X \times X^*$ that is coercive in both variables. By the first part of the proof, $H_{M_\lambda^*}$ is therefore a tempered antisymmetric Hamiltonian on X. But in view of Proposition 3.4, we have $M_\lambda^* = L \oplus T_\lambda$ and therefore $H_{M_\lambda^*} = H_L + H_{T_\lambda}$. Consequently, L itself is a tempered self-dual Lagrangian since both $H_{M_\lambda^*}$ and H_{T_λ} are.

11.2 Regular maps and antisymmetric Hamiltonians

We now introduce the following notion that extends considerably the class of Hamiltonians associated to self-dual Lagrangians.

Definition 11.2. Let E be a convex subset of a reflexive Banach space X. A functional $M : E \times E \to \overline{\mathbf{R}}$ is said to be *an antisymmetric Hamiltonian* (or an *AS-Hamiltonian*) on $E \times E$ if it satisfies the following conditions:

1. For every $x \in E$, the function $y \to M(x, y)$ is proper concave.
2. For every $y \in E$, the function $x \to M(x, y)$ is weakly lower semicontinuous.
3. $M(x, x) \leq 0$ for every $x \in E$.

We shall denote the class of antisymmetric Hamiltonians on a convex set E by $\mathscr{H}_B^{asym}(E)$. It is clear that $\mathscr{H}^{asym}(E)$ is a convex cone of functions on $E \times E$.

Examples of antisymmetric Hamiltonians

1. If L is a B-self-dual Lagrangian on a Banach space $X \times X^*$, then the Hamiltonian $M(x, y) = H_L(y, Bx)$ is clearly antisymmetric on $\mathrm{Dom}_1(L)$. Similarly, the Hamiltonian $N(p, q) = \tilde{H}_L(B^*p, q)$ is antisymmetric on $\mathrm{Dom}_2(L)$.
2. If $\Lambda : D(\Lambda) \subset X \to X^*$ is a non necessarily linear operator that is continuous on its domain for the weak topologies of X and X^* and such that $E \subset D(\Lambda)$, while the function $x \to \langle \Lambda x, x \rangle$ is weakly lower semicontinuous on E, then the Hamiltonian $H(x, y) = \langle x - y, \Lambda x \rangle$ is in $\mathscr{H}^{asym}(E)$.

More generally, we shall need the following definition.

Definition 11.3. Consider a pair of non necessarily linear operators $A : D(A) \subset Z \to X$ and $\Lambda : D(\Lambda) \subset Z \to X^*$, where Z and X are two Banach spaces.

1. The pair (A, Λ) is said to be *regular* if both are weak-to-weak continuous on their respective domains and if

$$u \to \langle Au, \Lambda u \rangle \text{ is weakly lower semicontinuous on } D(\Lambda) \cap D(A). \qquad (11.5)$$

2. The pair (A, Λ) is said to be *conservative* if it is regular and satisfies

$$\langle Au, \Lambda u \rangle = 0 \text{ for all } u \in D(\Lambda) \cap D(A). \qquad (11.6)$$

If A is the identity operator, we shall simply say that $\Lambda : D(\Lambda) \subset X \to X^*$ is a *regular* (resp., *conservative*) map.

The following proposition will be frequently used in the sequel.

Proposition 11.3. *Let L be a B-self-dual Lagrangian on a reflexive Banach space X, where $B : X \to X$ is a bounded linear operator, and let H_L (resp., $\tilde{H}_L$) be its corresponding Hamiltonian (resp., co-Hamiltonian). Suppose $A : D(A) \subset Z \to X$ and $\Lambda : D(\Lambda) \subset Z \to X^*$ are operators on a Banach space Z such that the pair $(A, B^*\Lambda)$ is regular.*

1. If A is linear, then the function

$$M(x, y) = \langle BAx - BAy, \Lambda x \rangle + H_L(Ay, BAx)$$

is an AS-Hamiltonian on any convex subset E of $D(A) \cap D(\Lambda)$.
2. If the operator Λ is linear, then the function

$$\tilde{M}(x, y) = \langle BAx, \Lambda x - \Lambda y \rangle + \tilde{H}_L(-B^*\Lambda x, -\Lambda y)$$

is an AS-Hamiltonian on any convex subset $\tilde{E}$ of $D(A) \cap D(\Lambda)$.

Proof. (1) For each $x \in E$, we have that $y \to M(x, y)$ is concave since $y \to \langle BAx - BAy, \Lambda x \rangle$ is clearly linear, while $y \to H_L(Ay, BAx)$ is concave.

For each $y \in E$, the function $x \to M(x, y)$ is weakly lower semicontinuous since $x \to \langle BAy, \Lambda x \rangle$ is weakly continuous by assumption, while $x \to \langle BAx, \Lambda x \rangle$ is weakly lower semicontinuous on E since the pair $(A, B^*\Lambda)$ is regular. Moreover, $x \to H_L(y, x)$ is the supremum of continuous affine functions on X and $x \to BAx$ is weakly continuous on E. Hence, $x \to H_L(Ay, BAx)$ is also weakly lower semicontinuous.

Finally, use Proposition 11.1 to infer that $H_L(Ax, BAx) \le 0$ and therefore $M(x, x) \le 0$ for all $x \in E$.

(2) Similarly, for each $x \in \tilde{E}$, we have that $y \to \tilde{M}(x, y)$ is concave since $y \to \langle BAx, \Lambda x - \Lambda y \rangle$ is clearly linear, while $y \to \tilde{H}_L(-B^*\Lambda x, -\Lambda y)$ is concave. Also, for each $y \in \tilde{E}$, the function $x \to \tilde{M}(x, y)$ is weakly lower semicontinuous since $x \to \langle BAx, \Lambda y \rangle$ is weakly continuous, while again $x \to \langle BAx, \Lambda x \rangle$ is weakly lower semicontinuous. The last term $x \to \tilde{H}_L(-B^*\Lambda x, -\Lambda y)$ is a composition of a convex lower semicontinuous function with a weakly continuous function and hence, is weakly lower semicontinuous. Finally, use again Proposition 11.1 to deduce that $\tilde{H}_L(-B^*\Lambda x, -\Lambda x) \le 0$ and therefore $\tilde{M}(x, x) \le 0$ for all $x \in E$.

Exercises 11.A. Examples of regular maps

1. Show that if $A : X \to X$ (resp., $\Lambda : X \to X^*$) are bounded linear operators such that $\langle Au, \Lambda u \rangle \geq 0$, then they form a *regular pair*, and that regular conservative maps include skew-symmetric bounded linear operators.

2. Show that the linear operator $\Lambda u = J\dot{u}$ is regular on the Sobolev space $H^1_{per}[0, T]$ of $\mathbf{R}^{2N}$-valued periodic functions on $[0, T]$, where J is the symplectic matrix.

3. Show that the nonlinear operator $u \to u \cdot \nabla u$ acting on the subspace of $H^1_0(\Omega, \mathbf{R}^n)$ consisting of divergence-free vector fields into its dual is regular as long as the dimension $n \leq 4$.

4. Show that completely continuous operators (i.e., those that map weakly compact sets in X into norm compact sets in X^*) are necessarily regular maps.

5. Let Ω be a smooth bounded domain in $\mathbf{R}^n$ ($n \geq 3$), and let $b : \Omega \times R \times \mathbf{R}^n \to R$ be a Caratheodory function such that for some positive constants c_1, c_2, q_1, q_2, we have for all $(u, p) \in R \times \mathbf{R}^n$ and almost all $x \in \Omega$,

$$|b(x, u, p)| \leq c_1 + c_2|u|^{q_1} + c_2|p|^{q_2}.$$

Show that the map Λ defined by $\Lambda u(x) := b(x, u(x), Du(x))$ is completely continuous from $H^1_0(\Omega)$ into $H^{-1}(\Omega)$, provided $q_1 < \frac{n+2}{n-2}$ and $q_2 < \frac{n+2}{n}$.
Hint: The embedding $i : H^1_0(\Omega) \to L^p(\Omega)$ is compact if $p < \frac{2n}{n-2}$, while $j : L^q(\Omega) \to H^{-1}(\Omega)$ is compact, provided $q > \frac{2n}{n+2}$.

6. Let $X \subset H \subset X^*$ be an evolution triple such that the injection $X \to H$ is compact, and let $\varphi : H \to R \cup \{+\infty\}$ be convex lower semicontinuous on H such that $X \subset \text{Dom}(\varphi)$. Show that, for each $\lambda > 0$, the map $\partial \varphi_\lambda$ is completely continuous from X to X^*, where

$$\varphi_\lambda(x) = \inf \left\{ \varphi(z) + \frac{\|x - z\|_H}{\lambda}; z \in H \right\}.$$

Hint: Show that $\|\partial \varphi_\lambda x - \partial \varphi_\lambda y\|_H \leq \frac{1}{\lambda}\|x - y\|_H$.

7. Suppose $f : [0, T] \times \mathbf{R}^n \to \mathbf{R}^n$ is a continuous map such that for every bounded set $B \subset \mathbf{R}^n$ there exists $C(B) > 0$ such that for $x, y \in B$

$$\|f(t, x) - f(t, y)\| \leq C(B)\|x - y\|.$$

Consider the space $\mathscr{X}_{p,q}[0, T]$, where $p > 1$ and $\frac{1}{p} + \frac{1}{q} = 1$, and show that the map $F : \mathscr{X}_{p,q}[0, T] \to L^q_{\mathbf{R}^n}[0, T]$ defined by $F(u)(t) := f(t, u(t))$ is completely continuous.

11.3 Self-dual functionals

Definition 11.4. Let $I : X \to R \cup \{+\infty\}$ be a functional on a Banach space X.

1. Say that I is a *self-dual functional on a convex set* $E \subset X$ if it is nonnegative and if there exists an antisymmetric Hamiltonian $M : E \times E \to R$ such that

$$I(x) = \sup_{y \in E} M(x, y) \text{ for every } x \in E.$$

2. Say that a self-dual functional I is *strongly coercive on* E if, for some $y_0 \in E$, the set $E_0 = \{x \in E; M(x, y_0) \leq 0\}$ is bounded in X, where M is a corresponding AS-Hamiltonian.

Note that this notion of coercivity is slightly stronger than the coercivity of I, which amounts to saying that $\lim\limits_{\|x\|\to+\infty} I(x) = \lim\limits_{\|x\|\to+\infty} \sup\limits_{y\in E} M(x,y) = +\infty.$

Basic examples of self-dual functionals

1. Every completely self-dual functional is clearly a self-dual functional. Indeed, if $I(x) = L(x,0)$, where L is a (partially) B-self-dual Lagrangian for some bounded linear operator $B : X \to X$, then

$$I(x) = L(x,0) = \sup\{H_L(y,Bx); y \in \mathrm{Dom}_1(L)\} \text{ for every } x \in \mathrm{Dom}_1(L),$$

where H_L is the Hamiltonian associated to L. It then follows that I is a self-dual functional on $\mathrm{Dom}_1(L)$ with $M(x,y) := H_L(y,Bx)$ being the corresponding AS-Hamiltonian. Strong coercivity for I is then implied by the condition that, for some $y_0 \in X$, we have $\lim\limits_{\|x\|\to+\infty} H_L(y_0,Bx) = +\infty.$

In the case where $L(x,p) = \varphi(x) + \varphi^*(p)$ for φ proper, convex, and lower semi-continuous, the strong coercivity is simply equivalent to the coercivity of φ.

2. More generally, consider a functional of the form $I(x) = L(x,\Gamma x)$, where L is a B-self-dual Lagrangian on the graph of an operator $\Gamma : X \to X^*$ such that $B^* \circ \Gamma$ is a skew-adjoint bounded linear operator. It is then easy to see that

$$M(x,y) = \langle Bx - By, \Gamma x \rangle + H_L(y,Bx)$$

is an *AS*-Hamiltonian for I on $\mathrm{Dom}_1(L)$.

In the following we show that this notion is much more encompassing, as it covers iterates of self-dual Lagrangians with operators that need not be skew-adjoint or even linear.

Proposition 11.4. *Let L be a B-self-dual Lagrangian on a reflexive Banach space X, where $B : X \to X$ is a bounded linear operator, and let $A : D(A) \subset Z \to X$ and $\Lambda : D(\Lambda) \subset Z \to X^*$ be two operators on a Banach space Z such that the pair $(A, B^*\Lambda)$ is regular. Consider on $D(A) \cap D(\Gamma)$ the functional*

$$I(x) = L(Ax, -\Lambda x) + \langle BAx, \Lambda x \rangle.$$

1. *If the operator A is linear and E is a convex subset of $D(\Lambda) \cap D(A)$ such that $\mathrm{Dom}_1(L) \subset A(E)$, then I is a self-dual functional on E with an AS-Hamiltonian given by*

$$M(x,y) = \langle BAx - BAy, \Lambda x \rangle + H_L(Ay, BAx),$$

   where H_L is the Hamiltonian associated to L.

2. *If the operator Λ is linear, and $\tilde{E}$ is a convex subset of $D(\Lambda) \cap D(A)$ such that $\mathrm{Dom}_2(L) \subset \Lambda(-\tilde{E})$, then I is a self-dual functional on $\tilde{E}$ with an AS-Hamiltonian*

given by

$$\tilde{M}(x,y) = \langle BAx, \Lambda x - \Lambda y \rangle + \tilde{H}_L(-B^*\Lambda x, -\Lambda y),$$

where $\tilde{H}_L$ is the co-Hamiltonian associated to L.

Proof. The B-self-duality of L yields that I is nonnegative on $D(\Lambda) \cap D(A)$. For 1), we use the fact that the Lagrangian L is B-self-dual and that $\mathrm{Dom}_1(L) \subset A(E)$ to write for each $x \in E$

$$
\begin{aligned}
I(x) &= L(Ax, -\Lambda x) + \langle BAx, \Lambda x \rangle \\
&= L^*(-B^*\Lambda x, BAx) + \langle BAx, \Lambda x \rangle \\
&= \sup\{\langle z, -B^*\Lambda x \rangle + \langle p, BAx \rangle - L(z, p); z \in X, p \in X^*\} + \langle BAx, \Lambda x \rangle \\
&= \sup\{\langle z, -B^*\Lambda x \rangle + \langle p, BAx \rangle - L(z, p); z \in \mathrm{Dom}_1(L), p \in X^*\} + \langle BAx, \Lambda x \rangle \\
&= \sup\{\langle Ay, -B^*\Lambda x \rangle + \sup\{\langle p, BAx \rangle - L(Ay, p); , p \in X^*\}; y \in E\} + \langle BAx, \Lambda x \rangle \\
&= \sup\{\langle BAx - BAy, \Lambda x \rangle + H_L(Ay, BAx); y \in E\} \\
&= \sup_{y \in E} M(x, y),
\end{aligned}
$$

where $M(x,y) = \langle BAx - BAy, \Lambda x \rangle + H_L(Ay, BAx)$ and where H_L is the Hamiltonian associated to L. By Proposition 11.3, M is an antisymmetric Hamiltonian.

Similarly, for (2) we use the fact that $\mathrm{Dom}_2(L) \subset \Lambda(-\tilde{E})$ to write for each $x \in \tilde{E}$

$$
\begin{aligned}
I(x) &= L(Ax, -\Lambda x) + \langle \Lambda x, BAx \rangle \\
&= L^*(-B^*\Lambda x, BAx) + \langle \Lambda x, BAx \rangle \\
&= \sup\{\langle z, -B^*\Lambda x \rangle + \langle p, BAx \rangle - L(z, p); z \in X, p \in X^*\} + \langle \Lambda x, BAx \rangle \\
&= \sup\{\langle z, -B^*\Lambda x \rangle + \langle p, BAx \rangle - L(z, p); z \in X, p \in \mathrm{Dom}_2(L)\} + \langle \Lambda x, BAx \rangle \\
&= \sup\{\langle -\Lambda y, BAx \rangle + \sup\{\langle z, -B^*\Lambda x \rangle - L(z, -\Lambda y); , z \in X\}; y \in \tilde{E}\} + \langle \Lambda x, BAx \rangle \\
&= \sup\{\langle BAx, \Lambda x - \Lambda y \rangle + \tilde{H}_L(-B^*\Lambda x, -\Lambda y); y \in \tilde{E}\} \\
&= \sup_{y \in \tilde{E}} \tilde{M}(x, y),
\end{aligned}
$$

where $\tilde{M}(x,y) = \langle BAx, \Lambda x - \Lambda y \rangle + \tilde{H}_L(-B^*\Lambda x, -\Lambda y)$ and where $\tilde{H}_L$ is the co-Hamiltonian associated to L. By Proposition 11.3, $\tilde{M}$ is an antisymmetric Hamiltonian on $\tilde{E}$.

Exercises 11.B. The class of *B*-skew-symmetric Hamiltonians

1. Say that a functional $H : X \times X \to \mathbf{R} \cup \{+\infty\} \cup \{-\infty\}$ is a *B-skew-symmetric Hamiltonian* if:

   a. For each $y \in X$, the function $x \to H(x, y)$ is concave.
   b. For each $x \in X$, the function $y \to H(x, y)$ is convex and lower semicontinuous.
   c. For each $x, y \in X$, we have $H(y, Bx) \leq -H(x, By)$.

The class of B-skew-symmetric Hamiltonians on X will be denoted by $\mathscr{H}_B^{skew}(X)$. The most basic skew-symmetric Hamiltonian is $H(x,y) = \|y\|^2 - \|x\|^2$ (Maxwell's Hamiltonian) or more generally $H(x,y) = \varphi(y) - \varphi(Bx)$ is a B-skew-symmetric Hamiltonian for any operator B on X. Show that for a bounded operator B on X, $\Gamma : X \to X^*$ a linear operator such that $B^*\Gamma$ is skew-adjoint, $f \in X^*$, and $\varphi : X \to \mathbf{R} \cup \{+\infty\}$ proper, convex, and lower semicontinuous, the Hamiltonian

$$H(x,y) = \begin{cases} \varphi(y) - \varphi(Bx) + \langle \Gamma x, y \rangle + \langle f, -Bx+y \rangle & \text{if } Bx \in \text{Dom}(\varphi) \\ -\infty & \text{if } Bx \notin \text{Dom}(\varphi) \end{cases} \qquad (11.7)$$

is B-skew-symmetric.

2. Show that scalar products $H(x,y) = \langle Bx, y \rangle$ on a Hilbert space are clearly B-skew-symmetric Hamiltonians for any skew-adjoint operator B on H. More generally, for any operator $\Gamma : X \to X^*$, the Hamiltonian $H(x,y) = \langle \Gamma x, y \rangle$ is B-skew-symmetric whenever $B : X \to X$ is an operator such that $B^*\Gamma$ is nonpositive (i.e., $\langle \Gamma x, Bx \rangle \le 0$ for all $x \in X$).

3. Deduce that this class does not characterize those functions $H : X \times X \to \bar{\mathbf{R}}$ such that there exists a self-dual Lagrangian L on $X \times X^*$ such that $H = H_L$.

4. Suppose $H : X \times X \to \bar{\mathbf{R}}$ is such that $H_2^*(x,p) = (-H)_1^*(p,x)$ for each $(x,p) \in X \times X^*$. Show that $L(x,p) := H_2^*(x,p) = (H)_1^*(p,x)$ is a self-dual Lagrangian such that $H = H_L$.

5. Show that the class of B-skew-symmetric Hamiltonians satisfies the following permanence properties:

   a. If H and K are in $\mathscr{H}_B^{skew}(X)$ and $\lambda > 0$, then the Hamiltonians $H + K$ (defined as $-\infty$ if the first variable is not in $\text{Dom}_1(H) \cap \text{Dom}_1(K)$) and $\lambda \cdot H$ also belong to $\mathscr{H}_B^{skew}(X)$.

   b. If $H_i \in \mathscr{H}_{B_i}^{skew}(X_i)$, where X_i is a reflexive Banach space for each $i \in I$, then the Hamiltonian $H := \Sigma_{i \in I} H_i$ defined by $H((x_i)_i, (y_i)_i) = \Sigma_{i \in I} H_i(x_i, y_i)$ is in $\mathscr{H}_{\bar{B}}^{skew}(\Pi_{i \in I} X_i)$, where $\bar{B}$ is the operator $\Pi_{i \in I} B_i$.

   c. If $H \in \mathscr{H}_B^{skew}(X)$ and $\Gamma : X \to X^*$ is a bounded linear operator such that $B^*\Gamma$ is skew-adjoint, then the Hamiltonian H_B defined by $H_B(x,y) = H(x,y) + \langle \Gamma x, y \rangle$ is also in $\mathscr{H}_B^{skew}(X)$.

   d. If $H \in \mathscr{H}_{B_1}^{skew}(X)$ and $K \in \mathscr{H}_{B_2}^{skew}(Y)$, then for any bounded linear operator $A : X \to Y^*$ such that $AB_1 = B_2^*A$, the Hamiltonian $H +_A K$ defined by

   $$(H +_A K)((x,y),(z,w)) = H(x,z) + K(y,w) + \langle A^* y, z \rangle - \langle Ax, w \rangle$$

   belongs to $\mathscr{H}_{(B_1,B_2)}^{skew}(X \times Y)$.

   e. Let φ be proper, convex, and lower semicontinuous function on $X \times Y$, B_1 an operator on X, and B_2 an operator on Y. If A is a bounded linear operator $A : X \to Y^*$ such that $AB_1 = B_2^*A$, then the Hamiltonian $H_{\varphi,A}$ defined by

   $$H_{\varphi,A}((x,y),(z,w)) = \varphi(z,w) - \varphi(B_1 x, B_2 y) + \langle A^* y, z \rangle - \langle Ax, w \rangle$$

   belongs to $\mathscr{H}_{(B_1,B_2)}^{skew}(X \times Y)$.

Chapter 12
Variational Principles for Self-dual Functionals and First Applications

We establish the basic variational principle for self-dual functionals which states that under appropriate coercivity conditions, the infimum of such a functional is attained and is equal to zero. Applying this to functionals of the form

$$I(x) = L(x, -\Lambda x) + \langle x, \Lambda x \rangle,$$

we obtain solutions to nonlinear equations of the form $0 \in \Lambda x + \overline{\partial} L(x)$. This allows the variational resolution of a large class of PDEs, in particular nonlinear Lax-Milgram problems of the type

$$\Lambda u + Au + f \in -\partial \varphi(u),$$

where φ is a convex lower semicontinuous functional, Λ is a nonlinear regular operator, and A is a linear – not necessarily bounded – positive operator.

Immediate applications include a variational resolution to various equations involving nonlinear operators, such as the stationary Navier-Stokes equation, as well as to equations involving nonlocal terms such as the generalized Choquard-Pekar Schrödinger equation.

12.1 Ky Fan's min-max principle

We start by proving the following important result that is due to Ky Fan [49].

Theorem 12.1. *Let E be a closed convex subset of a reflexive Banach space X, and let $M(x,y)$ be an antisymmetric Hamiltonian on $E \times E$ such that for some $y_0 \in E$, the set $E_0 = \{x \in E; M(x,y_0) \leq 0\}$ is bounded in X. Then, there exists $x_0 \in E$ such that*

$$\sup_{y \in E} M(x_0, y) \leq 0. \tag{12.1}$$

This will follow immediately from the following lemma.

Lemma 12.1. *Let $\emptyset \neq D \subset E \subset X$, where E is a closed convex set in a Banach space X, and consider $M : E \times conv(D) \to \mathbf{R} \cup \{\pm\infty\}$ to be a functional such that*

N. Ghoussoub, *Self-dual Partial Differential Systems and Their Variational Principles*, Springer Monographs in Mathematics, DOI 10.1007/978-0-387-84897-6_12,
© Springer Science+Business Media, LLC 2009

1. *For each $y \in D$, the map $x \mapsto M(x,y)$ is weakly lower semicontinuous on E.*
2. *For each $x \in E$, the map $y \mapsto M(x,y)$ is concave on conv(D).*
3. *There exists $\gamma \in \mathbf{R}$ such that $M(x,x) \leq \gamma$ for every $x \in$ conv(D).*
4. *There exists a nonempty subset $D_0 \subset D \cap B$, where B is a weakly compact convex subset of E such that $E_0 = \bigcap_{y \in D_0} \{x \in E : M(x,y) \leq \gamma\}$ is weakly compact.*

Then, there exists a point $x^* \in E$ such that $M(x^*,y) \leq \gamma$ for all $y \in D$.

Proof. Suppose first that E is weakly compact. By way of contradiction, assume that for every $x \in E$ there exists some point $y \in D$ such that

$$M(x,y) > \gamma. \tag{12.2}$$

It follows that $E \subset \bigcup_{y \in D} N(y)$, where each set $N(y) = \{x \in E : M(x,y) > \gamma\}$ is weakly open in E. Since the latter is weakly compact, $\{N(y)\}$ has a finite subcover $N(y_1), \ldots, N(y_m)$. Choose a partition of unity $\mu_j : E \to \mathbf{R}$ subordinate to $\{N(y)\}$, and define a map $T : E \to E$ by

$$T(x) = \sum_{j=1}^{m} \mu_j(x) y_j,$$

which is continuous and maps E into $S \equiv \mathrm{co}\{y_1, \ldots, y_m\}$. In particular, T maps S into itself, and therefore, by Brouwer's fixed-point theorem, there exists $x_\lambda \in S$ such that $T(x_\lambda) = x_\lambda$.

Letting $I = \{j : 1 \leq j \leq m, \mu_j(x_\lambda) > 0\}$, we get that $x_\lambda = \sum_{j \in I} \mu_j(x_\lambda) y_j$ and, for all $j \in I$, $x_\lambda \in N(y_j)$, meaning that $\varphi(x_\lambda, y_j) > \gamma$, which contradicts the concavity of $y \to M(x_\lambda, y)$.

Suppose now that E is not assumed to be weakly compact but only closed and convex. For each $y \in D$, we set $K(y) = \{x \in E : M(x,y) \leq \gamma\}$. We shall prove that the collection $\{E_0 \cap K(y) : y \in D\}$ has the finite intersection property.

Indeed, for any arbitrary finite subset $\{y_1, \ldots, y_m\}$ of D, we consider the set $D_1 = D_0 \cup \{y_1, \ldots, y_m\}$ and E_1 to be the convex hull of $B \cup \{y_1, \ldots, y_m\} \subset E$. Since B is weakly compact and convex, so is E_1, and we can apply the first part of the proof to $D_1 \subset E_1$ to find a vector $x' \in E_1$ such that $M(x',y) \leq \gamma$ for all $y \in D_1$, which means that $x' \in E_0 \cap \left[\bigcap_{j=1}^{m} K(y_j) \right]$.

The collection $\{E_0 \cap K(y) : y \in D\}$ therefore has the finite intersection property, and since E_0 is weakly compact and $K(y)$ is weakly closed, $E_0 \cap K(y)$ is also weakly compact, and hence, $\bigcap_{y \in D} [E_0 \cap K(y)] \neq \emptyset$. So there exists a vector $x^* \in E$ such that $x^* \in K(y)$ for all $y \in D$ and thus $M(x^*,y) \leq \gamma$ for all $y \in D$, and we are done.

Now we can deduce the main variational principle for self-dual functionals.

Theorem 12.2. *If $I : E \to \mathbf{R} \cup \{+\infty\}$ is a self-dual functional that is strongly coercive on a closed convex subset E of a reflexive Banach space X, then there exists $\bar{x} \in E$ such that*

$$I(\bar{x}) = \inf_{x \in E} I(x) = 0.$$

The following is a key variational principle for the superposition of self-dual vector fields with regular operators.

Theorem 12.3. *Let L be a B-self-dual Lagrangian on a reflexive Banach space X, where $B : X \to X$ is a bounded linear operator. Consider two operators $A : D(A) \subset Z \to X$ and $\Lambda : D(\Lambda) \subset Z \to X^*$, such that the pair $(A, B^*\Lambda)$ is regular from a reflexive Banach space Z into $X \times X^*$. Assume one of the two following situations:*

1. *The operator A is linear, E is a closed convex subset of $D(\Lambda) \cap D(A)$ such that $\mathrm{Dom}_1(L) \subset A(E)$, as well as the coercivity condition:*

$$\lim_{x \in E; \|x\| \to +\infty} H_L(0, BAx) + \langle BAx, \Lambda x \rangle = +\infty. \tag{12.3}$$

2. *The operator Λ is linear, $\tilde{E}$ is a closed convex subset of $D(\Lambda) \cap D(A)$ such that $\mathrm{Dom}_2(L) \subset \Lambda(-\tilde{E})$, as well as the coercivity condition:*

$$\lim_{x \in \tilde{E}; \|x\| \to +\infty} \tilde{H}_L(-B^*\Lambda x, 0) + \langle BAx, \Lambda x \rangle = +\infty. \tag{12.4}$$

Then, $I(x) := L(Ax, -\Lambda x) + \langle BAx, \Lambda x \rangle$ is a self-dual functional on E (resp., $\tilde{E}$) and there exists $\bar{x} \in D(\Lambda) \cap D(A)$ such that

$$I(\bar{x}) = \inf_{x \in X} I(x) = 0, \tag{12.5}$$

$$-\Lambda \bar{x} \in \bar{\partial}_B L(A\bar{x}). \tag{12.6}$$

Proof. Note first that $I(x) = L(Ax, -\Lambda x) + \langle \Lambda x, BAx \rangle \geq 0$ for all $x \in E$ (resp., $x \in \tilde{E}$). In the first case, the associated AS-Hamiltonian is

$$M(x, y) = \langle BAx - BAy, \Lambda x \rangle + H_L(Ay, BAx),$$

where H_L is the Hamiltonian associated to L. The strong coercivity follows from the fact that the set $E_0 = \{x \in E; M(x, 0) \leq 0\}$ is bounded in X since $M(x, 0) = H_L(0, BAx) + \langle \Lambda x, BAx \rangle$ and the latter goes to infinity with $\|x\|$.

In the second case, the AS-Hamiltonian is

$$\tilde{M}(x, y) = \langle BAx, \Lambda x - \Lambda y \rangle + \tilde{H}_L(-B^*\Lambda x, -\Lambda y).$$

The strong coercivity follows from the fact that $\tilde{M}(x, 0) = \tilde{H}_L(-B^*\Lambda x, 0) + \langle \Lambda x, BAx \rangle$ and the latter goes to infinity with $\|x\|$.

It follows from Theorem 12.2 that in either case there exists $\bar{x} \in D(\Lambda) \cap D(A)$ such that

$$I(\bar{x}) = \inf_{x \in X} I(x) = L(A\bar{x}, -\Lambda \bar{x}) + \langle \Lambda \bar{x}, BA\bar{x} \rangle = 0,$$

which means that $(-B^*\Lambda \bar{x}, B\bar{x}) \in \partial L(\bar{x}, -\Lambda \bar{x})$ or equivalently $-\Lambda \bar{x} \in \bar{\partial}_B(A\bar{x})$.

Remark 12.1. One can easily see that the hypothesis in Theorem 12.3 above can be relaxed in two ways:

1. The pair (A, Λ) need only be regular on the set E.
2. The Lagrangian L need only be B-self-dual on the graph $\{(Ax, \Lambda x); x \in E\}$ or $\{(Ax, \Lambda x); x \in \tilde{E}\}$.

In the case where A is the identity and $\Lambda \equiv 0$, this yields in particular the following refinement of Theorem 6.1.

Corollary 12.1. *If L is a partially B-self-dual Lagrangian on a reflexive Banach space such that* $\lim_{\|x\| \to +\infty} H_L(0, Bx) = +\infty$, *then there exists $\bar{x} \in X$ such that*

$$\begin{cases} L(\bar{x}, 0) = \inf_{x \in X} L(x, 0) = 0 \\ \quad 0 \in \partial_B L(\bar{x}). \end{cases} \tag{12.7}$$

Exercises 12.A. A more general Min-Max principle

Weaker hypotheses on the Hamiltonian M are known to be sufficient to obtain the same conclusion as in the Ky Fan min-max theorem above. For our purpose, this translates to only assuming that the operator Λ is *pseudoregular* in the sense that it only needs to satisfy the following property:

$$\text{If } x_n \rightharpoonup x \text{ and } \limsup_n \langle \Lambda x_n, x_n - x \rangle \leq 0, \quad \text{then } \liminf_n \langle \Lambda x_n, x_n - y \rangle \geq \langle \Lambda x, x - y \rangle. \tag{12.8}$$

1. Show that if $T : H \to H$ is a Lipschitz continuous monotone map such that $\langle Tx, x \rangle = 0$ for all $x \in H$, then T is pseudoregular.
2. Show that the same conclusion as in Theorem 12.1 will still hold if Λ is only assumed to be pseudoregular.
3. Show that the same conclusion as in Theorem 12.1 will still hold if L is only supposed to be a subself-dual Lagrangian.

12.2 Variational resolution for general nonlinear equations

We now give some of the most immediate applications of Theorem 12.3 to the variational resolution of various nonlinear systems that are not of Euler-Lagrange type. We shall actually apply it to the most basic self-dual Lagrangians of the form $L(x, p) = \varphi(x) + \varphi^*(p)$, where φ is a convex function. The following version will be used throughout this chapter.

Corollary 12.2. *Let φ be a bounded below convex lower semicontinuous function on a reflexive Banach space X, let $f \in X^*$, and suppose $A : D(A) \subset Z \to X$ and $\Lambda : D(\Lambda) \subset Z \to X^*$ are two operators, with A being linear. Consider $E \subset D(A) \cap D(\Lambda)$ to be a closed convex subset Z such that*

$$\text{Dom}(\varphi) \subset A(E), \tag{12.9}$$

$$\text{the pair } (A, \Lambda) \text{ is regular on the set } E, \tag{12.10}$$

and

$$\lim_{\substack{\|x\| \to \infty \\ x \in E}} \varphi(Ax) + \langle x, \Lambda x + f \rangle = +\infty. \tag{12.11}$$

Then, there exists a solution $\bar{x} \in E$ to the equation

$$0 \in f + \Lambda x + \partial \varphi(Ax). \tag{12.12}$$

It is obtained as a minimizer of the self-dual functional

$$I(x) = \varphi(Ax) + \varphi^*(-\Lambda x - f) + \langle Ax, \Lambda x + f \rangle, \tag{12.13}$$

which has zero as its minimal value over X.

Proof. Consider the convex and lower semicontinuous function $\psi(x) = \varphi(x) + \langle f, x \rangle$ and the Lagrangian $L(x, p) = \psi(x) + \psi^*(p)$, which is then self-dual. Since φ is bounded below, we have $\text{Dom}_1(L) \subset \text{Dom}(\varphi)$. Moreover, $H_L(x, y) = \varphi(y) - \varphi(x) + \langle f, x + y \rangle$ when $x \in \text{Dom}(\varphi)$, and Theorem 12.3 then applies to yield the claim.

Example 12.1. Sub-quadratic semilinear equations with advection

We now give some of the most immediate applications to semilinear equations of the type

$$\begin{cases} \mathbf{a} \cdot \nabla u - \Delta u + a_0 u = \alpha |u|^{p-1} u - \beta |u|^{q-1} u + f(x) & \text{on } \Omega \\ u = 0 & \text{on } \partial \Omega, \end{cases} \tag{12.14}$$

where Ω is a bounded smooth domain in $\mathbf{R}^n$ ($n \geq 3$), $f \in H^{-1}(\Omega)$, $\alpha, \beta \geq 0$, and $p, q > 0$. Throughout this section, $\mathbf{a} : \Omega \to \mathbf{R}^n$ will be a smooth vector field on a neighborhood of a bounded domain Ω of $\mathbf{R}^n$, and we shall consider the first-order linear operator $\mathbf{a} \cdot \nabla v = \Sigma_{i=1}^n a_i \frac{\partial v}{\partial x_i}$.

Theorem 12.4. *Let $f \in H^{-1}$, $\alpha \geq 0$, $\beta > 0$, $0 < q \leq \frac{n+2}{n-2}$, $0 < p < \max\{1, q\}$, and $\text{div}(\mathbf{a}) - 2a_0 < \lambda_1$ on $\bar{\Omega}$, where λ_1 is the first eigenvalue of the Laplacian on $H_0^1(\Omega)$. Consider the convex continuous functional on $H_0^1(\Omega)$,*

$$\psi(u) = \frac{1}{2} \int_\Omega |\nabla u|^2 \, dx + \frac{\beta}{q+1} \int_\Omega |u|^{q+1} \, dx - \int_\Omega f u \, dx.$$

The functional

$$I(u) = \psi(u) + \psi^*(-\mathbf{a} \cdot \nabla u - a_0 u + \alpha |u|^{p-1} u)$$
$$+ \frac{1}{2} \int_\Omega (2a_0 - \text{div}(\mathbf{a})) u^2 \, dx - \alpha \int_\Omega |u|^{p+1} \, dx$$

is then self-dual on $H_0^1(\Omega)$, has zero infimum, and the latter is attained at a solution $\bar{u}$ for equation (12.14).

Proof. First we check that the nonlinear operator $\Lambda u = \mathbf{a} \cdot \nabla u + a_0 u - \alpha |u|^{p-1} u$ is regular. Indeed, it is weak-to-weak continuous from $H_0^1(\Omega)$ into $H^{-1}(\Omega)$ since by the Sobolev embedding we have that $u_n \to u$ strongly in L^r for all $r < 2^* := \frac{2n}{n-2}$ whenever $u_n \to u$ weakly in $H_0^1(\Omega)$. Setting $\alpha = \frac{2n}{n+2}$, we have that $1 < \alpha p < \frac{2n}{n+2} \cdot \frac{n+2}{n-2} = \frac{2n}{n-2}$, and therefore $|u_n|^{p-1} u_n \to |u|^{p-1} u$ strongly in L^α.

Note that $\frac{1}{\alpha} + \frac{1}{2^*} = 1$, and therefore, by Hölder's inequality, we have for any $v \in H_0^1(\Omega) \subset L^{2^*}$ that

$$\left| \int_\Omega |u_n|^{p-1} u_n v - |u|^{p-1} u v \, dx \right| \leq \int_\Omega \left| |u_n|^{p-1} u_n - |u|^{p-1} u \right| |v| \, dx$$

$$\leq \left(\int_\Omega \left| |u_n|^{p-1} u_n^p - |u|^{p-1} u \right|^\alpha \right)^{\frac{1}{\alpha}} \left(\int_\Omega |v|^{2^*} \right)^{\frac{1}{2^*}}.$$

It follows that $|u_n|^{p-1} u_n \to |u|^{p-1} u$ weakly in $H^{-1}(\Omega)$. On the other hand,

$$u \to \langle \Lambda u, u \rangle = \int_\Omega \left(\mathbf{a} \cdot \nabla u + a_0 u - \alpha |u|^{p-1} u \right) u \, dx$$

$$= \frac{1}{2} \int_\Omega (2a_0 - \operatorname{div}(\mathbf{a})) |u|^2 dx - \alpha \int_\Omega |u|^{p+1} \, dx$$

is weakly continuous on $H_0^1(\Omega)$. Let now $\lambda := \sup_{x \in \bar{\Omega}} \operatorname{div} \mathbf{a}(x)$. We then have

$$\psi(u) + \langle \Lambda u, u \rangle = \frac{1}{2} \int_\Omega |\nabla u|^2 dx + \frac{\beta}{q+1} \int_\Omega |u|^{q+1} \, dx$$

$$+ \frac{1}{2} \int_\Omega (2a_0 - \operatorname{div}(\mathbf{a})) u^2 \, dx - \alpha \int_\Omega |u|^{p+1} \, dx$$

$$\geq \frac{1}{2} \left(1 - \frac{\lambda - 2a_0}{\lambda_1} \right) \int_\Omega |\nabla u|^2 dx + \frac{1}{q+1} \int_\Omega |u|^{q+1} \, dx$$

$$- \alpha \int_\Omega |u|^{p+1} \, dx,$$

which means that it is coercive since $q > p$, $0 < p < \max\{1, q\}$, and $2a_0 + \operatorname{div}(\mathbf{a}) > -\lambda_1$. Now we can apply Corollary 12.2 and get a solution for (12.14).

Example 12.2. A Schrödinger equation with a nonlocal term

Consider the generalized Choquard-Pekar equation

$$-\Delta u + V(x) u = \left(w * f(u) \right) g(u) + h(x), \tag{12.15}$$

where V, w, and h are real-valued functions on $\mathbf{R}^N$ and $w * f(u)$ denotes the convolution of $f(u)$ and w.

We consider here the case where $f(u) = |u|^p$ and $g(u) = |u|^{q-2}u$, but note first that if $p = q$, then (12.15) can be solved by standard variational methods since weak solutions are critical points of the energy function $\Phi(u) = \psi(u) - \varphi(u)$, where

$$\psi(u) = \tfrac{1}{2} \int_\Omega |\nabla u|^2 + \tfrac{1}{2} \int_\Omega V(x)u^2 - \int_\Omega h(x)u(x)dx$$

and

$$\varphi(u) = \int_\Omega \left(w * f(u)\right) f(u)\, dx.$$

However, as soon as $p \neq q$, (12.15) ceases to be an Euler-Lagrange equation, but we can, however, proceed in the following way.

Theorem 12.5. *Consider $h \in L^2(\mathbf{R}^N)$, $w \in L^1(\mathbf{R}^N)$, and V such that $V(x) \geq \delta > 0$ for $x \in \mathbf{R}^N$. Assume that either V and w are both radial or that $\lim\limits_{\|x\|\to\infty} V(x) = +\infty$.*

Assume also that one of the following conditions holds:

(A) $1 \leq p < \frac{2^*}{2}$, $1 < q < \frac{2^*}{2}$ *and* $w(x) \leq 0$ *on* $\mathbf{R}^N$.
(B) $1 \leq p < \frac{2^*}{2}$, $1 < q < \frac{2^*}{2}$ *and* $1 \leq pq < 2$.

The functional

$$I(u) = \psi(u) + \psi^*\left((w * |u|^p)|u|^{q-1}u\right) - \int_\Omega (w * |u|^p)|u|^q\, dx$$

is then self-dual on $H^1(\mathbf{R}^N)$ (resp., on $H_r^1(\mathbf{R}^N)$ in the radial case), and has zero as an infimum, which is attained at a solution of equation (12.15).

The proof uses the following standard facts:

- Let $w \in L^r(\mathbf{R}^N)$, $r \geq 1$, and $s = \frac{2r}{2r-1}$. The bilinear map $(u, v) \to (w * u)v$ is then well defined and continuous from $L^s \times L^s$ into L^1 and satisfies $\left|(w * u)v\right|_{L^1(\Omega)} \leq \|w\|_r \|u\|_s \|v\|_s$. Moreover, if (v_n) and $(u_n) \subseteq L^s(\mathbf{R}^N)$ are bounded and if either $u_n \to u$ in $L^s(\mathbf{R}^N)$ and $v_n \to v$ in $L_{Loc}^s(\mathbf{R}^N)$ or vice-versa, $u_n \to u$ in $L_{Loc}^s(\mathbf{R}^N)$ and $v_n \to v$ in $L^s(\mathbf{R}^N)$, then $(w * u_n)v_n \to (w * u)v$ in L^1.
- If $\limsup\limits_{|x|\to+\infty} V(x) = +\infty$, then the space $X = \{u \in H^1(\mathbf{R}^N) \mid \int_{\mathbf{R}^N} V(x)u^2\, dx < \infty\}$ embeds compactly in $L^k(\mathbf{R}^N)$, provided $2 \leq k < 2^*$.
- The space $H_r^1(\mathbf{R}^N) := \{u \in H^1(\mathbf{R}^N) \mid u \text{ is radial}\}$ also embeds compactly in $L^k(\mathbf{R}^N)$ for $2 \leq k < 2^*$.

We now show that the map $\Lambda : X \to X^*$ defined by $\Lambda u = -(w * |u|^p)|u|^{q-1}u$ is regular when X is either $H_r^1(\mathbf{R}^N)$ for the radial case or when $X = \{u \in H^1(\mathbf{R}^N) \mid \int_{\mathbf{R}^N} V(x)|u|^2\, dx < \infty\}$ for the case where $\lim\limits_{|x|\to+\infty} V(x) = +\infty$.

First note that $\Lambda : X \to X^*$ is well defined since by Young's inequality and then by Hölder's we have

$$|\langle \Lambda u, v \rangle| = \left| \int_{\mathbf{R}^N} (w * |u|^p) |u|^{q-2} uv \, dx \right|$$

$$\leq \|w\|_1 \||u|^p\|_2 \||u|^{q-1}|v|\|_2$$

$$\leq \|w\|_1 \|u\|_{2p}^p \|u\|_{2q}^{q-1} \|v\|_{2q} < \infty.$$

To show that Λ is weak-to-weak continuous, let $u_n \rightharpoonup u$ weakly in X so that $u_n \to u$ strongly in $L^r(\mathbf{R}^N)$ for $2 \leq r < 2^*$. It follows that $|u_n|^p \to |u|^p$ strongly in $L^2(\mathbf{R}^N)$, and $|u_n|^{q-2} u \to |u|^{q-2} u$ strongly in $L^{\frac{2q}{q-1}}(\mathbf{R}^N)$. For every $v \in L^{2q}$, the sequence $|u_n|^{q-2} uv$ then converges strongly to $|u|^{q-2} uv$ in $L^2(\mathbf{R}^N)$. Therefore, by Young's inequality, we get that $\langle \Lambda u_n, v \rangle \to \langle \Lambda u, v \rangle$ and consequently Λ is weak-to weak continuous. On the other hand, in case (A) we have

$$\langle \Lambda u, u \rangle = - \int_{\mathbf{R}^N} (w * |u|^p) |u|^{q+1} \, dx \geq 0,$$

so that the functional $\psi(u) + \langle \Lambda u, u \rangle$ is coercive. For case (B), even though $\langle \Lambda u, u \rangle$ may be negative, the functional $\varphi(u) + \langle \Lambda u, u \rangle$ does not lose its coercivity since $1 < pq < 2$. Corollary 12.2 then applies to yield the claimed result.

Example 12.3. A subquadratic nonlinear system

Consider the problem

$$\begin{cases} \operatorname{div} F(x, u) - \Delta u = -|u|^{p-2} u & \text{on} \quad \Omega \\ u = 0 & \text{on} \quad \partial\Omega, \end{cases} \tag{12.16}$$

on a smooth bounded domain Ω in $\mathbf{R}^n$, where $2 \leq p < \frac{2n}{n-2}$ and $F : \Omega \times R \to \mathbf{R}^n$ is a Caratheodory function that satisfies for some $0 < q < 1$ and $g \in L^2(\Omega)$ the following growth condition:

$$|F(x, u)| \leq g(x) + |u|^q. \tag{12.17}$$

Theorem 12.6. *Equation (12.16) has a solution that can be obtained by minimizing the self-dual functional on $H_0^1(\Omega)$*

$$I(u) = \Phi(u) + \Phi^*(-\Lambda u) + \langle u, \Lambda u \rangle,$$

where $\Phi(u) = \frac{1}{2} \int_\Omega |\nabla u|^2 \, dx + \frac{1}{p} \int_\Omega |u|^p \, dx$ and $\Lambda : H_0^1(\Omega) \to H^{-1}(\Omega)$ is the map defined by

$$\langle \Lambda u, v \rangle = \int_\Omega F(x, u) \cdot \nabla v \, dx.$$

Proof. Obviously Λ is well defined and to show that it is weak-to-weak continuous, we consider $u_n \rightharpoonup u$ weakly in $H_0^1(\Omega)$ so that $u_n \to u$ strongly in $L^r(\Omega)$ for $1 \leq r < 2^*$ and $u_n(x) \to u(x)$ for almost every $x \in \Omega$. Hence, $F(x, u_n(x)) \to F(x, u(x))$ for almost every $x \in \Omega$. But $\int_\Omega |F(x, u_n)|^2 \, dx \leq C + \int_\Omega |u_n|^2 \, dx$ and $u_n \to u$ strongly in

$L^2(\Omega)$, and so $F(x, u_n(x)) \to F(x, u(x))$ strongly in $L^2(\Omega)$. Therefore Λ is weak to weak continuous. Note also that

$$\int_\Omega F(x, u_n) \cdot \nabla u_n \, dx \to \int_\Omega F(x, u) \cdot \nabla u \, dx,$$

which means that Λ is regular.

On the other hand,

$$|\langle \Lambda u, u \rangle| = \left| \int_\Omega F(x, u) \cdot \nabla u \, dx \right| \leq \int_\Omega g(x) |\nabla u| \, dx + \int_\Omega |u|^q |\nabla u| \, dx,$$

and since $0 < q < 1$, we deduce that

$$\lim_{\|u\|_{H_0^1} \to +\infty} \frac{1}{2} \int_\Omega |\nabla u|^2 \, dx + \frac{1}{p} \int_\Omega |u|^p \, dx + \int_\Omega F(x, u) \cdot \nabla u \, dx = +\infty,$$

and Theorem 12.3 can now be applied to get our claim.

Example 12.4. A nonlinear biharmonic equation

Consider the problem

$$\begin{cases} -\Delta^2 u + f(x, u, \nabla u) = 0 & \text{on} \quad \Omega \\ u = \frac{\partial u}{\partial n} = 0 & \text{on} \quad \partial \Omega, \end{cases} \tag{12.18}$$

on a smooth bounded domain Ω in $\mathbf{R}^n$, where $f : \Omega \times \mathbf{R} \times \mathbf{R}^N \to \mathbf{R}$ is a nonnegative Caratheodory function that satisfies for some $0 \leq g \in L^2(\Omega)$, $r \geq 1$, and $1 \leq s < \frac{N}{N-2}$ the conditions

$$uf(x, u, p) \geq 0 \text{ and } f(x, u, p) \leq g(x) + |u|^r + |\nabla u|^s. \tag{12.19}$$

Theorem 12.7. *Assume that either $N > 4$ or that $2 < N \leq 4$ and $r < \frac{N}{N-4}$. Then, equation (12.18) has a weak solution that can be obtained by minimizing the self-dual functional*

$$I(u) = \Phi(u) + \Phi^*(-\Lambda u) + \langle u, \Lambda u \rangle,$$

on $H_0^2(\Omega)$, where $\Phi(u) = \frac{1}{2} \int_\Omega |\Delta u|^2 \, dx$ and $\Lambda : H_0^2(\Omega) \to H_0^2(\Omega)^*$ is the map defined by

$$\Lambda u = f(x, u, \nabla u).$$

Proof. Setting $X = H_0^2(\Omega)$, we first show that $\Lambda : X \to X^*$ is weak-to-weak continuous. Indeed, let $u_n \rightharpoonup u$ weakly in $H_0^2(\Omega)$, so that $u_n \to u$ and $\nabla u_n \to \nabla u$ for almost every $x \in \Omega$. Since f is a caratheodory function, we have

$$f(x, u_n, \nabla u_n) \to f(x, u, \nabla u) \text{ for } x \in \Omega. \tag{12.20}$$

By (12.19), we have $f(x,u_n,\nabla u_n)^2 \leq g(x)^2 + |u_n|^{2r} + |\nabla u_n|^{2s}$, and hence, $f(x,u_n,\nabla u_n)$ is uniformly bounded in $L^2(\Omega)$. It follows – modulo passing to a subsequence – that $f(x,u_n,\nabla u_n) \to f(x,u,\nabla u)$ weakly in $L^2(\Omega)$. On the other hand, $u_n \to u$ strongly in $L^2(\Omega)$, so that

$$\lim_{n \to +\infty} \int u_n f(x,u_n,\nabla u_n)dx = \int u f(x,u,\nabla u)dx,$$

and Λ is therefore regular.

Since now $uf(x,u,p) \geq 0$, we have that

$$\lim_{\|u\|_X \to +\infty} \frac{1}{2} \int_\Omega |\Delta u|^2 + uf(x,u,\nabla u)dx = +\infty,$$

and Theorem 12.3 can therefore be applied to get our claim.

Exercises 12.B. Other applications

1. Show that equation (12.14) can still have a solution when $\beta = 0$, provided $0 < p < 1$.
2. Consider the nonlocal semilinear elliptic equation on a smooth bounded domain Ω in $\mathbf{R}^n$

$$\begin{cases} -\Delta u + \partial \varphi(u) + f(x,u,\nabla u) + |u|^{p-2}u(\int_\Omega \psi(x,u,\nabla u)dx)^\alpha = 0, & x \in \Omega, \\ u = 0, & x \in \partial\Omega, \end{cases} \quad (12.21)$$

   where $\alpha \geq 0$, and $1 \leq p < 2^*$. Give conditions on f and ψ that make Theorem 12.3 applicable in order to find a weak solution in $H_0^1(\Omega)$.
3. Show that the equation

$$\begin{cases} -\Delta u + |u|^{p-2}u\{\int_\Omega |u|^q dx\}^\alpha = g(x) & x \in \Omega, \\ u = 0 & x \in \partial\Omega, \end{cases} \quad (12.22)$$

   has a a weak solution in $H_0^1(\Omega)$ whenever $1 \leq p,q < 2^*$ and $\alpha > 0$.
4. Give conditions on g_1, g_2, φ_1, and φ_2 that make Theorem 12.3 applicable in order to find a weak solution in $H_0^1(\Omega)$ for the system of semilinear elliptic equations

$$\begin{cases} -\Delta u + \partial \varphi_1(u) + g_1(x,u,v,Du,Dv) = 0 & x \in \Omega, \\ -\Delta v + \partial \varphi_2(v) + g_2(x,u,v,Du,Dv) = 0 & x \in \Omega, \\ u = v = 0 & x \in \partial\Omega. \end{cases} \quad (12.23)$$

5. Repeat Exercise 4 above for the nonlinear polyharmonic equation

$$\begin{cases} m,(-\Delta^m)u + \partial \varphi(u) + h(x,u,Du,\ldots D^m u) = 0 & x \in \Omega \\ \quad \text{for } |\alpha| \leq m-1 \qquad D^\alpha u = 0 \; x \in \partial\Omega. \end{cases} \quad (12.24)$$

6. Develop a self-dual variational formulation and resolution for the problem of existence of 2π-periodic solutions for the following nonlinear wave equation

$$\begin{cases} y_{tt}(t,x) - y_{xx}(t,x) + \partial \varphi(y(t,x)) = f(t,x) & \text{on } \mathbf{R} \times (0,\pi) \\ y(t + 2\pi,x) = y(t,x) & \text{on } \mathbf{R} \times (0,\pi) \\ y(t,0) = y(t,\pi) = 0 & \text{on } (0,\pi), \end{cases} \quad (12.25)$$

where f is a 2π-periodic function in $L^\infty((0,\pi) \times \mathbf{R})$, and φ is a convex function on $\mathbf{R}$ such that, for every $x \in \mathbf{R}$

$$\alpha_1|x|^2 + \beta_1 \le \varphi(x) \le \alpha_2|x|^2 + \beta_2$$

where $\beta_1 \le \beta_2$, and $0 < \alpha_1 < \alpha_2 < \frac{4}{3}$.

Hint: A solution can be obtained by minimizing the self-dual functional

$$\int_0^{2\pi} \int_0^\pi \{\varphi(y) + \varphi^*(\mathscr{W}y - f) - y(\mathscr{W}y - f)\}\, dxdt$$

over all functions y in $L^2([0,2\pi] \times [0,\pi])$, where $\mathscr{W}$ is the wave operator on $L^2([0,2\pi] \times [0,\pi])$ defined by

$$\int_0^{2\pi} \int_0^\pi \varphi \mathscr{W}y\, dxdt = \int_0^{2\pi} \int_0^\pi y(\varphi_{tt} - \varphi_{xx})\, dxdt$$

for every $\varphi \in C^2([0,2\pi] \times [0,\pi])$ such that $\varphi_t(0,x) = \varphi_t(2\pi,x)$, $\varphi(t,0) = \varphi(t,\pi) = 0$, and $\varphi(0,x) = \varphi(2\pi,x) = 0$ for all (t,x) in $[0,2\pi] \times [0,\pi]$.

12.3 Variational resolution for the stationary Navier-Stokes equations

Example 12.5. Variational resolution for the stationary Navier-Stokes equation

Consider the incompressible stationary Navier-Stokes equation on a bounded smooth domain Ω of $\mathbf{R}^3$

$$\begin{cases} (u \cdot \nabla)u + f = \alpha \Delta u - \nabla p & \text{on } \Omega, \\ \operatorname{div} u = 0 & \text{on } \Omega, \\ u = 0 & \text{on } \partial\Omega, \end{cases} \tag{12.26}$$

where $\alpha > 0$ and $f \in L^p(\Omega; \mathbf{R}^3)$. Let

$$\Phi(u) = \frac{\alpha}{2} \int_\Omega \Sigma_{j,k=1}^3 \left(\frac{\partial u_j}{\partial x_k}\right)^2 dx \tag{12.27}$$

be the convex and coercive function on the Sobolev subspace

$$X = \{u \in H_0^1(\Omega; \mathbf{R}^3); \operatorname{div} u = 0\}, \tag{12.28}$$

Its Legendre transform Φ^* on X^* can be characterized as $\Phi^*(v) = \frac{1}{2}\langle Sv, v\rangle$, where $S : X^* \to X$ is the bounded linear operator that associates to $v \in X^*$ the solution $\hat{v} = Sv$ of the Stokes problem

$$\begin{cases} \alpha \Delta \hat{v} + \nabla p = -v & \text{on } \Omega, \\ \operatorname{div} \hat{v} = 0 & \text{on } \Omega, \\ \hat{v} = 0 & \text{on } \partial\Omega. \end{cases} \tag{12.29}$$

It is easy to see that (12.26) can be reformulated as

$$\begin{cases} (u \cdot \nabla)u + f \in -\partial \Phi(u) = \alpha \Delta u - \nabla p, \\ \qquad\qquad u \in X. \end{cases} \tag{12.30}$$

Consider now the nonlinear operator $\Lambda : X \to X^*$ defined as

$$\langle \Lambda u, v \rangle = \int_\Omega \Sigma_{j,k=1}^3 u_k \frac{\partial u_j}{\partial x_k} v_j \, dx = \langle (u \cdot \nabla)u, v \rangle.$$

We can deduce the following known existence result.

Theorem 12.8. *Assume Ω is a bounded domain in $\mathbf{R}^3$, and consider $f \in L^p(\Omega; \mathbf{R}^3)$ for $p > \frac{6}{5}$. Then,*

$$I(u) = \Phi(u) + \Phi^*(-(u \cdot \nabla)u + f) - \int_\Omega \Sigma_{j=1}^3 f_j u_j$$

is a self-dual functional on X, its infimum is equal to zero, and the latter is attained at a solution of the Navier-Stokes equation (12.26).

Proof. To apply Corollary 12.2, it remains to show that Λ is a regular conservative operator. It is standard to show that $\langle \Lambda u, u \rangle = 0$ on X. For the weak-to weak continuity, assume that $u^n \to u$ weakly in $H^1(\Omega)$, and fix $v \in X$. We have that

$$\langle \Lambda u^n, v \rangle = \int_\Omega \Sigma_{j,k=1}^3 u_k^n \frac{\partial u_j^n}{\partial x_k} v_j \, dx = -\int_\Omega \Sigma_{j,k=1}^3 u_k^n \frac{\partial v_j}{\partial x_k} u_j^n \, dx$$

converges to $\langle \Lambda u, v \rangle = \int_\Omega \Sigma_{j,k=1}^3 u_k \frac{\partial v_j}{\partial x_k} u_j \, dx$. Indeed, the Sobolev embedding in dimension 3 implies that (u^n) converges strongly in $L^p(\Omega; \mathbf{R}^3)$ for $1 \le p < 6$. On the other hand, $\frac{\partial u_j}{\partial x_k}$ is in $L^2(\Omega)$ and the result follows from an application of Hölder's inequality.

Example 12.6. Variational resolution for a fluid driven by its boundary

We now deal with the Navier-Stokes equation with a boundary moving with a prescribed velocity:

$$\begin{cases} (u \cdot \nabla)u + f = \alpha \Delta u - \nabla p & \text{on } \Omega, \\ \qquad\qquad \mathrm{div} u = 0 & \text{on } \Omega, \\ \qquad\qquad u = u^0 & \text{on } \partial\Omega, \end{cases} \tag{12.31}$$

where $\int_{\partial\Omega} u^0 \cdot \mathbf{n} \, d\sigma = 0$, $\alpha > 0$, and $f \in L^p(\Omega; \mathbf{R}^3)$. Assuming that $u^0 \in H^{3/2}(\partial\Omega)$ and that $\partial\Omega$ is connected, a classical result of Hopf [156], then yields for each $\varepsilon > 0$, the existence of $v^0 \in H^2(\Omega)$ such that

$$v^0 = u^0 \text{ on } \partial\Omega, \quad \mathrm{div}\, v^0 = 0, \quad \text{and} \quad \int_\Omega \Sigma_{j,k=1}^3 u_k \frac{\partial v_j^0}{\partial x_k} u_j \, dx \le \varepsilon \|u\|_X^2 \text{ for } u \in X. \tag{12.32}$$

Setting $v = u + v^0$, solving (12.31) then reduces to finding a solution for

$$\begin{cases} (u \cdot \nabla)u + (v^0 \cdot \nabla)u + (u \cdot \nabla)v^0 + f - \alpha \Delta v^0 + (v^0 \cdot \nabla)v^0 = \alpha \Delta u - \nabla p & \text{on } \Omega \\ \text{div} u = 0 & \text{on } \Omega \\ u = 0 & \text{on } \partial \Omega. \end{cases} \tag{12.33}$$

This can be reformulated as the following equation in the space X

$$(u \cdot \nabla)u + (v^0 \cdot \nabla)u + (u \cdot \nabla)v^0 + g \in -\partial \Phi(u), \tag{12.34}$$

where Φ is again the convex functional $\Phi(u) = \frac{\alpha}{2} \int_\Omega \Sigma_{j,k=1}^3 (\frac{\partial u_j}{\partial x_k})^2 dx$ as above and

$$g := f - \alpha \Delta v^0 + (v^0 \cdot \nabla)v^0 \in X^*.$$

In other words, this is an equation of the form

$$\Lambda u + \Gamma u + g \in -\partial \Phi(u) \tag{12.35}$$

with $\Lambda u = (u \cdot \nabla)u$ a regular conservative operator and $\Gamma u = (v^0 \cdot \nabla)u + (u \cdot \nabla)v^0$ a bounded linear operator. Note that the component $u \to (v^0 \cdot \nabla)u$ is skew-symmetric, which means that Hopf's result yields the required coercivity condition:

$$\Psi(u) := \Phi(u) + \frac{1}{2} \langle \Gamma u, u \rangle \geq \frac{1}{2} (\alpha - \varepsilon) \|u\|^2 \quad \text{for all } u \in X.$$

In other words, Ψ is convex and coercive, and therefore we can apply Theorem 12.3 to deduce the following theorem.

Theorem 12.9. *Under the hypotheses above and letting A^a be the antisymmetric part of the operator $Au = (u \cdot \nabla)v^0$, the functional*

$$I(u) = \Psi(u) + \Psi^*(-(u \cdot \nabla)u - (v^0 \cdot \nabla)u - A^a u + g) - \int_\Omega \Sigma_{j=1}^3 g_j u_j$$

is self-dual on X, has zero infimum, and the is latter attained at a solution $\bar{u}$ for equation (12.33).

Example 12.7. Variational resolution for a fluid driven by a transport operator

Let $a \in C^\infty(\bar{\Omega}, \mathbf{R}^3)$ be a smooth vector field on a neighborhood of a C^∞ bounded open set $\Omega \subset \mathbf{R}^3$, let $a_0 \in L^\infty(\Omega)$, and consider again the space X as in (12.28) and the skew-adjoint transport operator $\Gamma : u \mapsto (a \cdot \nabla)u + \frac{1}{2} \text{div}(a)u$ from X into X^*. Consider now the following equation on the domain $\Omega \subset \mathbf{R}^3$

$$\begin{cases} (u \cdot \nabla)u + (a \cdot \nabla)u + a_0 u + |u|^{m-2}u + f = \alpha \Delta u - \nabla p & \text{on } \Omega, \\ \text{div} u = 0 & \text{on } \Omega, \\ u = 0 & \text{on } \partial \Omega, \end{cases} \tag{12.36}$$

where $\alpha > 0$, $6 \geq m \geq 1$, and $f \in L^q(\Omega;\mathbf{R}^3)$ for $q \geq \frac{6}{5}$. Suppose

$$\frac{1}{2}\mathrm{div}(\mathbf{a}) - a_0 \geq 0 \quad \text{on} \quad \Omega, \tag{12.37}$$

and consider the functional

$$\Psi(u) = \frac{\alpha}{2}\int_\Omega \Sigma_{j,k=1}^3 \left(\frac{\partial u_j}{\partial x_k}\right)^2 dx + \frac{1}{4}\int_\Omega (\mathrm{div}\,\mathbf{a} - 2a_0)|u|^2 dx + \frac{1}{m}\int_\Omega |u|^m dx + \int_\Omega uf dx,$$

which is convex and coercive function on X. Theorem 12.3 then applies to yield the following result.

Theorem 12.10. *Under the above hypotheses, the functional*

$$I(u) = \Psi(u) + \Psi^*\left(-(u \cdot \nabla)u - \mathbf{a} \cdot \nabla u - \frac{1}{2}\mathrm{div}(\mathbf{a})u\right)$$

is self-dual on X, has zero infimum, and the latter is attained at a solution $\bar{u}$ for (12.34).

12.4 A variational resolution for certain nonlinear systems

Corollary 12.3. *Let φ be a proper convex lower semicontinuous function on $X \times Y$, let $A : X \to Y^*$ be any bounded linear operator, let $B_1 : X \to X^*$ (resp., $B_2 : Y \to Y^*$) be two positive bounded linear operators, and assume $\Lambda := (\Lambda_1, \Lambda_2) : X \times Y \to X^* \times Y^*$ is a regular conservative operator. Assume that*

$$\lim_{\|x\|+\|y\| \to \infty} \frac{\varphi(x,y) + \frac{1}{2}\langle B_1 x, x\rangle + \frac{1}{2}\langle B_2 y, y\rangle}{\|x\| + \|y\|} = +\infty.$$

Then, for any $(f,g) \in X^* \times Y^*$, there exists $(\bar{x}, \bar{y}) \in X \times Y$ that solves the system

$$\begin{cases} -\Lambda_1(x,y) - A^*y - B_1 x + f \in \partial_1 \varphi(x,y) \\ -\Lambda_2(x,y) + Ax - B_2 y + g \in \partial_2 \varphi(x,y). \end{cases} \tag{12.38}$$

The solution is obtained as a minimizer on $X \times Y$ of the self-dual functional

$$I(x,y) = \psi(x,y) + \psi^*(-A^*y - B_1^a x - \Lambda_1(x,y), Ax - B_2^a y - \Lambda_2(x,y)),$$

where

$$\psi(x,y) = \varphi(x,y) + \frac{1}{2}\langle B_1 x, x\rangle + \frac{1}{2}\langle B_2 y, y\rangle - \langle f, x\rangle - \langle g, y\rangle,$$

and where B_1^a (resp., B_2^a) are the skew-symmetric parts of B_1 and B_2.

Proof. Consider the self-dual Lagrangian

$$L((x,y),(p,q)) = \psi(x,y) + \psi^*(-A^*y - B_1^a x + p, Ax - B_2^a y + q).$$

Theorem 12.3 yields that $I(x,y) = L((x,y), -\Lambda(x,y))$ attains its minimum at some point $(\bar{x}, \bar{y}) \in X \times Y$ and that the minimum is 0. In other words,

$$
\begin{aligned}
0 &= I(\bar{x}, \bar{y}) \\
&= \psi(\bar{x}, \bar{y}) + \psi^*(-A^*\bar{y} - B_1^a\bar{x} - \Lambda_1(\bar{x}, \bar{y}), A\bar{x} - B_2^a\bar{y} - \Lambda_2(\bar{x}, \bar{y})) \\
&= \psi(\bar{x}, \bar{y}) + \psi^*(-A^*\bar{y} - B_1^a\bar{x} - \Lambda_1(\bar{x}, \bar{y}), A\bar{x} - B_2^a\bar{y} - \Lambda_2(\bar{x}, \bar{y})) \\
&\quad - \langle (\bar{x}, \bar{y}), (-A^*\bar{y} - B_1^a\bar{x} - \Lambda_1(\bar{x}, \bar{y}), A\bar{x} - B_2^a\bar{y} - \Lambda_2(\bar{x}, \bar{y})) \rangle,
\end{aligned}
$$

from which it follows that

$$
\begin{cases}
-A^*y - B_1^a x - \Lambda_1(x, y) \in \partial_1 \varphi(x, y) + B_1^s(x) - f \\
Ax - B_2^a y - \Lambda_1(x, y) \in \partial_2 \varphi(x, y) + B_2^s(y) - g.
\end{cases}
\tag{12.39}
$$

Example 12.8. Doubly nonlinear coupled equations

Let $\mathbf{b_1}: \Omega \to \mathbf{R}^n$ and $\mathbf{b_2}: \Omega \to \mathbf{R}^n$ be two smooth vector fields on the neighborhood of a bounded domain Ω of $\mathbf{R}^n$, and let $B_1 v = \mathbf{b_1} \cdot \nabla v$ and $B_2 v = \mathbf{b_2} \cdot \nabla v$ be the corresponding first-order linear operators. Consider the Dirichlet problem

$$
\begin{cases}
\Delta(v + u) + \mathbf{b_1} \cdot \nabla u = |u|^{p-2} u + u^{m-1} v^m + f & \text{on } \Omega \\
\Delta(v - u) + \mathbf{b_2} \cdot \nabla v = |v|^{q-2} q - u^m v^{m-1} + g & \text{on } \Omega \\
u = v = 0 & \text{on } \partial\Omega.
\end{cases}
\tag{12.40}
$$

We can use Corollary 12.3 to get the following result.

Theorem 12.11. *Assume* $\mathrm{div}(\mathbf{b_1}) \geq 0$ *and* $\mathrm{div}(\mathbf{b_2}) \geq 0$ *on* Ω, $2 < p, q \leq \frac{2n}{n-2}$, *and* $1 < m < \frac{n+2}{n-2}$, *and consider on* $H_0^1(\Omega) \times H_0^1(\Omega)$ *the functional*

$$
\begin{aligned}
I(u, v) &= \Psi(u) + \Psi^*\left(\mathbf{b_1} . \nabla u + \frac{1}{2} \mathrm{div}(\mathbf{b_1}) u + \Delta v - u^{m-1} v^m\right) \\
&\quad + \Phi(v) + \Phi^*\left(\mathbf{b_2} . \nabla v + \frac{1}{2} \mathrm{div}(\mathbf{b_2}) v - \Delta u + u^m v^{m-1}\right),
\end{aligned}
$$

where

$$
\Psi(u) = \frac{1}{2} \int_\Omega |\nabla u|^2 dx + \frac{1}{p} \int_\Omega |u|^p dx + \int_\Omega fu\, dx + \frac{1}{4} \int_\Omega \mathrm{div}(\mathbf{b_1}) |u|^2 dx,
$$

$$
\Phi(v) = \frac{1}{2} \int_\Omega |\nabla v|^2 dx + \frac{1}{q} \int_\Omega |v|^q dx + \int_\Omega gv\, dx + \frac{1}{4} \int_\Omega \mathrm{div}(\mathbf{b_2}) |v|^2 dx
$$

and Ψ^* *and* Φ^* *are their Legendre transforms. Then, there exists* $(\bar{u}, \bar{v}) \in H_0^1(\Omega) \times H_0^1(\Omega)$ *such that*

$$
I(\bar{u}, \bar{v}) = \inf\{I(u, v); (u, v) \in H_0^1(\Omega) \times H_0^1(\Omega)\} = 0,
$$

and $(\bar{u}, \bar{v})$ *is a solution of* (12.40).

Proof. Let $A = \Delta$ on H_0^1, $B_1 = \mathbf{b_1}.\nabla$, $B_2 = \mathbf{b_2}.\nabla$. Let $X = H_0^1 \times H_0^1$, and consider on $X \times X^*$ the self-dual Lagrangian

$$L((u,v),(r,s))) = \Psi(u) + \Psi^* \left(\mathbf{b_1}.\nabla u + \frac{1}{2}\mathrm{div}(\mathbf{b_1}) u + \Delta v + r \right)$$
$$+ \Phi(v) + \Phi^* \left(\mathbf{b_2}.\nabla v + \frac{1}{2}\mathrm{div}(\mathbf{b_2}) v - \Delta u + s \right).$$

It is also easy to verify that the map $\Lambda : H_0^1 \times H_0^1 \to H^{-1} \times H^{-1}$ defined by

$$\Lambda(u,v) = (u^{m-1}v^m, -u^m v^{m-1})$$

is regular and conservative.

12.5 A nonlinear evolution involving a pseudoregular operator

We now apply Theorem 12.3 to solve nonlinear evolution equations of the form

$$\begin{cases} -\dot{v}(t) - \Lambda v(t) \in \overline{\partial}L(t,v(t)) & \text{for a.e. } t \in [0,T] \\ \quad\quad\quad\quad v(0) = v_0, \end{cases} \qquad (12.41)$$

where L is a time-dependent self-dual Lagrangian on phase space $X \times X^*$ and Λ is a regular map on path space. As seen in Part II, one can lift L to a self-dual Lagrangian $\mathscr{L}$ on either one of the two path spaces $\mathscr{X}_{2,2}([0,T]) \times \mathscr{X}_{2,2}([0,T])^*$ (Theorem 9.2) or $L_X^2[0,T] \times L_{X^*}^2[0,T]$ (Theorem 4.3). One can then try to get a solution for 12.41 by minimizing the functional

$$\mathscr{I}(u) = \mathscr{L}(u, -\Lambda u) + \int_0^T \langle \Lambda u(t), u(t) \rangle \, dt$$

on either $\mathscr{X}_{2,2}([0,T])$ or $L_X^2[0,T]$.

Now the space $\mathscr{X}_{2,2}$ presents the advantage of having a strong topology that increases the chance for a given map Λ to be regular on $\mathscr{X}_{2,2}$. On the other hand, the coercivity condition becomes harder to satisfy, and one needs to establish the conclusions of Theorem 12.3 under much weaker coercivity conditions. This is done in Chapters 17 and 18 in order to deal with the fact that the nonlinear operator $u \to u \cdot \nabla u$ – which appears in the Navier-Stokes equations – is only regular on $\mathscr{X}_{2,2}$ (at least in dimensions $n = 2$). On the other hand, by working on the space L_X^2, the coercivity condition on $\mathscr{I}$ is often easy to verify but, short of considering positive linear operators as in part II, it is harder to find operators that are regular on L_X^2.

In this section, we shall consider cases where this latter setting is applicable, provided the operator Λ can be regularized to become pseudoregular on L_X^2. First, we give a general result.

Theorem 12.12. *Let $X \subset H \subset X^*$ be an evolution triple, and consider a self-dual Lagrangian L on $[0,T] \times X \times X^*$ as well as a boundary Lagrangian ℓ on $H \times H$ satisfying $\ell^*(-x,p) = \ell(x,p)$. Assume the following conditions:*

$$u \to \int_0^T L(t,u(t),p(t))dt \text{ is bounded on the balls of } L_X^2 \text{ for each } p \in L_{X^*}^2, \quad (12.42)$$

$$\lim_{\|v\|_{L^2(X)} \to +\infty} \int_0^T H_L(t,0,v(t))dt = +\infty, \quad (12.43)$$

$$\ell(a,b) \leq C(1 + \|a\|_H^2 + \|b\|_H^2) \text{ for all } (a,b) \in H \times H. \quad (12.44)$$

(i) Then, for any pseudoregular map $\Lambda : L_X^2 \to L_{X^*}^2$, the functional

$$I_{\ell,L,\Lambda}(u) = \int_0^T \left\{ L(t,u(t),-\Lambda u(t) - \dot{u}(t)) + \langle \Lambda u(t), u(t) \rangle \right\} dt + \ell(u(0),u(T))$$

has zero infimum on L_X^2. Moreover, there exists $v \in \mathscr{X}_{2,2}$ such that

$$I_{\ell,L,\Lambda}(v) = \inf_{u \in \mathscr{X}_{2,2}} I_{\ell,L,\Lambda}(u) = 0, \quad (12.45)$$

$$(-v(0),v(T)) \in \partial \ell(v(0),v(T)), \quad (12.46)$$

$$-\dot{v}(t) - \Lambda v(t) \in \bar{\partial} L(t,v(t)). \quad (12.47)$$

(ii) In particular, for every $v_0 \in H$, the self-dual functional

$$I_{v_0,L,\Lambda}(u) = \int_0^T \left\{ L(t,u(t),-\Lambda u(t) - \dot{u}(t)) + \langle \Lambda u(t), u(t) \rangle \right\} dt$$
$$+ \frac{1}{2}\|u(0)\|^2 - 2\langle v_0, u(0) \rangle + \|v_0\|^2 + \frac{1}{2}\|u(T)\|^2$$

has zero infimum on $\mathscr{X}_{2,2}$. It is attained at a unique path v such that $v(0) = v_0$ and satisfying (12.45- 12.47). In particular, we have the following "conservation of energy type" formula: For every $t \in [0,T]$,

$$\|v(t)\|_H^2 = \|v_0\|^2 - 2\int_0^t \left\{ L(s,v(s),-\Lambda v(s) - \dot{v}(s)) + \langle v(s),\Lambda v(s) \rangle \right\} ds. \quad (12.48)$$

Proof. (i) We first apply Theorem 4.3 to get that the Lagrangian

$$\mathscr{L}(u,p) = \begin{cases} \int_0^T L(t,u(t),p(t) - \dot{u}(t))dt + \ell(u(0),u(T)) & \text{if } u \in \mathscr{X}_{2,2} \\ +\infty & \text{otherwise} \end{cases}$$

is self-dual on $L_X^2 \times L_{X^*}^2$. We then apply Theorem 12.3 (actually its extension to pseudoregular operators as of Exercise 12.A), with the space L_X^2, to conclude that the infimum of $\mathscr{L}(u,-\Lambda u) + \langle u,\Lambda u \rangle$ on L_X^2 is equal to 0 and is achieved. This yields claim (12.45).

Since now $L(t,v(t),\Lambda v(t) + \dot{v}(t)) + \langle v(t),\Lambda v(t) + \dot{v}(t) \rangle \geq 0$ for all $t \in [0,T]$, and $\ell(v(0),v(T)) \geq \frac{1}{2}(\|v(T)\|_H^2 - \|v(0)\|_H^2)$, claims (12.46) and (12.47) follow from the identity

$$0 = I_{\ell,L,\Lambda}(v) = \int_0^T \{L(t,v(t),-\Lambda v(t)-\dot{v}(t)) + \langle v(t),\Lambda v(t)+\dot{v}(t)\rangle\} dt$$
$$-\frac{1}{2}(\|v(T)\|_H^2 - \|v(0)\|_H^2) + \ell(v(0),v(T)).$$

It follows that

$$L(t,v(t),-\Lambda v(t)-\dot{v}(t)) + \langle v(t),\Lambda u(t)+\dot{v}(t)\rangle = 0 \quad \text{a.e. } t \in [0,T] \qquad (12.49)$$

and

$$\ell(v(0),v(T)) = \frac{1}{2}(\|v(T)\|_H^2 - \|v(0)\|_H^2), \qquad (12.50)$$

which imply claims (12.47) and (12.46) respectively.

For (ii) it suffices to apply the first part with the boundary Lagrangian

$$\ell(r,s) = \frac{1}{2}\|r\|^2 - 2\langle v_0,r\rangle + \|v_0\|^2 + \frac{1}{2}\|s\|^2.$$

We then get

$$I_{\ell,L,\Lambda}(u) = \int_0^T [L(t,u(t),-\Lambda u(t)-\dot{u}(t)) + \langle u(t),\Lambda u(t)+\dot{u}(t)\rangle] dt + \|u(0)-v_0\|^2.$$

Note also that (12.49) yields

$$\frac{d(|v(s)|^2)}{ds} = -2[L(s,v(s),-\Lambda v(s)-\dot{v}(s)) + \langle \Lambda v(s),v(s)\rangle],$$

which readily implies (12.48).

We now apply Theorem 12.12 to the particular class of self-dual Lagrangians of the form $L(x,p) = \varphi(x) + \varphi^*(Ax-p)$ to obtain variational formulations and resolutions of various nonlinear parabolic equations.

Corollary 12.4. *Let $X \subset H \subset X^*$ be an evolution triple, and consider for each $t \in [0,T]$ a bounded linear operator $A_t : X \to X^*$ and $\varphi : [0,T] \times X \to \mathbf{R}$ such that for each t the functional $\psi(t,x) := \varphi(t,x) + \frac{1}{2}\langle A_t x,x\rangle$ is convex, lower semicontinuous, and satisfies for some $C > 0$, $m,n > 1$ the following growth condition: For $x \in L_X^2$,*

$$\frac{1}{C}\left(\|x\|_{L_X^2}^m - 1\right) \le \int_0^T \{\varphi(t,x(t)) + \frac{1}{2}\langle A_t x(t),x(t)\rangle\} dt \le C\left(\|x\|_{L_X^2}^n + 1\right). \qquad (12.51)$$

If $\Lambda : L_X^2 \to L_{X^*}^2$ is a pseudoregular map, and if A_t^a is the antisymmetric part of the operator A_t, then for every $v_0 \in X$ the functional

$$I(x) = \int_0^T \{\psi(t,x(t)) + \psi^*(t,-\Lambda x(t) - A_t^a x(t) - \dot{x}(t)) + \langle \Lambda x(t),x(t)\rangle\} dt$$

$$+\frac{1}{2}(|x(0)|^2 + |x(T)|^2) - 2\langle x(0), v_0 \rangle + |v_0|^2$$

attains its infimum on L_X^2 at a path $v \in \mathscr{X}_{2,2}$ such that

$$I(v) = \inf_{x \in \mathscr{X}_{2,2}} I(x) = 0, \tag{12.52}$$

$$\begin{cases} -\dot{v}(t) - A_t v(t) - \Lambda v(t) \in \partial \varphi(t, v(t)) & \text{for a.e. } t \in [0,T] \\ \qquad\qquad v(0) = v_0. \end{cases} \tag{12.53}$$

Proof. $L(t,x,p) := \psi(t,x) + \psi^*(t, -A^a x + p)$ is a self-dual Lagrangian on $X \times X^*$, and it is easy to check that all the conditions of Theorem 12.12 are satisfied by L, ℓ, and Λ. Hence, there exists $v \in \mathscr{X}_{2,2}$ such that $I(v) = 0$. We obtain

$$0 = \int_0^T \left(\psi(t,v) + \psi^*(t, -\Lambda v - A_t^a v - \dot{v}) + \langle v(t), \Lambda v(t) + A_t v(t) + \dot{v}(t) \rangle \right) dt$$
$$+ \frac{1}{2}\|v(0) - v_0\|_H^2,$$

which yields since the integrand is nonnegative for each t and by the limiting case of Legendre-Fenchel duality, that

$$\begin{cases} -\dot{v}(t) - A_t^a v(t) - \Lambda v(t) \in \partial \varphi(t, v(t)) + A_t^s v(t) & \text{for a.e. } t \in [0,T] \\ \qquad\qquad v(0) = v_0. \end{cases} \tag{12.54}$$

Example 12.9. Complex Ginsburg-Landau evolutions

We consider the initial boundary value problem for the complex Ginzburg-Landau equation in $\Omega \subseteq \mathbf{R}^N$

$$\begin{cases} \dot{u}(t) - (\kappa + i\alpha)\Delta u + (\gamma + i\beta)|u|^{q-1}u - \omega u = 0, \\ \qquad\qquad\qquad\qquad u(x,0) = u_0, \end{cases} \tag{12.55}$$

where $\kappa \geq 0, \gamma \geq 0, q \geq 1, \alpha, \beta \in \mathbf{R}$.
  We apply Theorem 12.12 to establish the following theorem.

Theorem 12.13. *Let Ω be a bounded domain in $\mathbf{R}^N$, $\kappa > 0$, $\gamma \geq 0$, $\beta \in \mathbf{R}$, and $q > 1$. Let $H := L^2(\Omega)$, $X := H_0^1(\Omega)$, $V_1 := L_X^2$, $V_2 := L^{q+1}(0,T;L^{q+1}(\Omega))$, and $V := V_1 \cap V_2$. Then, for every $u_0 \in X$, there exists $u \in V$ with $\dot{u} \in V^*$ satisfying equation (12.55).*

We would like to apply Theorem 12.12 with the nonlinear operator

$$\Lambda u := -i\Delta u + i\beta|u|^{q-1}u - \omega u$$

and the convex functional

$$\Phi(x) := \frac{\kappa}{2} \int_\Omega |\nabla u|^2 dx + \frac{\gamma}{q+1} \int_\Omega |u|^{q+1} dx$$

with $X = H_0^1(\Omega)$ and $H := L^2(\Omega)$. However, the operator Λ is not necessarily regular from its domain into $L^2_{X^*}$. We shall therefore replace Λ by the pseudoregular operator

$$\Lambda_\lambda(u) := -i\Delta u + i\beta \partial \psi_\lambda(u) - \omega u,$$

where ψ_λ is the λ-regularization of $\psi(u) = \frac{\gamma}{q+1} \int |u|^{q+1} dx$ on $H := L^2(\Omega)$. In this case, Φ needs to be replaced by

$$\Phi_\lambda(x) := \frac{\kappa}{2} \int_\Omega |\nabla u|^2 dx + \psi_\lambda(u).$$

Indeed, we first prove the following lemma

Lemma 12.2. *Suppose* $\kappa, \gamma, \omega > 0$ *and* $u_0 \in X$. *For every* $0 < \lambda < \frac{1}{2\omega}$, *there exists a solution* $u_\lambda \in \mathscr{X}_{2,2}$ *of the* λ*-regularized problem*

$$\begin{cases} \dot{u}(t) - (\kappa + i\alpha)\Delta u + (\gamma + i\beta)\partial \psi_\lambda(u) - \omega u = 0 & \text{on } \Omega, \\ u(x,0) = u_0. \end{cases} \quad (12.56)$$

Proof. In order to apply Theorem 12.12, we need to show that Λ_λ is pseudoregular on L^2_X, and that the functional

$$\Phi_\lambda(u) + \langle \Lambda_\lambda u, u \rangle = \frac{\kappa}{2} \int_\Omega |\nabla u|^2 dx + \psi_\lambda(u) - \omega \|u\|_H^2$$

is coercive on H. For Λ_λ, we first note that the operator $-i\Delta u - \omega u$ is bounded and linear, and so clearly "lifts" to a regular operator from $L^2_X \to L^2_{X^*}$ since $-i\Delta$ is skew-adjoint and that $u \to -\omega u$ is compact from $L^2_X \to L^2_H$. So we only need to verify that $B_\lambda(u) := i\partial \psi_\lambda u : L^2_X \to L^2_{X^*}$ is pseudoregular. For that, suppose that $x_n \rightharpoonup x$ weakly in L^2_X. Since B_λ is Lipschitz continuous on L^2_H, we can assume that $B_\lambda x_n \rightharpoonup y$ weakly in $L^2_{X^*}$. Since $\langle u, B_\lambda(u) \rangle = 0$ for every $u \in X$, it therefore suffices to show that $y = B_\lambda x$ as long as $0 \le \langle x, y \rangle$.

Now, by the monotonicity property of $\partial \varphi_\lambda$, we have $\langle B_\lambda x_n - B_\lambda u, x_n - u \rangle \ge 0$ for every $u \in L^2_X$. It follows that

$$\langle y - B_\lambda u, x - u \rangle \ge \langle y, -u \rangle + \langle -B_\lambda u, x - u \rangle \quad (12.57)$$

$$\ge \lim_n \langle B_\lambda x_n, -u \rangle + \lim_n \langle -B_\lambda u, x_n - u \rangle \quad (12.58)$$

$$= \lim_n \langle B_\lambda x_n - B_\lambda u, x_n - u \rangle \quad (12.59)$$

$$\ge 0. \quad (12.60)$$

Hence, $\langle y - B_\lambda u, x - u \rangle \ge 0$ for all $u \in L^2_X$. For $w \in L^2_X$, set $u = x - tw$ with $t > 0$ in such a way that

$$0 \le \frac{1}{t}\langle y - B_\lambda u, x - u \rangle = \langle y - B_\lambda(x - tw), w \rangle. \quad (12.61)$$

Since B_λ is Lipschitz continuous on L_H^2, we have $\lim_{t\to 0}\langle B_\lambda(x-tw),w\rangle = \langle B_\lambda x,w\rangle$, which yields that $\langle y-B_\lambda(x),w\rangle \geq 0$ for every $w\in L_X^2$, and therefore $y=B_\lambda x$.

For the coercivity, it suffices to show that for every $\omega > 0$ with $0<\lambda<\frac{1}{2\omega}$, the functional $\psi_\lambda(u)-\omega\|u\|_H^2$ is coercive on H. For that, write for $u\in H$

$$
\begin{aligned}
\psi_\lambda(u)-\omega\|u\|_H^2 &= \inf_{v\in H}\left\{\psi(v)+\frac{\|u-v\|_H^2}{2\lambda}\right\}-\omega\|u\|_H^2 \\
&= \inf_{v\in H}\left\{\psi(v)+\frac{\|u\|_H^2}{2\lambda}+\frac{\|v\|_H^2}{2\lambda}-\frac{1}{\lambda}\langle u,v\rangle-\omega\|u\|_H^2\right\} \\
&= \left(\frac{1}{2\lambda}-\omega\right)\|u\|_H^2+\inf_{v\in H}\left\{\psi(v)+\frac{\|v\|_H^2}{2\lambda}-\frac{1}{\lambda}\langle u,v\rangle\right\} \\
&= \left(\frac{1}{2\lambda}-\omega\right)\|u\|_H^2-\sup_{v\in H}\left\{\frac{1}{\lambda}\langle u,v\rangle-\psi(v)-\frac{\|v\|_H^2}{2\lambda}\right\} \\
&\geq \left(\frac{1}{2\lambda}-\omega\right)\|u\|_H^2-\sup_{v\in H}\left\{\frac{1}{\lambda}\langle u,v\rangle-\psi(v)\right\} \\
&\geq \left(\frac{1}{2\lambda}-\omega\right)\|u\|_H^2-\psi^*\left(\frac{1}{\lambda}u\right) \\
&= \left(\frac{1}{2\lambda}-\omega\right)\|u\|_H^2-\frac{\gamma^{-1/q}}{p\lambda^p}\int_\Omega |u|^p\,dx,
\end{aligned}
$$

where $\frac{1}{p}+\frac{1}{q+1}=1$. Since $q+1>2$, we have $p<2$, which implies the required coercivity of $\psi_\lambda(u)-\omega\|u\|_H^2$ on H.

All conditions of Theorem 12.12 are therefore satisfied, and there exists then a solution $u_\lambda\in\mathscr{X}_{2,2}$ of (12.56). In order to complete the proof of Theorem 12.13, we need some estimates for u_λ. For that, we perform an inner product of u_λ with equation (12.56) to get for all $t\in[0,T]$

$$
\frac{1}{2}\frac{d}{dt}\|u_\lambda\|_H^2+\kappa\int_\Omega |\nabla u_\lambda|^2 dx+\psi_\lambda(u_\lambda)-\omega\|u_\lambda\|_H^2 \leq 0. \tag{12.62}
$$

Using Gronwall's inequality, we obtain that $\|u_\lambda\|_H$ is bounded and that consequently u_λ is bounded in L_X^2. It also follows from the above inequality that $\int_0^T\psi_\lambda(u_\lambda)\,dt$ is bounded. On the other hand, the regularization process gives for every $\lambda>0$ a unique $j_\lambda u_\lambda\in L_H^2$ such that, for some constant $C>0$,

$$
\int_0^T\psi_\lambda(u_\lambda)\,dt=\int_0^T\left\{\psi(j_\lambda u_\lambda)+\frac{\|u_\lambda-j_\lambda u_\lambda\|_H^2}{2\lambda}\right\}dt\leq C. \tag{12.63}
$$

We now claim that there exists $u\in V$ such that

$$
u_\lambda \rightharpoonup u \quad \text{weakly in} \quad L_H^2, \tag{12.64}
$$

$$
u_\lambda \to u \quad \text{a.e.} \quad \text{in} \quad [0,T]\times\Omega. \tag{12.65}
$$

Indeed, It follows from (12.62) and (12.63) that u_λ and $j_\lambda u_\lambda$ are bounded in V_1 and V_2, respectively. Since $\partial \psi$ and $-\Delta$ are duality maps, it follows that $-\Delta u_\lambda$ and $\partial \psi_\lambda(u_\lambda) = \partial \psi(j_\lambda u_\lambda)$ are bounded in V_1^* and V_2^*, respectively. Let $m \in \mathbf{N}$ with $m > N/2$ in such a way that $V_0 = W_0^{m,2}(\Omega)$ continuously embeds in $X \cap L^{q+1}(\Omega)$. It follows from equation (12.56) that $\{u_\lambda\}$ is bounded in the space

$$\mathscr{Y} := \{u \in L_X^2; \dot{u} \in L^p(0,T:V_0^*)\}, \text{ where } p = \frac{q+1}{q}.$$

Since $X \subseteq H \subseteq V_0^*$, where $X \subseteq H$ is compact and $H \subseteq V_0^*$ is continuous, it follows that the injection $\mathscr{Y} \subseteq L_H^2$ is compact, and therefore, there exists $u \in \mathscr{Y}$ such that $u_\lambda \to u$ in L_H^2 and $u_\lambda \to u$ a.e. in $[0,T] \times \Omega$. It follows that, up to a subsequence,

$$u_\lambda \rightharpoonup u \quad \text{weakly in } L_X^2, \tag{12.66}$$

$$j_\lambda u_\lambda \rightharpoonup u \quad \text{weakly in } V_2, \tag{12.67}$$

$$\partial \psi_\lambda(u_\lambda) \rightharpoonup \partial \psi(u) \quad \text{weakly in } V_2^*, \tag{12.68}$$

$$u_\lambda(T) \rightharpoonup a \quad \text{weakly in } H \text{ for some } a \in H. \tag{12.69}$$

Indeed, since (u_λ) is bounded in L_X^2 and since $u_\lambda \to u$ a.e. in $[0,T] \times \Omega$, we easily get (12.66), that $j_\lambda u_\lambda \to u$, and $\partial \psi_\lambda(u_\lambda) \to \partial \psi(u)$ a.e. in $[0,T] \times \Omega$, which together with the fact that $\partial \psi_\lambda(u_\lambda)$ is bounded in V_2^* imply (12.67) and (12.68). To prove (12.69), it suffices to note that $u_\lambda(T)$ is bounded in H and therefore $u_\lambda(T) \rightharpoonup a$ for some $a \in H$.

To complete the proof of Theorem 12.13, take any $v \in C^1([0,T];V)$ and deduce from equation (12.56) that

$$0 = \int_0^T \langle -(\kappa + i\alpha)\Delta u_\lambda + (\gamma + i\beta)\partial \psi_\lambda(u_\lambda) - \omega u_\lambda, v(t) \rangle \, dt$$
$$- \int_0^T \langle \dot{v}(t), u_\lambda(t) \rangle + \langle u_\lambda(T), v(T) \rangle - \langle u_0, v(0) \rangle.$$

Letting λ go to zero, it follows from (12.66)-(12.69) that

$$0 = \int_0^T \langle -(\kappa + i\alpha)\Delta u + (\gamma + i\beta)\partial \psi(u) - \omega u, v(t) \rangle \, dt$$
$$- \int_0^T \langle \dot{v}(t), u(t) \rangle + \langle a, v(T) \rangle - \langle u_0, v(0) \rangle.$$

Therefore $u \in V$ is a solution of equation (12.55).

Further comments

The min-max theorem of Ky Fan [49] can be found in several books (see for example, Aubin-Ekeland [8]). The version that covers pseudoregular operators can be found, for example, in Brézis-Nirenberg-Stampachia [31]. The selfdual variational principle (Theorem 12.3) was established by Ghoussoub in [56], where one can also find some of the applications mentioned in this chapter. The self-dual formulation to Navier-Stokes equations seems to have been considered first by Auchmuty [10]. A proof of the stationary case appeared in Auchmuty [13].

The global existence of unique strong solutions to (12.55) was first proved by Pecher and Von Wahl [128] under the conditions: $1 \leq q \leq \infty$ if $N = 1, 2$ and $1 \leq q \leq \frac{N+2}{N-2}$ for dimensions $3 \leq N \leq 8$. They conjectured that $\frac{N+2}{N-2}$ is the largest possible exponent (if $N > 2$) for the global existence of strong solutions (see [128], Remark 1.3). Shigeta managed in [144] to remove the restriction $N \leq 8$ on the dimension, but since the arguments in both [128] and [144] are based on the Gagliardo-Nirenberg inequality, they could not handle the case where $q > \frac{N+2}{N-2}$. This was done recently by Okazawa and Yokota [124], who proved the existence of strong solutions for all exponents $q \geq 1$. However, unlike the global argument above, their proof seems to work only for convex functions of power type.

Chapter 13
The Role of the Co-Hamiltonian in Self-dual Variational Problems

Self-dual functionals of the form $I(x) = L(Ax, -\Lambda x) + \langle Ax, \Lambda x \rangle$ have more than one antisymmetric Hamiltonian associated to them. In all previous examples, we have so far used the one involving the Hamiltonian H_L associated to the self-dual Lagrangian L. In this chapter, we shall see that the one corresponding to the co-Hamiltonian $\tilde{H}_L$ can be more suitable not only when the operator A is nonlinear but also in situations where we need a constrained minimization in order to obtain the appropriate boundary conditions. Furthermore, even if both A and Λ are linear, we shall see that the co-Hamiltonian representation can be more suitable for ensuring the required coercivity conditions.

Applications are given to provide variational solutions for semilinear equations in divergence form, Cauchy problems for Hamiltonian systems, doubly nonlinear evolutions, and gradient flows of certain nonconvex functionals.

13.1 A self-dual variational principle involving the co-Hamiltonian

An immediate application of Theorem 12.3 yields the following variational principle that will be used throughout this chapter.

Proposition 13.1. *Let L be a self-dual Lagrangian on a reflexive Banach space X, and consider a not necessarily linear operator $A : D(A) \subset Z \to X$ and a linear operator $\Lambda : D(\Lambda) \subset Z \to X^*$, where Z is a reflexive Banach space. Let E be a closed convex subset of $D(\Lambda) \cap D(A) \subset Z$ such that*

$$\text{Dom}_2(L) \subset \Lambda(E), \tag{13.1}$$

$$(A, \Lambda) \text{ is a regular pair on } E, \tag{13.2}$$

$$\lim_{x \in E; \|x\| \to +\infty} \tilde{H}_L(-B^*\Lambda x, 0) + \langle BAx, \Lambda x \rangle = +\infty. \tag{13.3}$$

Then, $I(x) := L(Ax, -\Lambda x) + \langle Ax, \Lambda x \rangle$ is a self-dual functional on E and there exists $\bar{x} \in D(\Lambda) \cap D(A)$ such that

N. Ghoussoub, *Self-dual Partial Differential Systems and Their Variational Principles*, Springer Monographs in Mathematics, DOI 10.1007/978-0-387-84897-6_13,
© Springer Science+Business Media, LLC 2009

$$I(\bar{x}) = \inf_{x \in X} I(x) = 0, \tag{13.4}$$

$$-\Lambda \bar{x} \in \bar{\partial}_B L(A\bar{x}). \tag{13.5}$$

In the particular case where L is a self-dual Lagrangian $L(x,p) = \varphi(x) + \varphi^*(p)$ associated to a bounded below, and lower semicontinuous convex functional φ, we then have $\mathrm{Dom}_2(L) \subset \mathrm{Dom}(\varphi^*)$, and $\tilde{H}_L(p,q) = \varphi^*(p) - \varphi^*(q)$ when $q \in \mathrm{Dom}(\varphi^*)$ and $-\infty$ elsewhere. One can then deduce the following useful corollary.

Corollary 13.1. *Let φ be a bounded below convex lower semicontinuous function on a reflexive Banach space X, and let $f \in X^*$. Consider $A : D(A) \subset Z \to X$ and $\Lambda : D(\Lambda) \subset Z \to X^*$, to be two operators, with Λ being linear. Let $E \subset D(A) \cap D(\Lambda)$ be a closed convex subset Z such that*

$$\mathrm{Dom}(\varphi^*) \subset \Lambda(E), \tag{13.6}$$

$$\text{the pair } (A,\Lambda) \text{ is regular on the set } E, \tag{13.7}$$

and

$$\lim_{\|x\| \to \infty, x \in E} \varphi^*(-\Lambda x) + \langle Ax, \Lambda x + f \rangle = +\infty. \tag{13.8}$$

Then, there exists a solution $\bar{x} \in E$ to the equation

$$0 \in f + \Lambda x + \partial \varphi(Ax). \tag{13.9}$$

It is obtained as a minimizer of the self-dual functional

$$I(x) = \varphi(Ax) + \varphi^*(-\Lambda x - f) + \langle Ax, \Lambda x + f \rangle, \tag{13.10}$$

which has zero as its minimal value over E.

13.2 The Cauchy problem for Hamiltonian flows

Example 13.1. Convex Hamiltonian systems

The first application of Corollary 13.1 deals with the following standard Cauchy problem for Hamiltonian systems. Given a time $T > 0$, we shall let $X = A^2_{\mathbf{R}^{2N}}([0,T])$ be the Sobolev space endowed with the norm $\|u\| = \left(\|u\|^2_{L^2} + \|\dot{u}\|^2_{L^2} \right)^{\frac{1}{2}}$.

Theorem 13.1. *Suppose $\varphi : \mathbf{R}^N \times \mathbf{R}^N \to \mathbf{R}$ is a proper, convex, lower semicontinuous function such that $\varphi(p,q) \to \infty$ as $|p| + |q| \to \infty$. Assume that*

$$-\alpha \leq \varphi(p,q) \leq \beta(|p|^r + |q|^r + 1) \qquad (1 < r < \infty), \tag{13.11}$$

where α, β are positive constants. Then, the infimum of the functional

$$I(p,q) := \int_0^T \left[\varphi\big(p(t),q(t)\big) + \varphi^*\big(-\dot{q}(t),\dot{p}(t)\big) + \dot{q}(t)\cdot p(t) - \dot{p}(t)\cdot q(t) \right] dt$$

on the set $E := \{(p,q) \in X; p(0) = p_0, q(0) = q_0\}$ is equal to zero and is attained at a solution of

$$\begin{cases} \dot{p}(t) \in \partial_2 \varphi\big(p(t),q(t)\big) \ t \in (0,T) \\ -\dot{q}(t) \in \partial_1 \varphi\big(p(t),q(t)\big) \ t \in (0,T) \\ (p(0),q(0)) = (p_0,q_0). \end{cases} \tag{13.12}$$

We shall first consider the subquadratic case $(1 < r < 2)$.

Proposition 13.2. *Assume φ is a proper, convex, and lower semicontinuous function on $\mathbf{R}^N \times \mathbf{R}^N$, and consider the following functional on $X \times X$:*

$$I(p,q) := \int_0^T \left[\varphi\big(p(t),q(t)\big) + \varphi^*\big(-\dot{q}(t),\dot{p}(t)\big) + \dot{q}(t)\cdot p(t) - \dot{p}(t)\cdot q(t) \right] dt$$

1. *If φ is subquadratic on $\mathbf{R}^N \times \mathbf{R}^N$, then for any $(p_0,q_0) \in \mathbf{R}^N \times \mathbf{R}^N$, I is a self-dual functional on the closed convex set $E := \{(p,q) \in X \times X, p(0) = p_0, q(0) = q_0\}$ with a corresponding antisymmetric Hamiltonian on $E \times E$ given by*

$$M(r,s;p,q) := \int_0^T \left[\varphi^*(-\dot{q},\dot{p}) - \varphi^*(-\dot{s},\dot{r}) + (\dot{r},-\dot{s})\cdot(q,p) + (\dot{q},-\dot{p})\cdot(p,q) \right] dt.$$

2. *The infimum of I on E is zero and is attained at a solution of system (13.12).*

Proof of Proposition 13.2: Fenchel-Legendre duality gives that $I(p,q) \geq 0$ for every $(p,q) \in X$. We shall apply Corollary 13.1 with $E \subset X$, $\Lambda u = J\dot{u}(t)$, the symplectic automorphism $J(p,q) = (-q,p)$, and the self-dual Lagrangian

$$L\big((p,q),(r,s)\big) = \int_0^T \left[\varphi\big(p(t),q(t)\big) + \varphi^*\big(r(t),s(t)\big) \right] dt \tag{13.13}$$

on $X \times X^*$. Note that the map $u \to \int_0^T \langle Ju, \dot{u}\rangle dt$ is weakly continuous, making Λ a regular linear operator. The functional I can then be written as: $I(z) = L(z,\Lambda z) - \langle Jz, \Lambda z\rangle$ over X, and the result then follows from Theorem 12.3. The corresponding AS-Hamiltonian is then

$$M(z,w) = \tilde{H}_L(\Lambda z, \Lambda w) + \langle z, \Lambda(z-w)\rangle, \tag{13.14}$$

which in this case is equal to (13.13).

Note now that the subquadraticity of φ ensures that I is strongly coercive, and we are then able to apply Theorem 12.3 to find $(\bar{p},\bar{q}) \in E$ such that $I(\bar{p},\bar{q}) = 0$.

In order to deal with the general case (that is, when (13.11) holds with $r > 2$), we shall use a variation of the standard inf-convolution procedure to reduce the problem to the subquadratic case where Proposition 13.2 applies. To do that, define for every $\lambda > 0$, the functional

$$\varphi_\lambda(p,q) := \inf_{u,v\in\mathbf{R}^N} \left\{ \varphi(u,v) + \frac{\|p-u\|_s^s}{s\lambda^s} + \frac{\|q-v\|_s^s}{s\lambda^s} \right\}, \tag{13.15}$$

where $s = \frac{r}{r-1}$. Obviously, $1 < s < 2$, and since φ is convex and lower semicontinuous, the infimum is clearly attained, so that for every $p,q \in \mathbf{R}^N$ there exist unique points $i(p), j(q) \in \mathbf{R}^N$ such that

$$\varphi_\lambda(p,q) = \varphi(i(p),j(q)) + \frac{\|p-i(p)\|_s^s}{s\lambda^s} + \frac{\|q-j(q)\|_s^s}{s\lambda^s}. \tag{13.16}$$

Lemma 13.1. *The regularized functional φ_λ satisfies the following properties:*

(i) $\varphi_\lambda(p,q) \to \varphi(p,q)$ as $\lambda \to 0^+$.
(ii) $\varphi_\lambda(p,q) \le \varphi(0,0) + \frac{\|q\|_s^s + \|p\|_s^s}{s\lambda^s}$.
(iii) $\varphi_\lambda^*(p,q) = \varphi^*(p,q) + \frac{\lambda^r}{r}(\|p\|_r^r + \|q\|_r^r)$.

Proof. (i) and (ii) are easy. For (iii), we have

$$\varphi_\lambda^*(p,q) = \sup_{u,v\in\mathbf{R}^N} \{u\cdot p + v\cdot q - \varphi_\lambda(u,v)\}$$

$$= \sup_{u,v\in\mathbf{R}^N} \left\{ u\cdot p + v\cdot q - \inf_{z,w\in\mathbf{R}^N} \left\{ \varphi(z,w) + \frac{\|z-u\|_s^s + \|w-v\|_s^s}{s\lambda^s} \right\} \right\}$$

$$= \sup_{u,v\in\mathbf{R}^N} \sup_{z,w\in\mathbf{R}^N} \left\{ u\cdot p + v\cdot q - \varphi(z,w) - \frac{\|z-u\|_s^s}{s\lambda^s} - \frac{\|w-v\|_s^s}{s\lambda_s} \right\}$$

$$= \sup_{z,w\in\mathbf{R}^N} \sup_{u,v\in\mathbf{R}^N} \left\{ (u-z)\cdot p + (v-w)\cdot q + z\cdot p + w\cdot q - \varphi(z,w) \right.$$
$$\left. - \frac{\|z-u\|_s^s + \|w-v\|_s^s}{s\lambda^s} \right\}$$

$$= \sup_{z,w\in\mathbf{R}^N} \sup_{u_1,v_1\in\mathbf{R}^N} \left\{ u_1\cdot p + v_1\cdot q - \frac{\|u_1\|_s^s}{s\lambda^s} - \frac{\|v_1\|_s^s}{s\lambda^s} \right.$$
$$\left. + z\cdot p + w\cdot q - \varphi(z,w) \right\}$$

$$= \sup_{u_1,v_1\in\mathbf{R}^N} \left\{ u_1\cdot p + v_1\cdot q - \frac{\|u_1\|_s^s}{s\lambda^s} - \frac{\|v_1\|_s^s}{s\lambda^s} \right\}$$
$$+ \sup_{z,w\in\mathbf{R}^N} \{z\cdot p + w\cdot q - \varphi(z,w)\}$$

$$= \frac{\lambda^r}{r}(\|p\|_r^r + \|q\|_r^r) + \varphi^*(p,q).$$

Lemma 13.2. *For every $\lambda > 0$, there exists $(p_\lambda,q_\lambda) \in X \times X$ such that $p_\lambda(0) = p_0, q_\lambda(0) = q_0$, and*

$$\begin{cases} \dot{p}_\lambda \in \partial_2\varphi(i_\lambda(p_\lambda), j_\lambda(q_\lambda)) \\ -\dot{q}_\lambda \in \partial_1\varphi(i_\lambda(p_\lambda), j_\lambda(q_\lambda)). \end{cases}$$

Proof. Consider the Cauchy problem associated to φ_λ. By Proposition 13.2, there exists $(p_\lambda, q_\lambda) \in X \times X$ such that $p_\lambda(0) = p_0, q_\lambda(0) = q_0$, and

$$I(p_\lambda, q_\lambda) = \int_0^T \left[\varphi_\lambda(p_\lambda, q_\lambda) + \varphi_\lambda^*(-\dot{q}_\lambda, \dot{p}_\lambda) + \dot{q}_\lambda \cdot p_\lambda - \dot{p}_\lambda \cdot q_\lambda \right] dt = 0 \quad (13.17)$$

yielding

$$\begin{cases} \dot{p}_\lambda \in \partial_2 \varphi_\lambda(p_\lambda, q_\lambda) \\ -\dot{q}_\lambda \in \partial_1 \varphi_\lambda(p_\lambda, q_\lambda) \\ p_\lambda(0) = p_0, \quad q_\lambda(0) = q_0, \end{cases} \quad (13.18)$$

while

$$\varphi_\lambda(p_\lambda, q_\lambda) = \varphi(i_\lambda(p_\lambda), j_\lambda(q_\lambda)) + \frac{\|p_\lambda - i_\lambda(p_\lambda)\|_s^s}{s\lambda^s} + \frac{\|q_\lambda - j_\lambda(q_\lambda)\|_s^s}{s\lambda^s}. \quad (13.19)$$

To relate (p_λ, q_λ) to the original Hamiltonian, use (13.17) and Legendre-Fenchel duality to write

$$\varphi_\lambda(p_\lambda, q_\lambda) + \varphi_\lambda^*(-\dot{q}_\lambda, \dot{p}_\lambda) + \dot{q}_\lambda \cdot p_\lambda - \dot{p}_\lambda \cdot q_\lambda = 0 \quad \forall t \in (0, T). \quad (13.20)$$

Part (iii) of Lemma 13.1, together with (13.18) and (13.19), gives

$$\begin{aligned} 0 = {}& \varphi(i_\lambda(p_\lambda), j_\lambda(q_\lambda)) + \frac{\|p_\lambda - i_\lambda(q_\lambda)\|_s^s + \|q_\lambda - j_\lambda(q_\lambda)\|_s^s}{s\lambda^s} \\ & + \varphi^*(-\dot{q}_\lambda, \dot{p}_\lambda) + \frac{\lambda^r}{r} \left(\|\dot{p}_\lambda\|_r^r + \|\dot{q}_\lambda\|_r^r \right) \\ & + \dot{q}_\lambda \cdot p_\lambda - \dot{p}_\lambda \cdot q_\lambda. \end{aligned} \quad (13.21)$$

Note that

$$\begin{aligned} p_\lambda \cdot \dot{q}_\lambda &= (p_\lambda - i_\lambda(p_\lambda)) \cdot \dot{q}_\lambda + i_\lambda(p_\lambda) \cdot \dot{q}_\lambda \\ \dot{p}_\lambda \cdot q_\lambda &= (q_\lambda - j_\lambda(q_\lambda)) \cdot \dot{p}_\lambda + (\dot{p}_\lambda \cdot j_\lambda(q_\lambda)). \end{aligned} \quad (13.22)$$

By Young's inequality, we have

$$\left| (p_\lambda - i_\lambda(p_\lambda)) \cdot \dot{q}_\lambda \right| \leq \frac{\|p_\lambda - i_\lambda(p_\lambda)\|_s^s}{s\lambda^s} + \frac{\lambda^r}{r} \|\dot{q}_\lambda\|_r^r, \quad (13.23)$$

$$\left| (q_\lambda - j_\lambda(q_\lambda)) \cdot \dot{p}_\lambda \right| \leq \frac{\|q_\lambda - j_\lambda(q_\lambda)\|_s^s}{s\lambda^s} + \frac{\lambda^r}{r} \|\dot{p}_\lambda\|_r^r. \quad (13.24)$$

Combining the last three inequalities gives

$$\begin{aligned} 0 = {}& \varphi(i_\lambda(p_\lambda), j_\lambda(q_\lambda)) + \frac{\|p_\lambda - i_\lambda(q_\lambda)\|_s^s + \|q_\lambda - j_\lambda(q_\lambda)\|_s^s}{s\lambda^s} \\ & + \varphi^*(-\dot{q}_\lambda, \dot{p}_\lambda) + \frac{\lambda^r}{r} \left(\|\dot{p}_\lambda\|_r^r + \|\dot{q}_\lambda\|_r^r \right) \\ & + (p_\lambda - i_\lambda(p_\lambda)) \cdot \dot{q}_\lambda + i_\lambda(p_\lambda) \cdot \dot{q}_\lambda - (q_\lambda - j_\lambda(q_\lambda) \cdot \dot{p}_\lambda - \dot{p}_\lambda \cdot j_\lambda(q_\lambda)) \end{aligned}$$

$$\geq \varphi\big(i_\lambda(p_\lambda), j_\lambda(q_\lambda)\big) + \frac{\|p_\lambda - i_\lambda(q_\lambda)\|_s^s + \|q_\lambda - j_\lambda(q_\lambda)\|_s^s}{s\lambda^s}$$

$$+\varphi^*(-\dot{q}_\lambda, \dot{p}_\lambda) + \frac{\lambda^r}{r}\big(\|\dot{p}_\lambda\|_r^r + \|\dot{q}_\lambda\|_r^r\big)$$

$$+i_\lambda(p_\lambda)\cdot\dot{q}_\lambda - \dot{p}_\lambda\cdot j_\lambda(q_\lambda) - \frac{\lambda^r}{r}\big(\|\dot{p}_\lambda\|_r^r + \|\dot{q}_\lambda\|_r^r\big)$$

$$-\frac{\|p_\lambda - i_\lambda(p_\lambda)\|_s^s + \|q_\lambda - j_\lambda(q_\lambda)\|_s^s}{s\lambda^s}$$

$$= \varphi\big(i_\lambda(p_\lambda), j_\lambda(q_\lambda)\big) + \varphi^*(-\dot{q}_\lambda, \dot{p}_\lambda) + i_\lambda(p_\lambda)\cdot\dot{q}_\lambda - \dot{p}_\lambda\cdot j_\lambda(q_\lambda).$$

On the other hand, by the definition of Fenchel-Legendre duality,

$$\varphi\big(i_\lambda(p_\lambda), j_\lambda(q_\lambda)\big) + \varphi^*(-\dot{q}_\lambda, \dot{p}_\lambda) + i_\lambda(p_\lambda)\cdot\dot{q}_\lambda - \dot{p}_\lambda\cdot j_\lambda(q_\lambda) \geq 0,$$

which means we have equality, so that

$$\begin{cases} \dot{p}_\lambda \in \partial_2\varphi\big(i_\lambda(p_\lambda), j_\lambda(q_\lambda)\big) \\ -\dot{q}_\lambda \in \partial_1\varphi\big(i_\lambda(p_\lambda), j_\lambda(q_\lambda)\big). \end{cases}$$

Lemma 13.3. *With the above notation, the following estimate holds:*

1. $\sup\limits_{t\in(0,T)} |q_\lambda - j_\lambda(q_\lambda)| + |p_\lambda - i_\lambda(p_\lambda)| \leq c\lambda$, *where c is a constant.*
2. *If $\varphi(p,q) \to \infty$ as $|p| + |q| \to \infty$, then $\sup\limits_{\substack{t\in(0,T)\\\lambda>0}} |q_\lambda| + |j_\lambda(q_\lambda)| + |p_\lambda| + |i_\lambda(p_\lambda)| < \infty.$*
3. $\sup\limits_{t\in[0,T],\lambda>0} |\dot{p}_\lambda(t)| + |\dot{q}_\lambda(t)| < +\infty.$

Proof. (1) For every $\lambda > 0$ and $t \in (0,T)$, multiply the first equation of (13.18) by $\dot{q}_\lambda$ and the second one by $\dot{p}_\lambda$ to get

$$\begin{cases} \dot{p}_\lambda\dot{q}_\lambda = \dot{q}_\lambda\partial_2\varphi_\lambda(p_\lambda, q_\lambda), \\ -\dot{q}_\lambda p_\lambda = \dot{p}_\lambda\partial_1\varphi_\lambda(p_\lambda, q_\lambda). \end{cases}$$

So $\frac{d}{dt}\varphi_\lambda(p_\lambda, q_\lambda) = 0$ and $\varphi_\lambda\big(p_\lambda(t), q_\lambda(t)\big) = \varphi_\lambda\big(p(0), q(0)\big) \leq \varphi\big(p(0), q(0)\big) :=$ $c < +\infty$. Hence, it follows from (13.19) that

$$\varphi\big(i_\lambda(p_\lambda(t)), j_\lambda(q_\lambda(t))\big) + \frac{\|p_\lambda - i_\lambda(p_\lambda)\|_s^s + \|q_\lambda - j_\lambda(q_\lambda)\|_s^s}{s\lambda^s} \leq c,$$

which yields $\sup_{t\in(0,T]} |q_\lambda - j_\lambda(q_\lambda)| + |p_\lambda - i_\lambda p_\lambda| \leq c\lambda$ and

$$\sup_{t\in(0,T]} \varphi\big(i_\lambda(p_\lambda(t)), j_\lambda(q_\lambda(t))\big) < +\infty.$$

(2) Since φ is coercive the last equation gives $\sup_{t\in(0,T)\lambda} |j_\lambda(q_\lambda)| + |i_\lambda p_\lambda| < \infty$, which together with (1) proves claim (2).

(3) Since $-\alpha < \varphi(p,q) \leq \beta|p|^r + \beta|q|^r + \beta$ with $r > 2$, an easy calculation shows that if $(p^*, q^*) \in \partial\varphi(p,q)$, then

$$|p^*| + |q^*| \leq \left\{ s(2\beta)^{\frac{r}{s}}(|p| + |q| + \alpha + \beta) + 1 \right\}^{r-1}. \tag{13.25}$$

Since by Lemma 13.1 we have $(\dot{p}_\lambda, -\dot{q}_\lambda) = \partial\varphi(i_\lambda(p_\lambda), j_\lambda(q_\lambda))$, it follows that

$$|\dot{p}_\lambda| + |\dot{q}_\lambda| \leq \left\{ s(2\beta)^{\frac{r}{s}}(|i_\lambda(p_\lambda)| + |j_\lambda(q_\lambda)| + \alpha + \beta) + 1 \right\}^{r-1}, \tag{13.26}$$

which together with (1) and (2) proves the desired result.

End of proof of Theorem 13.1: From Lemma 13.1, we have

$$\int_0^T \left[\varphi(i_\lambda(p_\lambda), j_\lambda(q_\lambda)) + \varphi^*(-\dot{q}_\lambda, \dot{p}_\lambda) + \dot{q}_\lambda \cdot i_\lambda(p_\lambda) - \dot{p}_\lambda \cdot j_\lambda(q_\lambda) \right] dt = 0, \tag{13.27}$$

while $p_\lambda(0) = p_0$ and $q_\lambda(0) = q_0$. By Lemma 13.3, $\dot{p}_\lambda$ and $\dot{q}_\lambda$ are bounded in $L^2(0,T;\mathbf{R}^N)$ so there exists $(p,q) \in X \times X$ such that $\dot{p}_\lambda \rightharpoonup \dot{p}$ and $\dot{q}_\lambda \rightharpoonup \dot{q}$ weakly in $L^2(0,T;\mathbf{R}^N)$ and $p_\lambda \to p$ and $q_\lambda \to q$ strongly in $L_\infty(0,T;\mathbf{R}^N)$. So again by Lemma 13.3, $i_\lambda(p_\lambda) \to p$ and $j_\lambda(q_\lambda) \to q$ strongly in $L_\infty(0,T;\mathbf{R}^N)$. Hence, by letting $\lambda \to 0$ in (13.27), we get $\int_0^T [\varphi(p,q) + \varphi^*(-\dot{q}, \dot{p}) + \dot{q} \cdot p - \dot{p} \cdot q] dt \leq 0$, which means $p(0) = p_0$ and $q(0) = q_0$ and (p,q) satisfies (13.12).

Exercises 13.A. More on co-Hamiltonians

1. Show directly that the functional I given in Proposition 13.2 is given by

$$\begin{aligned} I(p,q) &= \sup_{(r,s)\in E\times E} M(r,s;p,q) \\ &= \sup_{f,g\in L^2} \left\{ \int_0^T [(f,-g)\cdot(q,p) + \varphi^*(-\dot{q}, \dot{p}) - \varphi^*(-g,f) + \dot{q}p - \dot{p}\cdot q] dt \right\}. \end{aligned}$$

2. Show that the map $u \to \int_0^T \langle J\dot{u}(t), u(t) \rangle dt$ is weakly continuous on the space $A_{\mathbf{R}^{2N}}^2([0,T])$.

13.3 The Cauchy problem for certain nonconvex gradient flows

Example 13.2. A doubly nonlinear evolution

Consider the doubly nonlinear equation

$$\begin{cases} \partial\psi(-\dot{u}(t)) = DF(u(t)) \ t \in (0,T), \\ u(0) = u_0, \end{cases} \tag{13.28}$$

or equivalently

$$\begin{cases} -\dot{u}(t) \in \partial\varphi(DF(u(t))) \ t \in (0,T) \\ u(0) = u_0, \end{cases} \tag{13.29}$$

where $\varphi : X \to \mathbf{R}$ (resp., $\psi = \varphi^*$) is convex lower semicontinuous on a reflexive Banach space X (resp., X^*) and where $F : X^* \to \mathbf{R}$ is a Fréchet differentiable functional with differential $DF : X^* \to X$. We shall assume that $X \subset H \subset X^*$ is an evolution triple in such a way that the following formula holds on the space $\mathscr{X}_{p,q}[0,T]$, where $\frac{1}{p} + \frac{1}{q} = 1$.

$$\int_0^T \langle \dot{u}(t), DF(u(t))dt \rangle = F(u(T)) - F(u(0)). \tag{13.30}$$

Corollary 13.2. *Let φ be a convex function on a reflexive Banach space X such that for some $p > 1$ and $C > 0$ we have*

$$-C \leq \varphi(x) \leq C(1 + \|x\|^p) \text{ for all } x \in X. \tag{13.31}$$

Let F be a weakly lower semicontinuous Fréchet-differentiable functional on X^* such that the induced map $A : \mathscr{X}_{p,q}[0,T] \to L_X^q[0,T]$ defined by $(Au)(t) = DF(u(t))$ is weak-to-weak continuous. Then, the infimum of the self-dual functional

$$I(u) = \int_0^T \left\{ \varphi(DF(u(t))) + \varphi^*(-\dot{u}(t)) \right\} dt + F(u(T)) - F(u(0))$$

on the set $E = \{u \in \mathscr{X}_{p,q}[0,T]; u(0) = u_0\}$ is equal to zero and is attained at a solution of equation (13.28).

Proof. Indeed, apply the preceding theorem with the linear operator $\Lambda u = \dot{u}$ mapping $\mathscr{X}_{p,q}[0,T]$ to $L_X^q[0,T]$, and the nonlinear weakly continuous map A mapping $\mathscr{X}_{p,q}[0,T]$ to $L_{X^*}^q[0,T]$. Note that E is a closed convex set such that $\Lambda(E) = L_X^q[0,T]$, and since F is assumed to be weakly lower semicontinuous, we get from (13.30) that the pair (A,Λ) is regular on the set E. The coercivity condition is now

$$\lim_{u \in E, \|u\|_{\mathscr{X}_{p,q}} \to +\infty} \int_0^T \varphi^*(-\dot{u})dx + F(u(T)) = +\infty,$$

which holds since the growth assumption on φ implies that $\varphi^*(y) \geq C'(1 + \|y\|^q)$.

Example 13.3. A nonconvex gradient flow

Consider the initial-value problem

$$\begin{cases} -u_x = u|u|^{p-2} - u|u|^{q-2} + 1 \ x \in (0,1), \\ u(0) = 0, \end{cases} \tag{13.32}$$

where $p > 2 \geq q > 1$.

We reformulate the problem in the following way. Set

$$A(u) = u|u|^{q-2} = \partial F(u), \tag{13.33}$$

where $F(u) = \frac{1}{q} \int_0^1 |u|^q \, dx$, so that it becomes

$$-u_x = u|u|^{p-2} - A(u) + 1 = |A(u)|^{(q'-1)(p-1)-1} A(u) - A(u) + 1$$
$$= \partial \psi(A(u)) - A(u) + 1,$$

where $\frac{1}{q} + \frac{1}{q'} = 1$ and

$$\psi(u) = \frac{1}{(q'-1)(p-1)+1} \int_0^1 |u|^{(q'-1)(p-1)+1} \, dx. \tag{13.34}$$

Problem (13.32) is now equivalent to

$$\begin{cases} -u_x \in \partial \psi(A(u)) - A(u) + 1 \\ u(0) = 0. \end{cases} \tag{13.35}$$

Moreover, by considering the convex functional defined by

$$\varphi(u) = \frac{1}{(p-1)(q'-1)+1} \int_0^1 e^{2\tau} |e^{-\tau} u|^{(p-1)(q'-1)+1} \, dx + \frac{1}{2} \int_0^1 \tau_x |u|^2 \, dx + \int_0^1 u \, dx$$

for an appropriate function $\tau \in C^1([0,1])$, we get that any solution $\bar{u}$ to

$$\begin{cases} -u_x \in \partial \varphi(A(u)) \\ u(0) = 0 \end{cases} \tag{13.36}$$

yields a solution $\bar{v}(x) := e^{-\tau(x)} \bar{u}(x)$ to equation (13.32).

In order to solve (13.36), we set $p^{\#} = (q'-1)(p-1)+1$ and its conjugate $q^{\#} = 1 + \frac{1}{(q'-1)(p-1)}$, and consider the Banach space $X = L^{p^{\#}}[0,1]$, its dual $X^* = L^{q^{\#}}[0,1]$, and the space

$$Z = \{u \in L^{p^{\#}}[0,1]; u_x \in L^{q^{\#}}[0,1]\}.$$

Since $p > 2 \geq q > 1$, we have that $X \subset L^2 \subset X^*$, and it is therefore easy to prove that $A : Z \to X$ and $\Lambda u = \dot{u}$ from Z to X^* are weak-to-weak continuous. Moreover, we have

$$\int_0^1 \langle u_x, A u(x) dx \rangle = \int_0^1 u_x u |u|^{q-2} \, dx = \frac{|u(1)|^q}{q} - \frac{|u(0)|^q}{q} \tag{13.37}$$

and therefore weakly lower semicontinuous on Z. Finally, the set $E = \{u \in Z; u(0) = 0\}$ is closed and convex and $\Lambda(E) = X^* = L^{q^{\#}}[0,1]$. Now we can conclude since the functional

$$I(u) = \int_0^T \left\{ \varphi(\nabla F(u(t))) + \varphi^*(-\dot{u}(t)) \right\} dt + \frac{|u(1)|^q}{q}$$

is self-dual and coercive on E, and therefore its infimum is attained at $\bar{u}$ in such a way that $\bar{v} = e^{-\tau} \bar{u}$ is a solution of equation (13.32).

Exercises 13.B. More on co-Hamiltonians

1. Let Z be a reflexive Banach space, and let $\varphi : X \to \mathbf{R} \cup \{+\infty\}$ be proper, convex, and lower semicontinuous on a reflexive space X. Suppose $\Lambda : D(\Lambda) \subset Z \to X^*$ is a linear operator, and $(b_1, b_2) : D(b_1, b_2) \subset Z \times x \to H_1 \times H_2$ are linear operators into a Hilbert space $H_1 \times H_2$. $F : D(F) \subset Z \to X$, a not necessarily linear weak-to-weak continuous map, such that for every $x \in D(\Lambda) \cap D(F) \cap D(b_1, b_2)$,

$$\langle \Lambda x, F(x) \rangle = \psi(x) + \ell_2(b_2 x) - \ell_1(b_1 x),$$

where $\ell_2 : H_2 \to \mathbf{R} \cup \{+\infty\}$ (resp., $\ell_1 : H_1 \to \mathbf{R} \cup \{+\infty\}$) are bounded below weakly lower semicontinuous functions on H_1 (resp., H_2), and ψ is weakly lower semicontinuous on Z. Suppose E is a closed convex subset of Z such that for some $a \in H_1$ we have

$$E \subset D(\Lambda) \cap D(F) \cap D(b_1, b_2) \cap \{x \in X : b_1 x = a\}$$

and

$$\Lambda(E) \text{ is dense in } X^*.$$

   a. Show that the functional

$$I(z) = \varphi(F(z)) + \varphi^*(-\Lambda z) + \langle \Lambda z, F(z) \rangle$$

   is self-dual on E and write its antisymmetric Hamiltonian.
   b. Show that the problem

$$\begin{cases} -\Lambda u \in \partial \varphi(F(u)) \\ b_1(u) = a \end{cases}$$

   can be solved, provided $\frac{\varphi^*(-\Lambda z)}{\|z\|} \to +\infty$ as $\|z\|_Z \to \infty$.

2. Fill in the details in Example 13.3.
3. Develop a self-dual variational approach to initial-value problems of the form

$$f \in \frac{d}{dt} S_1(u) + S_2(u) \tag{13.38}$$

   where S_1 and S_2 are two maximal monotone operators.

Further comments

The self-dual approach to the Cauchy problem for Hamiltonian flows appeared in Ghoussoub-Moameni [62]. Doubly nonlinear evolutions have a long history starting probably with Barbu [16]. See also Arai [3], Senba [142] and Colli and Visentin

[39] for a Hilbertian framework, and Colli [38] and Visentin [159], [158] in Banach space settings. Fo more recent work, we refer to Stefanelli [151], [147], [152], and to Mielke et al. [111], [107], [102], [112], and [108]. For equations of the form (13.38), we refer to the recent work of Stefanelli [145], and [146] and the references therein.

Chapter 14
Direct Sum of Self-dual Functionals and Hamiltonian Systems

This chapter extends the results of Chapter 9. The context is similar, as we assume that a system of linear equations

$$\Gamma_i x = p_i, \ i = 1, ..., n,$$

with Γ_i being a linear operator from a Banach space Z into the dual of another one X_i, can be solved for any $p_i \in X^*$. We then investigate when one can solve variationally the semilinear system of equations

$$\Gamma_i x \in \partial \varphi_i (A_i x),$$

where each A_i is a bounded linear operator from Z to X_i. Unlike Chapter 9, where we required $\sum\limits_{i=1}^{n} \langle A_i z, \Gamma_i z \rangle$ to be identically zero, here we relax this assumption considerably by only requiring that the map

$$z \to \sum\limits_{i=1}^{n} \langle A_i z, \Gamma_i z \rangle \text{ be weakly upper semicontinuous}$$

as long as we have some control of the form

$$\left| \sum\limits_{i=1}^{n} \langle A_i z, \Gamma_i z \rangle \right| \leq \sum\limits_{i=1}^{n} \alpha_i \| \Gamma_i z \|^2$$

for some $\alpha_i \geq 0$. In this case, the growth of the potentials φ_i should not exceed a quadratic growth of factor $\frac{1}{2\alpha_i}$. This result is then applied to derive self-dual variational resolutions to periodic solutions of Hamiltonian systems as well as orbits connecting certain Lagrangian submanifolds.

14.1 Self-dual systems of equations

We first state the following general result.

N. Ghoussoub, *Self-dual Partial Differential Systems and Their Variational Principles*, Springer Monographs in Mathematics, DOI 10.1007/978-0-387-84897-6_14,

Theorem 14.1. *Consider n reflexive Banach spaces* $X_1, X_2, ..., X_n$, *bounded linear operators* B_i *acting on each* X_i, *and* B_i-*self-dual Lagrangians* L_i *on* $X_i \times X_i^*$ *for* $i = 1, ..., n$. *Let Z be a reflexive Banach space and let* $\Gamma_i : Z \to X_i^*$ *be bounded linear operators such that* $\Gamma := (\Gamma_1, \Gamma_2, ..., \Gamma_n)$ *maps a closed convex subset E of Z onto* $\text{Dom}_2(L_1) \times \text{Dom}_2(L_2) \times ... \times \text{Dom}_2(L_n) \subset \Pi_{i=1}^n X_i^*$. *Consider also bounded linear operators* $A_i : Z \to X_i$ *such that the map*

$$z \to \sum_{i=1}^n \langle B_i A_i z, \Gamma_i z \rangle \text{ is weakly upper semicontinuous on } E. \qquad (14.1)$$

1. The functional

$$I(z) = \sum_{i=1}^n L_i(A_i z, \Gamma_i z) - \langle B_i A_i z, \Gamma_i z \rangle \qquad (14.2)$$

is self-dual on E with corresponding AS-Hamiltonian

$$M(z, w) = \sum_{i=1}^n \tilde{H}_{L_i}(B_i^* \Gamma_i z, \Gamma_i w) + \langle B_i A_i z, \Gamma_i(w - z) \rangle, \qquad (14.3)$$

where for each $i = 1, ..., n$, $\tilde{H}_{L_i}$ *is the co-Hamiltonian on* $X_i^* \times X_i^*$ *associated to* L_i.

2. If the coercivity condition

$$\lim_{z \in E, \|z\| \to +\infty} \sum_{i=1}^n \tilde{H}_{L_i}(B_i^* \Gamma_i z, 0) - \langle B_i A_i z, \Gamma_i z \rangle = +\infty, \qquad (14.4)$$

holds, then the infimum of the functional I over E is zero and is attained at $\bar{x} \in E$, *which then solves the system of equations*

$$\Gamma_i \bar{x} \in \overline{\partial}_{B_i} L_i(A_i \bar{x}) \quad \text{for} \quad i = 1, ..., n. \qquad (14.5)$$

Proof. First, it is clear that I is nonnegative since each L_i is B_i-self-dual. Now write for any $z \in Z$

$$I(z) = \sum_{i=1}^n L_i(A_i z, \Gamma_i z) - \langle B_i A_i z, \Gamma_i z \rangle$$

$$= \sum_{i=1}^n L_i^*(B_i^* \Gamma_i z, B_i A_i z) - \langle B_i A_i z, \Gamma_i z \rangle$$

$$= \sum_{i=1}^n \sup \left\{ \langle B_i^* \Gamma_i z, x_i \rangle + \langle B_i A_i z, w_i \rangle - L_i(x_i, w_i); x_i \in X_i, w_i \in X_i^* \right\}$$

$$- \sum_{i=1}^n \langle B_i A_i z, \Gamma_i z \rangle$$

$$= \sum_{i=1}^n \sup \left\{ \tilde{H}_{L_i}(B_i^* \Gamma_i z, w_i) + \langle B_i A_i z, w_i \rangle; w_i \in X_i^* \right\} - \sum_{i=1}^n \langle B_i A_i z, \Gamma_i z \rangle$$

$$= \sum_{i=1}^{n} \sup \left\{ \tilde{H}_{L_i} (B_i^* \Gamma_i z, w_i) + \langle B_i A_i z, w_i \rangle ; w_i \in \text{Dom}_2(L_i) \right\} - \sum_{i=1}^{n} \langle B_i A_i z, \Gamma_i z \rangle$$

$$= \sup \left\{ \sum_{i=1}^{n} \tilde{H}_{L_i} (B_i^* \Gamma_i z, \Gamma_i w) + \langle B_i A_i z, \Gamma_i w \rangle ; w \in E \right\} - \sum_{i=1}^{n} \langle B_i A_i z, \Gamma_i z \rangle$$

$$= \sup \left\{ \sum_{i=1}^{n} \tilde{H}_{L_i} (B_i^* \Gamma_i z, \Gamma_i w) + \langle B_i A_i z, \Gamma_i (w - z) \rangle ; w \in E \right\}.$$

For (2), use the coercivity condition and Theorem 12.2 to deduce the existence of $\bar{z}$ such that

$$0 = \sum_{i=1}^{n} L_i(A_i \bar{z}, \Gamma_i \bar{z}) - \langle B_i A_i \bar{z}, \Gamma_i \bar{z} \rangle.$$

Again, since each term is nonnegative , we get that $L_i(A_i \bar{z}, \Gamma_i \bar{z}) - \langle B_i A_i \bar{z}, \Gamma_i \bar{z} \rangle = 0$ for each $i = 1, ..., n$, and therefore

$$(B_i^* \Gamma_i \bar{z}, B_i A_i \bar{z}) \in \partial L_i(A_i \bar{z}, \Gamma_i \bar{z}).$$

The following is a major improvement on Corollary 9.1.

Theorem 14.2. *For $i = 1, ..., n$, we consider $\Gamma_i : Z \to X_i^*$ to be bounded linear operators from a reflexive Banach space Z into the dual of a Banach space X_i such that $(\Gamma_i)_i$ is an isomorphism from Z onto $\Pi_{i=1}^{n} X_i^*$. Let $A_i : Z \to X_i$ be bounded linear operators and scalars $\alpha_i \geq 0$, $i = 1, ..., n$ such that*

$$z \to \sum_{i=1}^{n} \langle A_i z, \Gamma_i z \rangle \text{ is weakly upper semicontinuous on } Z \tag{14.6}$$

and

$$\left| \sum_{i=1}^{n} \langle A_i z, \Gamma_i z \rangle \right| \leq \sum_{i=1}^{n} \alpha_i \| \Gamma_i z \|^2 \text{ for } z \in Z. \tag{14.7}$$

Consider bounded below, convex, and lower semicontinuous functions Φ_i on X_i ($1 \leq i \leq n$) such that for $1 \leq i \leq k$ there are $\beta_i > 0$ so that

$$0 \leq \alpha_i < \frac{1}{2\beta_i} \quad \text{and} \quad \Phi_i(x) \leq \frac{\beta_i}{2}(\|x\|^2 + 1) \text{ for all } x \in X_i, \tag{14.8}$$

while for $k+1 \leq i \leq n$ we have

$$\Phi_i(0) < +\infty \quad \text{and} \quad \lim_{\|p\| \to \infty} \frac{\Phi_i^*(p)}{\|p\|^2} \geq \alpha_i. \tag{14.9}$$

Assuming the coercivity condition

$$\lim_{\|z\| \to +\infty} \sum_{i=1}^{n} \left[\Phi_i(A_i z) + \Phi_i^*(\Gamma_i z) + \alpha_i \| \Gamma_i z \|^2 \right] = +\infty, \tag{14.10}$$

then there exists a solution to the system

$$\Gamma_i z \in \partial \Phi_i(A_i z) \quad \textit{for all } i = 1, \ldots n. \tag{14.11}$$

It is obtained as the minimum of the nonnegative functional

$$I(z) = \sum_{i=1}^{n} \Phi_i(A_i z) + \Phi_i^*(\Gamma_i z) - \langle A_i z, \Gamma_i z \rangle \tag{14.12}$$

over Z, which is then equal to zero.

We start with the following lemma, which proves the theorem above under stronger boundedness conditions.

Lemma 14.1. *Let $\Gamma_i : Z \to X_i^*$ be bounded linear operators from a reflexive Banach space Z into the dual of a Banach space X_i such that $(\Gamma_i)_i$ is an isomorphism from Z onto $\Pi_{i=1}^{n} X_i^*$. Consider bounded linear operators $A_i : Z \to X_i$ such that (14.6) is satisfied. Let $(\alpha_i)_{i=1}^{n}$ be scalars such that for some $p > 1$*

$$\left| \sum_{i=1}^{n} \langle A_i z, \Gamma_i z \rangle \right| \le \sum_{i=1}^{n} \alpha_i \|\Gamma_i z\|_{X_i}^{p} \textit{ for } z \in Z. \tag{14.13}$$

Consider convex lower semicontinuous functions Φ_i on X_i verifying, for some $\varepsilon_i > 0$, $\beta_i > 0$, $r > 1$,

$$\varepsilon_i(\|x\|^r - 1) \le \Phi_i(x) \le \frac{\beta_i}{q}(\|x\|^q + 1) \textit{ for all } x \in X_i, \tag{14.14}$$

where $\frac{1}{p} + \frac{1}{q} = 1$. Assuming that

$$0 \le \alpha_i < \frac{\beta_i^{1-p}}{p} \textit{ for } i = 1, \ldots, n, \tag{14.15}$$

then there exists a solution to the system (14.11). It is obtained as the minimum over Z of the self-dual functional I given in (14.12), which is then equal to zero.

Proof. This is a direct application of Proposition 14.1 applied to the self-dual Lagrangians $L_i(x, p) = \Phi_i(x) + \Phi_i^*(p)$. The AS-Hamiltonian M associated to the self-dual functional I is then

$$M(z, w) = \sum_{i=1}^{n} \Phi_i^*(\Gamma_i z) - \Phi_i^*(\Gamma_i w) + \langle A_i z, \Gamma_i(w - z) \rangle. \tag{14.16}$$

In order to check the coercivity condition, we write

$$M(z, 0) = \sum_{i=1}^{n} \Phi_i^*(\Gamma_i z) - \Phi_i^*(0) - \langle A_i z, \Gamma_i z \rangle$$

$$\ge \sum_{i=1}^{n} \left(\frac{\beta_i^{1-p}}{p} \|\Gamma_i z\|^p - \frac{\beta_i}{q} \right) - \sum_{i=1}^{n} \Phi_i^*(0) - \sum_{i=1}^{n} \alpha_i \|\Gamma_i z\|_{X_i}^{p}$$

$$\geq \sum_{i=1}^{n} \left(\frac{\beta_i^{1-p}}{p} - \alpha_i \right) \|\Gamma_i z\|^p - C,$$

and hence, $M(z,0) \to +\infty$ whenever $\|z\| \to \infty$. It follows that there exists $\bar{z} \in Z$ such that

$$0 = I(\bar{z}) = \sum_{i=1}^{n} \Phi_i(A_i\bar{z}) + \Phi_i^*(\Gamma_i\bar{z}) - \langle A_i\bar{z}, \Gamma_i\bar{z} \rangle,$$

yielding $\Phi_i(A_i\bar{z}) + \Phi_i^*(\Gamma_i\bar{z}) - \langle A_i\bar{z}, \Gamma_i\bar{z} \rangle = 0$ for $i = 1, ...n$, and hence $\bar{z}$ satisfies system (14.11).

Proof of Theorem 14.2. We shall use inf-convolution to relax the boundedness condition on Φ_i in the lemma above. We first assume that, for all $1 \leq i \leq n$, we have

$$C_i \leq \Phi_i(x) \leq \frac{\beta_i}{2}(\|x\|^2 + 1) \text{ for all } x \in X_i, \text{ and } \alpha_i < \frac{1}{2\beta_i}. \tag{14.17}$$

In this case, the convex functions $\Phi_{i,\varepsilon}(x) = \Phi_i(x) + \frac{\varepsilon}{2}\|x\|^2$ clearly satisfy the conditions of Lemma 14.1, provided ε is chosen small enough so that for $1 \leq i \leq n$

$$0 \leq \alpha_i < \frac{1}{2(\beta_i + \varepsilon)},$$

and therefore we can find $z_\varepsilon \in Z$ such that the minimum of the self-dual functional

$$I_\varepsilon(z) = \sum_{i=1}^{n} \Phi_{i,\varepsilon}(A_i z) + \Phi_{i,\varepsilon}^*(\Gamma_i z) - \langle A_i z, \Gamma_i z \rangle \tag{14.18}$$

over Z is attained at z_ε and $I_\varepsilon(z_\varepsilon) = 0$. We now show that $(z_\varepsilon)_\varepsilon$ is bounded in Z. Indeed, the fact that $I_\varepsilon(z_\varepsilon) = 0$ coupled with (14.7) yields that

$$\sum_{i=1}^{n} \Phi_{i,\varepsilon}(A_i z_\varepsilon) + \Phi_{i,\varepsilon}^*(\Gamma_i z_\varepsilon) - \sum_{i=1}^{n} \alpha_i \|\Gamma_i z_\varepsilon\|^2 \leq 0. \tag{14.19}$$

This, combined with the bounds on $(\Phi_i)_{i=1}^{n}$, guarantees the existence of $C > 0$ independent of ε such that

$$\sum_{i=1}^{n} \left(\frac{1}{2(\beta_i + \varepsilon)} - \alpha_i \right) \|\Gamma_i z_\varepsilon\|^2 \leq C.$$

It follows that $(z_\varepsilon)_\varepsilon$ is bounded in Z, and therefore there exists $\bar{z}$ such that $z_\varepsilon \to \bar{z}$ weakly as $\varepsilon \to 0$. Write now

$$\Phi_{i,\varepsilon}^*(\Gamma_i z_\varepsilon) = \inf \left\{ \Phi_i^*(v) + \frac{1}{2\varepsilon} \|\Gamma_i z_\varepsilon - v\|^2; v \in X_i^* \right\},$$

where the infimum is attained at some $v_\varepsilon^i \in X_i^*$ in such a way that $\Phi_{i,\varepsilon}^*(\Gamma_i z_\varepsilon) = \Phi_i^*(v_\varepsilon^i) + \frac{1}{2\varepsilon}\|\Gamma_i z_\varepsilon - v_\varepsilon^i\|^2$. The boundedness of $(z_\varepsilon)_\varepsilon$ in Z yields that there is $C > 0$ independent of ε such that

$$\Phi_i^*(v_\varepsilon^i) + \frac{1}{2\varepsilon}\|\Gamma_i z_\varepsilon - v_\varepsilon^i\|^2 = \Phi_{i,\varepsilon}^*(\Gamma_i z_\varepsilon) \leq C,$$

and since Φ_i^* is bounded below, we get that $\|\Gamma_i z_\varepsilon - v_\varepsilon^i\| \leq C\varepsilon$, which means that $v_\varepsilon^i \to \Gamma_i \bar{z}$ weakly in X_i^* for every $i = 1, \dots, n$.

Since now each Φ_i^* is convex and (weakly) lower semicontinuous, we get

$$\Phi_i^*(\Gamma_i \bar{z}) \leq \liminf_{\varepsilon \to 0} \Phi_i^*(v_\varepsilon^i)$$

$$\leq \liminf_{\varepsilon \to 0} \Phi_i^*(v_\varepsilon^i) + \frac{1}{2\varepsilon}\|\Gamma_i z_\varepsilon - v_\varepsilon^i\|^2$$

$$= \liminf_{\varepsilon \to 0} \Phi_{i,\varepsilon}^*(\Gamma_i z_\varepsilon).$$

Similarly, we have

$$\Phi_i(A_i \bar{z}) \leq \liminf_{\varepsilon \to 0} \Phi_i(A_i z_\varepsilon) \leq \liminf_{\varepsilon \to 0} \Phi_i(A_i z_\varepsilon) + \frac{\varepsilon}{2}\|z_\varepsilon\|^2 = \liminf_{\varepsilon \to 0} \Phi_{i,\varepsilon}(A_i z_\varepsilon).$$

These, combined with the weak upper semicontinuity of $z \to \sum_{i=1}^{n}\langle A_i z, \Gamma_i z\rangle$ on Z, finally yields

$$I(\bar{z}) = \sum_{i=1}^{n} \Phi_i(A_i z) + \Phi_i^*(\Gamma_i z) - \langle A_i z, \Gamma_i z\rangle$$

$$\leq \sum_{i=1}^{n}\left[\liminf_{\varepsilon \to 0}\Phi_{i,\varepsilon}(A_i z_\varepsilon) + \liminf_{\varepsilon \to 0}\Phi_{i,\varepsilon}^*(\Gamma_i z_\varepsilon)\right] + \liminf_{\varepsilon \to 0}\sum_{i=1}^{n} -\langle A_i z_\varepsilon, \Gamma_i z_\varepsilon\rangle$$

$$\leq \liminf_{\varepsilon \to 0} I_\varepsilon(z_\varepsilon)$$

$$= 0.$$

On the other hand, $I(z) \geq 0$ for every $z \in X$, and the proof is complete under the assumption (14.17).

Now, for the general case, we only assume that for $1 \leq i \leq k$

$$C_i \leq \Phi_i(x) \leq \frac{\beta_i}{2}(\|x\|^2 + 1) \text{ for all } x \in X_i, \tag{14.20}$$

while for $k + 1 \leq i \leq n$ we have

$$\Phi_i(0) < +\infty \text{ and } \lim_{\|p\| \to \infty} \frac{\Phi_i^*(p)}{\|p\|^2} \geq \alpha_i. \tag{14.21}$$

In this case, we use inf-convolution on $\{\Phi_i; k + 1 \leq i \leq n\}$ and define for $\lambda > 0$ the convex functional $\Phi_{i,\lambda}(x) = \inf\left\{\Phi_i(y) + \frac{\lambda}{2}\|x - y\|^2; y \in X_i\right\}$ in such a way that their conjugates are $\Phi_{i,\lambda}^*(v) = \Phi_i^*(v) + \frac{2}{\lambda}\|v\|^2$. Note that $\Phi_{i,\lambda}$ are also bounded below, and since $\Phi_i(0) < +\infty$ for $i = k+1, ..., n$, we have that Φ_i^* and consequently $\Phi_{i,\lambda}^*$ is bounded below for all $i = k+1, ..., n$.

The first part of the proof applies to $\{\Phi_1, ..., \Phi_k, \Phi_{k+1,\lambda}, ..., \Phi_{n,\lambda}\}$ as long as λ is small enough so that $\alpha_i < \frac{2}{\lambda}$ for all $i = k+1, ..., n$. We then obtain for each $\lambda > 0$ a $z_\lambda \in Z$ such that the minimum of the self-dual functional

$$I_\lambda(z) = \sum_{i=1}^{k} \left[\Phi_i(A_i z) + \Phi_i^*(\Gamma_i z) - \langle A_i z, \Gamma_i z\rangle\right]$$

$$+ \sum_{i=k+1}^{n} \left[\Phi_{i,\lambda}(A_i z) + \Phi_{i,\lambda}^*(\Gamma_i z) - \langle A_i z, \Gamma_i z\rangle\right]$$

over Z is attained at z_λ and $I_\lambda(z_\lambda) = 0$. We shall show that $(z_\lambda)_\lambda$ is bounded in Z. Indeed, combining again the fact that $I_\lambda(z_\lambda) = 0$ and (14.7) yields that

$$\sum_{i=1}^{k} \left[\Phi_i(A_i z_\lambda) + \Phi_i^*(\Gamma_i z_\lambda)\right] \tag{14.22}$$

$$+ \sum_{i=k+1}^{n} \left[\Phi_{i,\lambda}(A_i z_\lambda) + \Phi_{i,\lambda}^*(\Gamma_i z_\lambda)\right] - \sum_{i=1}^{n} \alpha_i \|\Gamma_i z_\lambda\|^2 \leq 0.$$

From the lower bounds on Φ_i^* $(1 \leq i \leq k)$, we get

$$\sum_{i=1}^{k} \left[\Phi_i(A_i z_\lambda) + \left(\frac{1}{2\beta_i} - \alpha_i\right)\|\Gamma_i z_\lambda\|^2\right]$$

$$+ \sum_{i=k+1}^{n} \left[\Phi_{i,\lambda}(A_i z_\lambda) + \Phi_{i,\lambda}^*(\Gamma_i z_\lambda) - \alpha_i \|\Gamma_i z_\lambda\|^2\right] \leq 0. \tag{14.23}$$

This, combined with the lower bounds on $(\Phi_i)_{i=1}^{n}$, guarantees the existence of $C > 0$ independent of λ such that

$$\sum_{i=1}^{k} \left(\frac{1}{2\beta_i} - \alpha_i\right)\|\Gamma_i z_\lambda\|^2 + \sum_{i=k+1}^{n} \left[\Phi_{i,\lambda}^*(\Gamma_i z_\lambda) - \alpha_i \|\Gamma_i z_\lambda\|^2\right] \leq C.$$

It follows that $\sum_{i=k+1}^{n} \left[\Phi_i^*(\Gamma_i z_\lambda) + (\frac{2}{\lambda} - \alpha_i)\|\Gamma_i z_\lambda\|^2\right] \leq C$, and in view of (14.9) this means that $(\|\Gamma_i z_\lambda\|)_\lambda$ is bounded for each $i \in \{k, k+1, ..., n\}$. Since Φ_i^* is bounded below for all such i's, we get that $\sum_{i=1}^{k} \|\Gamma_i z_\lambda\|^2$ is also bounded. In other words, we have in view of (14.23) and the fact that $\Phi_{i,\lambda}$ and $\Phi_{i,\lambda}^*$ are bounded below, that

$$\sum_{i=1}^{k} \left[\Phi_i(A_i z_\lambda) + \alpha_i \|\Gamma_i z_\lambda\|^2\right] \leq C_1 \tag{14.24}$$

and

$$\sum_{i=k+1}^{n} \left[\Phi_{i,\lambda}(A_i z_\lambda) + \Phi_{i,\lambda}^*(\Gamma_i z_\lambda) \right] \le C_2. \tag{14.25}$$

It follows that $\sum_{i=1}^{n} \left[\Phi_i(A_i z_\lambda) + \Phi_i^*(\Gamma_i z_\lambda) + \alpha_i \|\Gamma_i z_\lambda\|^2 \right] < +\infty$. Using the coercivity condition (14.10), we conclude that $(z_\lambda)_\lambda$ is necessarily bounded in Z, and therefore there exists $\bar{z}$ such that $z_\lambda \to \bar{z}$ weakly as $\lambda \to 0$. Therefore

$$\Phi_i(A_i \bar{z}) + \Phi_i^*(\Gamma_i \bar{z}) \le \liminf_{\lambda \to 0} [\Phi_i(A_i z_\lambda) + \Phi_i^*(\Gamma_i z_\lambda)]$$

for $1 \le i \le k$, and from (14.25) combined with Lemma 3.3, we get for $k+1 \le i \le n$ that

$$\Phi_i(A_i \bar{z}) + \Phi_i^*(\Gamma_i \bar{z}) \le \liminf_{\lambda \to 0} [\Phi_{i,\lambda}(A_i z_\lambda) + \Phi_{i,\lambda}^*(\Gamma_i z_\lambda)].$$

The last two assertions, coupled with the weak upper semicontinuity of $z \to \sum_{i=1}^{n} \langle A_i z, \Gamma_i z \rangle$ on Z, give

$$\begin{aligned}
I(\bar{z}) &= \sum_{i=1}^{n} \Phi_i(A_i z) + \Phi_i^*(\Gamma_i z) - \langle A_i z, \Gamma_i z \rangle \\
&\le \sum_{i=1}^{k} \liminf_{\lambda \to 0} \left[\Phi_i(A_i z_\lambda) + \Phi_i^*(\Gamma_i z_\lambda) \right] + \sum_{i=k+1}^{n} \liminf_{\lambda \to 0} \left[\Phi_{i,\lambda}(A_i z_\lambda) + \Phi_{i,\lambda}^*(\Gamma_i z_\lambda) \right] \\
&\quad + \liminf_{\lambda \to 0} \sum_{i=1}^{n} -\langle A_i z_\lambda, \Gamma_i z_\lambda \rangle \\
&\le \liminf_{\lambda \to 0} I_\lambda(z_\lambda) \\
&= 0.
\end{aligned}$$

On the other hand, $I(z) \ge 0$ for all $z \in X$, and the proof of Theorem 14.2 is complete.

14.2 Periodic orbits of Hamiltonian systems

For a given Hilbert space H, we consider the subspace H_T^1 of A_H^2 consisting of all periodic functions, equipped with the norm induced by A_H^2. We also consider the space H_{-T}^1 consisting of all functions in A_H^2 that are antiperiodic, i.e. $u(0) = -u(T)$. The norm of H_{-T}^1 is given by $\|u\|_{H_{-T}^1} = (\int_0^T |\dot{u}|^2 \, dt)^{\frac{1}{2}}$. We start by establishing a few useful inequalities on H_{-T}^1 that can be seen as the counterparts of Wirtinger's inequality,

$$\int_0^T |u|^2 \, dt \le \frac{T^2}{4\pi^2} \int_0^T |\dot{u}|^2 \, dt \quad \text{for} \quad u \in H_T^1 \quad \text{and} \quad \int_0^T u(t) \, dt = 0, \tag{14.26}$$

and the Sobolev inequality on H^1_T,

$$\|u\|^2_\infty \le \frac{T}{12} \int_0^T |\dot u|^2 \, dt \quad \text{for} \quad u \in H^1_T \quad \text{and} \quad \int_0^T u(t) \, dt = 0. \tag{14.27}$$

Proposition 14.1. *If $u \in H^1_{-T}$, then*

$$\int_0^T |u|^2 \, dt \le \frac{T^2}{\pi^2} \int_0^T |\dot u|^2 \, dt \tag{14.28}$$

and

$$\|u\|^2_\infty \le \frac{T}{4} \int_0^T |\dot u|^2 \, dt. \tag{14.29}$$

Proof. Since $u(0) = -u(T)$, it has the Fourier expansion $u(t) = \sum\limits_{k=-\infty}^{\infty} u_k \exp(\frac{(2k-1)i\pi t}{T})$. The Parseval equality implies that

$$\int_0^T |\dot u|^2 \, dt = \sum_{k=-\infty}^{\infty} T((2k-1)^2 \pi^2 / T^2) |u_k|^2 \ge \frac{\pi^2}{T^2} \sum_{k=-\infty}^{\infty} T |u_k|^2 = \frac{\pi^2}{T^2} \int_0^T |u|^2 \, dt.$$

On the other hand, the Cauchy-Schwarz inequality yields for $t \in [0, T]$,

$$|u(t)|^2 \le \left(\sum_{k=-\infty}^{\infty} |u_k| \right)^2$$

$$\le \left[\sum_{k=-\infty}^{\infty} \frac{T}{\pi^2 (2k-1)^2} \right] \left[\sum_{k=-\infty}^{\infty} T((2k-1)^2 \pi^2 / T^2) |u_k|^2 \right]$$

$$= \frac{T}{\pi^2} \sum_{k=-\infty}^{\infty} \frac{1}{(2k-1)^2} \int_0^T |\dot u|^2 \, dt,$$

and we conclude by noting that $\sum_{k=-\infty}^{\infty} \frac{1}{(2k-1)^2} = \frac{\pi^2}{4}$.

Proposition 14.2. *Consider the space A^2_X where $X = H \times H$, and let J be the symplectic operator on X defined as $J(p, q) = (-q, p)$.*

1. If H is any Hilbert space, then for every $u \in A^2_X$

$$\left| \int_0^T \langle J\dot u, u \rangle \, dt + \left\langle J \frac{u(0) + u(T)}{2}, u(T) - u(0) \right\rangle \right| \le \frac{T}{2} \int_0^T |\dot u(t)|^2 \, dt.$$

2. If H is finite-dimensional, then

$$\left| \int_0^T \langle J\dot u, u \rangle \, dt + \left\langle J \frac{u(0) + u(T)}{2}, u(T) - u(0) \right\rangle \right| \le \frac{T}{\pi} \int_0^T |\dot u(t)|^2 \, dt.$$

Proof. For part (1), note that each $u \in A_X^2$ can be written as

$$u(t) = \frac{1}{2} \left(\int_0^t \dot{u}(s)\,ds - \int_t^T \dot{u}(s)\,ds \right) + \frac{u(0) + u(T)}{2},$$

where $v(t) = u(t) - \frac{u(0)+u(T)}{2} = \frac{1}{2} \left(\int_0^t \dot{u}(s)\,ds - \int_t^T \dot{u}(s)\,ds \right)$ clearly belongs to H_{-T}^1.
Multiplying both sides by $J\dot{u}$ and integrating over $[0, T]$, we get

$$\int_0^T \langle J\dot{u}, u \rangle\,dt = \frac{1}{2} \int_0^T \left\langle \int_0^t \dot{u}(s)\,ds - \int_t^T \dot{u}(s)\,ds, J\dot{u} \right\rangle dt$$
$$+ \left\langle \frac{u(0) + u(T)}{2}, \int_0^T J\dot{u}(t)\,dt \right\rangle.$$

Hence,

$$\int_0^T \langle J\dot{u}, u \rangle\,dt - \left\langle \frac{u(0) + u(T)}{2}, J\big(u(T) - u(0)\big) \right\rangle$$
$$= \frac{1}{2} \int_0^T \left\langle \int_0^t \dot{u}(s)\,ds - \int_t^T \dot{u}(s)\,ds, J\dot{u} \right\rangle dt,$$

and since J is skew-symmetric, we have

$$\int_0^T \langle J\dot{u}, u \rangle\,dt + \left\langle J\frac{u(0) + u(T)}{2}, u(T) - u(0) \right\rangle \qquad (14.30)$$
$$= \frac{1}{2} \int_0^T \left\langle \int_0^t \dot{u}(s)\,ds - \int_t^T \dot{u}(s)\,ds, J\dot{u} \right\rangle dt. \qquad (14.31)$$

Applying Hölder's inequality for the right-hand side, we get

$$\left| \int_0^T \langle J\dot{u}, u \rangle\,dt + \left\langle J\frac{u(0) + u(T)}{2}, u(T) - u(0) \right\rangle \right| \leq \frac{T}{2} \int_0^T |\dot{u}(t)|^2\,dt.$$

For part (2), set $v(t) = u(t) - \frac{u(0)+u(T)}{2}$ and note that

$$\int_0^T \langle J\dot{u}, u \rangle\,dt + \left\langle J\frac{u(0) + u(T)}{2}, u(T) - u(0) \right\rangle = \int_0^T \langle J\dot{v}, v \rangle\,dt. \qquad (14.32)$$

Since $v \in H_{-T}^1$, Hölder's inequality and Proposition 14.1 imply,

$$\left| \int_0^T \langle J\dot{v}, v \rangle\,dt \right| \leq \left(\int_0^T |v|^2\,dt \right)^{\frac{1}{2}} \left(\int_0^T |J\dot{v}|^2\,dt \right)^{\frac{1}{2}}$$
$$\leq \frac{T}{\pi} \left(\int_0^T |\dot{v}|^2\,dt \right)^{\frac{1}{2}} \left(\int_0^T |J\dot{v}|^2\,dt \right)^{\frac{1}{2}}$$
$$= \frac{T}{\pi} \int_0^T |\dot{v}|^2\,dt = \frac{T}{\pi} \int_0^T |\dot{u}|^2\,dt.$$

Combining this inequality with (14.30) yields the claimed inequality.

Lemma 14.2. *If $H = \mathbf{R}^N$ and $X = H \times H$, then the functional $F : A_X^2 \to \mathbf{R}$*

$$F(u) = \int_0^T \langle J\dot{u}, u \rangle \, dt + \left\langle u(T) - u(0), J\frac{u(T) + u(0)}{2} \right\rangle$$

is weakly continuous.

Proof. Let u_k be a sequence in A_X^2 that converges weakly to u in A_X^2. The injection A_X^2 into $C([0,T];X)$ with natural norm $\| \ \|_\infty$ is compact. Hence $u_k \to u$ strongly in $C([0,T];X)$ and specifically $u_k(T) \to u(T)$ and $u_k(0) \to u(0)$ strongly in X. Therefore

$$\lim_{k \to +\infty} \left\langle u_k(T) - u_k(0), J\frac{u_k(T) + u_k(0)}{2} \right\rangle = \left\langle u(T) - u(0), J\frac{u(T) + u(0)}{2} \right\rangle.$$

Since H is finite-dimensional, we use again that $u \to \int_0^T \langle J\dot{u}, u \rangle \, dt$ is weakly continuous on $A_X^2[0,T]$ (Proposition 1.2 in [96]), which together with (14.33) implies that F is weakly continuous on $A_X^2[0,T]$.

Periodic orbits for Hamiltonian systems

We now establish the following existence result.

Theorem 14.3. *Let $\varphi : [0,T] \times X \to \mathbf{R}$ be a time-dependent, convex, and lower semicontinuous function, and let $\psi : X \to \mathbf{R} \cup \{\infty\}$ be convex and lower semicontinuous on X such that the following conditions are satisfied. There exists $\beta > 0$, $\gamma, \alpha \in L^2(0,T;\mathbf{R}_+)$ such that:*

(B₁) $-\alpha(t) \le \varphi(t,u) \le \frac{\beta}{2}|u|^2 + \gamma(t)$ *for every $u \in H$ and all $t \in [0,T]$.*

(B₂) $\int_0^T \varphi(t,u) \, dt \to +\infty$ *as* $\|u\|_2 \to +\infty$.

(B₃) ψ *is bounded from below and $0 \in \mathrm{Dom}(\psi)$.*

(1) For any $T < \frac{\pi}{2\beta}$, the infimum on A_X^2 of the nonnegative functional

$$I_1(u) = \int_0^T \left[\varphi(t,u(t)) + \varphi^*(t,-J\dot{u}(t)) + \langle J\dot{u}(t), u(t) \rangle \right] dt \tag{14.33}$$

$$+ \left\langle u(T) - u(0), J\frac{u(0) + u(T)}{2} \right\rangle + \psi(u(T) - u(0)) + \psi^*\left(-J\frac{u(0) + u(T)}{2} \right)$$

is equal to zero and is attained at a solution of

$$\begin{cases} -J\dot{u}(t) & \in \partial\varphi(t,u(t)) \\ -J\frac{u(T)+u(0)}{2} & \in \partial\psi(u(T) - u(0)). \end{cases} \tag{14.34}$$

(2) Under the same assumptions, the infimum on A_X^2 of the functional

$$I_2(u) = \int_0^T \left[\varphi\big(t, u(t)\big) + \varphi^*\big(t, -J\dot{u}(t)\big) + \langle J\dot{u}(t), u(t) \rangle \right] dt$$
$$+ \langle Ju(0), u(T) \rangle + \psi\big(u(0)\big) + \psi^*\big(Ju(T)\big)$$

is also zero and is attained at a solution of

$$\begin{cases} -J\dot{u}(t) \in \partial\varphi\big(t, u(t)\big) \\ Ju(T) \quad \in \partial\psi\big(u(0)\big). \end{cases} \tag{14.35}$$

Proof. It suffices to apply Theorem 14.2 with $Z = A_X^2 = A_{H \times H}^2$, $X_1 = L_{H \times H}^2$, $X_2 = X = H \times H$, and the isomorphism $(\Gamma_1, \Gamma_2) : Z \to X_1^* \times X_2^*$, where

$$\Gamma_1(u) = -J\dot{u}(t) \quad \text{and} \quad \Gamma_2(u) := -J\frac{u(T)+u(0)}{2}, \tag{14.36}$$

while $A_1 : Z \to X_1$, and $A_2 : Z \to X_2$ are defined as

$$A_1(u) := u \quad \text{and} \quad A_2(u) := u(T) - u(0). \tag{14.37}$$

Consider now the functional Φ on $X_1 = L_{H \times H}^2$ defined by $\Phi_1(u) = \int_0^T \varphi(t, u(t)) \, dt$ and let $\Phi_2(x) = \psi(x)$ on $X_2 = H \times H$. The functional I on Z can be written as

$$I_1(u) = \Phi_1(A_1 u) + \Phi_1^*(\Gamma_1 u) - \langle A_1 u, \Gamma_1 u \rangle$$
$$+ \Phi_2(A_2 u) + \Phi_2^*(\Gamma_2 u) - \langle A_2 u, \Gamma_2 u \rangle. \tag{14.38}$$

From Lemma 14.2, we have that $u \to \langle A_1 u, \Gamma_1 u \rangle + \langle A_2 u, \Gamma_2 u \rangle$ is weakly continuous on Z, and from part (2) of Proposition 14.2, we have

$$\left| \langle A_1 u, \Gamma_1 u \rangle + \langle A_2 u, \Gamma_2 u \rangle \right| = \left| \int_0^T \langle J\dot{u}(t), u(t) \rangle \, dt + \left\langle u(T) - u(0), J\frac{u(T)+u(0)}{2} \right\rangle \right|$$
$$\leq \frac{T}{\pi} \int_0^T |\dot{u}(t)|^2 \, dt.$$

This means that (14.7) above applies with $\alpha_1 = \frac{T}{\pi}$ and $\alpha_2 = 0$. Moreover, since $\frac{1}{2\beta} - \frac{T}{\pi} > 0$, condition (14.14) is satisfied, and since $\psi(0)$ is finite, hypothesis (14.10) also holds. It remains to show that

$$\mathscr{L}(u) := \int_0^T \left[\varphi(t, u(t)) + \varphi^*(t, -J\dot{u}(t)) + \|\dot{u}(t)\|^2 \right] dt$$
$$+ \psi\big(u(T) - u(0)\big) + \psi^*\left(-J\frac{u(0)+u(T)}{2} \right)$$

is coercive on Z. But this follows from condition (B_2) since ψ and ψ^* are bounded below,

$$\mathscr{L}(u) \geq \int_0^T \left[\varphi(t, u(t)) + \|\dot{u}(t)\|^2 \right] dt - C,$$

and $\int_0^T \varphi(t, u(t))dt \to +\infty$ with $\int_0^T |u(t)|^2 dt$. It follows that there exists $\bar{u} \in X$ such that $I(\bar{u}) = 0$, which yields

$$
\begin{aligned}
I(\bar{u}) = &\int_0^T \left[\varphi(t, \bar{u}) + \varphi^*(t, -J\dot{\bar{u}}) + \langle \bar{u}, J\dot{\bar{u}} \rangle \right] dt \\
&+ \psi\left(\bar{u}(T) - \bar{u}(0) \right) + \psi^*\left(-J\frac{\bar{u}(0) + \bar{u}(T)}{2} \right) + \left\langle \bar{u}(T) - \bar{u}(0), J\frac{\bar{u}(0) + \bar{u}(T)}{2} \right\rangle \\
= &\, 0.
\end{aligned}
$$

The result now follows from the following identities and from the limiting case in Legendre-Fenchel duality:

$$
\varphi(t, \bar{u}(t)) + \varphi^*(t, -J\dot{\bar{u}}(t)) + (\bar{u}(t), J\dot{\bar{u}}(t)) = 0,
$$

$$
\psi\left(\bar{u}(T) - \bar{u}(0) \right) + \psi^*\left(-J\frac{\bar{u}(0) + \bar{u}(T)}{2} \right) + \left\langle \bar{u}(T) - \bar{u}(0), J\frac{\bar{u}(0) + \bar{u}(T)}{2} \right\rangle = 0.
$$

Example 14.1. Periodic solutions for Hamiltonian systems

As mentioned earlier, one can choose the boundary Lagrangian ψ appropriately to solve Hamiltonian systems of the form

$$
\begin{cases}
-J\dot{u}(t) \in \partial \varphi(t, u(t)) \\
u(0) = u_0, \text{ or } u(T) - u(0) \in K, \text{ or } u(T) = -u(0), \text{ or } u(T) = Ju(0).
\end{cases}
$$

For example:

- For an initial boundary condition $x(0) = x_0$ with a given $x_0 \in H$, use the functional I_1 with $\bar{\varphi}(t, x) = \varphi(t, x - x_0)$ and $\psi(x) = 0$ at 0 and $+\infty$ elsewhere.
- For periodic solutions $x(0) = x(T)$, or more generally $x(0) - x(T) \in K$, where K is a closed convex subset of $H \times H$, use the functional I_1 with ψ chosen as

$$
\psi(x) = \begin{cases} 0 & x \in K \\ +\infty & \text{elsewhere.} \end{cases}
$$

- For antiperiodic solutions $x(0) = -x(T)$, use the functional I_1 with $\psi(x) = 0$ for each $x \in H$.
- For skew-periodic solutions $x(0) = Jx(T)$, use the functional I_2 with $\psi(x) = \frac{1}{2}|x|^2$.

Corollary 14.1. *Let $\varphi : [0, T] \times X \to \mathbf{R}$ be a time-dependent convex and lower semicontinuous function, and let $\psi : X \to \mathbf{R} \cup \{\infty\}$ be convex and lower semicontinuous on X such that (B_2) and (B_3) are satisfied, while (B_1) is replaced by:*

(B_1') $\alpha(t) \leq \varphi(t, u) \leq \frac{\beta}{p}|u|^p + \gamma(t)$ *for every $u \in H$ and all $t \in \mathbf{R}^+$,* *for some $\beta > 0$, $1 \le p < 2$, and $\gamma, \alpha \in L^2([0,T]; \mathbf{R})$. Then, there is a solution for the system (14.35).*

Proof. It is an immediate application of Theorem 14.3 since for every $\varepsilon > 0$ there exists $C > 0$ such that $|x|^p \le \varepsilon |x|^2 + C$, leading to solutions for arbitrarily large T.

Example 14.2. Periodic solutions for second-order equations

One can also use the method to solve second order systems with convex potential and with prescribed nonlinear boundary conditions such as

$$\begin{cases} -\ddot{q}(t) = \partial \varphi(t, q(t)), \\ -\frac{q(0)+q(T)}{2} = \partial \psi_1(\dot{q}(T) - \dot{q}(0)), \\ \frac{\dot{q}(0)+\dot{q}(T)}{2} = \partial \psi_2(q(T) - q(0)), \end{cases} \tag{14.39}$$

and

$$\begin{cases} \ddot{q}(t) = \partial \varphi(t, q(t)), \\ -q(T) = \partial \psi_1(\dot{q}(0)), \\ \dot{q}(T) = \partial \psi_2(q(0)), \end{cases} \tag{14.40}$$

where ψ_1 and ψ_2 are convex lower semicontinuous. One can deduce the following

Corollary 14.2. *Let $\varphi : [0,T] \times X \to \mathbf{R}$ be a time-dependent convex and lower semicontinuous function, and let $\psi_i : H \to \mathbf{R} \cup \{\infty\}$, $i = 1, 2$ be convex and lower semicontinuous on H. Assume the following conditions. There exists $\beta \in (0, \frac{\pi}{2T})$ and $\gamma, \alpha \in L^2(0, T; \mathbf{R}_+)$ such that*

(A_1) $-\alpha(t) \le \varphi(t,q) \le \frac{\beta^2}{2}|q|^2 + \gamma(t)$ *for every $q \in H$ and a.e. $t \in [0,T]$.*
(A_2) $\int_0^T \varphi(t,q) \, dt \to +\infty$ *as* $|q| \to +\infty$.
(A_3) *ψ_1 and ψ_2 are bounded from below, and $0 \in \text{Dom}(\psi_i)$ for $i = 1, 2$.*

Then, equations (14.39) and (14.40) have at least one solution in A_H^2.

Proof. Define $\Psi : H \times H \to \mathbf{R} \cup \{+\infty\}$ by $\Psi(p,q) := \psi_1(p) + \psi_2(q)$ and $\Phi : [0,T] \times H \times H \to \mathbf{R}$ by $\Phi(t,u) := \frac{\beta}{2}|p|^2 + \frac{1}{\beta}\varphi(t, q(t))$, where $u = (p,q)$. It is easily seen that Φ is convex and lower semicontinuous in u and that

$$-\alpha(t) \le \Phi(t,u) \le \frac{\beta}{2}|u|^2 + \frac{\gamma(t)}{\beta} \quad \text{and} \quad \int_0^T \Phi(t,u) \, dt \to +\infty \quad \text{as} \quad |u| \to +\infty.$$

Also, from (A_3), the function Ψ is bounded from below and $0 \in \text{Dom}(\Psi)$. By Theorem 14.3, the infimum of the functional

$$I(u) := \int_0^T \left[\Phi(t, u(t)) + \Phi^*(t, -J\dot{u}(t)) + \langle J\dot{u}(t), u(t) \rangle \right] dt$$

$$+ \left\langle u(T) - u(0), J\frac{u(0)+u(T)}{2} \right\rangle + \Psi(u(T) - u(0)) + \Psi^*\left(-J\frac{u(0)+u(T)}{2} \right)$$

on A_X^2 is zero and is attained at a solution of

$$
\begin{cases}
-J\dot{u}(t) \in \partial\Phi(t,u(t)) \\
-J\frac{u(T)+u(0)}{2} = \partial\Psi(u(T)-u(0)).
\end{cases}
$$

Now, if we rewrite this problem for $u = (p,q)$, we get

$$
-\dot{p}(t) = \frac{1}{\beta}\partial\varphi(t,q(t)),
$$

$$
\dot{q}(t) = \beta p(t),
$$

$$
-\frac{q(T)+q(0)}{2} = \partial\psi(p(T)-p(0)),
$$

$$
\frac{p(T)+p(0)}{2} = \partial\psi(q(T)-q(0)),
$$

and hence, $q \in A_H^2$ is a solution of (14.39).

As in the case of Hamiltonian systems, one can then solve variationally the differential equation $-\ddot{q}(t) = \partial\varphi(t,q(t))$ with any one of the following boundary conditions:

(i)   periodic: $\dot{q}(T) = \dot{q}(0)$ and $q(T) = q(0)$;
(ii)   antiperiodic: $\dot{q}(T) = -\dot{q}(0)$ and $q(T) = -q(0)$;
(iii) Cauchy: $q(0) = q_0$ and $\dot{q}(0) = q_1$ for given $q_0, q_1 \in H$.

14.3 Lagrangian intersections

We consider again the Hamiltonian system

$$
\begin{cases}
\dot{p}(t) \in \partial_2\varphi(p(t),q(t)) & t \in (0,T) \\
-\dot{q}(t) \in \partial_1\varphi(p(t),q(t)) & t \in (0,T),
\end{cases}
\tag{14.41}
$$

where $\varphi : \mathbf{R}^N \times \mathbf{R}^N \to \mathbf{R}$ is a convex and lower semicontinuous function, but we now establish the existence of solutions that connect two Lagrangian submanifolds associated to given convex lower semicontinuous functions ψ_1 and ψ_2 on $\mathbf{R}^N$; that is, solutions satisfying

$$
q(0) \in \partial\psi_1(p(0)) \quad \text{and} \quad -p(T) \in \partial\psi_2(q(T)).
\tag{14.42}
$$

In other words, the Hamiltonian path must connect the graph of $\partial\psi_1$ to the graph of $-\partial\psi_2$, which are typical Lagrangian submanifolds in $\mathbf{R}^{2N}$. First, we consider the case of a convex Hamiltonian, and we extend it later to the semiconvex case.

Theorem 14.4. *Suppose $\varphi : \mathbf{R}^{2N} \to \mathbf{R}$ is a convex lower semicontinuous Hamiltonian such that, for some $0 < \beta < \frac{1}{2}$ and some constants α, γ, we have*

$$\alpha \le \varphi(p,q) \le \tfrac{\beta}{2}(|p|^2 + |q|^2) + \gamma \quad \text{for all } (p,q) \in \mathbf{R}^{2N}. \tag{14.43}$$

Let ψ_1 and ψ_2 be two convex lower semicontinuous and coercive functions on R^N such that one of them (say ψ_1) satisfies the following condition:

$$\liminf_{|p| \to +\infty} \frac{\psi_1(p)}{|p|^2} > 0. \tag{14.44}$$

Then, for any $T < \min\left\{\liminf_{|p| \to +\infty} \frac{\psi_1(p)}{2|p|^2}, \frac{1}{2\sqrt{\beta}}\right\}$, the minimum of the functional

$$I(p,q) := \int_0^T \left[\varphi(p(t), q(t)) + \varphi^*(-\dot{q}(t), \dot{p}(t)) + 2\dot{q}(t) \cdot p(t) \right] dt$$
$$+ \psi_2(q(T)) + \psi_2^*(-p(T)) + \psi_1(p(0)) + \psi_1^*(q(0))$$

on $Y = A_{\mathbf{R}^N}^2 \times A_{\mathbf{R}^N}^2$ is zero and is attained at a solution of

$$\begin{cases} \dot{p}(t) \in \partial_2 \varphi(p(t), q(t)) \ t \in (0, T) \\ -\dot{q}(t) \in \partial_1 \varphi(p(t), q(t)) \ t \in (0, T) \\ q(0) \in \partial \psi_1(p(0)) \\ -p(T) \in \partial \psi_2(q(T)). \end{cases} \tag{14.45}$$

Proof. We again apply Theorem 14.2 with $Z = A_{\mathbf{R}^N}^2 \times A_{\mathbf{R}^N}^2$, $X_1 = L_{\mathbf{R}^N}^2 \times L_{\mathbf{R}^N}^2$, $X_2 = \mathbf{R}^N$, $X_3 = \mathbf{R}^N$, and the isomorphism $(\Gamma_1, \Gamma_2, \Gamma_3) : Z \to X_1^* \times X_2^* \times X_3^*$, where

$$\Gamma_1(p(t), q(t)) := (-\dot{q}(t), \dot{p}(t)),$$
$$\Gamma_2(p(t), q(t)) := p(0),$$
$$\Gamma_3(p(t), q(t)) := q(T),$$

while $A_1 : Z \to X_1, A_2 : Z \to X_2, A_3 : Z \to X_3$ are defined as

$$A_1(p(t), q(t)) := (p, q), \ A_2(p(t), q(t)) := q(0), \text{ and } A_3(p(t), q(t)) := -p(T).$$

From Lemma 14.2, we have by setting $z = (p, q)$ that the functional

$$z \to \sum_{i=1}^3 \langle A_i z, \Gamma_i z \rangle = \int_0^T (-\dot{q} \cdot p + \dot{p} \cdot q) \, dt + p(0)q(0) - p(T)q(T)$$
$$= -2 \int_0^T \dot{q} \cdot p \, dt$$

is weakly continuous on $Z = A_{\mathbf{R}^N}^2 \times A_{\mathbf{R}^N}^2$. An easy calculation shows that

$$\|p\|_{L^2} \le T\|\dot{p}\|_{L^2} + \sqrt{T}|p(0)|, \quad \|q\|_{L^2} \le T\|\dot{q}\|_{L^2} + \sqrt{T}|q(T)|. \tag{14.46}$$

Using Hölder's inequality, we then have

$$\left| \int_0^T \dot{q} \cdot p \, dt \right| \le \frac{1}{2} \int_0^T |p|^2 \, dt + \frac{1}{2} \int_0^T |\dot{q}|^2 \, dt$$

$$\le T^2 \int_0^T |\dot{p}|^2 \, dt + T|p(0)|^2 + \frac{1}{2} \int_0^T |\dot{q}|^2 \, dt. \qquad (14.47)$$

It follows that

$$\left| \sum_{i=1}^3 \langle A_i z, \Gamma_i z \rangle \right| = 2 \left| \int_0^T \dot{q} \cdot p \, dt \right| \le \max\{2T^2, 1\} \int_0^T (|\dot{p}|^2 + |\dot{q}|^2 \, dt + 2T|p(0)|^2.$$

This means that estimate (14.7) holds with $\alpha_1 = \max\{2T^2, 1\}$, $\alpha_2 = 2T$, and $\alpha_3 = 0$. Now consider the convex functions $\Phi_1(p, q) = \int_0^T \varphi(p(t), q(t)) \, dt$ on $X_1 = L^2_{\mathbf{R}^N} \times L^2_{\mathbf{R}^N}$, $\Phi_2 = \psi_1^*$ on $X_2 = \mathbf{R}^N$, and $\Phi_3 = \psi_2^*$ on $X_3 = \mathbf{R}^N$. One can then write the functional I as

$$I(z) = \sum_{i=1}^3 \Phi_i(A_i z) + \Phi_i^*(\Gamma_i z) - \langle A_i z, \Gamma_i z \rangle \qquad (14.48)$$

over Z. We now show that all the hypotheses of Theorem 14.2 are satisfied. Indeed, it is clear that $\alpha_1 = \max\{2T^2, 1\} < \frac{1}{2\beta}$ and that

$$\liminf_{|p| \to +\infty} \frac{\Phi_2^*(p)}{|p|^2} = \liminf_{|p| \to +\infty} \frac{\psi_1(p)}{|p|^2} > 2T = \alpha_2.$$

Moreover, that the functional

$$\mathcal{L}(z) := \sum_{i=1}^3 \Phi_i(A_i z) + \Phi_i^*(\Gamma_i z) + \alpha_1 \|\Gamma_1 z\|^2 + \alpha_2 \|\Gamma_2 z\|^2$$

$$\ge \psi_2(q(T)) + \alpha_1 \int_0^T (|\dot{p}|^2 + |\dot{q}|^2) \, dt + \alpha_2 |p(0)|^2 - C$$

is coercive follows from the fact that Φ_1, $\Phi_2 = \psi_1^*$, $\Phi_3 = \psi_2^*$ and their conjugates are bounded below, while $\Phi_3^* = \psi_2$ is coercive.

Theorem 14.2 now applies, and we obtain $(\bar{p}, \bar{q}) \in Z = A^2_{\mathbf{R}^N} \times A^2_{\mathbf{R}^N}$ such that $I(\bar{p}, \bar{q}) = 0$. It follows that

$$0 = I(\bar{p}, \bar{q})$$

$$= \int_0^T \left[\varphi(\bar{p}(t), \bar{q}(t)) + \varphi^*(-\dot{\bar{q}}(t), \dot{\bar{p}}(t)) + 2\dot{\bar{q}}(t) \cdot \bar{p}(t) \right] dt$$

$$+ \psi_2(\bar{q}(T)) + \psi_2^*(-\bar{p}(T)) + \psi_1(\bar{p}(0)) + \psi_1^*(\bar{q}(0))$$

$$= \int_0^T \left[\varphi(\bar{p}(t), \bar{q}(t)) + \varphi^*(-\dot{\bar{q}}(t), \dot{\bar{p}}(t)) + \dot{\bar{q}}(t) \cdot \bar{p}(t) - \dot{\bar{p}}(t) \cdot \bar{q}(t) \right] dt$$

$$+ \left[\psi_2(\bar{q}(T)) + \psi_2^*(-\bar{p}(T)) + \bar{p}(T) \cdot \bar{q}(T) \right]$$

$$+ \left[\psi_1(\bar{p}(0)) + \psi_1^*(\bar{q}(0)) - \bar{p}(0) \cdot \bar{q}(0) \right].$$

The result is now obtained from the following three identities and from the limiting case in Legendre-Fenchel duality:

$$\varphi\big(\bar{p}(t),\bar{q}(t)\big) + \varphi^*\big(-\dot{\bar{q}}(t),\dot{\bar{p}}(t)\big) + \dot{\bar{q}}(t)\cdot\bar{p}(t) - \dot{\bar{p}}(t)\cdot\bar{q}(t) = 0,$$

$$\psi_2\big(\bar{q}(T)\big) + \psi_2^*\big(-\bar{p}(T)\big) + \bar{p}(T)\cdot\bar{q}(T) = 0,$$

$$\psi_1\big(\bar{p}(0)\big) + \psi_1^*\big(\bar{q}(0)\big) - \bar{p}(0)\cdot\bar{q}(0) = 0.$$

Remark 14.1. As in the preceding example, if the Hamiltonian is assumed to satisfy

$$-\alpha \leq H(p,q) \leq \beta(|p|^r + |q|^r + 1) \qquad (1 < r < 2)$$

where α, β are any positive constants, then we have existence for any $T > 0$.

14.4 Semiconvex Hamiltonian systems

In this section, we consider the system

$$\begin{cases} \dot{p}(t) \in \partial_2\varphi\big(p(t),q(t)\big) - \delta_1 q(t) & t \in (0,T) \\ -\dot{q}(t) \in \partial_1\varphi\big(p(t),q(t)\big) - \delta_2 p(t) & t \in (0,T) \\ q(0) \in \partial\psi_1\big(p(0)\big) \\ -p(T) \in \partial\psi_2\big(q(T)\big), \end{cases} \tag{14.49}$$

where $\delta_1, \delta_2 \in \mathbf{R}$. Note that if $\delta_i \leq 0$, then the problem reduces to the one studied in the previous section, with a new convex Hamiltonian $\tilde{\varphi}(p,q) = \varphi(p,q) - \frac{\delta_1}{2}|q|^2 - \frac{\delta_2}{2}|p|^2$. The case that concerns us here is when $\delta_i > 0$.

Theorem 14.5. *Suppose $\varphi : \mathbf{R}^{2N} \to \mathbf{R}$ is a convex lower semicontinuous Hamiltonian such that, for some $0 < \beta < \frac{1}{2}$ and some constants α, γ, we have*

$$\alpha \leq \varphi(p,q) \leq \frac{\beta}{2}(|p|^2 + |q|^2) + \gamma \quad \text{for all } (p,q) \in \mathbf{R}^{2N}, \tag{14.50}$$

and let ψ_1 and ψ_2 be convex lower semicontinuous functions on $\mathbf{R}^N$ satisfying

$$\liminf_{|p|\to+\infty} \frac{\psi_i(p)}{|p|^2} > 0 \quad \text{for } i = 1,2. \tag{14.51}$$

Then, for T small enough, the minimum of the functional

$$\begin{aligned} I(p,q) := & \int_0^T \Big[\varphi\big(p(t),q(t)\big) + \varphi^*\big(-\dot{q}(t) + \delta_2 p(t), \dot{p}(t) + \delta_1 q(t)\big)\Big]\,dt \\ & - \int_0^T \big(\delta_1|q|^2 + \delta_2|p|^2 - 2\dot{q}(t)\cdot p(t)\big)\,dt \\ & + \psi_2\big(q(T)\big) + \psi_2^*\big(-p(T)\big) + \psi_1\big(p(0)\big) + \psi_1^*\big(q(0)\big) \end{aligned}$$

on $Z = A_{\mathbf{R}^N}^2 \times A_{\mathbf{R}^N}^2$ is equal to zero and is attained at a solution of (14.49).

We shall need the following lemma.

Lemma 14.3. *For any* $f, g \in L^2(0, T; \mathbf{R}^N)$, $x, y \in \mathbf{R}^N$, *and* $\delta_1, \delta_2 \geq 0$, *there exists* $(r, s) \in Z$ *such that*

$$\begin{cases} \dot{r}(t) = -\delta_2 s(t) + f(t), \\ -\dot{s}(t) = -\delta_1 r(t) + g(t), \\ r(0) = x, \\ s(T) = y. \end{cases} \tag{14.52}$$

Proof. This is standard and is essentially a linear system of ordinary differential equations. Also, one can rewrite the problem as

$$\begin{cases} \dot{r}(t) - f(t) = -\partial_2 G(r(t), s(t)), \\ -\dot{s}(t) - g(t) = -\partial_1 G(r(t), s(t)), \\ r(0) = x, \\ s(T) = y, \end{cases}$$

where $G(r(t), s(t)) = \frac{\delta_1}{2} \int_0^T |r(t)|^2 \, dt + \frac{\delta_2}{2} \int_0^T |s(t)|^2 \, dt$. Hence,

$$G^*\left(-\dot{s}(t) - g(t), \dot{r}(t) - f(t)\right) = \frac{1}{2\delta_2} \int_0^T |\dot{s}(t) + g(t)|^2 + \frac{1}{2\delta_1} \int_0^T |\dot{r}(t) - f(t)|^2 \, dt.$$

One can show as in Theorem 14.4 that whenever $|\delta_i| < \frac{1}{2T}$, coercivity holds and the following infimum is achieved at a solution of (14.52).

$$0 = \inf_{(r,s) \in D \subseteq X \times X} G^*\left(-\dot{s}(t) - g(t), \dot{r}(t) - f(t)\right) + G(r(t), s(t))$$

$$+ \int_0^T \dot{r}(t) \cdot s(t) \, dt - \int_0^T \dot{s}(t) \cdot r(t) \, dt - \int_0^T \left(f(t) \cdot s(t) + r(t) \cdot g(t)\right) dt,$$

where $D = \{(r, s) \in X \times X \mid r(0) = x, s(T) = y\}$.

Proof of Theorem 14.5. Apply Theorem 14.2 with $Z = A_{\mathbf{R}^N}^2 \times A_{\mathbf{R}^N}^2$, $X_1 = L_{\mathbf{R}^N}^2 \times L_{\mathbf{R}^N}^2$, $X_2 = \mathbf{R}^N$, $X_3 = \mathbf{R}^N$, and the isomorphism $(\Gamma_1, \Gamma_2, \Gamma_3) : Z \to X_1^* \times X_2^* \times X_3^*$, where

$$\Gamma_1(p(t), q(t)) := (-\dot{q}(t) + \delta_2 p(t), \dot{p}(t) + \delta_1 q(t)),$$
$$\Gamma_2(p(t), q(t)) := p(0),$$
$$\Gamma_3(p(t), q(t)) := q(T),$$

while $A_1 : Z \to X_1$, $A_2 : Z \to X_2$, and $A_2 : Z \to X_2$ are defined as

$$A_1(p(t), q(t)) := (p, q), \quad A_2(p(t), q(t)) := q(0), \text{ and } A_3(p(t), q(t)) := -p(T).$$

Again, from Lemma 14.2, we have by setting $z = (p, q)$ that the functional

$$z \to \sum_{i=1}^{3} \langle A_i z, \Gamma_i z \rangle = \int_0^T \left[\dot{q}(t) \cdot p(t) - \dot{p}(t) \cdot q(t) - \delta_1 |q|^2 - \delta_2 |p|^2 \right] dt$$

$$+ p(T) \cdot q(T) - p(0) \cdot q(0)$$

is weakly continuous on $Z = A_{\mathbf{R}^N}^2 \times A_{\mathbf{R}^N}^2$.

It is easily seen that

$$|\dot{q} - \delta_2 p|^2 \geq \frac{1}{2} |\dot{q}|^2 - |\delta_2|^2 |p|^2 \text{ and } |\dot{p} + \delta_1 q|^2 \geq \frac{1}{2} |\dot{p}|^2 - |\delta_1|^2 |q|^2 \quad (14.53)$$

and, as above, we have

$$\frac{1}{2} \int_0^T |p(t)|^2 dt \leq T^2 \int_0^T |\dot{p}|^2 dt + T |p(0)|^2 \tag{14.54}$$

and

$$\frac{1}{2} \int_0^T |q(t)|^2 dt \leq T^2 \int_0^T |\dot{q}|^2 dt + T |q(T)|^2. \tag{14.55}$$

Combining the inequalities above gives

$$\int_0^T \left[|\dot{q} - \delta_2 p|^2 + |\dot{p} + \delta_1 q|^2 \right] dt \geq \int_0^T \left[\frac{1}{2} \left(|\dot{q}|^2 + |\dot{p}|^2 \right) - |\delta_2|^2 |p|^2 - |\delta_1|^2 |q|^2 \right] dt$$

$$\geq \int_0^T \left[\frac{1}{2} \left(|\dot{q}|^2 + |\dot{p}|^2 \right) - 2T^2 \left(|\delta_2|^2 |\dot{p}|^2 + |\delta_1|^2 |\dot{q}|^2 \right) \right] dt$$

$$- 2T \left(|\delta_2|^2 |p(0)|^2 + |\delta_1|^2 |q(T)|^2 \right)$$

$$\geq \int_0^T \frac{1}{2} \left[(1 - 4T^2 |\delta_2|^2) |\dot{p}|^2 + (1 - 4T^2 |\delta_1|^2) |\dot{q}|^2 \right] dt$$

$$- 2T \left(|\delta_2|^2 |p(0)|^2 + |\delta_1|^2 |q(T)|^2 \right)$$

$$= \int_0^T \frac{1}{2} \left[\varepsilon_1 |\dot{q}|^2 + \varepsilon_2 |\dot{p}|^2 \right] - 2T \left(|\delta_2|^2 |p(0)|^2 + |\delta_1|^2 |q(T)|^2 \right),$$

where $\varepsilon_i := 1 - 4T^2 |\delta_i|^2 > 0$ since $|\delta_i| < \frac{1}{2T}$.

We can now estimate

$$\left| \sum_{i=1}^{3} \langle A_i z, \Gamma_i z \rangle \right| = 2 \left| \int_0^T \left[\dot{q} \cdot p + \delta_1 |q|^2 + \delta_2 |p|^2 \right] dt \right|$$

$$\leq \max\{2T^2, 1\} \int_0^T \left(|\dot{p}|^2 + |\dot{q}| \right)^2 dt + 2T |p(0)|^2$$

$$+ 8\delta_1 T^2 \int_0^T |\dot{q}|^2 dt + 4\delta_1 T |q(T)|^2$$

$$+ 8\delta_2 T^2 \int_0^T |\dot{p}|^2 dt + 4\delta_2 T |p(0)|^2$$

$$\leq \left(\max\{2T^2, 1\} + 8\delta_2 T^2\right) \int_0^T |\dot{p}|^2 \, dt + \left(\max\{2T^2, 1\}\right.$$

$$\left. + 8\delta_1 T^2\right) \int_0^T |\dot{q}|^2 \, dt + 2T(1 + 2\delta_2)|p(0)|^2 + 4\delta_1 T |q(T)|^2$$

$$\leq \alpha_1 \int_0^T \left[|\dot{q} - \delta_2 p|^2 + |\dot{p} + \delta_1 q|^2\right] dt + \alpha_2 |p(0)|^2 + \alpha_3 |q(T)|^2,$$

where $\alpha_i := \alpha_i(T) \to 0$ as $T \to 0$ for $i = 1, 2, 3$.

Now consider again the convex functions $\Phi_1(p, q) = \int_0^T \varphi(p(t), q(t)) \, dt$ on $X_1 = L^2_{\mathbf{R}^N}$, $\Phi_2 = \psi_1^*$ on $X_2 = \mathbf{R}^N$, and $\Phi_3 = \psi_2^*$ on $X_3 = \mathbf{R}^N$, in such a way that

$$I(z) = \sum_{i=1}^3 \Phi_i(A_i z) + \Phi_i^*(\Gamma_i z) - \langle A_i z, \Gamma_i z \rangle. \tag{14.56}$$

Take now T small enough so that $0 < \alpha_1(T) < \frac{1}{2\beta}$,

$$\liminf_{|p| \to +\infty} \frac{\psi_1(p)}{|p|^2} > \alpha_2(T) > 0 \quad \text{and} \quad \liminf_{|p| \to +\infty} \frac{\psi_2(p)}{|p|^2} > \alpha_3(T) > 0. \tag{14.57}$$

All the hypotheses of Theorem 14.2 are then satisfied, and we conclude as above.

Further comments

The general existence results in this chapter come from [59]. They were motivated by the applications to Hamiltonian systems exhibited by Ghoussoub and Moameni [63]. They can be seen as extensions of results obtained by Clarke and Ekeland for subquadratic Hamiltonian systems. See for example Ekeland [46] and Mawhin-Willem [96].

Chapter 15
Superposition of Interacting Self-dual Functionals

We consider situations where functionals of the form

$$I(x) = L_1(A_1 x, \Lambda_1 x) - \langle A_1 x, \Lambda_1 x \rangle + L_2(A_2 x, \Lambda_2 x) - \langle A_2 x, \Lambda_2 x \rangle$$

are self-dual on a Banach space Z, considering that $L_i, i = 1, 2$ are self-dual Lagrangians on spaces $X_i \times X_i^*$, and $(\Lambda_1, \Lambda_2) : Z \to X_1^* \times X_2^*$ and $(A_1, A_2) : Z \to X_1 \times X_2$ are linear or nonlinear operators on Z. However, unlike in the previous chapter, the space Z is not necessarily isomorphic to $X_1 \times X_2$, and the functional is not a direct sum, but still certain compatibility relations between the operators $A_1, A_2, \Lambda_1, \Lambda_2$ are needed according to which one of the operators is linear. One also needs that the functional $x \to \langle A_1 x, \Lambda_1 x \rangle + \langle A_2 x, \Lambda_2 x \rangle$ be weakly upper semicontinuous on Z. Under a suitable coercivity condition, I will attain its zero infimum at an element $\bar{x}$ that solves the system

$$\begin{cases} \Lambda_1 \bar{x} \in \bar{\partial} L_1(A_1 \bar{x}) \\ \Lambda_2 \bar{x} \in \bar{\partial} L_2(A_2 \bar{x}). \end{cases}$$

The theorem is applied to semilinear Laplace equations and to a nonlinear Cauchy-Riemann problem. An application to Hamiltonian systems of PDEs will be given in Chapter 16.

15.1 The superposition in terms of the Hamiltonians

Theorem 15.1. *Consider three reflexive Banach spaces Z, X_1, X_2, and bounded operators B_1 on X_1 and B_2 on X_2. Let $\Lambda_1 : D(\Lambda_1) \subset Z \to X_1^*$ (resp., $\Lambda_2 : D(\Lambda_2) \subset Z \to X_2^*$) be – not necessarily linear – weak-to-weak continuous operators, and let $A_1 : D(A_1) \subset Z \to X_1$ (resp., $A_2 : D(A_2) \subset Z \to X_2$) be linear operators. Suppose E is a closed linear subspace of Z such that $E \subset D(A_1) \cap D(A_2) \cap D(\Lambda_1) \cap D(\Lambda_2)$ such that the following properties are satisfied:*

1. *The image of $E_0 := \mathrm{Ker}(A_2) \cap E$ by A_1 is dense in X_1.*
2. *The image of E by A_2 is dense in X_2.*
3. *$x \to \langle B_1 A_1 x, \Lambda_1 x \rangle + \langle B_2 A_2 x, \Lambda_2 x \rangle$ is weakly upper semicontinuous on E.*

N. Ghoussoub, *Self-dual Partial Differential Systems and Their Variational Principles*, Springer Monographs in Mathematics, DOI 10.1007/978-0-387-84897-6_15, © Springer Science+Business Media, LLC 2009

Let $L_i, i = 1, 2$ be a B_i-self-dual Lagrangian on $X_i \times X_i^*$ such that the Hamiltonians H_{L_i} are continuous in the first variable on X_i.

(i) The functional

$$I(x) = L_1(A_1 x, \Lambda_1 x) - \langle B_1 A_1 x, \Lambda_1 x \rangle + L_2(A_2 x, \Lambda_2 x) - \langle B_2 A_2 x, \Lambda_2 x \rangle \qquad (15.1)$$

is then self-dual on E. Its corresponding AS-Hamiltonian on $E \times E$ is

$$\begin{aligned}
M(x, v) = {} & \langle B_1 A_1 (v - x), \Lambda_1 x \rangle + H_{L_1}(A_1 v, B_1 A_1 x) \\
& + \langle B_2 A_2 (v - x), \Lambda_2 x \rangle + H_{L_2}(A_2 v, B_2 A_2 x).
\end{aligned} \qquad (15.2)$$

(ii) Consequently, if

$$\lim_{\|x\| \to +\infty} H_{L_1}(0, B_1 A_1 x) - \langle \Lambda_1 x, B_1 A_1 x \rangle + H_{L_2}(0, B_2 A_2 x) - \langle \Lambda_2 x, B_2 A_2 x \rangle = +\infty,$$

$$(15.3)$$

then I attains its minimum at a point $z \in E$ in such a way that

$$\begin{cases}
I(z) = \inf_{x \in E} I(x) = 0 \\
\Lambda_1 z \in \overline{\partial}_{B_1} L_1(A_1 z) \\
\Lambda_2 z \in \overline{\partial}_{B_2} L_2(A_2 z).
\end{cases} \qquad (15.4)$$

Proof. (i) Recall first from Proposition 11.3 that, for every $z \in D(A_i) \cap D(\Lambda_i) \subset Z$, we have

$$L_i(A_i z, \Lambda_i z) = \sup \{ \langle B_i r, \Lambda_i z \rangle + H_{L_i}(r, B_i A_i z); \ r \in X_i \}.$$

Let $x \in D(A_1) \cap D(A_2) \cap D(\Lambda_1) \cap D(\Lambda_2)$, and write

$$\begin{aligned}
\sup_{v \in E} M(x, v) = {} & \sup_{v \in E} \{ \langle B_1 A_1 (v - x), \Lambda_1 x \rangle + H_{L_1}(A_1 v, B_1 A_1 x) \\
& \qquad + \langle B_2 A_2 (v - x), \Lambda_2 x \rangle + H_{L_2}(A_2 v, B_2 A_2 x) \} \\
= {} & \sup_{v \in E} \{ \langle B_1 A_1 v, \Lambda_1 x \rangle + H_{L_1}(A_1 v, B_1 A_1 x) + \langle B_2 A_2 v, \Lambda_2 x \rangle \\
& \qquad + H_{L_2}(A_2 v, B_2 A_2 x) \} - \langle B_1 A_1 x, \Lambda_1 x \rangle - \langle B_2 A_2 x, \Lambda_2 x \rangle \\
= {} & \sup_{v \in E, v_0 \in E_0} \{ \langle B_1 A_1 v, \Lambda_1 x \rangle + H_{L_1}(A_1 v, B_1 A_1 x) \\
& \qquad + \langle B_2 A_2 (v + v_0), \Lambda_2 x \rangle + H_{L_2}(A_2 (v + v_0), B_2 A_2 x) \} \\
& \quad - \langle B_1 A_1 x, \Lambda_1 x \rangle - \langle B_2 A_2 x, \Lambda_2 x \rangle \\
= {} & \sup_{w \in E, v_0 \in E_0} \{ \langle B_1 A_1 (w - v_0), \Lambda_1 x \rangle + H_{L_1}(A_1 (w - v_0), B_1 A_1 x) \\
& \qquad + \langle B_2 A_2 w, \Lambda_2 x \rangle + H_{L_2}(A_2 w, B_2 A_2 x) \} \\
& \quad - \langle B_1 A_1 x, \Lambda_1 x \rangle - \langle B_2 A_2 x, \Lambda_2 x \rangle \\
= {} & \sup_{w \in E, r \in X_1} \{ \langle B_1 (A_1 w + r), \Lambda_1 x \rangle + H_{L_1}(A_1 w + r, B_1 A_1 x) \\
& \qquad + \langle B_2 A_2 w, \Lambda_2 x \rangle + H_{L_2}(A_2 w, B_2 A_2 x) \}
\end{aligned}$$

$$-\langle B_1A_1x, \Lambda_1x \rangle - \langle B_2A_2x, \Lambda_2x \rangle$$

$$= \sup_{w \in E, x_1 \in X_1} \{\langle B_1x_1, \Lambda_1x \rangle + H_{L_1}(x_1, B_1A_1x)$$

$$+ \langle B_2A_2w, \Lambda_2x \rangle + H_{L_2}(A_2w, B_2A_2x)\}$$

$$- \langle B_1A_1x, \Lambda_1x \rangle - \langle B_2A_2x, \Lambda_2x \rangle$$

$$= \sup_{x_1 \in X_1} \{\langle B_1x_1, \Lambda_1x \rangle + H_{L_1}(x_1, B_1A_1x)\}$$

$$+ \sup_{x_2 \in X_2} \{\langle B_2x_2, \Lambda_2x \rangle + H_{L_2}(x_2, B_2A_2x)\}$$

$$- \langle B_1A_1x, \Lambda_1x \rangle - \langle B_2A_2x, \Lambda_2x \rangle$$

$$= L_1(A_1x, \Lambda_1x) - \langle B_1A_1x, \Lambda_1x \rangle + L_2(A_2x, \Lambda_2x) - \langle B_2A_2x, \Lambda_2x \rangle$$

$$= I(x).$$

For (ii) it suffices to apply Theorem 12.2 to get that $I(z) = 0$ for some $z \in E$. Now use the fact that $L_1(A_1x, \Lambda_1x) - \langle B_1A_1x, \Lambda_1x \rangle \geq 0$ and $L_2(A_2x, \Lambda_2x) - \langle B_2A_2x, \Lambda_2x \rangle \geq 0$ to conclude.

Example 15.1. Nonlinear Laplace equation

Consider the equation

$$\begin{cases} \Delta u + \mathbf{a} \cdot \nabla u \in \partial \varphi(u) & \text{on} \quad \Omega \subset \mathbf{R}^n \\ -\frac{\partial u}{\partial n} \in \partial \psi(u) & \text{on} \quad \partial \Omega, \end{cases} \tag{15.5}$$

where $\mathbf{a}$ is a smooth vector field on Ω with compact support with $\mathrm{div}(\mathbf{a}) \geq 0$, and φ and ψ are convex functions on the real line. We shall apply the preceding theorem with the spaces $Z = H^1(\Omega)$, $X_1 = L^2(\Omega)$, $X_2 = H^{1/2}(\partial \Omega)$, and the linear operators $A_1(u) = u$, $\Gamma_1 u = \Delta u + \mathbf{a} \cdot \nabla u$, $A_2(u) = u_{|\partial \Omega}$, $\Gamma_2 u = -\frac{\partial u}{\partial n}$. Note that

$$\langle A_1u, \Lambda_1u \rangle + \langle A_2u, \Lambda_2u \rangle = \int_\Omega u \Delta u \, dx - \frac{1}{2} \int_\Omega \mathrm{div}(\mathbf{a}) |u|^2 dx - \int_{\partial \Omega} u \frac{\partial u}{\partial n} d\sigma$$

$$= -\int_\Omega \mathrm{div}(\mathbf{a}) |u|^2 dx - \int_\Omega |\nabla u|^2 dx,$$

and therefore $u \to \langle A_1u, \Lambda_1u \rangle + \langle A_2u, \Lambda_2u \rangle$ is weakly upper semicontinuous on $H^1(\Omega)$. Moreover, the other assumptions on $A_1, A_2, \Gamma_1, \Gamma_2$ are easily verified. Assuming now that φ and ψ satisfy the growth conditions

$$\varphi(0) < +\infty \quad \text{and} \quad \varphi(x) \geq C(|x|^2 - 1) \tag{15.6}$$

and

$$\psi(0) < +\infty \quad \text{and} \quad \psi \text{ bounded below}, \tag{15.7}$$

one can then easily prove that the coercivity condition (15.3) is verified. Theorem 15.1 applies, and we obtain the following.

Theorem 15.2. *Under the assumptions above, the minimum of the functional*

$$I(u) = \int_\Omega \left(\varphi(u) + \varphi^*(\Delta u + \mathbf{a} \cdot \nabla u) + \frac{1}{2}\mathrm{div}(\mathbf{a})|u|^2 + |\nabla u|^2\right)dx$$
$$+ \int_{\partial\Omega} \left(\psi(u) + \psi^*\left(-\frac{\partial u}{\partial n}\right)\right)d\sigma$$

on $H^1(\Omega)$ is zero and is attained at a solution of equation (15.15).

15.2 The superposition in terms of the co-Hamiltonians

Theorem 15.3. *Consider three reflexive Banach spaces Z, X_1, X_2 and bounded operators B_1 on X_1 and B_2 on X_2. Let $\Gamma_1 : D(\Gamma_1) \subset Z \to X_1^*$ (resp., $\Gamma_2 : D(\Gamma_2) \subset Z \to X_2^*$) be linear operators, and let $A_1 : D(A_1) \subset Z \to X_1$ (resp., $A_2 : D(A_2) \subset Z \to X_2$) be – not necessarily linear – weak-to-weak continuous operators. Suppose F is a closed linear subspace of Z such that $F \subset D(A_1) \cap D(A_2) \cap D(\Gamma_1) \cap D(\Gamma_2)$, while the following properties are satisfied:*

1. *The image of $F_0 := \mathrm{Ker}(\Gamma_2) \cap F$ by Γ_1 is dense in X_1^*.*
2. *The image of F by Γ_2 is dense in X_2^*.*
3. *$x \to \langle B_1 A_1 x, \Gamma_1 x \rangle + \langle B_2 A_2 x, \Gamma_2 x \rangle$ is weakly upper semicontinuous on F.*

Let $L_i, i = 1, 2$ be a B_i-self-dual Lagrangian on $X_i \times X_i^*$ such that the co-Hamiltonians $\tilde{H}_{L_i}$ are continuous in the second variable on X_i^*.

(i) The functional

$$I(x) := L_1(A_1 x, \Gamma_1 x) - \langle B_1 A_1 x, \Gamma_1 x \rangle + L_2(A_2 x, \Gamma_2 x) - \langle B_2 A_2 x, \Gamma_2 x \rangle \qquad (15.8)$$

is then self-dual on F. Its corresponding AS-Hamiltonian on $F \times F$ is

$$M(x, v) = \langle \Gamma_1 v - \Gamma_1 x, B_1 A_1 x \rangle + \tilde{H}_{L_1}(B_1^* \Gamma_1 x, \Gamma_1 v)$$
$$+ \langle \Gamma_2 v - \Gamma_2 x, B_2 A_2 x \rangle + \tilde{H}_{L_2}(B_2^* \Gamma_2 x, \Gamma_2 v).$$

(ii) Consequently, if

$$\lim_{\|x\| \to +\infty} \tilde{H}_{L_1}(B_1^* \Gamma_1 x, 0) - \langle \Gamma_1 x, B_1 A_1 x \rangle + \tilde{H}_{L_2}(B_2^* \Gamma_2 x, 0) - \langle \Gamma_2 x, B_2 A_2 x \rangle = +\infty,$$
$$(15.9)$$

then I attains its minimum on X at a point $z \in F$ in such a way that

$$\begin{cases} I(z) = \inf_{x \in F} I(x) = 0, \\ \Lambda_1 z \in \bar{\partial}_{B_1} L_1(A_1 z), \\ \Lambda_2 z \in \bar{\partial}_{B_2} L_2(A_2 z). \end{cases} \qquad (15.10)$$

Proof. i) Recall again from Proposition 11.3 that, for every $z \in D(A_i) \cap D(\Lambda_i) \subset Z$, we have

$$L_i(A_i z, \Gamma_i z) = \sup\{\langle r, B_i A_i z \rangle + \tilde{H}_{L_i}(B_i^* \Gamma_i z, r); r \in X_i^*\}.$$

Let $x \in D(\Gamma_1) \cap D(\Gamma_2) \cap D(A_1) \cap D(A_2)$, and write

$$\begin{aligned}
\sup_{v \in F} M(x, v) &= \sup_{v \in F} \big\{ \langle \Gamma_1 v - \Gamma_1 x, B_1 A_1 x \rangle + \tilde{H}_{L_1}(B_1^* \Gamma_1 x, \Gamma_1 v) \\
&\qquad + \langle \Gamma_2 v - \Gamma_2 x, B_2 A_2 x \rangle + \tilde{H}_{L_2}(B_2^* \Gamma_2 x, \Gamma_2 v) \big\} \\
&= \sup_{v \in F, v_0 \in F_0} \big\{ \langle \Gamma_1 v, B_1 A_1 x \rangle + \tilde{H}_{L_1}(B_1^* \Gamma_1 x, \Gamma_1 v) \\
&\qquad + \langle \Gamma_2(v + v_0), B_2 A_2 x \rangle + \tilde{H}_{L_2}(B_2^* \Gamma_2 x, \Gamma_2(v + v_0)) \big\} \\
&\qquad - \langle \Gamma_1 x, B_1 A_1 x \rangle - \langle \Gamma_2 x, B_2 A_2 x \rangle \\
&= \sup_{w \in F, v_0 \in F_0} \big\{ \langle \Gamma_1(w - v_0), B_1 A_1 x \rangle + \tilde{H}_{L_1}(B_1^* \Gamma_1 x, \Gamma_1(w - v_0)) \\
&\qquad + \langle \Gamma_2 w, B_2 A_2 x \rangle + \tilde{H}_{L_2}(B_2^* \Gamma_2 x, \Gamma_2 w) \big\} \\
&\qquad - \langle \Gamma_1 x, B_1 A_1 x \rangle - \langle \Gamma_2 x, B_2 A_2 x \rangle \\
&= \sup_{w \in F, r \in X_1^*} \big\{ \langle \Gamma_1 w + r, B_1 A_1 x \rangle + \tilde{H}_{L_1}(B_1^* \Gamma_1 x, \Gamma_1 w + r) \\
&\qquad + \langle \Gamma_2 w, B_2 A_2 x \rangle + \tilde{H}_{L_2}(B_2^* \Gamma_2 x, \Gamma_2 w) \big\} \\
&\qquad - \langle \Gamma_1 x, B_1 A_1 x \rangle - \langle \Gamma_2 x, B_2 A_2 x \rangle \\
&= \sup_{w \in F, x_1^* \in X_1^*} \big\{ \langle x_1^*, B_1 A_1 x \rangle + \tilde{H}_{L_1}(B_1^* \Gamma_1 x, x_1^*) \\
&\qquad + \langle \Gamma_2 w, B_2 A_2 x \rangle + \tilde{H}_{L_2}(B_2^* \Gamma_2 x, \Gamma_2 w) \big\} \\
&\qquad - \langle \Gamma_1 x, B_1 A_1 x \rangle - \langle \Gamma_2 x, B_2 A_2 x \rangle \\
&= \sup_{x_1^* \in X_1^*} \big\{ \langle x_1^*, B_1 A_1 x \rangle + \tilde{H}_{L_1}(B_1^* \Gamma_1 x, x_1^*) \big\} \\
&\qquad + \sup_{x_2^* \in X_2^*} \big\{ \langle x_2^*, B_2 A_2 x \rangle + \tilde{H}_{L_2}(B_2^* \Gamma_2 x, x_2^*) \big\} \\
&\qquad - \langle \Gamma_1 x, B_1 A_1 x \rangle - \langle \Gamma_2 x, B_2 A_2 x \rangle \\
&= L_1(A_1 x, \Gamma_1 x) + L_2(A_2 x, \Gamma_2 x) - \langle \Gamma_1 x, B_1 A_1 x \rangle - \langle \Gamma_2 x, B_2 A_2 x \rangle \\
&= I(x).
\end{aligned}$$

The proof of (ii) is the same as in the proof of Theorem 15.1.

Example 15.2. Cauchy-Riemann equations

Consider the following nonlinear Cauchy-Riemann equation on a bounded simply connected domain $\Omega \subset \mathbf{R}^2$,

$$\begin{cases} \left(\frac{\partial u}{\partial x} - \frac{\partial v}{\partial y}, \frac{\partial v}{\partial x} + \frac{\partial u}{\partial y}\right) \in \partial \varphi\left(\frac{\partial v}{\partial y} - \frac{\partial u}{\partial x}, -\frac{\partial v}{\partial x} - \frac{\partial u}{\partial y}\right) \\ u_{|\partial \Omega} \in \partial \psi\left(n_x \frac{\partial v}{\partial y} - n_y \frac{\partial v}{\partial x}\right) \end{cases} \tag{15.11}$$

where φ (resp., ψ) is a convex lower semicontinuous function on $\mathbf{R}^2$ (resp., $\mathbf{R}$), and $\mathbf{n} = (n_x, n_y)$ is the outer normal vector on $\partial\Omega$. Let Z be the closed subspace of $H^1(\Omega, \mathbf{R}^2)$ consisting of all functions (u, v) such that $\int_\Omega v(x, y)\, dx\, dy = 0$. Set $X_1 = L^2(\Omega, \mathbf{R}^2)$, $X_2 = L^2(\partial\Omega, \mathbf{R})$, and consider the linear operators $A_1 : Z \to X_1$ and $A_2 : Z \to X_2$ defined by

$$A_1(u, v) = \left(\frac{\partial v}{\partial y} - \frac{\partial u}{\partial x}, -\frac{\partial v}{\partial x} - \frac{\partial u}{\partial y}\right) \quad \text{and} \quad A_2(u, v) = n_x \frac{\partial v}{\partial y} - n_y \frac{\partial v}{\partial x}.$$

Consider also the operators $\Gamma_1 : Z \to X_1^*$ and $\Gamma_2 : Z \to X_2^*$ defined by

$$\Gamma_1(u, v) = \left(\frac{\partial u}{\partial x} - \frac{\partial v}{\partial y}, \frac{\partial v}{\partial x} + \frac{\partial u}{\partial y}\right) \quad \text{and} \quad \Gamma_2(u, v) = u_{|\partial\Omega}.$$

An immediate application of Stokes' formula gives

$$\int\int_\Omega \left(\frac{\partial u}{\partial x}\frac{\partial v}{\partial y} - \frac{\partial u}{\partial y}\frac{\partial v}{\partial x}\right) dx\, dy = \frac{1}{2}\int_{\partial\Omega} u\left(n_y \frac{\partial v}{\partial x} - n_x \frac{\partial v}{\partial y}\right) d\sigma.$$

We now estimate $\Gamma(u, v) = \langle A_1(u, v), \Gamma_1(u, v)\rangle + \langle A_2(u, v), \Gamma_2(u, v)\rangle$ as follows.

$$\begin{aligned}
\Gamma(u, v) = {} & 2\int\int_\Omega \left(\frac{\partial u}{\partial x}\frac{\partial v}{\partial y} - \frac{\partial u}{\partial y}\frac{\partial v}{\partial x}\right) dx\, dy \\
& - \int\int_\Omega \left(\left|\frac{\partial v}{\partial x}\right|^2 + \left|\frac{\partial v}{\partial y}\right|^2 + \left|\frac{\partial u}{\partial y}\right|^2 + \left|\frac{\partial v}{\partial x}\right|^2\right) dx\, dy \\
& - \int_{\partial\Omega} u\left(n_y \frac{\partial v}{\partial x} - n_x \frac{\partial v}{\partial y}\right) d\sigma \\
= {} & -\int\int_\Omega \left(\left|\frac{\partial v}{\partial x}\right|^2 + \left|\frac{\partial v}{\partial y}\right|^2 + \left|\frac{\partial u}{\partial y}\right|^2 + \left|\frac{\partial v}{\partial x}\right|^2\right) dx\, dy,
\end{aligned}$$

and it is therefore weakly upper semicontinuous on Z.

Note now that the linear Cauchy-Riemann problem gives that the image of $\text{Ker}(\Gamma_2)$ by Γ_1 is dense in $X_1^* = L^2(\Omega, \mathbf{R}^2)$. It is also clear that the range of Γ_2 is dense in $X_2^* = L^2(\partial\Omega, \mathbf{R}^2)$.

The following existence result now follows from Theorem 15.3.

Theorem 15.4. *Let φ be a bounded below convex lower semicontinuous function on $\mathbf{R}^2$ such that $\varphi(0) < +\infty$, and let ψ be a convex lower semicontinuous function on $\mathbf{R}$ such that $-C \leq \psi(s) \leq C(\|s\|^2 + 1)$. Then, the minimum of the self-dual functional*

$$\begin{aligned}
I(u, v) = {} & \int\int_\Omega \left\{\varphi\left(\frac{\partial v}{\partial y} - \frac{\partial u}{\partial x}, -\frac{\partial v}{\partial x} - \frac{\partial u}{\partial y}\right) + \varphi^*\left(\frac{\partial u}{\partial x} - \frac{\partial v}{\partial y}, \frac{\partial v}{\partial x} + \frac{\partial u}{\partial y}\right)\right\} dx\, dy \\
& + \int\int_\Omega \left\{\left|\frac{\partial v}{\partial x}\right|^2 + \left|\frac{\partial v}{\partial y}\right|^2 + \left|\frac{\partial u}{\partial y}\right|^2 + \left|\frac{\partial v}{\partial x}\right|^2\right\} dx\, dy \\
& + \int_{\partial\Omega} \left\{\psi\left(n_x \frac{\partial v}{\partial y} - n_y \frac{\partial v}{\partial x}\right) + \psi^*(u)\right\} d\sigma
\end{aligned}$$

on Z is equal to zero, and is attained at a solution of equation (15.11).

Proof. It suffices to apply Theorem 15.3 with the spaces Z, X_1, X_2 defined above, the operators $A_1, A_2, \Gamma_1, \Gamma_2$, and the self-dual Lagrangians

$$L_1((u,v),(p,q)) = \int \int_\Omega \left\{ \varphi(u(x),v(x)) + \varphi^*(p(x),q(x)) \right\} dxdy$$

and

$$L_2(u,p) = \int_{\partial\Omega} \left\{ \psi(u(x)) + \psi^*(p(x)) \right\} d\sigma.$$

The coercivity conditions are then guaranteed by the assumptions on φ and ψ.

15.3 The superposition of a Hamiltonian and a co-Hamilonian

Theorem 15.5. *Consider three reflexive Banach spaces Z, X_1, X_2 and bounded operators B_1 on X_1 and B_2 on X_2. Let $\Gamma_1 : D(\Gamma_1) \subset Z \to X_1^*$ (resp., $A_2 : D(A_2) \subset Z \to X_2$) be linear operators, and let $A_1 : D(A_1) \subset Z \to X_1$ (resp., $\Lambda_2 : D(\Lambda_2) \subset Z \to X_2$) be – not necessarily linear – weak-to-weak continuous operators. Suppose G is a closed linear subspace of Z such that $G \subset D(A_1) \cap D(A_2) \cap D(\Gamma_1) \cap D(\Lambda_2)$, while the following properties are satisfied:*

1. *The image of $G_0 := \mathrm{Ker}(A_2) \cap G$ by Γ_1 is dense in X_1^*.*
2. *The image of G by A_2 is dense in X_2.*
3. *$x \to \langle B_1 A_1 x, \Gamma_1 x \rangle + \langle B_2 A_2 x, \Lambda_2 x \rangle$ is weakly upper semicontinuous on G.*

Let $L_i, i = 1, 2$ be a B_i-self-dual Lagrangian on $X_i \times X_i^*$ such that the co-Hamiltonian $\tilde{H}_{L_1}$ (resp., the Hamiltonian H_{L_2}) is continuous in the second variable on X_1^* (resp., is continuous in the first variable on X_2).

(i) *The functional*

$$I(x) := L_1(A_1 x, \Gamma_1 x) - \langle B_1 A_1 x, \Gamma_1 x \rangle + L_2(A_2 x, \Lambda_2 x) - \langle B_2 A_2 x, \Lambda_2 x \rangle \qquad (15.12)$$

is then self-dual on G. Its corresponding AS-Hamiltonian on $G \times G$ is

$$\begin{aligned} M(x,v) = {} & \langle \Gamma_1 v - \Gamma_1 x, B_1 A_1 x \rangle + \tilde{H}_{L_1}(B_1^* \Gamma_1 x, \Gamma_1 v) \\ & + \langle B_2 A_2 (v - x), \Lambda_2 x \rangle + H_{L_2}(\Lambda_2 v, B_2 A_2 x). \end{aligned}$$

(ii) *Consequently, if*

$$\lim_{\|x\| \to +\infty} \tilde{H}_{L_1}(B_1^* \Gamma_1 x, 0) - \langle \Gamma_1 x, B_1 A_1 x \rangle + H_{L_2}(0, B_2 A_2 x) - \langle \Lambda_2 x, B_2 A_2 x \rangle = +\infty,$$

$$(15.13)$$

then I attains its minimum on X at a point $z \in G$ in such a way that

$$\begin{cases} I(z) = \inf_{x \in G} I(x) = 0, \\ \Lambda_1 z \in \bar{\partial}_{B_1} L_1(A_1 z), \\ \Lambda_2 z \in \bar{\partial}_{B_2} L_2(A_2 z). \end{cases} \qquad (15.14)$$

Proof. i) Again we use that, for every $z \in D(A_i) \cap D(\Lambda_i) \subset Z$, we have

$$L_1(A_1z, \Gamma_1z) = \sup\{\langle r, B_1A_1z\rangle + \tilde{H}_{L_1}(B_1^*\Gamma_1z, r); r \in X_1^*\}$$

and

$$L_2(A_2z, \Lambda_2z) = \sup\{\langle B_2r, \Lambda_2z\rangle + H_{L_2}(r, B_2A_2z); r \in X_2\}.$$

Let $x \in D(\Gamma_1) \cap D(\Lambda_2) \cap D(A_1) \cap D(A_2)$, and write

$$
\begin{aligned}
\sup_{v \in G} M(x, v) &= \sup_{v \in G} \Big\{ \langle \Gamma_1 v - \Gamma_1 x, B_1A_1x\rangle + \tilde{H}_{L_1}(B_1^*\Gamma_1 x, \Gamma_1 v) \\
&\qquad\qquad + \langle B_2A_2(v - x), \Lambda_2 x\rangle + H_{L_2}(A_2v, B_2A_2x) \Big\} \\
&= \sup_{v \in G, v_0 \in G_0} \Big\{ \langle \Gamma_1 v, B_1A_1x\rangle + \tilde{H}_{L_1}(B_1^*\Gamma_1 x, \Gamma_1 v) \\
&\qquad\qquad + \langle B_2A_2(v + v_0), \Lambda_2 x\rangle + H_{L_2}(A_2(v + v_0), B_2A_2x) \Big\} \\
&\qquad - \langle \Gamma_1 x, B_1A_1x\rangle - \langle \Gamma_2 x, B_2A_2x\rangle \\
&= \sup_{w \in G, v_0 \in G_0} \Big\{ \langle \Gamma_1(w - v_0), B_1A_1x\rangle + \tilde{H}_{L_1}(B_1^*\Gamma_1 x, \Gamma_1(w - v_0)) \\
&\qquad\qquad + \langle B_2A_2w, \Lambda_2 x\rangle + H_{L_2}(A_2w, B_2A_2x) \Big\} \\
&\qquad - \langle \Gamma_1 x, B_1A_1x\rangle - \langle \Gamma_2 x, B_2A_2x\rangle \\
&= \sup_{w \in G, r \in X_1^*} \Big\{ \langle \Gamma_1 w + r, B_1A_1x\rangle + \tilde{H}_{L_1}(B_1^*\Gamma_1 x, \Gamma_1 w + r) \\
&\qquad\qquad + \langle B_2A_2w, \Lambda_2 x\rangle + H_{L_2}(A_2w, B_2A_2x) \Big\} \\
&\qquad \langle \Gamma_1 x, B_1A_1x\rangle - \langle \Gamma_2 x, B_2A_2x\rangle \\
&= \sup_{w \in G, x_1^* \in X^*} \Big\{ \langle x_1^*, B_1A_1x\rangle + \tilde{H}_{L_1}(B_1^*\Gamma_1 x, x_1^*) \\
&\qquad\qquad + \langle B_2A_2w, \Lambda_2 x\rangle + H_{L_2}(A_2w, B_2A_2x) \Big\} \\
&\qquad - \langle \Gamma_1 x, B_1A_1x\rangle - \langle \Lambda_2 x, B_2A_2x\rangle \\
&= \sup_{x_1^* \in X_1^*} \Big\{ \langle x_1^*, B_1A_1x\rangle + \tilde{H}_{L_1}(B_1^*\Gamma_1 x, x_1^*) \Big\} \\
&\qquad + \sup_{x_2 \in X_2} \Big\{ \langle B_2x_2, \Lambda_2 x\rangle + H_{L_2}(x_2, B_2A_2x) \Big\} \\
&\qquad - \langle \Gamma_1 x, B_1A_1x\rangle - \langle \Lambda_2 x, B_2A_2x\rangle \\
&= L_1(A_1x, \Gamma_1x) + L_2(A_2x, \Gamma_2x) - \langle \Gamma_1 x, B_1A_1x\rangle - \langle \Lambda_2 x, B_2A_2x\rangle \\
&= I(x).
\end{aligned}
$$

The rest is similar to the previous sections.

Exercise 15.A.

1. Suppose that in Example 15.1, the advection term $\mathbf{a}$ does not have compact support in Ω, and consider the corresponding entrance set $\Sigma_- = \{x \in \partial\Omega; \mathbf{a}(x) \cdot n(x) < 0\}$, and its complement Σ_+ in $\partial\Omega$. Use a self-dual variational principle to construct a solution for the equation

$$\begin{cases} \Delta u + \mathbf{a} \cdot \nabla u \in \partial\varphi(u) & \text{on} \quad \Omega \subset \mathbf{R}^n \\ -\frac{\partial u}{\partial n} \in \partial\psi(u) & \text{on} \quad \Sigma_+ \\ u = 0 & \text{on} \quad \Sigma_-. \end{cases} \qquad (15.15)$$

2. Use a self-dual variational approach to find periodic solutions for the abstract Hamiltonian system

$$\begin{cases} \mathscr{A}^* p \in \partial_y \mathscr{H}(y, p) + f \\ \mathscr{A} y \in \partial_p \mathscr{H}(y, p) + g, \end{cases} \qquad (15.16)$$

on a product Hilbert space $\mathscr{X} \times \mathscr{X}$, where $\mathscr{H}$ is a convex continuous Hamiltonian, $\mathscr{A} : D(\mathscr{A}) \subset \mathscr{X} \to \mathscr{X}$ is a linear, densely defined closed operator with closed range $R(\mathscr{A})$, and f, g are fixed elements in $\mathscr{X}$.

Further comments

The results of this chapter are new and have not been published elsewhere. We have included sample applications for each of these variational principles, but many more equations and systems can be handled with this approach. Theorem 15.5 will be applied in Chapter 16 to obtain solutions for certain Hamiltonian systems of PDEs.

Part IV
PERTURBATIONS OF SELF-DUAL SYSTEMS

Hamiltonian systems of PDEs, nonlinear Schrödinger equations, and Navier-Stokes evolutions can be written in the form

$$0 \in Au + \Lambda u + \overline{\partial} L(u),$$

where L is a self-dual Lagrangian on $X \times X^*$, $A : D(A) \subset X \to X^*$ is a linear – possibly unbounded – operator, and $\Lambda : D(\Lambda) \subset X \to X^*$ is a – non necessarily linear – map. They can be solved by minimizing the functionals

$$I(u) = L(u, -Au - \Lambda u) + \langle Au + \Lambda u, u \rangle$$

on X and by showing that their infimum is attained and is equal to zero.

These functionals are not necessarily self-dual functionals on their spaces of definition, as we need to deal with the difficulties arising from the superposition of the operators A and Λ. We are then often led to use the linear operator A to strengthen the topology on X by defining a new energy space $D(A)$ equipped with the norm $\|u\|_{D(A)}^2 = \|u\|_X^2 + \|Au\|_X^2$. In some cases, this closes the domain of A and increases the chance for Λ to be regular on $D(A)$, but may lead to a loss of strong coercivity on the new space. We shall present in this part situations where compactness and regularity can be restored without altering the self-duality of the system:

- If Λ is linear and is *almost orthogonal to A* in a sense to be made precise in Chapter 16, one may be able to add to I another functional J in such a way that $\tilde{I} = I + J$ is self-dual and coercive. This is applied in the next chapter when dealing with Hamiltonian systems of PDEs.
- The functional I may satisfy what we call *the self-dual Palais-Smale property* on the space $D(A)$, a property that is much weaker than the strong coercivity required in Part III. This method is applied in Chapter 18 to deal with Navier-Stokes and other nonlinear evolutions.

Chapter 16
Hamiltonian Systems of Partial Differential Equations

While dealing with Hamiltonian systems of PDEs, we encounter the standard difficulty of having – unlike the case of finite-dimensional Hamiltonian systems – the cross product $u \to \int_0^T \langle u(t), J\dot{u}(t) \rangle \, dt$ not necessarily weakly continuous on the Sobolev space $H_X^1[0,T]$ of absolutely continuous paths valued in an infinite-dimensional Hilbert space $X := H \times H$. Such systems can often be written in the form

$$J\dot{u}(t) + J\mathscr{A}u(t) \in \overline{\partial}L(t, u(t)),$$

where J is the symplectic operator, $\mathscr{A}$ is an unbounded linear operator on X, and L is a time-dependent self-dual Lagrangian on $[0,T] \times X \times X$. The idea is to use the linear operator $\mathscr{A}$ to strengthen the topology on X by considering the space $D(\mathscr{A})$ equipped with the norm $\|u\|_{D(\mathscr{A})}^2 = \|u\|_X^2 + \|\mathscr{A}u\|_X^2$, and a corresponding path space $\mathscr{W}_T$. The operator $\Lambda = J\dot{u} + J\mathscr{A}$ becomes regular on the new path space, but the functional

$$I(u) = \mathscr{L}(u, J\dot{u} + J\mathscr{A}u) - \langle u, J\dot{u} + J\mathscr{A}u \rangle$$

may cease to be coercive on the new space. We propose here a way to restore coercivity by perturbing the functional I without destroying self-duality. It can be used because $J\dot{u}$ is *almost orthogonal to* $J\mathscr{A}$ in a sense described below. In this case, one adds to I another functional J in such a way that $\tilde{I} = I + J$ is self-dual and coercive on $D(\mathscr{A})$. This will be applied to deal with Hamiltonian systems of PDEs such as

$$\begin{cases} -\dot{v}(t) - \Delta(v+u) + \mathbf{b}\cdot\nabla v = \partial\varphi_1(t,u), \\ \dot{u}(t) - \Delta(u+v) + \mathbf{a}\cdot\nabla u = \partial\varphi_2(t,v), \end{cases}$$

with Dirichlet boundary conditions, as well as

$$\begin{cases} -\dot{v}(t) + \Delta^2 v - \Delta v = \partial\varphi_1(t,u), \\ \dot{u}(t) + \Delta^2 u + \Delta u = \partial\varphi_2(t,v), \end{cases}$$

with Navier state-boundary conditions, and where $\varphi_i, i = 1,2$ are convex functions on some L^p-space.

N. Ghoussoub, *Self-dual Partial Differential Systems and Their Variational Principles*, 287
Springer Monographs in Mathematics, DOI 10.1007/978-0-387-84897-6_16,
© Springer Science+Business Media, LLC 2009

16.1 Regularity and compactness via self-duality

One novelty in this chapter is the introduction of a way to perturb a self-dual functional so as to make it coercive in an appropriate space without destroying self-duality. We shall illustrate this procedure on the simplified example

$$\Gamma x + Ax = -\partial \varphi(x), \tag{16.1}$$

where φ is a convex lower semicontinuous function on a Hilbert space H and both $A : D(A) \subset H \to H$ and $\Gamma : D(\Gamma) \subset H \to H$ are linear operators. The basic self-dual functional associated to (16.1) is

$$I(x) = \varphi(x) + \varphi^*(-Ax - \Gamma x) + \langle x, Ax + \Gamma x \rangle. \tag{16.2}$$

The main ingredients that allow us to use Theorem 12.3 and show that the infimum is attained and is zero are:

1. the weak lower semicontinuity of the function $x \to \langle x, Ax + \Gamma x \rangle$.
2. a coercivity condition that implies, for example, that $\lim_{\|x\| \to +\infty} I(x) = +\infty$.

Now suppose that A is a closed self-adjoint operator that satisfies $\langle Ax, x \rangle \geq c_0 \|x\|^2$ for all $x \in D(A)$ and that A^{-1} is a compact operator. Then, one can strengthen the topology on the domain of the functional I by considering the Banach space Y_A that is the completion of $D(A)$ for the norm $\|u\|_{Y_A}^2 = \|Au\|_H^2$. We can also consider the Hilbert space X_A that is the completion of $D(A)$ for the norm $\|u\|_{X_A}^2 = \langle Au, u \rangle_H$ induced by the scalar product $\langle u, v \rangle_{X_A} = \langle u, Av \rangle_H$. Note that $\|x\|_H^2 \leq \frac{1}{c_0} \|x\|_{X_A}^2 \leq \frac{1}{c_0^2} \|x\|_{Y_A}^2$, and the injections

$$Y_A \to X_A \to H = H^* \to X_A^* \to Y_A^*$$

are therefore continuous, with the injection of X_A into H being compact (See for example Cazenave[37]). The map $x \to \langle x, Ax \rangle$ is then readily weakly continuous on X_A (and Y_A), and the function $x \to \langle x, \Gamma x \rangle$ has a better chance to be lower semicontinuous for the weak topology of Y_A. On the other hand, by considering I on the space Y_A, we often lose coercivity for the new norm, which is not guaranteed by the following subquadratic growth that we shall assume on φ,

$$-C \leq \varphi(x) \leq \frac{\beta}{2}(\|x\|^2 + 1) \text{ for } x \in H, \tag{16.3}$$

for some $\beta > 0$ and $C \in \mathbf{R}$. We also assume that for $c_1 > 0$ we have

$$|\langle x, Ax + \Gamma x \rangle| \leq c_1 \|Ax\|^2 \text{ for } x \in D(A) \cap D(\Gamma). \tag{16.4}$$

Now condition (16.3) yields

$$\varphi^*(-Ax - \Gamma x) \geq \frac{1}{2\beta}(\|Ax + \Gamma x\|^2 - 1) \geq \frac{1}{2\beta}\|Ax\|^2 + \frac{1}{\beta}\langle Ax, \Gamma x \rangle - \frac{1}{2\beta}$$

in such a way that

$$I(x) \geq -C + \left(\frac{1}{2\beta} - c_1\right)\|Ax\|^2 + \frac{1}{\beta}\langle Ax, \Gamma x\rangle - \frac{1}{2\beta}, \tag{16.5}$$

meaning that the functional $I(x) - \frac{1}{\beta}\langle Ax, \Gamma x\rangle$ is coercive for the norm of Y_A. However, this new functional is not self-dual anymore, and we would like to replace the term $-\frac{1}{\beta}\langle Ax, \Gamma x\rangle$ by a larger term, while keeping the sum self-dual. We should be able to do that, provided A and Γ are "almost orthogonal" in the following sense. Assume the cross product $\langle Ax, \Gamma x\rangle$ can be resolved via a Green-Stokes type formula of the form

$$\langle Ax, \Gamma x\rangle + \langle T\mathscr{B}_1 x, \mathscr{B}_2 x\rangle = 0 \text{ for all } x \in D(A), \tag{16.6}$$

where $\mathscr{B}_1, \mathscr{B}_2$ are bounded linear operators from Y_A into a boundary Hilbert space H_0, and T is a positive self-adjoint operator on H_0. We then consider a bounded below self-dual Lagrangian ℓ on the Hilbert space $K_0 \times K_0$ equipped with the scalar product $\langle a, b\rangle_{K_0} = \langle a, Tb\rangle_{H_0}$ in such a way that

$$\ell(a, b) \geq \langle Ta, b\rangle \text{ for all } a, b \in H_0. \tag{16.7}$$

The functional

$$J(x) = I(x) + \ell\left(\mathscr{B}_1 x, \frac{1}{\beta}\mathscr{B}_2 x\right) - \left\langle \mathscr{B}_1 x, \frac{1}{\beta}T\mathscr{B}_2 x\right\rangle$$

is then nonnegative, self-dual, and coercive on Y_A as long as $\beta < \frac{1}{2c_1}$ since

$$J(x) \geq -C + \left(\frac{1}{2\beta} - c_1\right)\|Ax\|^2 - \frac{1}{2\beta} + D,$$

where D is a lower bound for ℓ. The infimum of J on Y_A is then equal to zero and is attained at a point $u \in Y_A$ satisfying

$$\begin{cases} Au + \Gamma u \in -\partial\varphi(u) \\ T\mathscr{B}_2 u \in \beta\bar{\partial}\ell(\mathscr{B}_1 u). \end{cases} \tag{16.8}$$

It is worth noting that the required bound on β normally leads to a time restriction in evolution equations and often translates into local existence results as opposed to global ones. The relevance of this approach will be illustrated in the next section.

16.2 Hamiltonian systems of PDEs with self-dual boundary conditions

Consider the system

$$J\dot{U}(t) + J\mathscr{A}U(t) \in \bar{\partial}L(t, U(t)), \tag{16.9}$$

with L being a time-dependent self-dual Lagrangian on $[0,T] \times X \times X$, where $X :=$ $H \times H$ for some – possibly infinite-dimensional – Hilbert space H, $U := (p,q) \in X$, $\mathscr{A}U = \mathscr{A}(p,q) := (Ap, -Aq)$, where $A : D(A) \subseteq H \to H$ is a self-adjoint operator and J is the symplectic operator $JU = J(p,q) := (-q,p)$.

Denote by $\tilde{A}$ the operator (A,A) on the product space $X = H \times H$, and consider the Hilbert space $X_A \subseteq X$, which is the completion of $D(\tilde{A})$ for the norm induced by the inner product

$$\langle U,V \rangle_{X_A} := \langle U, \tilde{A}V \rangle_X.$$

We also consider the Banach space $Y_A := \{U \in X; \tilde{A}U \in X\}$ equipped with norm $\|U\|_{Y_A} = \|U\|_X + \|\tilde{A}U\|_X$.

Assuming that $\langle Au,u \rangle \geq c_0 \|u\|_H^2$ on $D(A)$ for some $c_0 > 0$ and that A^{-1} is compact, we have the following diagram of continuous injections

$$Y_A \to X_A \to X = X^* \to X_A^* \to Y_A^*,$$

with the map from X_A into X assumed to be compact. The path space

$$\begin{aligned}
\mathscr{W}_T := \mathscr{W}_A[0,T] &= \{U \in L_X^2[0,T]; \dot{U} \text{ and } \tilde{A}U \in L_X^2[0,T]\} \\
&= \{U \in L_{Y_A}^2[0,T]; \dot{U} \in L_X^2[0,T]\}
\end{aligned}$$

is also a Hilbert space once equipped with the norm $\|U\|_{\mathscr{W}_T} = \left(\|\tilde{A}U\|_{L_X^2}^2 + \|\dot{U}\|_{L_X^2}^2 \right)^{\frac{1}{2}}$. The embedding $\mathscr{W}_T \to C([0,T];X)$ is then continuous, i.e.,

$$\|U\|_{C([0,T];X)} \leq c\|U\|_{\mathscr{W}_T}, \tag{16.10}$$

for some constant $c > 0$, while the injection $\mathscr{W}_T \to L^2([0,T];X)$ is compact.

We consider (16.9) with a boundary condition of the form

$$R\frac{U(T) + U(0)}{2} \in \bar{\partial}_A \ell (U(T) - U(0)), \tag{16.11}$$

where ℓ is a self-dual Lagrangian on $X_A \times X_A$ and R is the automorphism $R(p,q) = (p,-q)$ on X. The notation $\bar{\partial}_A$ means that the duality is taken in the space X_A. The following is our main result for Hamiltonian systems of PDEs.

Theorem 16.1. *Let* $L : [0,T] \times X \times X \to \mathbf{R} \cup \{+\infty\}$ *be a time-dependent self-dual Lagrangian on* $X \times X$, *and let* ℓ *be a self-dual Lagrangian on* $X_A \times X_A$. *Assume the following conditions:*

(C_1) *There exists* $0 < \beta < \frac{1}{4c\sqrt{T}}$ *and* $\gamma, \alpha \in L^2(0,T;\mathbf{R}_+)$ *such that, for every* $U \in X$ *and a.e.* $t \in [0,T]$,

$$-\alpha(t) \leq L(t,U,0) \leq \frac{\beta}{2}\|U\|_X^2 + \gamma(t).$$

(C_2) ℓ *is bounded below, and* $0 \in \text{Dom}(\ell)$.

Then, the infimum of the functional

$$I(U) = \int_0^T \left\{ L(t, U(t), J\dot{U}(t) + J\mathscr{A}U(t)) - \langle J\dot{U}(t) + J\mathscr{A}U(t), U(t)\rangle \right\} dt \quad (16.12)$$

$$+ \ell\left(U(T) - U(0), R\frac{U(T) + U(0)}{2\beta} \right) - \left\langle U(T) - U(0), \tilde{A}R\frac{U(0) + U(T)}{2\beta} \right\rangle$$

on $\mathscr{W}_T$ is equal to zero and is attained at some $U \in \mathscr{W}_T$ that is a solution of

$$\begin{cases} J\dot{U}(t) + J\mathscr{A}U(t) \in \overline{\partial}L(t, U(t)) \\ R\frac{U(T) + U(0)}{2} \in \beta\overline{\partial}_A\ell(U(T) - U(0)). \end{cases} \quad (16.13)$$

We start by establishing the following proposition that assumes a stronger condition on both Lagrangians L and ℓ.

Proposition 16.1. *Let $L : [0, T] \times X \times X \to \mathbf{R} \cup \{+\infty\}$ be a time-dependent self-dual Lagrangian, and let ℓ be a self-dual Lagrangian on $X_A \times X_A$. Assume the following conditions:*

(C_1') *There exists $\lambda > 0$, $0 < \beta < \frac{1}{4c\sqrt{T}}$, and $\gamma, \alpha \in L^2(0, T; \mathbf{R}_+)$ such that, for every (U, P) in $X \times X$ and a.e. $t \in [0, T]$,*

$$-\alpha(t) \le L(t, U, P) \le \tfrac{\beta}{2}\|U\|_X^2 + +\lambda\|P\|_X^2 + \gamma(t).$$

(C_2') *There exist positive constants $\alpha_1, \beta_1, \gamma_1 \in \mathbf{R}$ such that, for every $(U, P) \in X_A \times X_A$,*

$$-\alpha_1 \le \ell(U, P) \le \frac{\beta_1}{2}\left(\|U\|_{X_A}^2 + \|P\|_{X_A}^2\right) + \gamma_1.$$

The functional I given by (16.12) is then self-dual on $\mathscr{W}_T$, and its corresponding antisymmetric Hamiltonian on $\mathscr{W}_T \times \mathscr{W}_T$ is

$$M(U, V) = \int_0^T \left\{ \langle J\dot{V} + J\mathscr{A}V, U\rangle - \tilde{H}_L(t, J\dot{V} + J\mathscr{A}V, J\dot{U} + J\mathscr{A}U) \right\} dt$$

$$- \int_0^T \langle J\dot{U}(t) + J\mathscr{A}U(t), U(t)\rangle \, dt - \left\langle U(T) - U(0), \tilde{A}R\frac{U(T) + U(0)}{2\beta} \right\rangle$$

$$+ \left\langle V(T) - V(0), \tilde{A}R\frac{U(T) + U(0)}{2\beta} \right\rangle + H_\ell^A(V(T) - V(0), U(T) - U(0)),$$

where $H_\ell^A(V, W) = \sup\{\langle A(P), W\rangle_X - \ell(V, P); P \in X_A\}$ is the Hamiltonian of ℓ on the space $X_A \times X_A$. Moreover, the infimum of I on $\mathscr{W}_T$ is equal to zero and is attained at a solution U of equation (16.13).

The proof requires a few preliminary lemmas. We first establish the self-duality of the functional *I*.

Lemma 16.1. *With the above notation, we have:*

1. $I(U) \geq 0$ *for every $U \in \mathscr{W}_T$.*
2. *M is an antisymmetric Hamiltonian on $\mathscr{W}_T \times \mathscr{W}_T$.*
3. $I(U) = \sup_{V \in \mathscr{W}_T} M(U,V)$ *for every $U \in \mathscr{W}_T$.*

Proof. 1) Since L is a self-dual Lagrangian, we have, for any $U \in \mathscr{W}_T$,

$$L(t,U(t),J\dot{U}(t)+J\mathscr{A}U(t)) - \langle J\dot{U}(t)+J\mathscr{A}U(t),U(t)\rangle \geq 0 \qquad \text{for } t \in [0,T].$$

Also, since ℓ is self-dual on $X_A \times X_A$, we have

$$\ell\left(U(T)-U(0),\frac{R(U(T)+U(0))}{2\beta})\right) - \left\langle U(T)-U(0),\tilde{A}R\frac{U(0)+U(T)}{2\beta}\right\rangle \geq 0,$$

from which we obtain $I(U) \geq 0$.

2) The fact that M is an antisymmetric Hamiltonian on $\mathscr{W}_T \times \mathscr{W}_T$ is straightforward. Indeed, the weak lower semicontinuity of $U \to M(U,V)$ for any $V \in \mathscr{W}_T$ follows from the fact that the embedding $\mathscr{W}_T \subseteq L_X^2$ is compact and $\mathscr{W}_T \subseteq C(0,T;X)$ is continuous. It follows that if $U \in \mathscr{W}_T$ and $\{U_n\}$ is a bounded sequence in $\mathscr{W}_T$ such that $U_n \rightharpoonup U$ weakly in $\mathscr{W}_T$, then

$$\lim_{n\to\infty} \int_0^T \langle J\dot{U}_n + J\mathscr{A}U_n(t),U_n\rangle \, dt = \int_0^T \langle J\dot{U} + J\mathscr{A}U(t),U\rangle \, dt,$$

$$\lim_{n\to\infty} \langle U_n(T)-U_n(0),\tilde{A}R(U_n(0)+U_n(T))\rangle = \langle U(T)-U(0),\tilde{A}R(U(0)+U(T))\rangle.$$

3) Apply the superposition principle in Theorem 15.5 with $Z = \mathscr{W}$, $X_1 = L_X^2[0,T]$, and $X_2 = X_A$. The operators are $(A_1,\Gamma_1) : Z \to X_1 \times X_1^*$ defined by

$$A_1 U = U \quad \text{and} \quad \Gamma_1 U := J\dot{U}(t) + J\mathscr{A}U(t)),$$

while $A_2 : Z \to X_2$ and $\Lambda_2 : Z \to X_2^*$ are defined as

$$A_2 U := U(0) - U(T) \quad \text{and} \quad \Lambda_2 U := R\frac{U(T)+U(0)}{2}.$$

We consider the self-dual Lagrangian L_1 on $X_1 \times X_1^* = L_X^2[0,T] \times L_X^2[0,T]$,

$$L_1(U,P) = \int_0^T L(t,U(t),P(t))dt,$$

as well as the self-dual Lagrangian L_2 on $X_2 \times X_2^* = X_A \times X_A$,

$$L_2(X,Y) = \ell(X,Y).$$

The linear theory yields that the image of the space

$$G_0 := \mathrm{Ker}(A_2) = \{U \in \mathscr{W} : U(0) = U(T)\}$$

by Γ_1 is dense in $X_1^* = L_X^2[0,T]$. Moreover, for each $(a,b) \in Y_A \times Y_A$, there is $w \in \mathscr{W}$ such that $w(0) = a$ and $w(T) = b$, namely the linear path $w(t) = \frac{T-t}{T}a + \frac{t}{T}b$. Since Y_A is also dense in X_A, it follows that the image of $\mathscr{W}$ by A_2 is dense in $X_A \times X_A$.

It also follows from (C_1') and (C_2') that $\tilde{H}_L(t,.,.)$ is continuous in both variables on $L_X^2 \times L_X^2$ and that ℓ is continuous in both variables on $X_A \times X_A$. Theorem 15.5 then applies and gives the claim.

The following three lemmas are dedicated to the proof of the coercivity of $U \to M(U,0)$ on $\mathscr{W}$.

Lemma 16.2. *For any* $U \in \mathscr{W}_T$, *we have*

$$\int_0^T \langle J\dot{U}(t), J\mathscr{A}U(t)\rangle dt = \left\langle \tilde{A}U(T) - \tilde{A}U(0), R\frac{U(T)+U(0)}{2}\right\rangle. \tag{16.14}$$

Proof. Indeed, for $U = (p,q)$, we have

$$\int_0^T \langle J\dot{U}(t), J\mathscr{A}U(t)\rangle dt = \int_0^T \langle (-\dot{q},\dot{p}),(Aq,Ap)\rangle dt$$

$$= -\int_0^T \langle \dot{q}, Aq\rangle dt + \int_0^T \langle \dot{p}, Ap\rangle dt$$

$$= -\frac{1}{2}\int_0^T \frac{d}{dt}\|A^{\frac{1}{2}}q\|_{X_A}^2 dt + \frac{1}{2}\int_0^T \frac{d}{dt}\|A^{\frac{1}{2}}p\|_{X_A}^2 dt$$

$$= -\frac{1}{2}\|A^{\frac{1}{2}}q(T)\|_{X_A}^2 + \frac{1}{2}\|A^{\frac{1}{2}}q(0)\|_{X_A}^2$$

$$\quad + \frac{1}{2}\|A^{\frac{1}{2}}p(T)\|_{X_A}^2 - \frac{1}{2}\|A^{\frac{1}{2}}p(0)\|_{X_A}^2$$

$$= -\left\langle Aq(T) - Aq(0), \frac{q(0)+q(T)}{2}\right\rangle$$

$$\quad + \left\langle Ap(T) - Ap(0), \frac{p(0)+p(T)}{2}\right\rangle$$

$$= \left\langle \tilde{A}U(T) - \tilde{A}U(0), R\frac{U(T)+U(0)}{2}\right\rangle.$$

Lemma 16.3. *For each* $U \in \mathscr{W}_T$, *the following estimate holds:*

$$\left|\int_0^T \langle J\dot{U}(t) + J\mathscr{A}U(t), U(t)\rangle dt\right| \leq 2c\sqrt{T}\|U\|_{\mathscr{W}_T}^2.$$

Proof. Indeed, we have

$$\left|\int_0^T \langle J\dot{U}(t) + J\mathscr{A}U(t), U(t)\rangle dt\right| \leq \|U\|_{L_X^2}(\|\dot{U}\|_{L_X^2} + \|\mathscr{A}U\|_{L_X^2})$$

$$\leq \sqrt{T}\|U\|_{C(0,T;X)}(\|\dot{U}\|_{L_X^2} + \|\mathscr{A}U\|_{L_X^2})$$

$$\leq 2c\sqrt{T}\|U\|_{\mathscr{W}_T}^2.$$

Lemma 16.4. *There exists a constant* $C \geq 0$ *such that for any* $U \in \mathscr{W}_T$,

$$M(U,0) \geq \left(\frac{1}{2\beta} - 2c\sqrt{T}\right)\|U\|_{\mathscr{W}_T}^2 - C. \tag{16.15}$$

Proof. Note first that

$$-\int_0^T \tilde{H}_L(t,0,J\dot{U}(t)+J\mathscr{A}U(t))\,dt$$

$$= \int_0^T \tilde{H}_L(t,J\dot{U}(t)+J\mathscr{A}U(t),0)\,dt$$

$$= \sup_{P\in L_X^2} \int_0^T \left[\langle P(t),J\dot{U}(t)+J\mathscr{A}U(t)\rangle - L(t,P(t),0)\right]\,dt$$

$$\geq \sup_{P\in L_X^2} \int_0^T \left[\langle P(t),-J\dot{U}(t)-J\mathscr{A}U(t)\rangle - \frac{\beta}{2}\|P(t)\|_X^2 - \gamma(t)\right]\,dt$$

$$\geq \frac{1}{2\beta}\int_0^T \|J\dot{U}(t)+J\mathscr{A}U(t)\|_X^2\,dt - \int_0^T \gamma(t)\,dt$$

$$= \frac{1}{2\beta}\int_0^T \left(\|\dot{U}(t)\|_X^2 + \|\mathscr{A}U(t)\|_X^2\right)\,dt + \frac{1}{\beta}\int_0^T \langle J\dot{U}(t),J\mathscr{A}U(t)\rangle\,dt$$

$$- \int_0^T \gamma(t)\,dt. \tag{16.16}$$

It follows from Lemma 16.2, Lemma 16.3, and formula (16.16) that

$$M(U,0) = \int_0^T -\tilde{H}_L\big(t,0,J\dot{U}(t)+J\mathscr{A}U(t)\big)\,dt - \int_0^T \langle J\dot{U}(t)+J\mathscr{A}U(t),U(t)\rangle\,dt$$

$$- \left\langle U(T)-U(0),\tilde{A}R\frac{U(T)+U(0)}{2\beta}\right\rangle + H_\ell^A\big(0,U(T)-U(0)\big)$$

$$\geq \frac{1}{2\beta}\int_0^T \left(\|\dot{U}(t)\|_X^2 + \|\mathscr{A}U(t)\|_X^2\right)\,dt + \frac{1}{\beta}\int_0^T \langle J\dot{U}(t),J\mathscr{A}U(t)\rangle\,dt$$

$$- \int_0^T \gamma(t)\,dt - \int_0^T \langle J\dot{U}(t)+J\mathscr{A}U(t),U(t)\rangle\,dt$$

$$- \left\langle U(T)-U(0),\tilde{A}R\frac{U(T)+U(0)}{2\beta}\right\rangle + \frac{1}{2\beta_1}\|U(T)-U(0)\|_{X_A}^2 - \gamma_1$$

$$\geq \frac{1}{2\beta}\int_0^T \left(\|\dot{U}(t)\|_X^2 + \|\mathscr{A}U(t)\|_X^2\right)\,dt - 2c\sqrt{T}\|U\|_{\mathscr{W}_T}^2 - C$$

$$+ \frac{1}{\beta}\int_0^T \langle J\dot{U}(t),J\mathscr{A}U(t)\rangle\,dt - \frac{1}{\beta}\left\langle \tilde{A}(U(T)-U(0)),\frac{R(U(T)+U(0))}{2}\right\rangle$$

$$\geq \left(\frac{1}{2\beta} - 2c\sqrt{T}\right)\|U\|_{\mathscr{W}_T}^2.$$

Proof of Proposition 16.1. It follows from (C_1') and (C_2') that $\mathscr{L}$ is finite on $\mathscr{W}_T \times \mathscr{W}_T$ and from Lemma 16.1 that I is self-dual on $\mathscr{W}_T$. In view of the coercivity guaranteed by Lemma 16.4, we can apply Theorem 1.10 to get $V \in \mathscr{W}_T$ such that $I(V) = 0$. It follows that

$$L(t,V(t),J\dot{V}(t)+J\mathscr{A}V(t)) - \langle J\dot{V}(t)+J\mathscr{A}V(t),V(t)\rangle = 0 \qquad \text{for} \quad t \in [0,T]$$

and

$$\ell\left(V(T)-V(0),R\frac{V(T)+V(0)}{2\beta}\right) - \left\langle V(T)-V(0),\tilde{A}R\frac{V(T)+V(0)}{2\beta}\right\rangle = 0.$$

Since $V(0)$ and $V(T)$ are in $Y_A \subset X_A$, then $R\frac{V(T)+V(0)}{2} \in \beta\overline{\partial}_A\ell(V(0)-V(T))$.

Proof of Theorem 16.1. We now consider L and ℓ satisfying (C_1) and (C_2). We just need to show that the result of Proposition 16.1 still holds if one replaces (C_1') and (C_2') with (C_1) and (C_2) respectively. Indeed, for $0 < \lambda < \frac{1}{4c\sqrt{T}} - \beta$, we replace L with L_λ^2 in such a way that

$$L_\lambda^2(U,P) = \inf\left\{L(U,Q) + \frac{\|P-Q\|^2}{2\lambda} + \frac{\lambda\|U\|^2}{2};Q \in X\right\} \qquad (16.17)$$

$$\leq L(U,0) + \frac{\|P\|^2}{2\lambda} + \frac{\lambda\|U\|^2}{2}$$

$$\leq \frac{\lambda+\beta}{2}\|U\|^2 + \frac{\|P\|^2}{2\lambda} + \gamma(t), \qquad (16.18)$$

and therefore it satisfies (C_1') since $\lambda + \beta < \frac{1}{4c\sqrt{T}}$.

As to the boundary Lagrangian ℓ, we shall replace it by the Lagrangian $\ell_\lambda^{1,2}$ defined on $X_A \times X_A$ as

$$\ell_\lambda^{1,2}(U,P) = \inf\left\{\ell(V,Q) + \frac{1}{2\lambda}\|U-V\|_{X_A}^2 + \frac{\lambda}{2}\|P\|_{X_A}^2 + \frac{1}{2\lambda}\|Q-P\|_{X_A}^2 + \frac{\lambda}{2}\|V\|_{X_A}^2\right\}$$

over all $V \in X_A, Q \in X_A$. By Lemma 3.2, $\ell_\lambda^{1,2}$ is a self-dual Lagrangian on $X_A \times X_A$ and it is easy to see that it satisfies condition (C_2'). We can now apply Proposition 16.1 to L_λ^2 and $\ell_\lambda^{1,2}$ and find $U_\lambda \in \mathscr{W}_T$ with

$$I_\lambda(U_\lambda) = \int_0^T \left\{L_\lambda^2(t,U_\lambda(t),J\dot{U}_\lambda(t)+J\mathscr{A}U_\lambda(t)) - \langle J\dot{U}_\lambda(t)+J\mathscr{A}U_\lambda(t),U_\lambda(t)\rangle\right\}dt$$

$$+\ell_\lambda^{1,2}\left(U_\lambda(T)-U_\lambda(0),R\frac{U_\lambda(T)+U_\lambda(0)}{2(\lambda+\beta)}\right)$$

$$-\left\langle\tilde{A}(U_\lambda(T)-\tilde{U}_\lambda(0)),R\frac{U_\lambda(0)+U_\lambda(T)}{2(\lambda+\beta)}\right\rangle$$

$$= 0. \qquad (16.19)$$

From (16.17) and part (1) of Lemma 3.5, we have

$$\int_0^T L_\lambda^2(t, U_\lambda, J\dot{U}_\lambda + J\mathscr{A}U_\lambda)\, dt \geq \frac{1}{2(\lambda + \beta)}\|J\dot{U}_\lambda + J\mathscr{A}U_\lambda\|_{L_X^2}^2 - C_2. \quad (16.20)$$

From (16.20), (16.19), and Lemma 16.3, we get that

$$\frac{1}{2(\lambda + \beta)}\|J\dot{U}_\lambda(t) + J\mathscr{A}U_\lambda(t)\|_{L_X^2}^2 - 2c\sqrt{T}\|U_\lambda\|_{\mathscr{W}_T}^2$$

$$+\ell_\lambda^{1,2}\left(U_\lambda(0) - U_\lambda(T), R\frac{U_\lambda(T) + U_\lambda(0)}{2(\beta + \lambda)}\right)$$

$$-\left\langle U_\lambda(T) - \tilde{U}_\lambda(0), \tilde{A}R\frac{U_\lambda(0) + U_\lambda(T)}{2(\lambda + \beta)}\right\rangle \leq C,$$

where C is a constant independent of λ. Again using Lemma 16.2 and the fact that ℓ (and hence $\ell^{1,2}$) is bounded below, we obtain

$$\left(\frac{1}{2(\lambda + \beta)} - 2c\sqrt{T}\right)\|U_\lambda\|_{\mathscr{W}_T}^2 \leq C, \quad (16.21)$$

which ensures the boundedness of U_λ in $\mathscr{W}_T$. Assuming $U_\lambda \rightharpoonup U$ weakly in $\mathscr{W}_T$, it follows from Lemma 3.3 that $I(U) \leq \liminf_\lambda I_\lambda(U_\lambda) = 0$. Since, on the other hand, $I(U) \geq 0$, the latter is therefore equal to zero, and U is a solution of (16.13). The rest follows as in the proof of Proposition 16.1.

We now apply Theorem 16.1 to find solutions for the system

$$\begin{cases} J\dot{V}(t) + J\mathscr{A}V(t) \in \partial\varphi(t, V(t)) & \text{a.e. on } [0, T] \\ \tilde{A}R\frac{V(T) + V(0)}{2} \in \beta\partial\psi(V(T) - V(0)), \end{cases} \quad (16.22)$$

where $\varphi : [0, T] \times X \to \mathbf{R}$ is a time-dependent convex lower semicontinuous function on X, and ψ is a convex lower semicontinuous function on X. While the choice of the self-dual Lagrangian $L(t, U, p) = \varphi(t, U) + \varphi^*(t, p)$ on $X \times X$ is obvious, this is not the case for the boundary Lagrangian. We shall therefore need the following.

Lemma 16.5. *Let ψ be a bounded below convex lower semicontinuous function on X such that $0 \in \text{Dom}(\psi)$, and consider the following functions on X:*

$$\tilde{\psi}(U) = \begin{cases} \psi(U) & U \in X_A, \\ +\infty & U \in X \setminus X_A, \end{cases} \quad (16.23)$$

and

$$\psi^o(P) = \sup\{\langle P, \tilde{A}(U)\rangle - \psi(U); U \in X_A\}. \quad (16.24)$$

1. *The functional $\ell(U, P) = \psi(U) + \psi^o(P) = \psi(U) + (\tilde{\psi})^*(\tilde{A}(P))$ is then a self-dual Lagrangian on $X_A \times X_A$.*
2. *Assume $\psi^* = (\tilde{\psi})^*$. Then, $P \in \overline{\partial}_A\ell(U)$ if and only if $P \in X_A$ and $\tilde{A}(P) \in \partial\psi(U)$.*

The following is now a straightforward application of Theorem 16.1.

Corollary 16.1. *Let A, H, and X be as above and let ψ be a convex lower semi-continuous function on X that is bounded below and such that $0 \in \mathrm{Dom}(\psi)$ and $\psi^* = (\tilde{\psi})^*$. Let $\varphi : [0,T] \times X \to \mathbf{R}$ be a time-dependent, convex, lower semicontinuous function on X satisfying for some $\beta > 0$, $\gamma, \alpha \in L^2(0,T;\mathbf{R}_+)$,*

$$-\alpha(t) \le \varphi(t,U) \le \tfrac{\beta}{2}\|U\|_X^2 + \gamma(t) \text{ for every } U \in X \text{ and a.e. } t \in [0,T]. \quad (16.25)$$

Assume that

$$0 < T < \frac{1}{16c^2\beta^2}. \quad (16.26)$$

Then, the infimum on $\mathscr{W}_T$ of the functional

$$I(U) = \int_0^T \left\{ \varphi\big(t,U(t)\big) + \varphi^*\big(J\dot{U}(t) + J\mathscr{A}U(t)\big) - \langle J\dot{U}(t) + J\mathscr{A}U(t), U(t)\rangle \right\} dt$$

$$+ \psi\big(U(T) - U(0)\big) + \psi^\circ\left(R\frac{U(T)+U(0)}{2\beta}\right)$$

$$- \left\langle U(T) - U(0), \tilde{A}R\frac{U(0)+U(T)}{2\beta}\right\rangle$$

is equal to zero and is attained at some $V \in \mathscr{W}_T$ that is a solution of system (16.22).

Choices for boundary conditions. Here again, the general boundary condition above will allow us to obtain periodic and other types of solutions. Indeed:

- For periodic solutions $V(0) = V(T)$, then ψ is chosen as:

$$\psi(W) = \begin{cases} 0 & W = 0 \\ +\infty & \text{elsewhere.} \end{cases}$$

- For antiperiodic solutions $V(0) = -V(T)$, then $\psi \equiv 0$.

Note that in both cases above, we have that $\tilde{\psi} = \psi$.

- For the linking condition $p(0) = p_0$ and $q(T) = q_0$ for a given $p_0, q_0 \in X_A$, let $V_0 = (-p_0, q_0)$ and choose $\psi(W) = \tfrac{1}{4\beta}\langle A(W), W\rangle - \tfrac{1}{\beta}\langle W, A(V_0)\rangle$ on X. Since ψ is continuous on X and X_A is dense in X, we have that $(\tilde{\psi})^* = \psi^*$ on X and therefore $\partial\psi(U) = \partial\tilde{\psi}(U)$ for every $U \in X_A$. It follows that when

$$\tilde{A}R\frac{V(T)+V(0)}{2} = \beta\partial\big[\psi(V(T) - V(0))\big] = \frac{\tilde{A}V(T) - \tilde{A}V(0)}{2} - \tilde{A}V_0,$$

we get by setting $V = (p,q)$ that

$$\left(A\frac{p(T)+p(0)}{2}, -A\frac{q(T)+q(0)}{2}\right) = \tilde{A}R\left(\frac{p(T)+p(0)}{2}, \frac{q(T)+q(0)}{2}\right)$$

$$= \tilde{A}R\frac{V(T)+V(0)}{2}$$

$$= \tilde{A}\frac{V(T) - V(0)}{2} - \tilde{A}(V_0)$$

$$= \left(A\frac{p(T) - p(0)}{2} + p_0, A\frac{q(T) - q(0)}{2} - q_0\right),$$

from which we obtain $Ap(0) = Ap_0$ and $Aq(T) = Aq_0$.

Example 16.1. Periodic solutions for a coercive purely diffusive Hamiltonian system involving the Laplacian

We start with the following simple Hamiltonian system of PDEs:

$$\begin{cases} -\dot{v}(t) - \Delta v = |u|^{p-2}u + g(t,x) & (t,x) \in (0,T) \times \Omega, \\ \dot{u}(t) - \Delta u = |v|^{q-2}v + f(t,x) & (t,x) \in (0,T) \times \Omega, \\ u = 0 & (t,x) \in [0,T] \times \partial\Omega, \\ v = 0 & (t,x) \in [0,T] \times \partial\Omega. \end{cases} \tag{16.27}$$

It can be written as $J\dot{U}(t) + J\mathscr{A}U(t) = \overline{\partial}L(t,U(t))$, where $\mathscr{A}(u,v) = (-\Delta u, \Delta v)$, and $L(t,U,V) = \Phi(t,U) + \Phi^*(t,V)$ with

$$\Phi(t,U) = \frac{1}{p}\int_\Omega |u|^p\,dx + \langle u, f(t,x)\rangle + \frac{1}{q}\int_\Omega |v|^q\,dx + \langle v, g(t,x)\rangle.$$

Here $H = L^2(\Omega)$, $X = L^2(\Omega) \times L^2(\Omega)$, and $X_A = H_0^1(\Omega) \times H_0^1(\Omega)$. Corollary 16.1 yields the following existence result.

Corollary 16.2. *Suppose $f,g \in L_H^2$ and $1 < p,q < 2$. Then, for any $T > 0$ there exists a path $(u,v) \in \mathscr{W}_T$ satisfying (16.27) and one of the following boundary conditions:*

- *periodic solutions $u(0) = u(T)$ and $v(0) = v(T)$.*
- *antiperiodic solutions $u(0) = -u(T)$ and $v(0) = -v(T)$.*
- *linking condition $u(0) = u_0$ and $v(T) = v_0$ for any given $v_0, u_0 \in H$.*

Example 16.2. Periodic solutions for a coercive purely diffusive Hamiltonian System involving the bi-Laplacian

Let Ω be a bounded domain in $\mathbf{R}^N$, and consider the Hamiltonian System,

$$\begin{cases} -\dot{v}(t) + \Delta^2 v = \partial\varphi_1(t,u) & (t,x) \in (0,T) \times \Omega, \\ \dot{u}(t) + \Delta^2 u = \partial\varphi_2(t,v) & (t,x) \in (0,T) \times \Omega, \\ u = \Delta u = 0 & (t,x) \in [0,T] \times \partial\Omega, \\ v = \Delta v = 0 & (t,x) \in [0,T] \times \partial\Omega, \end{cases} \tag{16.28}$$

where $\varphi_i, i = 1,2$ are two convex lower semicontinuous functions on $H := H_0^1(\Omega)$ considered as a Hilbert space with the inner product $\langle u_1, u_2\rangle = \int_\Omega \nabla u_1 \cdot \nabla u_2\,dx$.

The system can be written as $J\dot{U}(t) + J\mathscr{A}U(t) = \overline{\partial}L(t,U(t))$, where $L(t,U,V) = \Phi(t,U) + \Phi^*(t,J\mathscr{B}U + V)$ with $\Phi(t,U) = \varphi_1(t,u) + \varphi_2(t,v)$ and $Au = \Delta^2 u$ so that for $U = (u,v)$, $\mathscr{A}U = \mathscr{A}(u,v) = (\Delta^2 u, -\Delta^2 v)$. Here $X = H_0^1(\Omega) \times H_0^1(\Omega)$, $H_A = \{u \in H_0^1(\Omega); \Delta u \in H_0^1(\Omega)\}$ equipped with the norm $\|u\|_{X_A}^2 = \int_\Omega |\nabla \Delta u|^2 \, dx$, and $X_A = H_A \times H_A$. Corollary 16.1 yields the following.

Corollary 16.3. *Suppose φ_1 and φ_2 satisfy the condition*

$$\gamma_i(t) \leq \varphi_i(t,u) \leq \alpha_i(t) + C_i \|u\|_{H_0^1(\Omega)}^2 \qquad i = 1,2, \tag{16.29}$$

where $\gamma_i, \alpha_i \in L^2([0,T])$ and $c_i, C_i > 0$. Then, for T small enough, there exist $U := (u,v) \in \mathscr{W}_T$ solution of the system (16.28) with one of the boundary conditions stated in Corollary 16.2.

16.3 Nonpurely diffusive Hamiltonian systems of PDEs

We now consider systems of the form

$$\begin{cases} J\dot{U}(t) + J\mathscr{A}U(t) + J\mathscr{B}U(t) \in \partial\varphi(t,U(t)) & \text{a.e on } [0,T] \\ \tilde{A}R\frac{U(T)+U(0)}{2} \in \beta\partial\psi(U(T) - U(0)), \end{cases} \tag{16.30}$$

where A, φ, and ψ are as in the last section but where $\mathscr{B}$ is an additional linear operator on X that is itself skew-adjoint or is such that $J\mathscr{B}$ is skew-adjoint.

Coercive nonpurely diffusive Hamiltonian systems of PDEs

In the case where $J\mathscr{B}$ is a skew-adjoint operator, we can directly apply Theorem 16.1 to the self-dual Lagrangian $L(t,U,p) = \varphi(t,U,p) + \varphi^*(t,U,J\mathscr{B}U + p)$ on $X \times X$ to obtain the following result.

Corollary 16.4. *Let A, H, and X be as in Section 16.2, and let $\mathscr{B}$ be a bounded linear operator on X such that $J\mathscr{B}$ is skew-adjoint. Let ψ be a convex lower semicontinuous function on X that is bounded below and such that $0 \in \mathrm{Dom}(\psi)$ and $\psi^* = (\tilde{\psi})^*$. Let $\varphi : [0,T] \times X \to \mathbf{R}$ be a time-dependent convex lower semicontinuous function on X such that, for some $\beta > 0$, $\gamma, \alpha \in L^2(0,T;\mathbf{R}_+)$, we have for all $U \in X$ and a.e. $t \in [0,T]$*

$$-\alpha(t) \leq \varphi(t,U) + \varphi^*(t,J\mathscr{B}U) \leq \tfrac{\beta}{2}\|U\|_X^2 + \gamma(t). \tag{16.31}$$

Assume that

$$0 < T < \frac{1}{16c^2\beta^2}. \tag{16.32}$$

Then, the infimum on $\mathscr{W}_T$ of the functional

$$I(U) = \int_0^T \left\{ \varphi(t, U(t)) + \varphi^* \left(J\dot{U}(t) + J\mathscr{A}U(t) + J\mathscr{B}U(t) \right) \right\} dt$$

$$- \int_0^T \langle J\dot{U}(t) + J\mathscr{A}U(t), U(t) \rangle \, dt$$

$$+ \psi(U(T) - U(0)) + \psi^o \left(R \frac{U(T) + U(0)}{2\beta} \right)$$

$$- \left\langle U(T) - U(0), \tilde{A}R \frac{U(0) + U(T)}{2\beta} \right\rangle$$

is equal to zero and is attained at a $V \in \mathscr{W}_T$ that is a solution of the system (16.30).

Example 16.3. Periodic solutions for a coercive nonpurely diffusive Hamiltonian system involving the bi-Laplacian

Let Ω be a bounded domain in $\mathbf{R}^N$, and consider the Hamiltonian system,

$$\begin{cases} -\dot{v}(t) + \Delta^2 v - \Delta v = \partial\varphi_1(t, u) & (t, x) \in (0, T) \times \Omega, \\ \dot{u}(t) + \Delta^2 u + \Delta u = \partial\varphi_2(t, v) & (t, x) \in (0, T) \times \Omega, \\ u = \Delta u = 0 & (t, x) \in [0, T] \times \partial\Omega, \\ v = \Delta v = 0 & (t, x) \in [0, T] \times \partial\Omega, \end{cases} \tag{16.33}$$

where $\varphi_i, i = 1, 2$ are two convex lower semicontinuous functions on $H := H_0^1(\Omega)$ considered as a Hilbert space with the inner product $\langle u_1, u_2 \rangle = \int_\Omega \nabla u_1 \cdot \nabla u_2 \, dx$. The system can be written as $J\dot{U}(t) + J\mathscr{A}U(t) = \bar{\partial}L(t, U(t))$, where $Au = \Delta^2 u$ so that for $U = (u, v)$, $\mathscr{A}U = \mathscr{A}(u, v) = (\Delta^2 u, -\Delta^2 v)$ and $L(t, U, V) = \Phi(t, U) + \Phi^*(t, V)$ with $\Phi(t, U) = \varphi_1(t, u) + \varphi_2(t, v)$ and $\mathscr{B}U = (\Delta u, \Delta v)$ in such a way that $J\mathscr{B}U = (-\Delta v, \Delta u)$ is skew-adjoint on $X = H_0^1(\Omega) \times H_0^1(\Omega)$. We consider again $H_A = \{u \in H_0^1(\Omega); \Delta u \in H_0^1(\Omega)\}$ equipped with the norm $\|u\|_{X_A}^2 = \int_\Omega |\nabla \Delta u|^2 \, dx$ and $X_A = H_A \times H_A$. Corollary 16.4 yields the following.

Corollary 16.5. *Suppose φ_1 and φ_2 satisfy the condition*

$$\gamma_i(t) + c_i\|u\|_{L^2(\Omega)}^2 \leq \varphi_i(t, u) \leq \alpha_i(t) + C_i\|u\|_{H_0^1(\Omega)}^2 \qquad i = 1, 2, \tag{16.34}$$

where $\gamma_i, \alpha_i \in L^2([0, T])$ and $c_i, C_i > 0$. Then, for T small enough, there exist $U := (u, v) \in \mathscr{W}_T$ a solution to (16.33) with either one of the boundary conditions stated in Corollary 16.2.

Proof. We just need to show that L satisfies condition (16.25) in Corollary 16.4. For that we let $C = \max\{C_1, C_2\}$, $c = \min\{c_1, c_2\}$, $\gamma(t) = \min\{\gamma_1(t), \gamma_2(t)\}$ and $\alpha(t) = \max\{\alpha_1(t), \alpha_2(t)\}$. It follows from (16.34) that

$$\gamma(t) + c\|U\|_{L^2(\Omega)}^2 \leq \Phi(t, U) \leq \alpha(t) + C\|U\|_{H_0^1(\Omega)}^2,$$

and therefore

$$-\alpha(t) + \frac{1}{4C}\|U\|^2_{H^1_0(\Omega)} \le \Phi^*(t,U) \le -\gamma(t) + \frac{1}{4c}\|\nabla(-\Delta)^{-1}U\|^2_{L^2(\Omega)} - \gamma(t),$$

from which we obtain

$$\gamma(t) - \alpha(t) \le L(t,U,0) \le \alpha(t) + C\|U\|^2_{H^1_0(\Omega)} + \frac{1}{4c}\|\nabla(-\Delta)^{-1}J\mathscr{B}U\|^2_{L^2(\Omega)}$$

$$= \alpha(t) - \gamma(t) + C\|U\|^2_{H^1_0(\Omega)} + \frac{1}{4c}\|\nabla U\|^2_{L^2(\Omega)}$$

$$= \alpha(t) - \gamma(t) + \left(C + \frac{1}{4c}\right)\|U\|^2_{H^1_0(\Omega)}.$$

Hence, for T small enough, Corollary 16.4 applies to yield our claim.

Noncoercive and nonpurely diffusive Hamiltonian systems of PDEs

Under a certain commutation property, we can relax the boundedness condition (16.25), provided one settles for periodic solutions up to an isometry.

Theorem 16.2. *Let $L : [0,T] \times X \times X \to \mathbf{R} \cup \{+\infty\}$, $\ell : X_A \times X_A \to \mathbf{R} \cup \{+\infty\}$, and $A : D(A) \subset H \to H$ be as in Theorem 16.1, let $\mathscr{B}$ be a skew-adjoint operator on $H \times H$ such that $\mathscr{A}\mathscr{B} = \mathscr{B}\mathscr{A}$ on $D(\mathscr{A})$, and let $(S_t)_t$ be its corresponding C_0-unitary group of operators on X. Then, the infimum of the functional*

$$I(U) = \int_0^T \{L(t, S_t U, JS_t\dot{U} + J\mathscr{A}S_t U) - \langle JS_t\dot{U} + J\mathscr{A}S_t U, S_t U\rangle\} \, dt$$

$$+\ell\left(U(T) - U(0), R\frac{U(T) + U(0)}{2\beta}\right) - \left\langle \tilde{A}(U(T) - \tilde{U}(0)), R\frac{U(0) + U(T)}{2\beta}\right\rangle$$

on $\mathscr{W}_T$ is equal to zero and is attained at some $U \in \mathscr{W}_T$ in such a way that $V(t) := S_t U(t)$ is a solution of

$$\begin{cases} J\dot{V}(t) + J\mathscr{A}V(t) + J\mathscr{B}V(t) \in \overline{\partial}L(t, V(t)) \\ R\frac{S_{(-T)}V(T) + V(0)}{2} \in \beta\overline{\partial}_A\ell(S_{(-T)}V(T) - V(0)). \end{cases} \quad (16.35)$$

Proof. It follows from Proposition 3.4 that $L_S(t,U,P) := L(t, S_t U, S_t P)$ is a self-dual Lagrangian on $[0,T] \times X \times X$. Since S_t is norm preserving, assumption (C_1) holds for the new Lagrangian L_S. Therefore, there exists $U \in \mathscr{W}_T$ such that $I(U) = 0$ and U is a solution of

$$\begin{cases} J\dot{U}(t) + J\mathscr{A}U(t) \in \overline{\partial}L_S(t, U(t)) \\ R\frac{U(T) + U(0)}{2} \in \beta\overline{\partial}_A\ell((U(T) - U(0)). \end{cases} \quad (16.36)$$

Note that $\overline{\partial} L_S(t, U(t)) = S_t^* \overline{\partial} L(t, S_t U(t))$, which together with (16.36) implies that

$$S_t \left(J\dot{U}(t) + J\mathscr{A} U(t) \right) = \overline{\partial} L(t, S_t U(t)).$$

Since $\mathscr{A}\mathscr{B} = \mathscr{B}\mathscr{A}$ on $D(\mathscr{A})$, we have $S_t \mathscr{A} U(t) = \mathscr{A} S_t U(t)$ and therefore

$$\begin{cases} JS_t \dot{U}(t) + J\mathscr{A} S_t U(t) \in \overline{\partial} L(t, S_t U(t)) \\ R\frac{U(T) + U(0)}{2} \in \beta \overline{\partial}_A \ell (U(T) - U(0)). \end{cases} \tag{16.37}$$

It is now clear that $V(t) := S(t) U(t)$ is a solution of problem (16.35).

By applying the above to the Lagrangian $L(t, U, P) = \varphi(t, U, P) + \varphi^*(t, P)$ on $X \times X$, we get the following.

Corollary 16.6. *Let $A : D(A) \subset H \to H$ be as in Theorem 16.1, let $\mathscr{B}$ be a skew-adjoint operator on $H \times H$ such that $\mathscr{A}\mathscr{B} = \mathscr{B}\mathscr{A}$ on $D(\mathscr{A})$, and let $(S_t)_t$ be its corresponding C_0-unitary group of operators on X. Let ψ be a convex lower semicontinuous function on X that is bounded below, such that $0 \in \mathrm{Dom}(\psi)$ and $\psi^* = (\tilde{\psi})^*$. Let $\varphi : [0, T] \times X \to \mathbf{R}$ be a time-dependent Gâteaux-differentiable convex function on X satisfying, for some $\beta > 0$, $\gamma, \alpha \in L^2(0, T; \mathbf{R}_+)$,*

$$-\alpha(t) \le \varphi(t, U) \le \tfrac{\beta}{2} \|U\|_X^2 + \gamma(t) \text{ for every } U \in X \text{ and a.e. } t \in [0, T]. \tag{16.38}$$

Assuming that

$$0 < T < \frac{1}{16c^2\beta^2}, \tag{16.39}$$

then the infimum on $\mathscr{W}_T$ of the functional

$$\bar{I}(U) = \int_0^T \left\{ \varphi \left(t, S_t U \right) + \varphi^* \left(JS_t \dot{U} + J\mathscr{A} S_t U \right) - \langle JS_t \dot{U} + J\mathscr{A} S_t U, S_t U \rangle \right\} dt$$

$$+ \psi \left(U(T) - U(0) \right) + \psi^o \left(R\frac{U(T) + U(0)}{2\beta} \right)$$

$$- \left\langle U(T) - U(0), \tilde{A} R\frac{U(0) + U(T)}{2\beta} \right\rangle$$

is equal to zero and is attained at some $U \in \mathscr{W}_T$ so that $V(t) := S_t U(t)$ is a solution of the system

$$\begin{cases} J\dot{V}(t) + J\mathscr{A} V(t) + J\mathscr{B} V(t) = \partial \varphi(t, V(t)) & \text{a.e on } [0, T] \\ \tilde{A} R\frac{S_{(-T)} V(T) + V(0)}{2} \in \beta \partial \psi \left(S_{(-T)} V(T) - V(0) \right). \end{cases} \tag{16.40}$$

Example 16.4. Periodic solutions up to an isometry for a noncoercive Hamiltonian system involving the bi-Laplacian

Consider the following Hamiltonian system,

$$\begin{cases} -\dot{v}(t)+\Delta^2 v-\Delta u=\partial\varphi_1(t,u) & (t,x)\in(0,T)\times\Omega, \\ \dot{u}(t)+\Delta^2 u-\Delta v=\partial\varphi_2(t,v) & (t,x)\in(0,T)\times\Omega, \\ u=\Delta u=0 & (t,x)\in[0,T]\times\partial\Omega, \\ v=\Delta v=0 & (t,x)\in[0,T]\times\partial\Omega, \end{cases} \quad (16.41)$$

where again $\varphi_i, i=1,2$ are two convex lower semicontinuous functions on $H:=H_0^1(\Omega)$. Corollary 16.6 yields the following existence result.

Corollary 16.7. *Suppose φ_1 and φ_2 satisfy the condition*

$$\gamma_i(t)\leq\varphi_i(t,u)\leq\alpha_i(t)+C_i\|u\|^2_{H_0^1(\Omega)} \qquad i=1,2, \qquad (16.42)$$

where $\gamma_i,\alpha_i\in L^2([0,T])$ and $c_i,C_i>0$. Then, for T small enough, there exist $U:=(u,v)\in\mathscr{W}_T$ satisfying (16.41) with either one of the following boundary conditions:

- *periodic solutions up to an isometry.*
- *antiperiodic solutions up to an isometry.*
- *mixed boundary condition $u(0)=u_0$ and $v(T)=v_0$ for a given $v_0,u_0\in H$.*

Proof. Let again $Au=\Delta^2 u$ in such a way that, for $U=(u,v)$, $\mathscr{A}U=\mathscr{A}(u,v)=(\Delta^2 u,-\Delta^2 v)$. Consider, however, the skew-adjoint operator $\mathscr{B}U=(-\Delta v,\Delta u)$ in such a way that $J\mathscr{B}U=(-\Delta u,-\Delta v)$.

Problem (16.41) can be rewritten as

$$J\dot{U}(t)+J\mathscr{A}U(t)+J\mathscr{B}U(t)=\partial\Phi(t,U(t)), \qquad (16.43)$$

where $\Phi(t,U)=\varphi_1(t,u)+\varphi_2(t,v)$.

Again $X_A=H_A\times H_A$, where $H_A=\{u\in H_0^1(\Omega);\Delta u\in H_0^1(\Omega)\}$ is equipped with the norm $\|u\|^2_{X_A}=\int_\Omega|\nabla\Delta u|^2\,dx$. In order to show that L satisfies condition (16.38) in Corollary 16.6, it suffices to note that

$$\gamma_1(t)+\gamma_2(t)\leq\Phi(t,U)\leq\alpha_1(t)+\alpha_2(t)+C\|U\|^2_{H_0^1(\Omega)}.$$

Example 16.5. Periodic solutions up to an isometry for a noncoercive Hamiltonian system involving the Laplacian and transport

Consider the Hamiltonian system of PDEs:

$$\begin{cases} -\dot{v}(t)-\Delta(v+u)+\mathbf{b}\cdot\nabla v=|u|^{p-2}u+g(t,x) & (t,x)\in(0,T)\times\Omega, \\ \dot{u}(t)-\Delta(u+v)+\mathbf{a}\cdot\nabla u=|v|^{q-2}v+f(t,x) & (t,x)\in(0,T)\times\Omega, \end{cases} \quad (16.44)$$

where $a,b\in\mathbf{R}^N$ are two constant vectors. Let $H=L^2(\Omega)$ and $X_A=H_0^1(\Omega)$.

Corollary 16.8. *Suppose $f,g\in L_H^2$ and $1<p,q<2$. Then, for any $T>0$, there exists a solution $U:=(u,v)\in\mathscr{W}_T$ for (16.44) with either one of the boundary conditions stated in Corollary 6.2*

Proof. Equation (16.44) can be rewritten as

$$J\dot{U}(t) + J\mathscr{A}U(t) + J\mathscr{B}U(t) = \partial\Phi(t, U(t)), \qquad (16.45)$$

where $\mathscr{A}(u,v) = (-\Delta u, \Delta v)$, $\mathscr{B}(u,v) = (-\Delta v + \mathbf{a}\cdot\nabla u, \Delta u - \mathbf{b}\cdot\nabla v)$ and

$$\Phi(t,U) = \frac{1}{p}\int_\Omega |u|^p\,dx + \langle u, f(t,x)\rangle + \frac{1}{q}\int_\Omega |v|^q\,dx + \langle v, g(t,x)\rangle.$$

It is clear that all hypotheses of Theorem 16.2 are satisfied.

Further comments

Infinite-dimensional Hamiltonian systems of the above type have been considered in the literature (See Barbu [18], [19]) but only in the case of linear boundary conditions. The results in this chapter are taken from the paper of Ghoussoub and Moameni [64].

Chapter 17
The Self-dual Palais-Smale Condition for Noncoercive Functionals

We extend the nonlinear variational principle for self-dual functionals of the form $I(x) = L(x, -\Lambda x) + \langle x, \Lambda x \rangle$ to situations where I does not satisfy the strong coercivity condition required in (1.46), but a much weaker notion of a *self-dual Palais-Smale property* on the functional I. This condition states that a sequence $(u_n)_n$ is bounded in X, provided it satisfies

$$\Lambda u_n + \overline{\partial} L(u_n) = -\varepsilon_n D u_n$$

for some $\varepsilon_n \to 0$. Here $D : X \to X^*$ is the duality map $\langle Du, u \rangle = \|u\|^2$.

We then deal again with the superposition of an unbounded linear operator $A : D(A) \subset X \to X^*$ with the – possibly nonlinear – map Λ, when trying to solve an equation of the form

$$0 \in Au + \Lambda u + \overline{\partial} L(u),$$

by minimizing the functional $I(u) = L(u, -Au - \Lambda u) + \langle u, Au + \Lambda u \rangle$. Unlike the previous chapter, we consider here the case when A is either positive, or skew-adjoint (possibly modulo a boundary operator).

Now the basic self-dual variational principle may not apply if the topology of X is not strong enough to make $A + \Lambda$ regular on X, and/or to keep the closure of $\text{Dom}_1(L)$ contained in the domain of $A + \Lambda$. We are then led to use the linear operator A to strengthen the topology on X by working with the space $Y_A := D(A)$ equipped with the norm $\|u\|_{Y_A}^2 = \|u\|_X^2 + \|Au\|_X^2$. This has the advantage of closing the domain of A and increases the chance for Λ to be regular on Y_A. This may, however, lead to a loss of strong coercivity on the new space, but there are instances where the functional I satisfies the *self-dual Palais-Smale* condition on the space Y_A, which then allows us to conclude.

A similar approach is used when A is skew-adjoint modulo a boundary operator. This is particularly relevant for the resolution of nonlinear evolution equations, and will be considered in detail in the next chapter.

N. Ghoussoub, *Self-dual Partial Differential Systems and Their Variational Principles*,   305
Springer Monographs in Mathematics, DOI 10.1007/978-0-387-84897-6_17,
© Springer Science+Business Media, LLC 2009

17.1 A self-dual nonlinear variational principle without coercivity

In this section, we show that the ideas behind the nonlinear self-dual variational principles can be extended in two different ways. For one, and as noted in Chapter 12, the hypothesis of regularity on the operator Λ in Theorem 12.3 can be weakened to pseudoregularity. We shall also relax the strong coercivity condition that proved prohibitive in the case of evolution equations.

We say that an operator Λ – linear or not – is *bounded* if it maps bounded sets into bounded sets. We denote by $D : X \to X^*$ the duality map $\langle Du, u \rangle = \|u\|^2$, and we assume that D is linear and continuous, which can always be done in the case where X is a reflexive Banach space, since then X can be equipped with an equivalent norm that is locally uniformly convex (see [42]).

The following is a useful extension of Theorem 12.3.

Theorem 17.1. *Let L be a self-dual Lagrangian on a reflexive Banach space X such that $0 \in \text{Dom}(L)$. Let $\Lambda : D(\Lambda) \subset X \to X^*$ be a bounded pseudoregular map such that $\overline{\text{Dom}_1(L)} \subset D(\Lambda)$ and*

$$\langle \overline{\partial} L(x) + \Lambda x, x \rangle \geq -C(\|x\| + 1) \text{ for large } \|x\|. \tag{17.1}$$

Then, for any $\lambda > 0$, the self-dual functional

$$I_\lambda(x) = L(x, -\Lambda x - \lambda Dx) + \langle \Lambda x + \lambda Dx, x \rangle$$

attains its infimum at $x_\lambda \in X$ in such a way that $I_\lambda(x_\lambda) = \inf_{x \in X} I_\lambda(x) = 0$ and x_λ is a solution of the differential inclusion

$$0 \in \Lambda x_\lambda + \lambda Dx_\lambda + \overline{\partial} L(x_\lambda). \tag{17.2}$$

For the proof, we shall need the following lemma.

Lemma 17.1. *Let L be a self-dual Lagrangian on a reflexive Banach space X, let $\Lambda : D(\Lambda) \subseteq X \to X^*$ be a pseudoregular map and let $F : D(F) \subseteq X \to X^*$ be a regular map. Assume $(x_n)_n$ is a sequence in $D(\Lambda) \cap D(F)$ such that $x_n \rightharpoonup x$ and $\Lambda x_n \rightharpoonup p$ weakly for some $x \in X$ and $p \in X^*$. If $0 \in \Lambda x_n + F x_n + \overline{\partial} L(x_n)$ for large $n \in \mathbf{N}$, then necessarily $0 \in \Lambda x + F x + \overline{\partial} L(x)$.*

Proof. We have

$$\limsup_n \langle \Lambda x_n, x_n - x \rangle \leq \lim_{n \to \infty} \langle \Lambda x_n, -x \rangle + \limsup_n \left\{ -L(x_n, -\Lambda x_n - F x_n) - \langle F x_n, x_n \rangle \right\}$$

$$= \langle p, -x \rangle - \liminf_n \left\{ L(x_n, -\Lambda x_n - F x_n) + \langle F x_n, x_n \rangle \right\}. \tag{17.3}$$

Since L is weakly lower semicontinuous and F is regular, we have

$$L(x, -p - F x) + \langle F x, x \rangle \leq \liminf_n \left\{ L(x_n, -\Lambda x_n - F x_n) + \langle F x_n, x_n \rangle \right\},$$

which together with (17.3) imply

$$\limsup_n \langle \Lambda x_n, x_n - x \rangle \leq \langle p, -x \rangle - L(x, -p - Fx) - \langle Fx, x \rangle$$
$$= \langle p + Fx, -x \rangle - L(x, -p - Fx).$$

L being a self-dual Lagrangian, we have $L(x, -p - Fx) \geq \langle p + Fx, -x \rangle$, and therefore

$$\limsup_n \langle \Lambda x_n, x_n - x \rangle \leq 0.$$

Now, since Λ is pseudoregular, we have $p = \Lambda x$ and $\liminf_n \langle \Lambda x_n, x_n \rangle \geq \langle \Lambda x, x \rangle$, from which we deduce that

$$L(x, -\Lambda x - Fx) + \langle \Lambda x + Fx, x \rangle \leq \liminf_n L(x_n, -\Lambda x_n - Fx_n) + \langle \Lambda x_n + Fx_n, x_n \rangle = 0.$$

On the other hand, since L is a self-dual Lagrangian, we have the reverse inequality $L(x, -\Lambda x - Fx) + \langle \Lambda x + Fx, x \rangle \geq 0$, which implies that the latter is equal to zero.

Remark 17.1. It is clear that under the hypotheses of the lemma above, one still gets the same conclusion, provided we have for large n, that

$$0 \in \Lambda x_n + Fx_n + \varepsilon_n T x_n + \overline{\partial} L(x_n), \tag{17.4}$$

where $T : X \to X^*$ is a bounded operator and $\varepsilon_n \downarrow 0$.

Proof of Theorem 17.1: Let $w(r) = \sup\{\|\Lambda u\|_* + 1; \|u\| \leq r\}$, set $Tu := w(\|u\|)Du$, and consider the λ-regularization of L with respect to the second variable,

$$L_\lambda^2(x, p) := \inf \left\{ L(x, q) + \frac{\|p - q\|_*^2}{2\lambda} + \frac{\lambda}{2}\|x\|^2; q \in X^* \right\}.$$

Since $0 \in \text{Dom}(L)$, the Lagrangian L and consequently L_λ^2 and therefore $H_{L_\lambda^2}(0, .)$ are bounded from below. Also, we have

$$\lim_{\|x\| \to +\infty} H_{L_\lambda^2}(0, -x) + \langle \Lambda x + \varepsilon Tx, x \rangle = +\infty$$

since $\langle \Lambda x + \varepsilon Tx, x \rangle \geq -w(\|x\|)\|x\| + \varepsilon w(\|x\|)\|x\|^2$.

Moreover, the map $\Lambda + \varepsilon T$ is pseudoregular, and therefore, from Theorem 12.3 and the remark following it, there exists $x_{\varepsilon,\lambda}$ such that

$$L_\lambda^2(x_{\varepsilon,\lambda}, -\Lambda x_{\varepsilon,\lambda} - \varepsilon T x_{\varepsilon,\lambda}) + \langle \Lambda x_{\varepsilon,\lambda} + \varepsilon T x_{\varepsilon,\lambda}, x_{\varepsilon,\lambda} \rangle = 0,$$

which means that $\Lambda x_{\varepsilon,\lambda} + \varepsilon T x_{\varepsilon,\lambda} \in -\overline{\partial} L_\lambda^2(x_{\varepsilon,\lambda})$, and therefore

$$\Lambda x_{\varepsilon,\lambda} + \varepsilon T x_{\varepsilon,\lambda} + \lambda D x_{\varepsilon,\lambda} \in -\overline{\partial} L(x_{\varepsilon,\lambda}). \tag{17.5}$$

This together with (17.1) implies $\langle \varepsilon T x_{\varepsilon,\lambda} + \lambda D x_{\varepsilon,\lambda}, x_{\varepsilon,\lambda} \rangle \leq C \|x_{\varepsilon,\lambda}\|$, thereby giving

$$\varepsilon w(\|x_{\varepsilon,\lambda}\|) \|x_{\varepsilon,\lambda}\|^2 + \lambda \|x_{\varepsilon,\lambda}\|^2 \leq C \|x_{\varepsilon,\lambda}\|,$$

which in turn implies that $(T x_{\varepsilon,\lambda})_\varepsilon$ and $(x_{\varepsilon,\lambda})_\varepsilon$ are bounded. Since now Λ is a bounded operator, we get that $(\Lambda x_{\varepsilon,\lambda})_\varepsilon$ is bounded in X^*. Suppose, up to a subsequence, that $x_{\varepsilon,\lambda} \rightharpoonup x_\lambda$ and $\Lambda x_{\varepsilon,\lambda} \rightharpoonup p_\lambda$. Then, it follows from Lemma 17.1 and since D is also regular that, for every $\lambda > 0$, we have

$$0 \in \Lambda x_\lambda + \lambda D x_\lambda + \overline{\partial} L(x_\lambda).$$

Remark 17.2. Note that we do not really need Λ to be a bounded operator but rather a weaker condition of the form $\|\Lambda x\| \leq C H_L(0,x) + w(\|x\|)$ for some nondecreasing function w and some constant $C > 0$.

The theorem above justifies the following weakened notions of coercivity. It can be seen as a self-dual version of the classical Palais-Smale condition in standard variational problems. Indeed, if I is a self-dual functional of the form $I(x) = L(x, -\Lambda x) + \langle x, \Lambda x \rangle$, then its *stationary points* correspond to when $I(\bar{x}) = \inf_{x \in D(\Lambda)} I(x) = 0$, in

which case they satisfy the equation $0 \in \overline{\partial} L(\bar{x}) + \Lambda \bar{x}$. So by analogy to classical variational theory, we introduce the following.

Definition 17.1. Given a map $\Lambda : D(\Lambda) \subset X \to X^*$ and a Lagrangian L on $X \times X^*$:

1. Say that $(x_n)_n$ is a *self-dual Palais-Smale sequence* for the functional $I_{L,\Lambda}(x) = L(x, -\Lambda x) + \langle x, \Lambda x \rangle$ if for some $\varepsilon_n \to 0$ it satisfies

$$\Lambda x_n + \overline{\partial} L(x_n) = -\varepsilon_n D x_n. \tag{17.6}$$

2. The functional $I_{L,\Lambda}$ is said to satisfy the *self-dual Palais-Smale condition* (self-dual-PS) if every self-dual Palais-Smale sequence for $I_{L,\Lambda}$ is bounded in X.
3. The functional $I_{L,\Lambda}$ is said to be *weakly coercive* if

$$\lim_{\|x_n\| \to +\infty} L\left(x_n, -\Lambda x_n - \frac{1}{n} D x_n\right) + \langle x_n, \Lambda x_n \rangle + \frac{1}{n} \|x_n\|^2 = +\infty. \tag{17.7}$$

Remark 17.3. (1) It is clear that a weakly coercive functional necessarily satisfies the self-dual Palais-Smale condition.

(2) On the other hand, a strongly coercive self-dual functional is necessarily weakly coercive. Indeed, recall that strong coercivity means that

$$\lim_{\|x\| \to +\infty} H_L(0,x) + \langle \Lambda x, x \rangle = +\infty, \tag{17.8}$$

and so in order to show that condition (17.8) is stronger than (17.7), write for each $(x,p) \in X \times X^*$

$$L(x,p) = \sup\{\langle y, p \rangle - H_L(x,y); y \in X\} \geq -H_L(x,0) \geq H_L(0,x)$$

in such a way that if $\|x_n\| \to +\infty$, then

$$\lim_{n\to+\infty} L\left(x_n, -\Lambda x_n - \frac{1}{n}Dx_n\right) + \langle x_n, \Lambda x_n\rangle + \frac{1}{n}\|x_n\|^2 \geq \lim_{n\to+\infty} H_L(0, x_n) + \langle \Lambda x_n, x_n\rangle = +\infty.$$

Corollary 17.1. *Under the conditions of Theorem 17.1, if the functional $I_{L,\Lambda}$ also satisfies the self-dual Palais-Smale condition, then it attains its infimum at $\bar{x} \in D(\Lambda)$ in such a way that $I(\bar{x}) = \inf_{x \in D(\Lambda)} I(x) = 0$ and $0 \in \Lambda\bar{x} + \overline{\partial}L(\bar{x})$.*

Proof. Since $I_{L,\Lambda}$ has the self-dual Palais-Smale condition, the family $(x_\lambda)_\lambda$ obtained in Theorem 17.1 is bounded in X and therefore converges weakly – up to a subsequence – to $\bar{x} \in X$. Again, since Λ is a bounded operator, Λx_λ is also bounded in X^*, and then Lemma 17.1 yields $L(\bar{x}, -\Lambda\bar{x}) + \langle \Lambda\bar{x}, \bar{x}\rangle = 0$, which means that $-\Lambda\bar{x} \in \overline{\partial}L(\bar{x})$.

Remark 17.4. Note that Theorem 17.1 is indeed an extension of Theorem 12.3 since we have that for large $\|x\|$

$$\langle \overline{\partial}L(x) + \Lambda x, x\rangle = L(x, \overline{\partial}L(x)) + \langle \Lambda x, x\rangle \geq H_L(0, x) + \langle \Lambda x, x\rangle \geq -C(\|x\| + 1),$$

which means that condition (17.1) is also implied by (17.8).

17.2 Superposition of a regular map with an unbounded linear operator

We now deal with the difficulties arising from the superposition of an unbounded linear operator $A : D(A) \subset X \to X^*$ with another – possibly nonlinear – map $\Lambda : D(\Lambda) \subset X \to X^*$ when trying to resolve an equation of the form

$$Au + \Lambda u \in -\overline{\partial}L(u) \tag{17.9}$$

by considering the functional

$$I(u) = L(u, -Au - \Lambda u) + \langle u, Au + \Lambda u\rangle. \tag{17.10}$$

If now A is skew-adjoint, then the functional can be written as

$$I(x) = L_A(x, -\Lambda x) + \langle x, \Lambda x\rangle, \tag{17.11}$$

where – under the appropriate conditions – L_A is the self-dual Lagrangian $L_A(x, p) = L(x, -Ax + p)$ when $x \in D(A)$ and $+\infty$ elsewhere. However, if A is an unbounded operator, the domain $\text{Dom}_1(L_A)$ of the new Lagrangian L_A is not necessarily closed, even when $\text{Dom}_1(L)$ is, and its closure may not be contained in $D(\Lambda)$ and as such Theorem 17.1 could not apply.

If now A has a closed graph, then we can strengthen the topology on X by considering the space Y_A, which is the completion of $D(A)$ for the norm $\|u\|_{Y_A}^2 =$

$\|u\|_X^2 + \|Au\|_{X^*}^2$. This then makes $\mathrm{Dom}_1(L_A)$ closed in Y_A and increases the chance for Λ to be regular on the new space Y_A. Moreover, L_A will still induce a self-dual Lagrangian on $Y_A \times Y_A^*$. On the other hand, this may lead to a loss of the *strong coercivity* on the new space since it requires that

$$\lim_{\|x\| + \|Ax\| \to +\infty} H_L(0, -x) + \langle Ax + \Lambda x, x \rangle = +\infty. \tag{17.12}$$

However, there are cases where the approach is still manageable, such as when the functional is weakly coercive or when it satisfies the *self-dual (PS)-condition* on Y_A. The following is a situation where this may happen.

Theorem 17.2. *Let $A : D(A) \subset X \to X^*$ be a closed linear operator on a reflexive Banach space X with a dense domain, and let Λ be a map from $D(A)$ into X^* that induces a pseudoregular operator $\Lambda : Y_A \to X^*$, while satisfying for some constant $0 < k < 1$, and a nondecreasing function w,*

$$\|\Lambda x\|_{X^*} \le k\|Ax\|_{X^*} + w(\|x\|_X) \tag{17.13}$$

and

$$x \to \langle Ax + \Lambda x, x \rangle \text{ is bounded below on } D(A). \tag{17.14}$$

Suppose L is a self-dual Lagrangian on $X \times X^*$ such that for some $C_1, C_2 > 0$ and $r_1 \ge r_2 > 1$ we have

$$C_1(\|x\|_X^{r_2} - 1) \le L(x, 0) \le C_2(1 + \|x\|_X^{r_1}) \text{ for all } x \in X. \tag{17.15}$$

The functional $I(x) = L(x, -Ax - \Lambda x) + \langle x, Ax + \Lambda x \rangle$ then attains its minimum at some $\bar{x} \in D(A)$ such that $I(\bar{x}) = \inf_{x \in Y_A} I(x) = 0$ and

$$0 \in \Lambda \bar{x} + A\bar{u} + \bar{\partial} L(\bar{x}). \tag{17.16}$$

We shall need the following.

Lemma 17.2. *Let $A : D(A) \subset X \to X^*$ be a closed linear operator on a reflexive Banach space X with a dense domain, and let Λ be a map from $D(A)$ into X^* that induces a bounded pseudoregular operator $\Lambda : Y_A \to Y_A^*$. Suppose L is a self-dual Lagrangian on $X \times X^*$ that satisfies the following conditions:*

For each $p \in \mathrm{Dom}_2(L)$, the functional $x \to L(x, p)$ is continuous on X, (17.17)

$$x \to L(x, 0) \text{ is bounded on the unit ball of } X, \tag{17.18}$$

and

$$\langle \bar{\partial} L(x) + Ax + \Lambda x, x \rangle \ge -C(1 + \|x\|_{Y_A}). \tag{17.19}$$

Then, for every $\lambda > 0$, there exists $x_\lambda \in Y_A$ that satisfies the equation

$$0 \in \Lambda x_\lambda + Ax_\lambda + \bar{\partial} L(x_\lambda) + \lambda D_{Y_A} x_\lambda. \tag{17.20}$$

Proof. Note first that $Y_A \subseteq X \subseteq X^* \subseteq Y_A^*$, and the injections are all continuous with dense range. We first show that the Lagrangian

$$\mathscr{M}(u,p) := \begin{cases} L(u,p), & p \in X^* \\ +\infty & p \in Y_A^* \setminus X^* \end{cases}$$

is a self-dual Lagrangian on $Y_A \times Y_A^*$. Indeed, if $q \in X^*$, use the fact that $D(A)$ is dense in X and that the functional $x \to L(x,p)$ is continuous on X to write

$$\begin{aligned} \mathscr{M}^*(q,v) &= \sup\{\langle u,q \rangle + \langle v,p \rangle - \mathscr{M}(u,p); (u,p) \in Y_A \times Y_A^*\} \\ &= \sup\{\langle u,q \rangle + \langle v,p \rangle - L(u,p); (u,p) \in X \times X^*\} \\ &= L^*(q,v) = L(v,q) = \mathscr{M}(v,q). \end{aligned}$$

If now $q \in Y_A^* \setminus X^*$, then there exists $\{x_n\}_n \subseteq Y_A$ with $\|x_n\|_X \leq 1$ such that $\lim_{n \to +\infty} \langle x_n, q \rangle \to +\infty$. Since $\{L(x_n, 0)\}_n$ is bounded, it follows that

$$\begin{aligned} \mathscr{M}^*(q,v) &= \sup\{\langle u,q \rangle + \langle v,p \rangle - \mathscr{M}(u,p); (u,p) \in Y_A \times X^*\} \\ &\geq \sup\{\langle x_n, q \rangle - L(x_n, 0)\} \\ &= +\infty = \mathscr{M}(v,q), \end{aligned}$$

and $\mathscr{M}$ is therefore self-dual on $Y_A \times Y_A^*$. Since $A + \Lambda$ is now pseudoregular on Y_A, we can apply Theorem 17.1 and obtain the claimed result.

Proof of Theorem 17.2: All the hypotheses in Lemma 17.2 are readily satisfied except condition (17.19). For that, note that (17.15) yields via Proposition 6.1 that $\partial L(0) \neq \emptyset$. It then follows from the monotonicity of ∂L and (17.14) that

$$\begin{aligned} \langle \partial L(x) + \Lambda x + Ax, x \rangle &\geq \langle \partial L(0), x \rangle + \langle \Lambda x + Ax, x \rangle \\ &\geq -C(1 + \|x\|_X) \\ &\geq -C(1 + \|x\|_{Y_A}). \end{aligned}$$

By Lemma 17.2, there exists then a self-dual Palais-Smale sequence for I on Y_A, and it remains to show that I is weakly coercive on Y_A. For that we assume that $(x_n)_n$ is a sequence in Y_A such that

$$L\left(x_n, -Ax_n - \Lambda x_n - \frac{1}{n}D_{Y_A}x_n\right) + \langle x_n, Ax_n + \Lambda x_n \rangle + \frac{1}{n}\|x_n\|_{Y_A}^2$$

is bounded, where here $D_A : Y_A \to Y_A^*$ is the duality map. It follows from (17.14) that $\langle Ax_n + \Lambda x_n, x_n \rangle + \frac{1}{n}\|x_n\|_{Y_A}^2$ is bounded below, and therefore the sequence $L(x_n, -Ax_n - \Lambda x_n - \frac{1}{n}D_Ax_n)$ is bounded from above. From (17.15), we get via Lemma 3.5 that there exist $D_1, D_2 > 0$ such that

$$D_1(\|p\|_X^{s_1} + \|x\|_X^{r_2} - 1) \leq L(x,p) \leq D_2(1 + \|x\|_X^{r_1} + \|p\|_X^{s_2}), \tag{17.21}$$

where $\frac{1}{r_i} + \frac{1}{s_i} = 1$ for $i = 1, 2$. It follows that both $(x_n)_n$ and $(Ax_n + \Lambda x_n + \frac{1}{n}D_{Y_A}x_n)_n$ are bounded in X and X^*, respectively. This, coupled with (17.13), yields that

$$\|Ax_n\|_{X^*} \leq \|\Lambda x_n + Ax_n + \frac{1}{n}D_{Y_A}x_n\|_{X^*} + \|\Lambda x_n + \frac{1}{n}D_{Y_A}x_n\|_{X^*}$$

$$\leq C + \|\Lambda x_n\|_{X^*} + \frac{1}{n}\|D_{Y_A}x_n\|_{X^*}$$

$$\leq C + k\|Ax_n\|_{X^*} + w(\|x_n\|_X) + \frac{1}{n}\|x_n\|_X + \frac{1}{n}\|Ax_n\|_{X^*}.$$

Hence, $(1 - k - \frac{1}{n})\|Ax_n\|_{X^*} \leq C + w(\|x_n\|_X) + \frac{1}{n}\|x_n\|_X$, and therefore $\|Ax_n\|_{X^*}$ is bounded, which implies the boundedness of $\{x_n\}$ in Y_A, and therefore I is weakly coercive on Y_A.

In the case of the basic self-dual Lagrangian $L(x, p) = \varphi(x) + \varphi^*(p)$, one can relax the strong boundedness condition on L and obtain the following useful corollary.

Corollary 17.2. *Let $A : D(A) \subset X \to X^*$ be a closed linear operator on a reflexive Banach space X with a dense domain, and let Λ be a map from $D(A)$ into X^* that induces a pseudoregular operator $\Lambda : Y_A \to X^*$, while verifying conditions (17.13) and (17.14). Let φ be a proper, convex, lower semicontinuous function such that φ is coercive as well as bounded on the bounded sets of X^*.*

Then, a solution $\bar{x} \in D(A)$ to the equation $0 \in \Lambda x + Ax + \partial \varphi(x)$ can be obtained as a minimizer on $D(A)$ of the functional

$$I(x) = \varphi(x) + \varphi^*(-\Lambda x - Ax) + \langle x, \Lambda x + Ax \rangle. \tag{17.22}$$

Proof. We assume without loss of generality that $0 \in \text{Dom}(\partial \varphi)$ and use the fact that $\partial \varphi$ is monotone, and that $\langle \Lambda x + Ax, x \rangle$ is bounded from below, to write

$$\langle \partial \varphi(x) + \Lambda x + Ax, x \rangle \geq \langle \partial \varphi(0), x \rangle + \langle \Lambda x + Ax, x \rangle$$

$$\geq -C(1 + \|x\|_X)$$

$$\geq -C(1 + \|x\|_{Y_A}).$$

The corollary is now a consequence of Lemma 17.2 applied to the self-dual Lagrangian $L(x, p) = \varphi(x) + \varphi^*(p)$, provided we prove that I is weakly coercive on Y_A. For that, we suppose $\{x_n\}_n \subseteq Y_A$ is such that $\|x_n\|_{Y_A} \to \infty$. We show that

$$\varphi(x_n) + \varphi^*\left(-\Lambda x_n - Ax_n - \frac{1}{n}D_{Y_A}x_n\right) + \langle x_n, \Lambda x_n + Ax_n \rangle + \frac{1}{n}\|x_n\|_{Y_A} \to \infty.$$

Indeed if not, and since $\langle x_n, \Lambda x_n + Ax_n \rangle + \frac{1}{n}\|x_n\|_X$ is bounded below, we have that $\varphi(x_n) + \varphi^*(-\Lambda x_n - Ax_n - \frac{1}{n}D_{Y_A}x_n)$ is bounded from above. The coercivity of φ on X then ensures the boundedness of $\{\|x_n\|_X\}_n$. In order to show that $\{x_n\}$ is actually bounded in Y_A, we use that φ^* is coercive in X^* to get that

$$\|\Lambda x_n + Ax_n + \frac{1}{n}D_{Y_A}x_n\|_{X^*} \leq C$$

for some constant $C > 0$. This combined with (17.13) yields that

$$\|Ax_n\|_{X^*} \leq \|\Lambda x_n + Ax_n + \frac{1}{n}D_{Y_A}x_n\|_{X^*} + \|\Lambda x_n + \frac{1}{n}D_{Y_A}x_n\|_{X^*}$$

$$\leq C + \|\Lambda x_n\|_{X^*} + \frac{1}{n}\|D_{Y_A}x_n\|_{X^*}$$

$$\leq C + k\|Ax_n\|_{X^*} + w(\|x_n\|_X) + \frac{1}{n}\|x_n\|_X + \frac{1}{n}\|Ax_n\|_{X^*}.$$

Hence, $(1 - k - \frac{1}{n})\|Ax_n\|_{X^*} \leq C + w(\|x_n\|_X) + \frac{1}{n}\|x_n\|_X$, and therefore $\|Ax_n\|_{X^*}$ is bounded, which implies the boundedness of $\{x_n\}$ in Y_A, and we are done.

The following is an application to the resolution of certain nonlinear systems.

Corollary 17.3. *Let X_1 and X_2 be two reflexive Banach spaces, and consider a convex lower semicontinuous function φ on $X_1 \times X_2$ that is bounded on the balls. Let $A : D(A) \subset X_1 \to X_2^*$ be a linear operator, and consider the spaces*

$$\tilde{X}_1 := \{x \in X_1; Ax \in X_2^*\} \text{ and } \tilde{X}_2 := \{x \in X_2; A^*x \in X_1^*\}$$

equipped with the norms $\|x\|_{\tilde{X}_1} = \|x\|_{X_1} + \|A_1x\|_{X_2^*}$ and $\|p\|_{\tilde{X}_2} = \|p\|_{X_2^*} + \|A^*p\|_{X_1}$.

Let $A_1 : D(A_1) \subset X_1 \to X_1^*$ and $A_2 : D(A_2) \subset X_2 \to X_2^*$ be two positive linear operators, and consider for $i = 1, 2$ the Banach spaces

$$Y_i := \{x \in \tilde{X}_i; A_ix \in X_i^*\}$$

equipped with the norm $\|y\|_{Y_i} = \|y\|_{\tilde{X}_i} + \|A_iy\|_{X_i^*}$.

Assume $\Lambda := (\Lambda_1, \Lambda_2) : Y_1 \times Y_2 \to Y_1^* \times Y_2^*$ is a pseudoregular operator such that

$$\lim_{\|x\|_{X_1} + \|y\|_{X_2} \to \infty} \frac{\varphi(x, y) + \langle A_1x, x \rangle + \langle A_2y, y \rangle + \langle \Lambda(x, y), (x, y) \rangle}{\|x\|_{X_1} + \|y\|_{X_2}} = +\infty \quad (17.23)$$

and

$$\|(\Lambda_1, \Lambda_2)(x, y)\|_{X_1^* \times X_2^*} \leq k\|(A_1, A_2)(x, y)\|_{X_1^* \times X_2^*} + w(\|(x, y)\|_{X_1 \times X_2}) \quad (17.24)$$

for some continuous and nondecreasing function w and some constant $0 < k < 1$. Then, for any $(f, g) \in Y_1^* \times Y_2^*$, there exists $(\bar{x}, \bar{y}) \in Y_1 \times Y_2$, which solves the system

$$\begin{cases} -\Lambda_1(x, y) - A^*y - A_1x + f \in \partial_1\varphi(x, y) \\ -\Lambda_2(x, y) + Ax - A_2y + g \in \partial_2\varphi(x, y). \end{cases} \quad (17.25)$$

The solution is obtained as a minimizer on $Y_1 \times Y_2$ of the functional

$$I(x, y) = \psi(x, y) + \psi^*(-A^*y - A_1x - \Lambda_1(x, y), Ax - A_2y - \Lambda_2(x, y))$$
$$+ \langle A_1x, x \rangle + \langle A_2y, y \rangle + \langle \Lambda(x, y), (x, y) \rangle,$$

where $\psi(x, y) = \varphi(x, y) - \langle f, x \rangle - \langle g, y \rangle$.

Proof. Let $X = X_1 \times X_2$ in such a way that $\tilde{A}(x,y) = (-A^*y, Ax) : X_1 \times X_2 \to X^*$ is a skew-adjoint operator that is bounded on the space $Y_{\tilde{A}} = \tilde{X} = \tilde{X}_1 \times \tilde{X}_2$. Consider the following self-dual Lagrangian on $\tilde{X} \times \tilde{X}^*$:

$$L((x,y),(p,q)) = \psi(x,y) + \psi^*(-A^*y + p, Ax + q).$$

Setting $\mathscr{A} := (A_1, A_2)$, we have that $Y_1 \times Y_2 = \tilde{X}_{\mathscr{A}}$, and the operator $\mathscr{A} + \Lambda$ is regular on $Y_1 \times Y_2$. Corollary 17.2 yields that

$$I(x,y) = L((x,y), -\Lambda(x,y) - \tilde{A}(x,y)) + \langle \Lambda(x,y) + \tilde{A}(x,y), (x,y) \rangle$$

attains its minimum at some point $(\bar{x}, \bar{y}) \in Y_1 \times Y_2$ and that the minimum is 0. In other words,

$$
\begin{aligned}
0 &= I(\bar{x}, \bar{y}) \\
&= \psi(\bar{x}, \bar{y}) + \psi^*(-A^*\bar{y} - A_1\bar{x} - \Lambda_1(\bar{x}, \bar{y}), A\bar{x} - A_2\bar{y} - \Lambda_2(\bar{x}, \bar{y})) \\
&\quad + \langle \Lambda(\bar{x}, \bar{y}) + A(\bar{x}, \bar{y}), (\bar{x}, \bar{y}) \rangle \\
&= \psi(\bar{x}, \bar{y}) + \psi^*(-A^*\bar{y} - A_1\bar{x} - \Lambda_1(\bar{x}, \bar{y}), A\bar{x} - A_2\bar{y} - \Lambda_2(\bar{x}, \bar{y})) \\
&\quad + \langle (\Lambda_1(\bar{x}, \bar{y}) + A_1\bar{x} - A^*\bar{y}, \Lambda_2(\bar{x}, \bar{y}) + A_2\bar{y} + A\bar{x}), (\bar{x}, \bar{y}) \rangle,
\end{aligned}
$$

from which it follows that

$$
\begin{cases}
-A^*y - A_1 x - \Lambda_1(x,y) \in \partial_1 \varphi(x,y) - f \\
Ax - A_2 y - \Lambda_2(x,y) \in \partial_2 \varphi(x,y) - g.
\end{cases}
\tag{17.26}
$$

Example 17.1. A variational resolution for doubly nonlinear coupled equations

Let $\mathbf{b_1} : \Omega \to \mathbf{R^n}$ and $\mathbf{b_2} : \Omega \to \mathbf{R^n}$ be two compactly supported smooth vector fields on the neighborhood of a bounded domain Ω of $\mathbf{R^n}$. Consider the Dirichlet problem

$$
\begin{cases}
\Delta v + \mathbf{b_1} \cdot \nabla u = |u|^{p-2}u + u^{m-1}v^m + f & \text{on} \quad \Omega, \\
-\Delta u + \mathbf{b_2} \cdot \nabla v = |v|^{p-2}v - u^m v^{m-1} + g & \text{on} \quad \Omega, \\
\qquad\qquad u = v = 0 & \text{on} \quad \partial\Omega.
\end{cases}
\tag{17.27}
$$

We can use Corollary 17.3 to get the following.

Theorem 17.3. *Assume f, g in L^p, $p \geq 2$, $\mathrm{div}(\mathbf{b_1}) \geq 0$, $\mathrm{div}(\mathbf{b_2}) \geq 0$ on Ω, and that $1 \leq m < \frac{p-1}{2}$. Let $Y = \{u \in H_0^1(\Omega); u \in L^p(\Omega) \text{ and } \Delta u \in L^q(\Omega)\}$, and consider on $Y \times Y$ the functional*

$$I(u,v) = \Psi(u) + \Psi^*(\mathbf{b_1}.\nabla u + \Delta v - u^{m-1}v^m) + \Phi(v) + \Phi^*(\mathbf{b_2}.\nabla v - \Delta u + u^m v^{m-1})$$
$$+ \frac{1}{2}\int_\Omega \mathrm{div}(\mathbf{b_1})\,|u|^2 dx + \frac{1}{2}\int_\Omega \mathrm{div}(\mathbf{b_1})\,|v|^2 dx,$$

where

$$\Psi(u) = \frac{1}{p}\int_\Omega |u|^p dx + \int_\Omega f u dx \quad \text{and} \quad \Phi(v) = \frac{1}{p}\int_\Omega |v|^p dx + \int_\Omega g v dx$$

are defined on $L^p(\Omega)$ *and* Ψ^* *and* Φ^* *are their respective Legendre transforms in* $L^q(\Omega)$. *Then, there exists* $(\bar{u}, \bar{v}) \in Y \times Y$ *such that*

$$I(\bar{u}, \bar{v}) = \inf\{I(u, v); (u, v) \in Y \times Y\} = 0$$

and $(\bar{u}, \bar{v})$ *is a solution of problem* (17.27).

Proof. With the notation of Corollary 17.3, set $X_1 = X_2 = L^p(\Omega)$, $A = \Delta : D(A) \subset L^p(\Omega) \to L^q(\Omega)$ in such a way that $\tilde{X}_1 = \tilde{X}_2 = Y$. We note that the operators $A_1 u := \mathbf{b_1}.\nabla u$ and $A_2 u := \mathbf{b_2}.\nabla u$ are bounded on $\tilde{X}_i$ in such a way that $Y_i = \tilde{X}_i = Y$.

The functions Φ and Ψ are continuous and coercive on X_i, and so it remains to verify condition (17.24). Indeed, by Hölder's inequality, for $q = \frac{p}{p-1} \leq 2$ we obtain

$$\|u^m v^{m-1}\|_{L^q(\Omega)} \leq \|u\|^m_{L^{2mq}(\Omega)} \|v\|^{(m-1)}_{L^{2(m-1)q}(\Omega)},$$

and since $m < \frac{p-1}{2}$, we have $2mq < p$ and therefore

$$\|u^m v^{m-1}\|_{L^q(\Omega)} \leq C\left(\|u\|^{2m}_{L^p(\Omega)} + \|v\|^{2(m-1)}_{L^p(\Omega)}\right). \tag{17.28}$$

Also since $q \leq 2$,

$$\|\mathbf{b_1}.\nabla u\|_{L^q(\Omega)} \leq C\|\mathbf{b_1}\|_{L^\infty(\Omega)}\|\nabla u\|_{L^2(\Omega)}$$

$$\leq C\|\mathbf{b_1}\|_{L^\infty(\Omega)}\left(\int \langle -\Delta u, u\rangle dx\right)^{\frac{1}{2}} \leq C\|\mathbf{b_1}\|_{L^\infty(\Omega)}\|u\|^{\frac{1}{2}}_{L^p(\Omega)}\|\Delta u\|^{\frac{1}{2}}_{L^q(\Omega)}$$

$$\leq k\|\Delta u\|_{L^q(\Omega)} + C(k)\|\mathbf{b_1}\|^2_{L^\infty(\Omega)}\|u\|_{L^p(\Omega)} \tag{17.29}$$

for some $0 < k < 1$. Condition (17.24) now follows from (17.28) and (17.29).

Finally, it is easy to verify that the nonlinear operator

$$\Lambda(u, v) = (-u^{m-1}v^m + \mathbf{b_1}.\nabla u, u^m v^{m-1} + \mathbf{b_2}.\nabla v)$$

is regular from $Y \times Y \to L^q(\Omega) \times L^q(\Omega)$. It is worth noting that there is no restriction here on the power p, which can well be beyond the critical Sobolev exponent.

17.3 Superposition of a nonlinear map with a skew-adjoint operator modulo boundary terms

Consider now equations of the form

$$\begin{cases} Ax + \Lambda x \in -\bar{\partial}L(x) \\ R\mathcal{B}x \in \partial\ell(\mathcal{B}x), \end{cases} \tag{17.30}$$

where $\Lambda : D(\Lambda) \subset X \to X^*$ is a nonlinear operator, $A : D(A) \subset X \to X^*$ is a linear skew-adjoint operator modulo a boundary triplet $(H,R,\mathscr{B})$, where R is an automorphism on a boundary space H, and $\mathscr{B} : D(\mathscr{B}) \subset X \to H$ is a boundary operator. Here again L is a self-dual Lagrangian on phase space $X \times X^*$ and ℓ is an R-self-dual function on the boundary space H.

We consider the nonnegative functional

$$I(x) = \begin{cases} L(x, -Ax - \Lambda x) + \langle x, \Lambda x \rangle + \ell(\mathscr{B}x) & \text{if } x \in D(A) \cap D(\mathscr{B}) \\ +\infty & \text{if } x \notin D(A) \cap D(\mathscr{B}) \end{cases} \qquad (17.31)$$

and note that it suffices to show that there exists $\bar{x} \in D(\Lambda) \cap D(A) \cap D(\mathscr{B})$ such that $I(\bar{x}) = 0$ since then

$$0 = L(\bar{x}, -A\bar{x} - \Lambda\bar{x}) + \langle \bar{x}, \Lambda\bar{x} \rangle + \langle \bar{x}, A\bar{x} \rangle - \langle \bar{x}, A\bar{x} \rangle + \ell(\mathscr{B}\bar{x})$$

$$= L(\bar{x}, -A\bar{x} - \Lambda\bar{x}) + \langle \bar{x}, A\bar{x} + \Lambda\bar{x} \rangle - \frac{1}{2}\langle \mathscr{B}\bar{x}, R\mathscr{B}\bar{x} \rangle + \ell(\mathscr{B}\bar{x}).$$

Since $L(x, p) \geq \langle x, p \rangle$ and $\ell(s) \geq \frac{1}{2}\langle s, Rs \rangle$, we get

$$\begin{cases} L(\bar{x}, -A\bar{x} - \Lambda\bar{x}) = -\langle \bar{x}, A\bar{x} + \Lambda\bar{x} \rangle \\ \ell(\mathscr{B}\bar{x}) = \frac{1}{2}\langle \mathscr{B}\bar{x}, R\mathscr{B}\bar{x} \rangle, \end{cases} \qquad (17.32)$$

and are done with solving equation (17.30).

In order to apply Theorem 12.3, we note that I can also be written as

$$I(u) = L_{A,\ell}(u, -\Lambda u) + \langle x, \Lambda x \rangle,$$

where

$$L_{A,\ell}(x, p) = \begin{cases} L(x, -Ax + p) + \ell(\mathscr{B}x) & \text{if } x \in D(\Gamma) \cap D(\mathscr{B}) \\ +\infty & \text{if } x \notin D(\Gamma) \cap D(\mathscr{B}). \end{cases} \qquad (17.33)$$

According to Proposition 4.2 and under the right conditions on L and A, $L_{A,\ell}$ is again a self-dual Lagrangian on $X \times X^*$. However, the basic self-dual variational principle may not apply if the topology of X is not strong enough to make Λ regular on X, or/and to keep the closure of $\mathrm{Dom}_1(L_{A,\ell})$ contained in the domain of Λ.

So again, we consider a space $Y_{A,\mathscr{B}}$ with a strong enough topology to make $D(A) \cap D(\mathscr{B})$ closed and at the same time increase the chance for Λ to be regular on the new space $Y_{A,\mathscr{B}}$. Moreover, it is expected that, just as in Lemma 17.2, the functional I remains self-dual on $Y_{A,\mathscr{B}}$. The main problem, however, is that this may again lead to a loss of coercivity on the new space, but there are situations where the functional satisfies the self-dual Palais-Smale condition on $Y_{A,\mathscr{B}}$, which allows us to conclude.

We will not carry on this analysis in full generality here, but the approach will be considered in detail in the next chapter, while dealing with nonlinear evolution equations, in which case $Au = \dot{u}$.

Exercise 17.A. More on perturbations

Establish a counterpart of Theorem 17.2 in the case where the linear operator A is skew-adjoint modulo boundary terms, and establish a self-dual variational principle for the boundary value problem (17.30) under minimal hypotheses.

Further comments

This chapter is reminiscent of the work of Brézis [27], Brézis-Crandall-Pazy [28], and many others dealing with the sum of certain unbounded operators, or with perturbations of maximal monotone operators.

Chapter 18
Navier-Stokes and other Self-dual Nonlinear Evolutions

The nonlinear self-dual variational principle established in Chapter 12 – though good enough to be readily applicable in many stationary nonlinear partial differential equations – did not, however, cover the case of nonlinear evolutions such as the Navier-Stokes equations. One of the reasons is the prohibitive coercivity condition that is not satisfied by the corresponding self-dual functional on the path space $\mathscr{X}_{p,q}$. We show here that such a principle still holds for functionals of the form

$$I(u) = \int_0^T \left[L(t, u(t), -\dot{u}(t) - \Lambda u(t)) + \langle \Lambda u(t), u(t) \rangle \right] dt$$
$$+ \ell \left(u(0) - u(T), -\frac{u(T) + u(0)}{2} \right),$$

where L (resp., ℓ) is a self-dual Lagrangian on state space (resp., boundary space) and Λ is an appropriate nonlinear operator on path space. As a consequence, we provide a variational formulation and resolution to evolution equations involving nonlinear operators such as the Navier-Stokes equation (in dimensions 2 and 3) with various boundary conditions. In dimension 2, we recover the well-known weak solutions for the corresponding initial-value problem as well as periodic and antiperiodic ones, while in dimension 3 we get Leray solutions for the initial-value problems but also solutions satisfying $u(0) = \delta u(T)$ for any given δ in $(-1, 1)$. The approach is quite general and applies to certain nonlinear Schrödinger equations and many other evolutions.

18.1 Elliptic perturbations of self-dual functionals

Let $X \subset H \subset X^*$ be an evolution pair, $p, q > 1$ such that $\frac{1}{p} + \frac{1}{q} = 1$, and consider a time-dependent self-dual Lagrangian L on $[0, T] \times X \times X^*$ such that:

For each $r \in L_{X^*}^q$, the map $u \to \int_0^T L(t, u(t), r(t)) dt$ is continuous on L_X^p.   (18.1)

The map $u \to \int_0^T L(t, u(t), 0) dt$ is bounded on the unit ball of L_X^p.   (18.2)

N. Ghoussoub, *Self-dual Partial Differential Systems and Their Variational Principles*,   319
Springer Monographs in Mathematics, DOI 10.1007/978-0-387-84897-6_18,
© Springer Science+Business Media, LLC 2009

Let ℓ be a self-dual Lagrangian on $H \times H$ such that

$$-C \leq \ell(a,b) \leq C(1 + \|a\|_H^2 + \|b\|_H^2) \text{ for all } (a,b) \in H \times H. \tag{18.3}$$

Recall from Proposition 4.3 that the Lagrangian

$$\mathscr{L}(u,r) = \begin{cases} \int_0^T L(t,u(t),r(t) - \dot{u}(t))dt + \ell\left(u(0) - u(T), -\frac{u(T)+u(0)}{2}\right) & \text{if } u \in \mathscr{X}_{p,q} \\ +\infty & \text{otherwise} \end{cases}$$

is then self-dual on $L_X^p \times L_{X^*}^q$.

Consider now the convex lower semicontinuous function on L_X^p

$$\psi(u) = \begin{cases} \frac{1}{q} \int_0^T \|\dot{u}(t)\|_{X^*}^q \, dt & \text{if } u \in \mathscr{X}_{p,q} \\ +\infty & \text{if } u \in L_X^p \setminus \mathscr{X}_{p,q}, \end{cases} \tag{18.4}$$

and, for any $\mu > 0$, we let Ψ_μ be the self-dual Lagrangian on $L_X^p \times L_{X^*}^q$ defined by

$$\Psi_\mu(u,r) = \mu \psi(u) + \mu \psi^*\left(\frac{r}{\mu}\right). \tag{18.5}$$

Recall that, for each $(u,r) \in L_X^p \times L_{X^*}^q$, the Lagrangian

$$\mathscr{L} \oplus \Psi_\mu(u,r) := \inf_{s \in L_{X^*}^q} \{\mathscr{L}(u,s) + \Psi_\mu(u,r-s)\} \tag{18.6}$$

is self-dual on $L_X^p \times L_{X^*}^q$.

Lemma 18.1. *Let L and ℓ be two self-dual Lagrangians verifying (18.1), (18.2) and (18.3), and let $\mathscr{L}$ be the corresponding self-dual Lagrangian on path space $L_X^p \times L_{X^*}^q$. Suppose Λ is regular from $\mathscr{X}_{p,q}$ into $L_{X^*}^q$. Then,*

1. The functional

$$I_\mu(u) = \mathscr{L} \oplus \Psi_\mu(u, -\Lambda u) + \int_0^T \langle \Lambda u(t), u(t)\rangle \, dt$$

is self-dual on $\mathscr{X}_{p,q}$, and its corresponding antisymmetric Hamiltonian on $\mathscr{X}_{p,q} \times \mathscr{X}_{p,q}$ is

$$M_\mu(u,v) := \int_0^T \langle \Lambda u(t), u(t) - v(t)\rangle \, dt + H_{\mathscr{L}}(v,u) + \mu \psi(u) - \mu \psi(v),$$

where $H_{\mathscr{L}}(v,u) = \sup_{r \in L_{X^*}^q} \{\int_0^T \langle r,u\rangle \, dt - \mathscr{L}(v,r)\}$ is the Hamiltonian of $\mathscr{L}$ on $L_X^p \times L_X^p$.

2. If in addition $\lim_{\|u\|_{\mathscr{X}_{p,q}} \to +\infty} \int_0^T \langle \Lambda u(t), u(t)\rangle dt + H_{\mathscr{L}}(0,u) + \mu \psi(u) = +\infty$, then there exists $u \in \mathscr{X}_{p,q}$ with $\partial \psi(u) \in L_{X^*}^q$ such that

$$\dot{u}(t) + \Lambda u(t) + \mu \partial \psi(u(t)) \in -\overline{\partial}L(t,u(t)) \tag{18.7}$$

$$\frac{u(T)+u(0)}{2} \in -\overline{\partial}\ell(u(0)-u(T)) \tag{18.8}$$

$$\dot{u}(T) = \dot{u}(0) = 0. \tag{18.9}$$

Proof. First note that since $\mathscr{L}$ and Ψ_μ are both self-dual Lagrangians, we have that $\mathscr{L}\oplus\Psi_\mu(u,-r)+\langle u,r\rangle \geq 0$ for all $(u,r) \in L_X^p \times L_{X^*}^q$, and therefore $I(u) \geq 0$ on $\mathscr{X}_{p,q}$. Now we have, for any $(u,r) \in L_X^p \times L_{X^*}^q$,

$$\mathscr{L}\oplus\Psi_\mu(u,-r) = \sup_{v\in L_X^p}\left\{\int_0^T \langle -r,v\rangle\,dt + H_\mathscr{L}(v,u) + \mu\psi(u) - \mu\psi(v)\right\}.$$

But for $u \in \mathscr{X}_{p,q}$ and $v \in L_X^p \setminus \mathscr{X}_{p,q}$, we have

$$H_\mathscr{L}(v,u) = \sup_{r\in L_{X^*}^q}\left\{\int_0^T \langle r,u\rangle\,dt - \mathscr{L}(v,r)\right\} = -\infty,$$

and therefore for any $u \in \mathscr{X}_{p,q}$, we have

$$\sup_{v\in\mathscr{X}_{p,q}} M_\mu(u,v) = \sup_{v\in L_X^p} M_\mu(u,v)$$

$$= \int_0^T \langle \Lambda u(t),u(t)\rangle\,dt$$

$$\quad + \sup_{v\in L_X^p}\int_0^T \langle -\Lambda u(t),v(t)\rangle\,dt + H_\mathscr{L}(v,u) + \mu\psi(u) - \mu\psi(v)$$

$$= \int_0^T \langle \Lambda u(t),u(t)\rangle\,dt + \mathscr{L}\oplus\Psi_\mu(u,-\Lambda u)$$

$$= I(u).$$

It follows from Theorem 12.2 that there exists $u_\mu \in \mathscr{X}_{p,q}$ such that

$$I_\mu(u_\mu) = \mathscr{L}\oplus\Psi_\mu(u_\mu,-\Lambda u_\mu) + \int_0^T \langle \Lambda u_\mu(t),u_\mu(t)\rangle\,dt = 0. \tag{18.10}$$

Since $\mathscr{L}\oplus\Psi_\mu$ is convex and coercive in the second variable, there exists $\bar{r} \in L_{X^*}^q$ such that

$$\mathscr{L}\oplus\Psi_\mu(u_\mu,-\Lambda u_\mu) = \mathscr{L}(u_\mu,\bar{r}) + \Psi_\mu(u_\mu,-\Lambda u_\mu - \bar{r}). \tag{18.11}$$

It follows that

$$0 = \mathscr{L}(u_\mu,\bar{r}) + \Psi_\mu(u_\mu,-\Lambda u_\mu - \bar{r}) + \int_0^T \langle \Lambda u_\mu(t),u_\mu(t)\rangle\,dt$$

$$= \int_0^T \left[L(t,u_\mu(t),-\dot{u}_\mu(t)+\bar{r}(t)) - \langle u_\mu(t),\bar{r}(t)\rangle\right]dt$$

$$+\ell\Big(u_\mu(0)-u_\mu(T),-\frac{u_\mu(T)+u_\mu(0)}{2}\Big)$$

$$+\Psi_\mu(u_\mu,-\Lambda u_\mu-\bar r)+\int_0^T\langle\Lambda u_\mu(t)+\bar r(t),u_\mu(t)\rangle\,dt$$

$$=\int_0^T\Big[L(t,u_\mu(t),-\dot u_\mu(t)+\bar r(t))+\langle u_\mu(t),\dot u_\mu(t)-\bar r(t)\rangle\Big]\,dt$$

$$-\frac{1}{2}\|u_\mu(T)\|^2+\frac{1}{2}\|u_\mu(0)\|^2+\ell\Big(u_\mu(0)-u_\mu(T),-\frac{u_\mu(T)+u_\mu(0)}{2}\Big)$$

$$+\Psi_\mu(u_\mu,-\Lambda u_\mu-\bar r)+\int_0^T\langle\Lambda u_\mu+\bar r,u_\mu(t)\rangle\,dt.$$

Since this is the sum of three nonnegative terms, we get the three identities

$$\int_0^T\Big[L(t,u_\mu(t),-\dot u_\mu(t)+\bar r(t))+\langle u_\mu,\dot u_\mu-\bar r\rangle\Big]\,dt=0,\qquad(18.12)$$

$$\Psi_\mu(u_\mu,-\Lambda u_\mu-\bar r)+\int_0^T\langle\Lambda u_\mu+\bar r,u_\mu(t)\rangle\,dt=0,\qquad(18.13)$$

$$\ell\Big(u_\mu(0)-u_\mu(T),-\frac{u_\mu(T)+u_\mu(0)}{2}\Big)-\frac{1}{2}\|u_\mu(T)\|^2+\frac{1}{2}\|u_\mu(0)\|^2=0.\quad(18.14)$$

It follows from the limiting case of Fenchel duality that

$$\dot u_\mu(t)+\Lambda u_\mu(t)+\mu\partial\psi(u_\mu(t))\in-\overline\partial L(t,u_\mu(t))\ \text{a.e. }t\in[0,T],$$

$$\frac{u_\mu(T)+u_\mu(0)}{2}\in-\overline\partial\ell(u_\mu(0)-u_\mu(T)).$$

Since $u:=u_\mu\in\mathscr X_{p,q}$, we have that

$$-\mu\partial\psi(u(t))=\dot u(t)+\Lambda u(t)+\overline\partial L(t,u(t))\in L_{X^*}^q.$$

It follows that $\partial\psi(u(t)))=-\frac{d}{dt}(\|\dot u\|_*^{q-2}D^{-1}\dot u)$, where D is the duality map between X and X^*. Hence, for each $v\in\mathscr X_{p,q}$, we have

$$0=\int_0^T\Big[\langle\dot u(t)+\Lambda u(t)+\overline\partial L(t,u(t)),v\rangle+\mu\langle\|\dot u\|_*^{q-2}D^{-1}\dot u,\dot v\rangle\Big]\,dt$$

$$=\int_0^T\Big\langle\dot u(t)+\Lambda u(t)-\mu\frac{d}{dt}(\|\dot u\|_*^{q-2}D^{-1}\dot u)+\overline\partial L(t,u(t)),v\Big\rangle\,dt$$

$$+\mu\langle\|\dot u(T)\|_*^{q-2}D^{-1}\dot u(T),v(T)\rangle-\mu\langle\|\dot u(0)\|_*^{q-2}D^{-1}\dot u(0),v(0)\rangle,$$

from which we deduce that

$$\dot u(t)+\Lambda u(t)-\frac{d}{dt}(\|\dot u\|^{q-2}D^{-1}\dot u(t))\in-\overline\partial L(t,u(t))$$

$$\dot u(T)=\dot u(0)=0.$$

18.2 A self-dual variational principle for nonlinear evolutions

This section is dedicated to the proof of a general variational principle for nonlinear evolutions. We shall make use of the following self-dual Palais-Smale property for functionals on path space.

Definition 18.1. Let L be a time-dependent self-dual Lagrangian on $[0,T] \times X \times X^*$, ℓ a self-dual Lagrangian on $H \times H$, and $\Lambda : \mathscr{X}_{p,q} \to L^q_{X^*}$ a given map. Let $D : X \to X^*$ be a duality map, and consider the functional

$$
I_{L,\ell,\Lambda}(u) = \int_0^T \left[L(t,u(t),-\dot{u}(t) - \Lambda u(t)) + \langle \Lambda u(t), u(t) \rangle \right] dt
$$
$$
+ \ell\left(u(0) - u(T), -\frac{u(T) + u(0)}{2} \right).
$$

1. The functional $I_{L,\ell,\Lambda}$ is said to satisfy the *self-dual Palais-Smale condition* on $\mathscr{X}_{p,q}$ if any sequence $\{u_n\}_{n=1}^{\infty} \subseteq \mathscr{X}_{p,q}$ satisfying for some $\varepsilon_n \to 0$,

$$
\begin{cases}
\dot{u}_n(t) + \Lambda u_n(t) - \varepsilon_n \|u_n\|^{p-2} Du_n(t) \in -\bar{\partial}L(t,u_n(t)) \ t \in [0,T] \\
\frac{u_n(0)+u_n(T)}{2} \in -\bar{\partial}\ell\left(u_n(0) - u_n(T) \right)
\end{cases}
\tag{18.15}
$$

is necessarily bounded in $\mathscr{X}_{p,q}$.
2. The functional $I_{L,\ell,\Lambda}$ is said to be *weakly coercive* on $\mathscr{X}_{p,q}$ if

$$
\int_0^T \left[L\left(t,u_n,\dot{u}_n + \Lambda u_n + \frac{1}{n}\|u_n\|^{p-2}Du_n \right) + \langle u_n, \Lambda u_n \rangle + \frac{1}{n}\|u_n\|^p \right] dt
$$

goes to $+\infty$ when $\|u_n\|_{\mathscr{X}_{p,q}} \to +\infty$.

Theorem 18.1. *Let $X \subset H \subset X^*$ be an evolution triple, where X is a reflexive Banach space and H is a Hilbert space. Let L be a time-dependent self-dual Lagrangian on $[0,T] \times X \times X^*$ such that, for some $C > 0$ and $r > 0$, we have*

$$
\int_0^T L(t,u(t),0)dt \le C(1 + \|u\|^r_{L^p_X}) \text{ for every } u \in L^p_X.
\tag{18.16}
$$

Let ℓ be a self-dual Lagrangian on $H \times H$ that is bounded below with $0 \in \mathrm{Dom}(\ell)$, and consider $\Lambda : \mathscr{X}_{p,q} \to L^q_{X^*}$ to be a regular map such that

$$
\|\Lambda u\|_{L^q_{X^*}} \le k\|\dot{u}\|_{L^q_{X^*}} + w(\|u\|_{L^p_X}) \text{ for every } u \in \mathscr{X}_{p,q},
\tag{18.17}
$$

where w is a nondecreasing continuous real function and $0 < k < 1$. Assume one of the following two conditions:

(A) $\left| \int_0^T \langle \Lambda u(t), u(t) \rangle dt \right| \le w(\|u\|_{L^p_X})$ *for every $u \in \mathscr{X}_{p,q}$.*

(B) *For each $r \in L^q_{X^*}$, the functional $u \to \int_0^T L(t,u(t),r(t)) dt$ is continuous on L^p_X, and there exists $C > 0$ such that for every $u \in L^p_X$ we have*

$$\|\overline{\partial}L(t,u)\|_{L^q_{X^*}} \leq w(\|u\|_{L^p_X}), \tag{18.18}$$

$$\int_0^T \langle \overline{\partial}L(t,u(t)) + \Lambda u(t), u(t) \rangle \, dt \geq -C(\|u\|_{L^p_X} + 1). \tag{18.19}$$

If the functional $I_{L,\ell,\Lambda}$ is weakly coercive on $\mathscr{X}_{p,q}$, then it attains its minimum at $v \in \mathscr{X}_{p,q}$ in such a way that $I_{L,\ell,\Lambda}(v) = \inf\limits_{u \in \mathscr{X}_{p,q}} I_{L,\ell,\Lambda}(u) = 0$ and

$$\begin{cases} -\Lambda v(t) - \dot{v}(t) \in \overline{\partial}L(t,v(t)) \quad \text{on } [0,T] \\ -\dfrac{v(0)+v(T)}{2} \in \overline{\partial}\ell\big(v(0) - v(T)\big). \end{cases} \tag{18.20}$$

We start with the following lemma, in which we consider a regularization (coercivization) of the self-dual Lagrangian $\mathscr{L}$ by the self-dual Lagrangian Ψ_μ and also a perturbation of Λ by the operator

$$Ku = w(\|u\|_{L^p_X})Du + \|u\|_{L^p_X}^{p-1}Du, \tag{18.21}$$

which is regular from $\mathscr{X}_{p,q}$ into $L^q_{X^*}$.

Lemma 18.2. *Let Λ be a regular map from $\mathscr{X}_{p,q}$ into $L^q_{X^*}$ satisfing (18.17). Let L be a time-dependent self-dual Lagrangian on $[0,T] \times X \times X^*$ satisfying conditions (18.1) and (18.2), and let ℓ be a self-dual Lagrangian on $H \times H$ satisfying condition (18.3). Then, for any $\mu > 0$, the functional*

$$I_\mu(u) = \mathscr{L} \oplus \Psi_\mu(u, -\Lambda u - Ku) + \int_0^T \langle \Lambda u(t) + Ku(t), u(t) \rangle \, dt$$

is self-dual on $\mathscr{X}_{p,q} \times \mathscr{X}_{p,q}$.
Moreover, there exists $u_\mu \in \{u \in \mathscr{X}_{p,q}; \partial \psi(u) \in L^q_{X^*}, \dot{u}(T) = \dot{u}(0) = 0\}$ such that

$$\dot{u}_\mu(t) + \Lambda u_\mu(t) + Ku_\mu(t) + \mu \partial \psi(u_\mu(t)) \in -\overline{\partial}L(t, u_\mu(t)), \tag{18.22}$$

$$\ell\left(u_\mu(0) - u_\mu(T), -\frac{u_\mu(T) + u_\mu(0)}{2}\right) = \int_0^T \langle \dot{u}_\mu(t), u_\mu(t) \rangle \, dt. \tag{18.23}$$

Proof. It suffices to apply Lemma 18.1 to the regular operator $\Gamma = \Lambda + K$, provided we show the required coercivity condition $\lim\limits_{\|u\|_{\mathscr{X}_{p,q}} \to +\infty} \mathscr{M}(u,0) = +\infty$, where

$$\mathscr{M}(u,0) = \int_0^T \langle \Lambda u(t) + Ku(t), u(t) \rangle \, dt + H_{\mathscr{L}}(0,u) + \mu \psi(u).$$

Note first that it follows from (18.17) that for $\varepsilon < \frac{\mu}{q}$ there exists $C(\varepsilon)$ such that

$$\int_0^T \langle \Lambda u(t), u(t) \rangle \, dt \leq k\|u\|_{L^p_X}\|\dot{u}\|_{L^q_{X^*}} + w(\|u\|_{L^p_X})\|u\|_{L^p_X}$$

$$\leq \varepsilon\|\dot{u}\|_{L^q_{X^*}}^q + C(\varepsilon)\|u\|_{L^p_X}^p + w(\|u\|_{L^p_X})\|u\|_{L^p_X}.$$

On the other hand, by the definition of K, we have

$$\int_0^T \langle Ku(t), u(t) \rangle \, dt = w(\|u\|_{L_X^p}) \|u\|_{L_X^p}^2 + \|u\|_{L_X^p}^{p+1}.$$

Therefore, the coercivity follows from the estimate

$$
\begin{aligned}
\mathscr{M}(u,0) &= \int_0^T \left[\langle \Lambda u(t) + Ku(t), u(t) \rangle \right] dt + H_{\mathscr{L}}(0,u) + \mu \frac{1}{q} \|\dot{u}\|_{L_{X^*}^q}^q \\
&\geq -\varepsilon \|\dot{u}\|_{L_{X^*}^q}^q - C(\varepsilon) \|u\|_{L_X^p}^p - w(\|u\|_{L_X^p}) \|u\|_{L_X^p} + w(\|u\|_{L_X^p}) \|u\|_{L_X^p}^2 + \|u\|_{L_X^p}^{p+1} \\
&\quad - \mathscr{L}(0,0) + \mu \frac{1}{q} \|\dot{u}\|_{L_{X^*}^q}^q \\
&\geq \left(\frac{\mu}{q} - \varepsilon \right) \|\dot{u}\|_{L_{X^*}^q}^q + \|u\|_{L_X^p}^{p+1} \left(1 + o(\|u\|_{L_X^p}) \right).
\end{aligned}
$$

In the following lemma, we get rid of the regularizing diffusive term $\mu \psi(u)$ and prove the theorem with Λ replaced by the operator $\Lambda + K$, but under the additional assumption that ℓ satisfies the boundedness condition (18.3).

Lemma 18.3. *Let L be a time-dependent self-dual Lagrangian as in Theorem 18.1 satisfying either condition (A) or (B), and assume that ℓ is a self-dual Lagrangian on $H \times H$ that satisfies condition (18.3). Then, there exists $u \in \mathscr{X}_{p,q}$ such that*

$$
\begin{aligned}
0 &= \int_0^T \left[L(t, u(t), -\dot{u}(t) - \Lambda u(t) - Ku(t)) \, dt + \langle \Lambda u(t) + Ku(t), u(t) \rangle \right] dt \\
&\quad + \ell \left(u(0) - u(T), -\frac{u(T) + u(0)}{2} \right).
\end{aligned}
$$

Proof under condition (B). Note first that in this case L satisfies both conditions (18.1) and (18.2) of Lemma 18.1, which then yields for every $\mu > 0$ an element $u_\mu \in \mathscr{X}_{p,q}$ satisfying

$$\dot{u}_\mu(t) + \Lambda u_\mu(t) + Ku_\mu(t) + \mu \partial \psi(u_\mu(t)) \in -\bar{\partial} L(t, u_\mu(t)), \qquad (18.24)$$

and

$$\ell \left(u_\mu(0) - u_\mu(T), -\frac{u_\mu(T) + u_\mu(0)}{2} \right) = \int_0^T \langle \dot{u}_\mu(t), u_\mu(t) \rangle \, dt. \qquad (18.25)$$

We now establish upper bounds on the norm of u_μ in $\mathscr{X}_{p,q}$. Multiplying (18.24) by u_μ and integrating over $[0,T]$, we obtain

$$\int_0^T \langle \dot{u}_\mu + \Lambda u_\mu + Ku_\mu + \mu \partial \psi(u_\mu), u_\mu \rangle \, dt = -\int_0^T \langle \bar{\partial} L(t, u_\mu), u_\mu \rangle \, dt. \quad (18.26)$$

It follows from (18.19) and the above equality that

$$\int_0^T \langle \dot{u}_\mu(t) + K u_\mu(t) + \mu \partial \psi(u_\mu(t)), u_\mu(t) \rangle \leq C(1 + \|u_\mu\|_{L_X^p}). \quad (18.27)$$

Taking into account (18.25), (18.24) and the fact that $\int_0^T \langle \partial \psi(u_\mu(t)), u_\mu(t) \rangle \geq 0$, it follows that

$$\ell\left(u_\mu(0) - u_\mu(T), -\frac{u_\mu(T) + u_\mu(0)}{2}\right) + \int_0^T \langle K u_\mu(t), u_\mu(t) \rangle \leq C(1 + \|u_\mu\|_{L_X^p}).$$

Since ℓ is bounded from below (say by C_1), the above inequality implies that $\|u_\mu\|_{L_X^p}$ is bounded since we then have

$$C_1 + w(\|u_\mu\|_{L_X^p})\|u_\mu\|_{L_X^p}^2 + \|u_\mu\|_{L_X^p}^{p+1} \leq C\|u_\mu\|_{L_X^p}.$$

Now we show that $\|\dot{u}_\mu\|_{L_{X^*}^q}$ is also bounded. For that, we multiply (18.24) by $D^{-1}\dot{u}_\mu$ to get that

$$\|\dot{u}_\mu\|_{L_{X^*}^q}^2 + \int_0^T \langle \Lambda u_\mu + K u_\mu + \mu \partial \psi(u_\mu) + \bar{\partial}L(t, u_\mu), D^{-1}\dot{u}_\mu \rangle \, dt = 0. \quad (18.28)$$

The last identity and the fact that $\int_0^T \langle \partial \psi(u_\mu(t)), D^{-1}\dot{u}_\mu(t) \rangle \, dt = 0$ imply that

$$\|\dot{u}_\mu\|_{L_{X^*}^q}^2 \leq \|\Lambda u_\mu\|_{L_{X^*}^q}\|\dot{u}_\mu\|_{L_{X^*}^q} + \|K u_\mu\|_{L_{X^*}^q}\|\dot{u}_\mu\|_{L_{X^*}^q} + w(\|u\|_{L_X^p})\|\dot{u}_\mu\|_{L_{X^*}^q}.$$

It follows from the above inequality and (18.17) that

$$\|\dot{u}_\mu\|_{L_{X^*}^q} \leq \|\Lambda u_\mu\|_{L_{X^*}^q} + \|K u_\mu\|_{L_{X^*}^q} + w(\|u\|_{L_X^p})$$
$$\leq k\|\dot{u}_\mu\|_{L_{X^*}^q} + 2w(\|u\|_{L_X^p}) + \|K u_\mu\|_{L_{X^*}^q},$$

from which we obtain that $(1-k)\|\dot{u}_\mu\|_{L_{X^*}^q} \leq 2w(\|u_\mu\|_{L_X^p}) + \|K u_\mu\|_{L_{X^*}^q}$, which means that $\|\dot{u}_\mu\|_{L_{X^*}^q}$ is bounded.

Consider now $u \in \mathscr{X}_{p,q}$ such that $u_\mu \rightharpoonup u$ weakly in L_X^p and $\dot{u}_\mu \rightharpoonup \dot{u}$ in $L_{X^*}^q$. From (18.24) and (18.25), we have

$$J_\mu(u_\mu) := \int_0^T L(t, u_\mu(t), -\dot{u}_\mu(t) - \Lambda u_\mu(t) - K u_\mu(t) - \mu \partial \psi(u_\mu(t))) \, dt$$
$$+ \int_0^T \langle \Lambda u_\mu(t) + K u_\mu(t), u_\mu(t) \rangle \, dt + \ell\left(u_\mu(0) - u_\mu(T), -\frac{u_\mu(T) + u_\mu(0)}{2}\right)$$
$$\leq \int_0^T L(t, u_\mu(t), -\dot{u}_\mu(t) - \Lambda u_\mu(t) - K u_\mu(t) - \mu \partial \psi(u_\mu(t))) \, dt$$
$$+ \int_0^T \langle \Lambda u_\mu(t) + K u_\mu(t) - \mu \partial \psi(u_\mu(t)), u_\mu(t) \rangle$$
$$+ \ell\left(u_\mu(0) - u_\mu(T), -\frac{u_\mu(T) + u_\mu(0)}{2}\right)$$
$$= I_\mu(u_\mu) = 0.$$

Since $\Lambda + K$ is regular, $\langle \partial \psi(u_\mu), u_\mu \rangle = \|\dot{u}_\mu\|_{X^*}^q$ is uniformly bounded and L is weakly lower semicontinuous on $X \times X^*$, we get by letting $\mu \to 0$ that

$$\int_0^T \left[\langle \Lambda u(t) + Ku(t), u(t) \rangle + L(t, u(t), -\dot{u}(t) - \Lambda u(t) - Ku(t)) \right] dt$$
$$+ \ell \left(u(T) - u(0), -\frac{u(T) + u(0)}{2} \right) \leq 0.$$

The reverse inequality is true for any $u \in \mathcal{X}_{p,q}$ since L and ℓ are self-dual Lagrangians.

Proof of Lemma 18.3 under condition (A). Note first that condition (18.16) implies that there is a $D > 0$ such that

$$\int_0^T L(t, u(t), v(t)) dt \geq D(\|v\|_{L^q_{X^*}}^s - 1) \text{ for every } v \in L^q_{X^*}, \tag{18.29}$$

where $\frac{1}{r} + \frac{1}{s} = 1$. But since L is not supposed to satisfy condition (18.1), we first replace it by its λ-regularization $L^1_{\lambda,p}$ with respect to its first variable as defined in Section 3.5, so as to satisfy all properties of Lemma 18.2. Therefore, there exists $u_{\mu,\lambda} \in \mathcal{X}_{p,q}$ satisfying

$$\dot{u}_{\mu,\lambda} + \Lambda u_{\mu,\lambda} + Ku_{\mu,\lambda} + \mu \partial \psi(u_{\mu,\lambda}) = -\overline{\partial} L^1_{\lambda,p}(t, u_{\mu,\lambda}) \tag{18.30}$$

and

$$\ell \left(u_{\mu,\lambda}(T) - u_{\mu,\lambda}(0), -\frac{u_{\mu,\lambda}(T) + u_{\mu,\lambda}(0)}{2} \right) = \int_0^T \langle \dot{u}_{\mu,\lambda}, u_{\mu,\lambda} \rangle dt. \tag{18.31}$$

We shall first find bounds for $u_{\mu,\lambda}$ in $\mathcal{X}_{p,q}$ that are independent of μ. Multiplying (18.30) by $u_{\mu,\lambda}$ and integrating, we obtain

$$\int_0^T \langle \dot{u}_{\mu,\lambda}(t) + \Lambda u_{\mu,\lambda}(t) + Ku_{\mu,\lambda}(t) + \mu \partial \psi(u_{\mu,\lambda}(t)), u_{\mu,\lambda}(t) \rangle dt$$
$$= -\int_0^T \langle \overline{\partial} L^1_{\lambda,p}(t, u_{\mu,\lambda}(t)), u_{\mu,\lambda}(t) \rangle dt. \tag{18.32}$$

Since $\overline{\partial} L^1_{\lambda,p}(t, .)$ is monotone, we have

$$\int_0^T \langle \overline{\partial} L^1_{\lambda,p}(t, u_{\mu,\lambda}(t)) - \overline{\partial} L^1_{\lambda,p}(t, 0), u_{\mu,\lambda}(t) - 0 \rangle dt \geq 0,$$

and therefore

$$\int_0^T \langle \overline{\partial} L^1_{\lambda,p}(t, u_{\mu,\lambda}(t)), u_{\mu,\lambda}(t) \rangle dt \geq \int_0^T \langle \overline{\partial} L^1_{\lambda,p}(t, 0), u_{\mu,\lambda}(t) \rangle dt. \tag{18.33}$$

Taking into account (18.31), (18.33), and the fact that $\int_0^T \langle \partial \psi(u_\mu(t)), u_\mu(t) \rangle \geq 0$, it follows from (18.32) that

$$\ell\left(u_{\mu,\lambda}(0) - u_{\mu,\lambda}(T), -\frac{u_{\mu,\lambda}(T) + u_{\mu,\lambda}(0)}{2}\right) + \int_0^T \langle \Lambda u_{\mu,\lambda}(t) + K u_{\mu,\lambda}(t), u_{\mu,\lambda}(t) \rangle\, dt$$

$$\leq -\int_0^T \langle \partial L^1_{\lambda,p}(t,0), u_{\mu,\lambda}(t) \rangle\, dt.$$

This implies $\{u_{\mu,\lambda}\}_\mu$ is bounded in L^p_X, and by the same argument as under condition (B), one can prove that $\{\dot{u}_{\mu,\lambda}\}_\mu$ is also bounded in $L^q_{X^*}$. Consider $u_\lambda \in \mathscr{X}_{p,q}$ such that $u_{\mu,\lambda} \rightharpoonup u_\lambda$ weakly in L^p_X and $\dot{u}_{\mu,\lambda} \rightharpoonup \dot{u}_\lambda$ in $L^q_{X^*}$. It follows just as in the proof under condition (B) that

$$\int_0^T \left[\langle \Lambda u_\lambda + K u_\lambda, u_\lambda \rangle + L^1_{\lambda,p}(t, u_\lambda, -\dot{u}_\lambda - \Lambda u_\lambda - K u_\lambda) \right] dt$$

$$+\ell\left(u_\lambda(0) - u_\lambda(T), -\frac{u_\lambda(T) + u_\lambda(0)}{2}\right) = 0, \qquad (18.34)$$

and therefore

$$\dot{u}_\lambda(t) + \Lambda u_\lambda(t) + K u_\lambda(t) \in -\bar{\partial} L^1_{\lambda,p}(t, u_\lambda(t)). \qquad (18.35)$$

Now we obtain estimates on u_λ in $\mathscr{X}_{p,q}$. Since ℓ and L^1_λ are bounded from below, it follows from (18.34) that $\int_0^T \left[\langle \Lambda u_\lambda(t) + K u_\lambda(t), u_\lambda(t) \rangle \right] dt$ is bounded and therefore u_λ is bounded in L^p_X since

$$\int_0^T \left[\langle \Lambda u_\lambda(t) + K u_\lambda(t), u_\lambda(t) \rangle\, dt \geq -C(\|u\|_{L^p_X} + 1) - \int_0^T \langle \bar{\partial} L(t, u(t)), u(t) \rangle dt \right.$$

$$+ \int_0^T \langle K u(t), u(t) \rangle\, dt$$

$$\geq -C(\|u\|_{L^p_X} + 1) - w(\|u\|_{L^p_X})\|u\|_{L^p_X}$$

$$+ w(\|u\|_{L^p_X})\|u\|^2_{L^p_X} + \|u\|^{p+1}_{L^p_X}.$$

Setting $v_\lambda(t) := \dot{u}_\lambda(t) + \Lambda u_\lambda(t) + K u_\lambda(t)$, we get from (18.35) that

$$-v_\lambda(t) = \bar{\partial} L^1_\lambda(t, u_\lambda(t)) = \bar{\partial} L(t, u_\lambda(t)) + \lambda^{q-1}\|v_\lambda(t)\|^{q-2}_* D^{-1} v_\lambda(t).$$

This together with (18.34) implies that

$$\int_0^T L\left(t, u_\lambda(t) + \lambda\|v_\lambda(t)\|^{q-2}_* D^{-1} v_\lambda(t), -\dot{u}_\lambda(t) - \Lambda u_\lambda(t) - K u_\lambda(t)\right) dt$$

$$+ \int_0^T \left[\langle \Lambda u_\lambda(t) + K u_\lambda(t), u_\lambda(t) \rangle + \lambda\|v_\lambda(t)\|^q \right] dt$$

$$+\ell\left(u_\lambda(0) - u_\lambda(T), -\frac{u_\lambda(T) + u_\lambda(0)}{2}\right) = 0. \qquad (18.36)$$

It follows that $\int_0^T L(t, u_\lambda + \lambda \|v_\lambda\|_*^{q-2} D^{-1} v_\lambda, -\dot{u}_\lambda - \Lambda u_\lambda - K u_\lambda)\, dt$ is bounded from above. In view of (18.29), there exists then a constant $C > 0$ such that

$$\|\dot{u}_\lambda(t) + \Lambda u_\lambda(t) + K u_\lambda(t)\|_{L_{X*}^q}\, dt \leq C. \tag{18.37}$$

It follows that

$$\|\dot{u}_\lambda\|_{L_{X*}^q} \leq \|\Lambda u_\lambda\|_{L_{X*}^q} + \|K u_\lambda\|_{L_{X*}^q} + C \leq k \|\dot{u}_\lambda\|_{L_{X*}^q} + w(\|u\|_{L_X^p}) + \|K u_\lambda\|_{L_{X*}^q},$$

from which we obtain

$$(1 - k)\|\dot{u}_\lambda\|_{L_{X*}^q} \leq w(\|u_\lambda\|_{L_X^p}) + \|K u_\lambda\|_{L_{X*}^q},$$

which means that $\|\dot{u}_\lambda\|_{L_{X*}^q}$ is bounded. By letting λ go to zero in (18.36), we obtain

$$\int_0^T \left[\langle \Lambda u(t) + K u(t), u(t) \rangle + L(t, u(t), -\dot{u}(t) - \Lambda u(t) - K u(t)) \right] dt$$
$$+ \ell\left(u(0) - u(T), -\frac{u(T) + u(0)}{2} \right) = 0,$$

where u is a weak limit of $(u_\lambda)_\lambda$ in $\mathscr{X}_{p,q}$.

Proof of Theorem 18.1. First we assume that ℓ satisfies condition (18.3), and we shall work toward eliminating the perturbation K. Let $L_{\lambda,p}^2$ be the (λ, p)-regularization of L with respect to the second variable as defined in Section 3.5, in such a way that $L_{\lambda,p}^2$ satisfies (18.19). Indeed,

$$\int_0^T \langle \bar{\partial} L_{\lambda,p}^2(t, u) + \Lambda u, u \rangle\, dt = \int_0^T \langle \bar{\partial} L(t, u) + \Lambda u + \lambda^{p-1} \|u\|^{p-2} D u, u \rangle\, dt$$
$$\geq \int_0^T \langle \bar{\partial} L(t, u(t)) + \Lambda u(t), u(t) \rangle\, dt$$
$$\geq -C \|u\|_{L_X^p}. \tag{18.38}$$

Moreover, we have in view of (18.16) that

$$\int_0^T L_{\lambda,p}^2(t, u, v)\, dt \geq -D + \frac{\lambda^{p-1}}{p} \|u\|_{L_X^p}^p. \tag{18.39}$$

From Lemma 18.3, we get for each $\varepsilon > 0$ a $u_{\varepsilon,\lambda} \in \mathscr{X}_{p,q}$ such that

$$\int_0^T L_{\lambda,p}^2(t, u_{\varepsilon,\lambda}, -\dot{u}_{\varepsilon,\lambda} - \Lambda u_{\varepsilon,\lambda} - \varepsilon K u_{\varepsilon,\lambda})\, dt$$
$$+ \int_0^T \langle \Lambda u_{\varepsilon,\lambda} + \varepsilon K u_{\varepsilon,\lambda}, u_{\varepsilon,\lambda} \rangle$$
$$+ \ell\left(u_{\varepsilon,\lambda}(0) - u_{\varepsilon,\lambda}(T), -\frac{u_{\varepsilon,\lambda}(T) + u_{\varepsilon,\lambda}(0)}{2} \right) = 0 \tag{18.40}$$

and

$$\dot{u}_{\varepsilon,\lambda}(t) + \Lambda u_{\varepsilon,\lambda}(t) + \varepsilon K u_{\varepsilon,\lambda}(t) \in -\bar{\partial} L_{\lambda,p}^2(t, u_{\varepsilon,\lambda}(t)). \tag{18.41}$$

We shall first find bounds for $u_{\varepsilon,\lambda}$ in $\mathscr{X}_{p,q}$ that are independent of ε. Multiplying (18.41) by $u_{\varepsilon,\lambda}$ and integrating, we obtain

$$\int_0^T \langle \dot{u}_{\varepsilon,\lambda} + \Lambda u_{\varepsilon,\lambda} + \varepsilon K u_{\varepsilon,\lambda}, u_{\varepsilon,\lambda} \rangle \, dt = - \int_0^T \langle \bar{\partial} L_{\lambda,p}^2(t, u_{\varepsilon,\lambda}), u_{\varepsilon,\lambda} \rangle \, dt. \tag{18.42}$$

It follows from (18.38) and the above equality that

$$\int_0^T \langle \dot{u}_{\varepsilon,\lambda}(t) + \varepsilon K u_{\varepsilon,\lambda}(t), u_{\varepsilon,\lambda}(t) \rangle \le C \| u_{\varepsilon,\lambda} \|_{L_X^p}, \tag{18.43}$$

and therefore

$$\ell \left(u_{\varepsilon,\lambda}(0) - u_{\varepsilon,\lambda}(T), -\frac{u_{\varepsilon,\lambda}(T) + u_{\varepsilon,\lambda}(0)}{2} \right) + \int_0^T \langle \varepsilon K u_{\varepsilon,\lambda}(t), u_{\varepsilon,\lambda}(t) \rangle \le C \| u_{\varepsilon,\lambda} \|_{L_X^p},$$

which in view of (18.40) implies that

$$\left| \int_0^T L_\lambda^2(t, u_{\varepsilon,\lambda}(t), -\dot{u}_{\varepsilon,\lambda}(t) - \Lambda u_{\varepsilon,\lambda}(t) - \varepsilon K u_{\varepsilon,\lambda}(t)) \, dt \right| \le C \| u_{\varepsilon,\lambda} \|_{L_X^p}.$$

By (18.39), we deduce that $\{u_{\varepsilon,\lambda}\}_\mu$ is bounded in L_X^p. The same reasoning as above then shows that $\{\dot{u}_{\varepsilon,\lambda}\}_\mu$ is also bounded in $L_{X^*}^q$. Again, the regularity of Λ and the lower semicontinuity of L yields the existence of $u_\lambda \in \mathscr{X}_{p,q}$ such that

$$\int_0^T \left[L_\lambda^2(t, u_\lambda, -\dot{u}_\lambda - \Lambda u_\lambda) + \langle \Lambda u_\lambda, u_\lambda \rangle \right] dt$$
$$+ \ell \left(u_\lambda(0) - u_\lambda(T), -\frac{u_\lambda(T) + u_\lambda(0)}{2} \right) = 0.$$

In other words,

$$\int_0^T L\left(t, u_\lambda, -\dot{u}_\lambda - \Lambda u_\lambda + \lambda^{p-1} \| u_\lambda \|^{p-2} D u_\lambda \right) + \lambda^{p-1} \| u_\lambda \|^p \, dt$$
$$+ \int_0^T \langle \Lambda u_\lambda, u_\lambda \rangle \, dt + \ell \left(u_\lambda(0) - u_\lambda(T), -\frac{u_\lambda(T) + u_\lambda(0)}{2} \right) = 0. \tag{18.44}$$

Now since I satisfies the self-dual Palais-Smale condition, we get that $(u_\lambda)_\lambda$ is bounded in $\mathscr{X}_{p,q}$. Suppose $u_\lambda \rightharpoonup \bar{u}$ in L_X^p and $\dot{u}_\lambda \rightharpoonup \dot{\bar{u}}$ in $L_{X^*}^q$. It follows from (18.17) that Λu_λ is bounded in $L_{X^*}^q$. Again, we deduce that

$$\ell \left(\bar{u}(T) - \bar{u}(0), -\frac{\bar{u}(T) + \bar{u}(0)}{2} \right) + \int_0^T \left[L(t, \bar{u}, -\dot{\bar{u}} - \Lambda \bar{u}) + \langle \Lambda \bar{u}, \bar{u} \rangle \right] dt = 0.$$

Now, we show that we can do without assuming that ℓ satisfies (18.3), but that it is bounded below, while $(0,0) \in \text{Dom}(\ell)$. Indeed, let $\ell_\lambda := \ell_\lambda^{1,2}$ be the λ-regularization of the self-dual Lagrangian ℓ in both variables. Then, ℓ_λ satisfies (18.3) and therefore there exists $x_\lambda \in \mathcal{X}_{p,q}$ such that

$$\int_0^T \left[L(t, x_\lambda, -\dot{x}_\lambda - \Lambda x_\lambda) + \langle \Lambda x_\lambda, x_\lambda \rangle \right] dt$$
$$+ \ell_\lambda \left(x_\lambda(T) - x_\lambda(0), -\frac{x_\lambda(T) + x_\lambda(0)}{2} \right) = 0. \quad (18.45)$$

Since ℓ is bounded from below, so is ℓ_λ. This together with (18.45) implies that the family $\int_0^T \left[\langle \Lambda x_\lambda(t), x_\lambda(t) \rangle + L(t, x_\lambda(t), -\dot{x}_\lambda(t) - \Lambda x_\lambda(t)) \right] dt$ is bounded above. Again, since $I_{L,\ell,\Lambda}$ is weakly coercive, we obtain that $(x_\lambda)_\lambda$ is bounded in $\mathcal{X}_{p,q}$. The continuity of the injection $\mathcal{X}_{p,q} \subseteq C([0,T];H)$ also ensures the boundedness of $(x_\lambda(T))_\lambda$ and $(x_\lambda(0))_\lambda$ in H. Consider $\bar{x} \in \mathcal{X}_{p,q}$ such that $x_\lambda \rightharpoonup \bar{x}$ in L_X^p and $\dot{x}_\lambda \rightharpoonup \dot{\bar{x}}$ in $L_{X^*}^q$. It follows from the regularity of Λ and the lower semicontinuity of ℓ and L that

$$\int_0^T \left[\langle \Lambda \bar{x}, \bar{x} \rangle + L(\bar{x}, -\dot{\bar{x}} - \Lambda \bar{x}) \right] dt + \ell \left(\bar{x}(T) - \bar{x}(0), -\frac{\bar{x}(T) + \bar{x}(0)}{2} \right) = 0,$$

and therefore $\bar{x}$ satisfies equation (18.20).

Remark 18.1. Note that the hypothesis that $I_{L,\ell,\Lambda}$ is weakly coercive, is only needed in the last part of the proof to deal with the case where ℓ is not assumed to satisfy (18.3). Otherwise, the hypothesis that $I_{L,\ell,\Lambda}$ satisfies the self-dual Palais-Smale condition would have been sufficient. This will be useful in the application to Schrödinger equations mentioned below.

18.3 Navier-Stokes evolutions

We now consider the most basic time-dependent self-dual Lagrangians $L(t, x, p) = \varphi(t, x) + \varphi^*(t, -p)$, where for each t the function $x \to \varphi(t, x)$ is convex and lower semicontinuous on X. Let now $\psi : H \to \mathbf{R} \cup \{+\infty\}$ be another convex lower semicontinuous function that is bounded from below and such that $0 \in \text{Dom}(\psi)$, and set $\ell(a, b) = \psi(a) + \psi^*(-b)$. The above principle then yields that if for some $C_1, C_2 > 0$ we have

$$C_1 \left(\|x\|_{L_X^p}^p - 1 \right) \leq \int_0^T \varphi(t, x(t)) \, dt \leq C_2 \left(\|x\|_{L_X^p}^p + 1 \right) \text{ for all } x \in L_X^p,$$

then for every regular map Λ satisfying (18.17) and either condition (A) or (B) in Theorem 18.1, the infimum of the functional

$$I(x) = \int_0^T \left[\varphi(t, x(t)) + \varphi^*(t, -\dot{x}(t) - \Lambda x(t)) + \langle \Lambda x(t), x(t) \rangle \right] dt$$
$$+ \psi(x(0) - x(T)) + \psi^* \left(-\frac{x(0) + x(T)}{2} \right) \quad (18.46)$$

on $\mathscr{X}_{p,q}$ is zero and is attained at a solution $x(t)$ of the equation

$$\begin{cases} -\dot{x}(t) - \Lambda x(t) \in \partial\varphi(t,x(t)) & \text{for all } t \in [0,T] \\ -\frac{x(0)+x(T)}{2} \in \partial\psi(x(0)-x(T)). \end{cases}$$

As noted before, the boundary condition above is quite general and includes as a particular case the more traditional ones such as initial-value problems, periodic and antiperiodic orbits. It suffices to choose $\ell(a,b) = \psi(a) + \psi^*(b)$ accordingly.

- For the initial boundary condition $x(0) = x_0$ for a given $x_0 \in H$, we choose $\psi(x) = \frac{1}{4}\|x\|_H^2 - \langle x,x_0\rangle$.
- For periodic solutions $x(0) = x(T)$, ψ is chosen as

$$\psi(x) = \begin{cases} 0 & x = 0 \\ +\infty & \text{elsewhere.} \end{cases}$$

- For antiperiodic solutions $x(0) = -x(T)$, it suffices to choose $\psi(x) = 0$ for each $x \in H$.

As a consequence of the above theorem, we provide a variational resolution to evolution equations involving nonlinear operators such as the Navier-Stokes equation with various time-boundary conditions,

$$\begin{cases} \frac{\partial u}{\partial t} + (u \cdot \nabla)u + f = \alpha\Delta u - \nabla p & \text{on } \Omega, \\ \operatorname{div} u = 0 & \text{on } [0,T] \times \Omega, \\ u = 0 & \text{on } [0,T] \times \partial\Omega, \end{cases} \qquad (18.47)$$

where Ω is a smooth domain of $\mathbf{R}^n$, $f \in L_{X^*}^2([0,T])$, $\alpha > 0$.

Indeed, setting $X = \{u \in H_0^1(\Omega;\mathbf{R}^n); \operatorname{div} u = 0\}$ and $H = L^2(\Omega)$, we write the problem above in the form

$$\begin{cases} \frac{\partial u}{\partial t} + \Lambda u \in -\partial\Phi(t,u) \\ \frac{u(0)+u(T)}{2} \in -\partial\psi(u(0) - u(T)), \end{cases} \qquad (18.48)$$

where ψ is a convex lower semicontinuous function on H, while the functional Φ and the nonlinear operator Λ are defined by

$$\Phi(t,u) = \frac{\alpha}{2}\int_\Omega \Sigma_{j,k=1}^3 (\frac{\partial u_j}{\partial x_k})^2\,dx + \langle u, f(t,x)\rangle \quad \text{and} \quad \Lambda u := (u \cdot \nabla)u. \qquad (18.49)$$

Note that $\Lambda : X \to X^*$ is regular as long as the dimension $N \leq 4$. On the other hand, when Λ lifts to path space, we have the following lemma.

Lemma 18.4. *With the above notation, we have*

1. *If $N = 2$, then the operator $\Lambda : \mathscr{X}_{2,2} \to L_{X^*}^2$ is regular.*
2. *If $N = 3$, then the operator Λ is regular from $\mathscr{X}_{4,\frac{4}{3}} \to L_{X^*}^{\frac{4}{3}}$ as well as from the space $\mathscr{X}_{2,\frac{4}{3}} \cap L^\infty(0,T;H)$ to $L_{X^*}^{\frac{4}{3}}$.*

Proof. First note that the three embeddings $\mathcal{X}_{2,2} \subseteq L_H^2$, $\mathcal{X}_{4,\frac{4}{3}} \subseteq L_H^2$, and $\mathcal{X}_{2,\frac{4}{3}} \subseteq L_H^2$ are compact.

Assuming first that $N = 3$, let $u^n \rightharpoonup u$ weakly in $\mathcal{X}_{4,\frac{4}{3}}$, and fix $v \in C^1([0,T] \times \Omega)$. We have that

$$\int_0^T \langle \Lambda u^n, v \rangle = \int_0^T \int_\Omega \Sigma_{j,k=1}^3 u_k^n \frac{\partial u_j^n}{\partial x_k} v_j \, dx \, dt = -\int_0^T \int_\Omega \Sigma_{j,k=1}^3 u_k^n \frac{\partial v_j}{\partial x_k} u_j^n \, dx.$$

Therefore

$$\left| \int_0^T \langle \Lambda u^n - \Lambda u, v \rangle \right| = \left| \Sigma_{j,k=1}^3 \int_0^T \int_\Omega (u_k^n \frac{\partial v_j}{\partial x_k} u_j^n - u_k \frac{\partial v_j}{\partial x_k} u_j) \, dx \, dt \right|$$

$$\leq \|v\|_{C^1([0,T] \times \Omega)} \sum_{j,k=1}^3 \int_0^T \int_\Omega |u_k^n u_j^n - u_k u_j| \, dx \, dt. \quad (18.50)$$

Also

$$\int_0^T \int_\Omega |u_k^n u_j^n - u_k u_j| \, dx \, dt \leq \int_0^T \int_\Omega |u_k^n u_j^n - u_k u_j^n| \, dx \, dt + \int_0^T \int_\Omega |u_k u_j^n - u_k u_j| \, dx \, dt$$

$$\leq \|u_j^n\|_{L_H^2} \|u_k^n - u_k\|_{L_H^2} + \|u_k\|_{L_H^2} \|u_j^n - u_j\|_{L_H^2} \to 0.$$

Moreover, we have for $N = 3$ the following standard estimate (see for example [156]):

$$\|\Lambda u^n\|_{X^*} \leq c |u^n|_H^{\frac{1}{2}} \|u^n\|_X^{\frac{3}{2}}. \quad (18.51)$$

Since $\mathcal{X}_{4,\frac{4}{3}} \subseteq C(0,T;H)$ is continuous, we obtain

$$\|\Lambda u^n\|_{L_{X^*}^{\frac{4}{3}}} \leq c |u^n|_{C(0,T;H)}^{\frac{1}{2}} \|u^n\|_{L_X^2}^{\frac{3}{4}} \leq c \|u^n\|_{\mathcal{X}_{4,\frac{4}{3}}}^{\frac{1}{2}} \|u^n\|_{L_X^2}^{\frac{3}{4}}, \quad (18.52)$$

from which we conclude that Λu^n is a bounded sequence in $L_{X^*}^{\frac{4}{3}}$, and therefore the weak convergence of $\langle \lambda u^n, v \rangle$ to $\langle \lambda u, v \rangle$ holds for each $v \in L_X^4$.

Now, since $\mathcal{X}_{2,2} \subseteq C(0,T;H)$ is also continuous, the same argument works for $N = 2$, the only difference being that we have the following estimate, which is better than (18.51):

$$\|\Lambda u^n\|_{X^*} \leq c |u^n|_H \|u^n\|_X. \quad (18.53)$$

To consider the case $\Lambda : \mathcal{X}_{2,\frac{4}{3}} \cap L^\infty(0,T;H) \to L_{X^*}^{\frac{4}{3}}$, we note that relations (18.50) and (18.51) still hold if $u_n \rightharpoonup u$ weakly in $\mathcal{X}_{2,\frac{4}{3}}$. We also have estimate (18.51). However, unlike the above, one cannot deduce (18.52) since we do not necessarily have a continuous embedding from $\mathcal{X}_{2,\frac{4}{3}} \subseteq C(0,T;H)$. However, if (u_n) is also assumed to be bounded in $L^\infty(0,T;H)$, then we get from (18.51) the estimate

$$\|\Lambda u^n\|_{L_{X^*}^{\frac{4}{3}}} \le c|u^n|_{L^\infty(0,T;H)}^{\frac{1}{2}}\|u^n\|_{L_X^2}^{\frac{3}{4}}, \tag{18.54}$$

which ensures the boundedness of Λu^n in $L_{X^*}^{\frac{4}{3}}$.

Corollary 18.1. *Assuming $N = 2$, f in $L_{X^*}^2([0,T])$, and that ℓ is a self-dual Lagrangian on $H \times H$ that is bounded from below, then the infimum of the functional*

$$I(u) = \int_0^T \left[\Phi(t,u) + \Phi^*(t, -\dot{u} - (u \cdot \nabla)u)\right] dt + \ell\left(u(0) - u(T), \frac{u(0) + u(T)}{2}\right)$$

on $\mathscr{X}_{2,2}$ is zero and is attained at a solution u of (18.47) that satisfies the following time-boundary condition:

$$-\frac{u(0) + u(T)}{2} \in \bar{\partial}\ell(u(0) - u(T)). \tag{18.55}$$

Moreover, u verifies the following "energy identity": For every $t \in [0,T]$,

$$\|u(t)\|_H^2 + 2\int_0^t \left[\Phi(t, u(t)) + \Phi^*(t, -\dot{u}(t) - (u \cdot \nabla)u(t))\right] dt = \|u(0)\|_H^2. \tag{18.56}$$

In particular, with appropriate choices for the boundary Lagrangian ℓ, the solution u can be chosen to verify either of the following boundary conditions:

- *an initial-value problem: $u(0) = u_0$ where u_0 is a given function in X;*
- *a periodic orbit : $u(0) = u(T)$;*
- *an antiperiodic orbit : $u(0) = -u(T)$.*

Proof. By Lemma 18.4, one can verify that the operator $\Lambda : \mathscr{X}_{2,2} \to L_{X^*}^2$ satisfies condition (18.17) and (A) of Theorem 18.1. Therefore, the infimum of the functional

$$I(u) = \int_0^T \left[\Phi(t, u(t)) + \Phi^*(t, -\dot{u}(t) - (u \cdot \nabla)u(t))\right] dt + \ell\left(u(0) - u(T), \frac{u(0) + u(T)}{2}\right)$$

on $\mathscr{X}_{2,2}$ is zero and is attained at a solution $u(t)$ of (18.47).

However, in the 3-dimensional case, we have to settle for the following result.

Corollary 18.2. *Assume $N = 3$, f in $L_{X^*}^2([0,T])$, and consider ℓ to be a self-dual Lagrangian on $H \times H$ that is now coercive in both variables. Then, there exists $u \in \mathscr{X}_{2,\frac{4}{3}}$ such that*

$$I(u) = \int_0^T \left[\Phi(t,u) + \Phi^*(t, -\dot{u} - (u \cdot \nabla)u)\right] dt + \ell\left(u(0) - u(T), \frac{u(0) + u(T)}{2}\right) \le 0,$$

and u is a weak solution of (18.47) that satisfies the time-boundary condition (18.55). Moreover, u verifies the "energy inequality"

$$\frac{\|u(T)\|_H^2}{2} + \int_0^T \left[\Phi(t, u(t)) + \Phi^*(t, -\dot{u}(t) - (u \cdot \nabla)u(t))\right] dt \le \frac{\|u(0)\|_H^2}{2}. \tag{18.57}$$

In particular, with appropriate choices for the boundary Lagrangian ℓ, the solution u will verify either one of the following boundary conditions:

- *an initial-value problem: $u(0) = u_0$.*
- *a periodicity condition of the form $u(0) = \delta u(T)$ for any given δ with $-1 < \delta < 1$.*

Proof. We start by considering the functional on the space $\mathscr{X}_{4,\frac{4}{3}}$.

$$I_\varepsilon(u) := \int_0^T \left[\Phi_\varepsilon(t,u) + \Phi_\varepsilon^*(t, -\dot{u} - (u \cdot \nabla)u)\right] dt + \ell\left(u(0) - u(T), -\frac{u(0) + u(T)}{2}\right),$$

where $\Phi_\varepsilon(t,u) = \Phi(t,u) + \frac{\varepsilon}{4}\|u\|_X^4$. In view of the preceding lemma, the operator $\Lambda u := (u \cdot \nabla)u$ and Φ_ε satisfy all properties of Theorem 18.1. In particular, we have the estimate

$$\|\Lambda u\|_{X^*} \leq c|u|_H^{1/2}\|u\|_X^{3/2} \qquad \text{for every } u \in X. \tag{18.58}$$

It follows from Theorem 18.1 that there exists $u_\varepsilon \in \mathscr{X}_{4,\frac{4}{3}}$ with $I_\varepsilon(u_\varepsilon) = 0$. This implies that

$$\begin{cases} \frac{\partial u_\varepsilon}{\partial t} + (u_\varepsilon \cdot \nabla)u_\varepsilon + f(t,x) = \alpha \Delta u_\varepsilon + \text{div}\left(\varepsilon\|u_\varepsilon\|^2 \nabla u_\varepsilon\right) - \nabla p_\varepsilon & \text{on } [0,T] \times \Omega, \\ \text{div}u_\varepsilon = 0 & \text{on } [0,T] \times \Omega, \\ u_\varepsilon = 0 & \text{on } [0,T] \times \partial\Omega, \\ -\frac{u_\varepsilon(0) + u_\varepsilon(T)}{2} \in \bar{\partial}\ell(u_\varepsilon(0) - u_\varepsilon(T)). \end{cases} \tag{18.59}$$

Now, we show that $(u_\varepsilon)_\varepsilon$ is bounded in $\mathscr{X}_{2,4/3}$. Indeed, multiply (18.59) by u_ε to get

$$\frac{d}{dt}\frac{|u_\varepsilon(t)|^2}{2} + \alpha\|u_\varepsilon(t)\|_X^2 + \varepsilon\|u_\varepsilon(t)\|_X^4 = \langle f(t), u_\varepsilon(t)\rangle \leq \frac{\alpha}{2}\|u_\varepsilon(t)\|_X^2 + \frac{2}{\alpha}\|f(t)\|_{X^*}^2$$

so that

$$\frac{d}{dt}\frac{|u_\varepsilon(t)|^2}{2} + \frac{\alpha}{2}\|u_\varepsilon(t)\|_X^2 + \varepsilon\|u_\varepsilon(t)\|_X^4 \leq \frac{2}{\alpha}\|f(t)\|_{X^*}^2. \tag{18.60}$$

Integrating (18.60) over $[0,s]$, $(s < T)$, we obtain

$$\frac{|u_\varepsilon(s)|^2}{2} - \frac{|u_\varepsilon(0)|^2}{2} + \frac{\alpha}{2}\int_0^s \|u_\varepsilon\|_X^2 + \varepsilon\int_0^s \|u_\varepsilon\|_X^4 \leq \frac{2}{\alpha}\int_0^s \|f(t)\|_{X^*}^2. \tag{18.61}$$

On the other hand, it follows from (18.59) that $\ell(u_\varepsilon(0) - u_\varepsilon(T), -\frac{u_\varepsilon(0) + u_\varepsilon(T)}{2}) = \frac{|u_\varepsilon(T)|^2}{2} - \frac{|u_\varepsilon(0)|^2}{2}$. Considering this together with (18.61) with $s = T$, we get

$$\ell\left(u_\varepsilon(0) - u_\varepsilon(T), \frac{u_\varepsilon(0) + u_\varepsilon(T)}{2}\right) + \frac{\alpha}{2}\int_0^T \|u_\varepsilon\|_X^2\, dt + \varepsilon\int_0^T \|u_\varepsilon\|_X^4\, dt$$
$$\leq \frac{2}{\alpha}\int_0^T \|f(t)\|_{X^*}^2.$$

Since ℓ is bounded from below and is coercive in both variables, it follows from the above that $(u_\varepsilon)_\varepsilon$ is bounded in L_X^2, that $(u_\varepsilon(T))_\varepsilon$ and $(u_\varepsilon(0))_\varepsilon$ are bounded in H, and that $\varepsilon \int_0^T \|u_\varepsilon(t)\|^4$ is also bounded. It also follows from (18.61) coupled with the boundedness of $(u_\varepsilon(0))_\varepsilon$ that u_ε is bounded in $L^\infty(0,T;H)$. Estimate (18.58) combined with the boundedness of $(u_\varepsilon)_\varepsilon$ in $L^\infty(0,T;H) \cap L_X^2$ implies that $(\Lambda u_\varepsilon)_\varepsilon$ is bounded in $L_X^{4/3}$. We also have the estimate

$$\left\| \alpha \Delta u_\varepsilon + \operatorname{div}\left(\varepsilon \|u_\varepsilon\|^2 \nabla u_\varepsilon\right)\right\|_{X^*} \leq \alpha \|u_\varepsilon\| + \varepsilon \|u_\varepsilon\|^3,$$

which implies that $\alpha \Delta u_\varepsilon + \operatorname{div}\left(\varepsilon \|u_\varepsilon\|^2 \nabla u_\varepsilon\right)$ is bounded in $L_{X^*}^{4/3}$.

It also follows from (18.59) that for each $v \in L_X^4$ we have

$$\int_0^T \left\langle \frac{\partial u_\varepsilon}{\partial t}, v \right\rangle dt = \int_0^T \left\langle \alpha \Delta u_\varepsilon - (u_\varepsilon \cdot \nabla)u_\varepsilon - f(t,x) + \operatorname{div}\left(\varepsilon \|u_\varepsilon\|^2 \nabla u_\varepsilon\right), v \right\rangle dt. \quad (18.62)$$

Since the right-hand side is uniformly bounded with respect to ε, so is the left-hand side, which implies that $\frac{\partial u_\varepsilon}{\partial t}$ is bounded in $L_{X^*}^{4/3}$. Therefore, there exists $u \in \mathscr{X}_{2,4/3}$ such that

$$u_\varepsilon \rightharpoonup u \quad \text{weakly in } L_X^2, \tag{18.63}$$

$$\frac{\partial u_\varepsilon}{\partial t} \rightharpoonup \frac{\partial u_\varepsilon}{\partial t} \quad \text{weakly in } L_{X^*}^{4/3}, \tag{18.64}$$

$$\operatorname{div}\left(\varepsilon \|u_\varepsilon\|^2 \nabla u_\varepsilon\right) \rightharpoonup 0 \quad \text{weakly in } L_{X^*}^{4/3}, \tag{18.65}$$

$$u_\varepsilon(0) \rightharpoonup u(0) \quad \text{weakly in } H, \tag{18.66}$$

$$u_\varepsilon(T) \rightharpoonup u(T) \quad \text{weakly in } H. \tag{18.67}$$

Letting ε approach zero in (18.62), it follows from (18.63)-(18.67) that

$$\int_0^T \left\langle \frac{\partial u}{\partial t}, v \right\rangle dt = \int_0^T \left\langle -(u \cdot \nabla)u - f(t,x) + \alpha \Delta u, v \right\rangle dt. \tag{18.68}$$

Also it follows from (18.66), (18.67) and (18.59), and the fact that $\overline{\partial}\ell$ is monotone, that

$$-\frac{u(0) + u(T)}{2} \in \overline{\partial}\ell(u(0) - u(T)). \tag{18.69}$$

(18.68) and (18.69) yield that u is a weak solution of

$$\begin{cases} \frac{\partial u}{\partial t} + (u \cdot \nabla)u + f(t,x) = \alpha \Delta u - \nabla p & \text{on } [0,T] \times \Omega, \\ \operatorname{div} u = 0 & \text{on } [0,T] \times \Omega, \\ u = 0 & \text{on } [0,T] \times \partial \Omega, \\ -\frac{u(0)+u(T)}{2} \in \overline{\partial}\ell(u(0) - u(T)). \end{cases} \tag{18.70}$$

Now we prove inequality (18.57). Since $I_\varepsilon(u_\varepsilon) = 0$, a standard argument (see the proof of Theorem 18.1) yields that $I(u) \leq \liminf_\varepsilon I_\varepsilon(u_\varepsilon) = 0$, thereby giving that

$$
I_\varepsilon(u) := \int_0^T \left[\Phi(t, u(t)) + \Phi^*(t, -\dot{u}(t) - (u \cdot \nabla)u(t)) \right] dt
$$
$$
+ \ell\left(u(0) - u(T), \frac{u(0) + u(T)}{2} \right)
$$
$$
\leq 0.
$$

On the other hand, it follows from (18.69) that

$$
\ell\left(u(0) - u(T), -\frac{u(0) + u(T)}{2} \right) = \frac{|u(T)|^2}{2} - \frac{|u(0)|^2}{2}.
$$

This together with the above inequality gives

$$
\frac{|u(T)|^2}{2} + \int_0^T \left[\Phi(t, u(t)) + \Phi^*(t, -\dot{u}(t) - (u \cdot \nabla)u(t)) \right] dt \leq \frac{|u(0)|^2}{2}.
$$

Corollary 18.3. *In dimension $N = 3$, there exists, for any given δ with $|\delta| < 1$, a weak solution of the equation*

$$
\begin{cases}
\frac{\partial u}{\partial t} + (u \cdot \nabla)u + f(t, x) = \alpha \Delta u - \nabla p & \text{on } [0, T] \times \Omega, \\
\text{div} u = 0 & \text{on } [0, T] \times \Omega, \\
u = 0 & \text{on } [0, T] \times \partial\Omega, \\
u(0) = \delta u(T).
\end{cases}
$$

Proof. For each δ with $|\delta| < 1$, there exists $\lambda > 0$ such that $\delta = \frac{\lambda-1}{\lambda+1}$. Now consider $\ell(a, b) = \psi_\lambda(a) + \psi_\lambda^*(b)$, where $\psi_\lambda(a) = \frac{\lambda}{4}|a|^2$.

Example 18.1. Navier-Stokes evolutions driven by their boundary

We now consider the evolution equation

$$
\begin{cases}
\frac{\partial u}{\partial t} + (u \cdot \nabla)u + f = \alpha \Delta u - \nabla p & \text{on } [0, T] \times \Omega \\
\text{div} u = 0 & \text{on } [0, T] \times \Omega, \\
u(t, x) = u^0(x) & \text{on } [0, T] \times \partial\Omega, \\
u(0, x) = \delta u(T, x) & \text{on } \Omega,
\end{cases}
\tag{18.71}
$$

where $\int_{\partial\Omega} u^0 \cdot \mathbf{n} \, d\sigma = 0$, $\alpha > 0$, and $f \in L_{X^*}^p$. Assuming that $u^0 \in H^{3/2}(\partial\Omega)$ and that $\partial\Omega$ is connected, Hopf's extension theorem again yields the existence of $v^0 \in H^2(\Omega)$ such that $v^0 = u^0$ on $\partial\Omega$, div $v^0 = 0$, and

$$
\int_\Omega \Sigma_{j,k=1}^n u_k \frac{\partial v_j^0}{\partial x_k} u_j \, dx \leq \varepsilon \|u\|_X^2 \quad \text{for all } u \in V,
\tag{18.72}
$$

where $V = \{u \in H^1(\Omega; \mathbf{R}^n); \mathrm{div}\, u = 0\}$. Setting $v = u + v^0$, solving equation (18.71) reduces to finding a solution in the path space $\mathscr{X}_{2,2}$ corresponding to the Banach space $X = \{u \in H^1_0(\Omega; \mathbf{R}^n); \mathrm{div}\, v = 0\}$ and the Hilbert space $H = L^2(\Omega)$ for

$$\frac{\partial u}{\partial t} + (u \cdot \nabla)u + (v^0 \cdot \nabla)u + (u \cdot \nabla)v^0 \in -\partial \Phi(u) \tag{18.73}$$

$$u(0) - \delta u(T) = (\delta - 1)v^0,$$

where $\Phi(t, u) = \frac{\alpha}{2} \int_\Omega \sum_{j,k=1}^3 (\frac{\partial u_j}{\partial x_k})^2 \, dx + \langle g, u \rangle$ and

$$g := f - \alpha \Delta v^0 + (v^0 \cdot \nabla)v^0 \in L^p_{V^*}.$$

In other words, this is an equation of the form

$$\frac{\partial u}{\partial t} + \Lambda u \in -\partial \Phi(t, u), \tag{18.74}$$

where $\Lambda u := (u \cdot \nabla)u + (v^0 \cdot \nabla)u + (u \cdot \nabla)v^0$ is the nonlinear regular operator for $N = 2$ or $N = 3$.

Now recalling the fact that the component $Bu := (v^0 \cdot \nabla)u$ is skew-symmetric, it follows from Hopf's estimate that

$$C\|u\|_V^2 \geq \Phi(t, u) + \langle \Lambda u, u \rangle \geq (\alpha - \varepsilon)\|u\|^2 + \langle g, u \rangle \quad \text{for all } u \in X.$$

As in Corollary 18.2, we have the following.

Corollary 18.4. *Assume $N = 3$ and consider ℓ to be a self-dual Lagrangian on $H \times H$ that is coercive in both variables. Then, there exists $u \in \mathscr{X}_{2,\frac{4}{3}}$ such that*

$$I(u) = \int_0^T \left[\Phi(t, u(t)) + \Phi^*(t, -\dot{u}(t) - \Lambda u(t)) + \langle u(t), \Lambda u(t) \rangle \right] dt$$

$$+ \ell \left(u(0) - u(T), -\frac{u(0) + u(T)}{2} \right)$$

$$\leq 0,$$

and u is a weak solution of (18.71).

To obtain the boundary condition given in (18.73) that is $u(0) - \delta u(T) = (\delta - 1)v^0$, consider $\ell(a, b) = \psi_\lambda(a) + \psi_\lambda^*(-b)$, where $\delta = \frac{\lambda - 1}{\lambda + 1}$ and $\psi_\lambda(a) = \frac{\lambda}{4}|a|^2 - 4\langle a, v^0 \rangle$.

Example 18.2. An equation in magneto-hydrodynamics

Let Ω be an open bounded domain of $\mathbf{R}^N$, and let $\mathbf{n}$ be the unit outward normal on its boundary Γ. Consider the evolution equation for the pair $(V, B); \Omega \to \mathbf{R}^N \times \mathbf{R}^N$

$$\begin{cases} \dfrac{\partial V}{\partial t} + (v \cdot \nabla)V - \dfrac{1}{R_e}\Delta V - S(B \cdot \nabla)B + \nabla\left(P + \dfrac{SB^2}{2}\right) = g(x) \text{ on } \Omega, \\ \dfrac{\partial B}{\partial t} + (v \cdot \nabla)B + \dfrac{1}{R_m}\operatorname{curl}(\operatorname{curl} B) - (B \cdot \nabla)V = 0 \text{ on } \Omega, \\ \operatorname{div}(V) = 0, \\ \operatorname{div}(B) = 0, \\ V(t,x) = 0 \text{ on } \Gamma, \\ B \cdot \mathbf{n} = 0 \text{ on } \Gamma, \\ \operatorname{curl} B \times \mathbf{n} = 0 \text{ on } \Gamma, \end{cases} \qquad (18.75)$$

with the initial conditions

$$V(x,0) = V_0(x) \text{ and } B(x,0) = B_0(x) \quad \text{for } x \in \Omega,$$

where R_e, R_m, and S are positive constants. We define an appropriate Banach space $X = X_1 \times X_2$ as

$$X_1 = \left\{U \in H_0^1(\Omega)^N; \operatorname{div}(U) = 0\right\},$$
$$X_2 = \left\{B \in H^1(\Omega)^N; \operatorname{div}(B) = 0, \ B \cdot \mathbf{n} = 0 \text{ on } \Gamma\right\},$$

where

$$(V_1, V_2)_{X_1} = \sum_{i=1}^N \left\langle \frac{\partial V_1}{\partial x_i}, \frac{\partial V_2}{\partial x_j} \right\rangle, \quad \|V\|_{X_1} = (V,V)_{X_1}^{\frac{1}{2}},$$

$$(B_1, B_2)_{X_2} = \langle \operatorname{curl}(B_1), \operatorname{curl}(B_2) \rangle, \quad \|B\|_{X_2} = (B,B)_{X_2}^{\frac{1}{2}},$$

are the scalar products and norms on X_1 and X_2, respectively. The scalar product on $H := L^2(\Omega)^N$ will be denoted by $\langle \, , \, \rangle$.

It is known [157] that the norm $\| \; \|_{X_2}$ is equivalent to that induced by $H^1(\Omega)^N$, and we equip the space X with the following scalar product (resp., norm): For a pair $u_i = (V_i, B_i) \in X$, $i = 1, 2$,

$$(u_1, u_2)_X = (V_1, V_2)_{X_1} + S(B_1, B_2)_{X_2} \quad \text{and} \quad \|u\| = \{(u_1, u_2)\}_X^{\frac{1}{2}}.$$

We define on $X \times X \times X$ a trilinear form

$$b(u_1, u_2, u_3) = b_1(V_1, V_2, V_3) - Sb_1(B_1, B_2, V_3) + Sb_1(V_1, B_2, B_3) - Sb_1(B_1, V_2, B_3),$$

where

$$b_1(\varphi, \psi, \theta) = \sum_{i,j=1}^N \int_\Omega \varphi_i \frac{\partial \psi_j}{\partial x_i} \theta_j \, dx, \quad \forall \varphi, \psi, \theta \in X,$$

in such a way that

$$b_1(\varphi, \psi, \psi) = 0 \quad \forall \varphi, \psi \in X.$$

Define the bilinear form

$$B : X \times X \to X^*$$
$$\big(B(u,u),v\big) = b(u,u,v),$$

and set $\tilde{g} = \begin{pmatrix} g \\ 0 \end{pmatrix} \in L^2(\Omega)^N \times L^2(\Omega)^N$. Equation (18.75) can then be written as

$$\frac{du}{dt} + B(u) - f \in -\partial \Phi(u)$$

in V, where $B(u) = B(u,u)$ is a regular map on $\mathscr{X}_{2,2}$ and $\Phi : X \to \mathbf{R}$ is the convex energy functional defined on L^2_X by

$$\Phi(u) = \Phi((V,B)) = \frac{1}{2R_e} \int_\Omega \sum_{i,j=1}^N \left| \frac{\partial V_i}{\partial x_j} \right|^2 dx + \frac{S}{2R_m} \int_\Omega |\mathrm{curl}\, B|^2 \, dx.$$

Using Theorem 18.1, one can prove the following existence result.

Theorem 18.2. *Suppose $N = 2$, $\tilde{g} \in L^2(\Omega)^N \times L^2(\Omega)^N$, and $(V_0, B_0) \in X$. The infimum of the functional*

$$I(u) = \int_0^T \left\{ \Phi\big(u(t)\big) + \Phi^* \left(-\frac{du}{dt} - B(u) + \tilde{g} \right) + \langle \tilde{g}, u \rangle \right\} dt$$

$$+ \int_\Omega \left\{ \frac{1}{2} \left(|V(0,x)|^2 + |B(0,x)|^2 + |V(x,T)|^2 + |B(x,T)|^2 \right) \right\} dx$$

$$+ \int_\Omega \left\{ |V_0(x)|^2 + |B_0(x)|^2 - 2V(0,x) \cdot V_0(x) - 2B(0,x) \cdot B_0(x) \right\} dx$$

on $\mathscr{X}_{2,2}$ is then equal to zero and is attained at a solution of the problem (18.75).

18.4 Schrödinger evolutions

Consider the nonlinear Schrödinger equation

$$iu_t + \Delta u - |u|^{r-1} u = -i \bar{\partial} L(t,u) \qquad (t,x) \in [0,T] \times \Omega, \qquad (18.76)$$

where Ω is a bounded domain in $\mathbf{R}^N$ and L is a time-dependent self-dual Lagrangian on $[0,T] \times H_0^1(\Omega) \times H^{-1}(\Omega)$. Equation (18.76) can be rewritten as

$$u_t + \Lambda u = -\bar{\partial} L(t,u) \qquad (t,x) \in [0,T] \times \Omega,$$

where $\Lambda u = -i\Delta + i|u|^{r-1} u$. We can then deduce the following existence result.

Theorem 18.3. *Suppose $1 \leq r \leq \frac{N}{N-2}$. Let $p = 2r$, and assume that L satisfies*

$$-C \leq \int_0^T L(t,u(t),0) dt \leq C(1 + \|u\|^r_{L^p_{H_0^1}}) \text{ for every } u \in L^p_{H_0^1}[0,T], \qquad (18.77)$$

$$\langle \overline{\partial} L(u), -\Delta u + |u|^{r-1}u \rangle \geq 0 \text{ for each } u \in H^2(\Omega). \tag{18.78}$$

Let $u_0 \in H^2(\Omega)$ and $\ell(a,b) = \frac{1}{4}\|a\|_H^2 - \langle a, u_0 \rangle + \|b - u_0\|_H^2$. The functional

$$I(u) = \int_0^T \left[L(u, -\dot{u} - \Lambda u) + \langle \Lambda u, u \rangle \right] dt + \ell\left(u(0) - u(T), \frac{u(T) + u(0)}{2} \right) \tag{18.79}$$

then attains its minimum at $v \in \mathscr{X}_{p,q}$ in such a way that $I(v) = \inf_{u \in \mathscr{X}_{p,q}} I(u) = 0$ and

$$\begin{cases} \dot{v}(t) - i\Delta v(t) + i|v(t)|^{r-1}v(t) = -\overline{\partial} L(t, v(t)), \\ \hspace{6em} v(0) = u_0. \end{cases} \tag{18.80}$$

Proof. Let $X = H_0^1(\Omega)$ and $H = L^2(\Omega)$. Taking into account Theorem 18.1, we just need to verify conditions (18.17) and (A) and prove that I satisfies the self-dual Palais-Smale condition on $\mathscr{X}_{p,q}$. Condition (A) follows from the fact that $\langle \Lambda u, u \rangle = 0$. To prove (18.17), note that

$$\|\Lambda u\|_{H^{-1}} = \| -\Delta u + |u|^{r-1}u \|_{H^{-1}} \leq \| -\Delta u \|_{H^{-1}} + C\||u|^{r-1}u\|_{L^q(\Omega)}$$
$$= \|u\|_{H_0^1} + C\|u\|_{L^{rq}}^r.$$

Since $p \geq 2$, we have $qr \leq 2r \leq \frac{2N}{N-2}$. It follows from the Sobolev inequality and the above that

$$\|\Lambda u\|_{H^{-1}} \leq \|u\|_{H_0^1} + C\|u\|_{H_0^1}^r,$$

from which we obtain

$$\|\Lambda u\|_{L_{H^{-1}}^q} \leq \|u\|_{L_{H_0^1}^q} + C\|u\|_{L_{H_0^1}^{rq}}^r \leq C(\|u\|_{L_{H_0^1}^p} + \|u\|_{L_{H_0^1}^p}^r).$$

To show that I satisfies the self-dual Palais-Smale condition on $\mathscr{X}_{p,q}$, we assume that u_n is a sequence in $\mathscr{X}_{p,q}$ and $\varepsilon_n \to 0$ are such that

$$\begin{cases} -\dot{u}_n + i\Delta u_n - i|u_n|^{r-1}u_n = -\varepsilon_n \|u_n\|^{p-2}\Delta u_n + \overline{\partial} L(u_n), \\ \hspace{10em} u_n(0) = u_0. \end{cases} \tag{18.81}$$

Since $u_0 \in H^2(\Omega)$, it is standard that at least $u_n \in H^2(\Omega)$. Now multiply both sides of the above equation by $\Delta u_n(t) - |u_n(t)|^{r-1}u_n(t)$, and taking into account (12.37) we have

$$\langle \dot{u}_n(t), -\Delta u_n(t) + |u_n(t)|^{r-1}u_n(t) \rangle \leq 0,$$

from which we obtain

$$\frac{1}{2}\|u_n(t)\|_{H_0^1}^2 + \frac{1}{r+1}\|u_n(t)\|^{r+1} \leq \frac{1}{2}\|u(0)\|_{H_0^1}^2 + \frac{1}{r+1}\|u(0)\|^{r+1},$$

which, once combined with (18.81), gives the boundedness of $(u_n)_n$ in $\mathscr{X}_{p,q}$.

Example 18.3. Initial-value Schrödinger evolutions

Here are two typical examples of self-dual Lagrangians satisfying the assumptions of the preceding Theorem:

- $L(u, p) = \varphi(u) + \varphi^*(p)$, where $\varphi \equiv 0$. This leads to a solution of

$$\begin{cases} i\dot{u}(t) + \Delta u(t) - |u(t)|^{r-1} u(t) = 0, \\ \qquad\qquad\qquad\qquad\qquad u(0) = u_0. \end{cases} \tag{18.82}$$

- $L(u, p) = \varphi(u) + \varphi^*(\mathbf{a} \cdot \nabla u + p)$, where $\varphi(u) = \frac{1}{2} \int_{\Omega} |\nabla u|^2 \, dx$ and $\mathbf{a}$ is a divergence-free vector field on Ω with compact support. In this case, we have a solution for

$$\begin{cases} i\dot{u}(t) + \Delta u(t) - |u(t)|^{r-1} u(t) = i\mathbf{a} \cdot \nabla u + i\Delta u(t), \\ \qquad\qquad\qquad\qquad\qquad\qquad\quad u(0) = u_0. \end{cases} \tag{18.83}$$

18.5 Noncoercive nonlinear evolutions

We now assume that there is a symmetric linear duality map D between X and X^*.

Theorem 18.4. *Let $(\bar{S}_t)_{t \in \mathbf{R}}$ be a C_0-unitary group of operators associated to a skew-adjoint operator A on the Hilbert space X^*, and let $S_t := D^{-1} \bar{S}_t D$ be the corresponding group on X. For $p > 1$ and $q = \frac{p}{p-1}$, assume that $\Lambda : \mathscr{X}_{p,q} \to L_{X^*}^q$ is a regular map such that for some nondecreasing continuous real function w and $0 \le k < 1$ it satisfies*

$$\|\Lambda S_t x\|_{L_{X^*}^q} \le k \|\dot{x}\|_{L_{X^*}^q} + w(\|x\|_{L_X^p}) \text{ for every } x \in \mathscr{X}_{p,q} \tag{18.84}$$

and

$$\left| \int_0^T \langle \Lambda x(t), x(t) \rangle \, dt \right| \le w(\|x\|_{L_X^p}) \text{ for every } x \in \mathscr{X}_{p,q}. \tag{18.85}$$

Let ℓ be a self-dual Lagrangian on $H \times H$ that is bounded below with $0 \in \mathrm{Dom}(\ell)$, and let L be a time-dependent self-dual Lagrangian on $[0, T] \times X \times X^*$ such that for some $C > 0$ and $r > 1$, we have

$$-C \le \int_0^T L(t, u(t), 0) dt \le C(1 + \|u\|_{L_X^p}^r) \text{ for every } u \in L_X^p. \tag{18.86}$$

Assume that the functional

$$I(u) = \int_0^T \left[L(t, S_t u, -\bar{S}_t \dot{u} - \Lambda S_t u) + \langle \Lambda S_t u, S_t u \rangle \right] dt$$

$$+ \ell\left(u(0) - u(T), -\frac{u(T) + u(0)}{2} \right)$$

satisfies the self-dual Palais-Smale condition on $\mathscr{X}_{p,q}$. Then, it attains its minimum at $u \in \mathscr{X}_{p,q}$ in such a way that

$$I(u) = \inf_{w \in \mathscr{X}_{p,q}} I(w) = 0.$$

Moreover, if $S_t = \bar{S}_t$ on X, then $v(t) = S_t u(t)$ is a solution of

$$\begin{cases} \Lambda v(t) + Av(t) + \dot{v}(t) \in -\bar{\partial}L(t, v(t)), \\ \frac{v(0) + S_{(-T)}v(T)}{2} \in -\bar{\partial}\ell\big(v(0) - S_{(-T)}v(T)\big). \end{cases} \tag{18.87}$$

Proof. Define the nonlinear map $\Gamma : \mathscr{X}_{p,q} \to L^q_{X^*}$ by $\Gamma(u) = S_t^* \Lambda S_t(u)$. This map is also regular in view of the regularity of Λ. It follows from Remark 3.3 that the self-dual Lagrangian L_S also satisfies (18.16). It remains to show that Γ satisfies conditions (18.17) and (A) of Theorem 18.1. Indeed, for $x \in \mathscr{X}_{p,q}$, we have

$$\|\Gamma x\|_{L^q_{X^*}} = \|S_t^* \Lambda S_t x\|_{L^q_{X^*}} = \|\Lambda S_t x\|_{L^q_{X^*}} \le k\|\dot{x}\|_{L^q_{X^*}} + w(\|x\|_{L^p_X}) \tag{18.88}$$

and

$$\left| \int_0^T \langle \Gamma x(t), x(t) \rangle \, dt \right| = \left| \int_0^T \langle \Lambda S_t x(t), S_t x(t) \rangle \, dt \right| \le w(\|S_t x\|_{L^p_X}) = w(\|x\|_{L^p_X}).$$

This means that all the hypotheses in Theorem 18.1 are satisfied. Hence, there exists $u \in \mathscr{X}_{p,q}$ such that $I(u) = 0$ and as in the proof of Theorem 10.1, $v(t) = S_t u(t)$ is a solution of (18.87).

Example 18.4. Variational resolution for a fluid driven by $-i\Delta^2$

Consider the problem of finding "periodic-type" solutions for the following equation

$$\begin{cases} \frac{\partial u}{\partial t} + (u \cdot \nabla)u - i\Delta^2 u + f = \alpha\Delta u - \nabla p & \text{on } \Omega \subset \mathbf{R}^n, \\ \operatorname{div} u = 0 & \text{on } \Omega, \\ u = 0 & \text{on } \partial\Omega, \end{cases} \tag{18.89}$$

where $u = (u_1, u_2)$ and $i\Delta^2 u = (\Delta^2 u_2, -\Delta^2 u_1)$ with

$$\operatorname{Dom}(i\Delta^2) = \{u \in H_0^1(\Omega); \Delta u \in H_0^1(\Omega) \text{ and } u = \Delta u = 0 \text{ on } \partial\Omega\}.$$

Theorem 18.5. *Let $(S_t)_{t \in \mathbf{R}}$ be the C_0-unitary group of operators associated to the skew-adjoint operator $i\Delta^2$. Assuming $N = 2$, f in $L^2_{X^*}([0,T])$, and that ℓ is a self-dual Lagrangian on $H \times H$ that is bounded from below, then the infimum of the functional*

$$I(u) = \int_0^T \big[\Phi(t, S_t u) + \Phi^*(t, -S_t \dot{u} - S_t^* \Lambda S_t u)\big] \, dt + \ell\left(u(0) - u(T), -\frac{u(0) + u(T)}{2}\right)$$

on $\mathscr{X}_{2,2}$ is zero and is attained at $u(t)$ in such a way that $v(t) = S_t u(t)$ is a solution of (18.89) that satisfies the following time-boundary condition:

$$-\frac{v(0) + S_{(-T)}v(T)}{2} \in \overline{\partial}\ell\big(v(0) - S_{(-T)}v(T)\big). \qquad (18.90)$$

Moreover, u verifies the following "energy identity": For every $t \in [0,T]$,

$$\|u(t)\|_H^2 + 2\int_0^t \big[\Phi(s, S_s u) + \Phi^*(s, -S_s \dot{u} - S_s^* \Lambda S_s u)\big]\, ds = \|u(0)\|_H^2. \qquad (18.91)$$

In particular, with appropriate choices for the boundary Lagrangian ℓ, the solution v can be chosen to verify one of the following boundary conditions:

- *an initial-value problem: $v(0) = v_0$, where v_0 is a given function in H;*
- *a periodic orbit: $v(0) = S_{(-T)}v(T)$;*
- *an antiperiodic orbit: $v(0) = -S_{(-T)}v(T)$.*

Proof. The duality map between X and X^* is $D = -\Delta$ and is therefore linear and symmetric. Also, we have $S_t = e^{it\Delta^2}$ and therefore $S_t D = D S_t$. Now the result follows from Theorem 18.4.

Exercises 18.A.

1. Show that the norm $\|u\| = \|\mathrm{curl}(u)\|_2$ is equivalent to the H^1-norm on $H^1(\Omega; \mathbf{R}^2)$.
2. Show that the operator Λ defined in Example 18.2, is indeed regular.
3. Complete the proof of Theorem 18.75, and investigate whether it can be extended to higher dimensions.

Further comments

The existence of weak solutions for the Navier-Stokes evolutions is of course well known (see for example Leray [85], Masuda [94], or the books of Temam [156] and [157]). Our approach – developed in [65] – is new since it is variational, and it applies to many other nonlinear evolutions. What is remarkable is that, in dimension 2 (resp., dimension 3), it gives Leray solutions that satisfy the energy identity (resp., inequality), which appears like another manifestation of the concept of self-duality in the context of evolution equations.

References

1. R. A. Adams, *Sobolev spaces*, Academic Press, New York, 1975.
2. L. A. Ambrosio, N. Gigli, G. Savare, *Gradient Flows in Metric Spaces and in the Wasserstein Space of Probability Measures*, Lecture Notes in Mathematics, Birkhäuser, Boston, 2005.
3. T. Arai, *On the existence of the solution for* $\partial\varphi(u'(t)) + \partial\psi(u(t)) \ni f(t)$, J. Fac. Sci. Univ. Tokyo Sect. IA, Math., 26(1):75–96, 1979.
4. E. Asplund, *Averaged norms*, Isr. J. Math., 5:227–233, 1967.
5. E. Asplund, *Topics in the theory of convex functions*, in *Theory and Applications of Monotone Operators (Proceedings of the NATO Advanced Study Institute, Venice, 1968)*, pages 1–33, Edizioni "Oderisi", Gubbio, 1969.
6. J.-P. Aubin, *Variational principles for differential equations of elliptic, parabolic and hyperbolic type*, in *Mathematical Techniques of Optimization, Control and Decision*, pages 31–45, Birkhäuser, Boston, 1981.
7. J.-P. Aubin, A. Cellina, *Differential inclusions*, Grundlehren der Mathematischen Wissenschaften, volume 264, Springer-Verlag, Berlin, 1984.
8. J. P. Aubin, I. Ekeland, *Applied Nonlinear Analysis*, reprint of the 1984 original, Dover Publications, Inc., Mineola, NY, 2006.
9. G. Auchmuty, *Duality for nonconvex variational principles*, J. Differential Equations, 50(1):80 –145, 1983.
10. G. Auchmuty, *Variational principles for operator equations and initial-value problems*, Nonlinear Anal. Theory Methods Appl., 12(5):531–564, 1988.
11. G. Auchmuty, *Variational principles for variational inequalities*, Numer. Funct. Anal. Optim. 10(9-10): 863–874, 1989.
12. G. Auchmuty, *Saddle points and existence-uniqueness for evolution equations*, Differential Integral Equations, 6:1161–1171, 1993.
13. G. Auchmuty, *Min-max problems for nonpotential operator equations*, Optimization methods in partial differential equations (South Hadley, MA, 1996), 19–28, Contemp. Math., 209, Amer. Math. Soc., Providence, RI, 1997.
14. C. Baiocchi, A. Capelo, *Variational and Quasivariational Inequalities: Applications to Free Boundary Value Problems,* Wiley, New York, 1984.
15. E. J. Balder, *A general approach to lower semicontinuity and lower closure in optimal control theory*, SIAM J. Control Optim., 22(4):570–598, 1984.
16. V. Barbu. *Existence theorems for a class of two point boundary problems.* J. Differential Equations, 17:236–257, 1975.
17. V. Barbu, *Nonlinear Semigroups and Differential Equations in Banach spaces*, Noordhoff International Publishing, Leyden, 1976.
18. V. Barbu, *Optimal Control of Variational Inequalities*, Research Notes in Mathematics, volume 100, Pitman, Boston, 1984.

19. V. Barbu, *Abstract periodic Hamiltonian systems*, Adv. Differential Equations, 1(4): 675–688, 1996.

20. V. Barbu, K. Kunisch, *Identification of nonlinear parabolic equations*, Control Theor. Adv. Tech., 10(4, part 5):1959–1980, 1995.

21. C. Bardos, *Problèmes aux limites pour les equations aux dérivées partielles du premier ordre a coefficients réels; Théorèmes d'approximation; Application à l'équation de transport*, Ann. sci. Ecole Normale Superieure, Ser. 4, 3: 185–233, 1970.

22. H. H. Bauschke, X. Wang, *The kernel average of two convex functions and its applications to the extension and representation of monotone operators*, preprint, 2007.

23. M. A. Biot, *Variational principles in irreversible thermodynamics with application to viscoelasticity*, Phys. Rev. 97(2):1463–1469, 1955.

24. Y. Brenier, *Order preserving vibrating strings and applications to electrodynamics and magnetohydrodynamics*, Methods Appl. Anal., 11(4):515–532, 2004.

25. H. Brézis, *Opérateurs maximaux monotones et semigroupes de contractions dans les espaces de Hilbert*, North-Holland, Amsterdam, 1973.

26. H. Brézis, *Analyse Fonctionnelle, Theorie et applications*, Masson, Paris, 1987.

27. H. Brézis, *Nonlinear perturbations of monotone operators*, Technical Report 25, University of Kansas, Lawrence, 1972.

28. H. Brézis, M. G. Crandall, A. Pazy, *Perturbations of nonlinear maximal monotone sets in Banach space*, Commun. Pure Appl. Math., 23:123–144, 1970.

29. H. Brézis, I. Ekeland, *Un principe variationnel associé à certaines equations paraboliques. Le cas independant du temps*, C.R. Acad. Sci. Paris Sér. A, 282:971–974, 1976.

30. H. Brézis, I. Ekeland, *Un principe variationnel associé à certaines equations paraboliques. Le cas dependant du temps*, C. R. Acad. Sci. Paris Sér. A, 282:1197–1198, 1976.

31. H. Brézis, L. Nirenberg, G. Stampacchia, *A remark on Ky Fan's minimax principle*, Boll. U. M., 1:293–300, 1972.

32. H. Brézis, A. Pazy, *Semigroups of nonlinear contractions on convex sets*, J. Funct. Anal., 6:237–281, 1970.

33. F. Browder, *Problèmes non linéaires*, Presses de l'Université de Montréal, Montréal, 1966.

34. F. Browder, *Nonlinear maximal monotone operators in Banach space*, Math. Ann., 175: 89–113, 1968.

35. R.S. Burachik, B. F. Svaiter, *Maximal monotonicity, conjugation and the duality product*, Proc. Am. Math. Soc., 131 (8): 2379–2383, 2003.

36. C. Castaing, M. Valadier, *Convex Analysis and Measurable Multifunctions*, Springer-Verlag, New York, 1977.

37. T. Cazenave, *Semilinear Schrödinger Equations*, Courant Lecture Notes in Mathematics, volume 10. New York University, Courant Institute of Mathematical Sciences, New York, American Mathematical Society, Providence, RI, 2003.

38. P. Colli, *On some doubly nonlinear evolution equations in Banach spaces,* Jpn. J. Ind. Appl. Math., 9(2):181–203, 1992.

39. P. Colli and A. Visintin, *On a class of doubly nonlinear evolution equations,* Commun. Partial Differential Equations, 15(5):737–756, 1990.

40. G. Dal Maso, *An Introduction to Γ-Convergence*, Progress in Nonlinear Differential Equations and their Applications, volume 8, Birkhäuser, Boston, 1993.

41. R. Dautray, J. L. Lions. *Mathematical Analysis and Numerical Methods for Science and Technology, volume 2, Functional and Variational Methods*, Springer-Verlag, New york, 1988.

42. J. Diestel, *Banach Spaces – Selected Topics*, Lecture Notes in Mathematics, Volume 485, Springer-Verlag, Berlin, 1975.

43. J. Diestel, J. J. Uhl, *Vector Measures*, with a foreword by B. J. Pettis, Mathematical Surveys, No. 15, American Mathematical Society, Providence, R. I., 1977.

44. G. Duvaut, J.-L. Lions, *Inequalities in Mechanics and Physics*, Springer, Berlin, 1976.

45. M. A. Efendiev, A. Mielke, *On the rate-independent limit of systems with dry friction and small viscosity*, J. Convex Anal., 13(1):151–167, 2006.

46. I. Ekeland, *Convexity Methods in Hamiltonian Mechanics*, Springer-Verlag, Berlin, 1990.
47. I. Ekeland, R. Temam, *Convex Analysis and Variational Problems*, Classics in Applied Mathematics, volume 28, SIAM, Philadelphia, 1999.
48. L. C. Evans, *Partial Differential Equations*, Graduate Studies in Mathematics, volume 19, American Mathematical Society, Providence, RI, 1998.
49. Ky Fan, *Minimax theorems*, Proc. Natl. Acad. Sci. USA., 39:42–47, 1953.
50. S. P. Fitzpatrick, *Representing monotone operators by convex functions*, Proc. Centre Math. Anal. 20:59–65, 1989.
51. N. Ghoussoub, *Duality and Perturbation Methods in Critical Point Theory*, Cambridge Tracts in Mathematics, Cambridge University Press, Cambridge, 1993.
52. N. Ghoussoub, *A theory of antiself-dual Lagrangians: Stationary case*, C. R. Acad. Sci. Paris, Ser. I 340:245–250, 2005.
53. N. Ghoussoub, *A theory of antiself-dual Lagrangians: Dynamical case*, C. R. Acad. Sci. Paris, Ser. I 340:325–330, 2005.
54. N. Ghoussoub, *A variational principle for nonlinear transport equations*, Commun. Pure Appl. Anal., 4(4):735–742, 2005.
55. N. Ghoussoub, *Antiself-dual Lagrangians: Variational resolutions of non self-adjoint equations and dissipative evolutions*, Ann. Inst. Henri Poincaré, Analyse non-linéaire, 24: 171–205, 2007.
56. N. Ghoussoub, *Antisymmetric Hamiltonians: Variational resolution of Navier-Stokes equations and other nonlinear evolutions*, Commun. Pure Appl. Math., 60(5):619–653, 2007.
57. N. Ghoussoub, *Maximal monotone operators are self-dual vector fields and vice-versa*, Preprint available at: http://www.birs.ca/ nassif/, 2006.
58. N. Ghoussoub, *Superposition of self-dual functionals for nonhomogeneous boundary value problems and differential systems*, J. Discrete Contin. Dyn. Syst., 21(1):71–104, 2008.
59. N. Ghoussoub, *Hamiltonian systems as self-dual equations*, Front. Math. China, 3(2): 167–193, 2008.
60. N. Ghoussoub, *A variational theory for monotone vector fields*, J. Fixed Point Theor. Appli., (in press).
61. N. Ghoussoub, R. McCann, *A least action principle for steepest descent in a non-convex landscape*, Contemp. Math., 362:177-187, 2004.
62. N. Ghoussoub, A. Moameni, *On the existence of Hamiltonian paths connecting Lagrangian submanifolds*, Math. Rep. Acad. Sci. R. Soc. Can. (in press).
63. N. Ghoussoub, A. Moameni, *Self-dual variational principles for periodic solutions of Hamiltonian and other dynamical systems*, Commun. Partial Differential Equations, 32: 771–795, 2007.
64. N. Ghoussoub, A. Moameni, *Hamiltonian systems of PDEs with self-dual boundary conditions* (submitted), Preprint available at: http://www.birs.ca/ nassif/, 2008.
65. N. Ghoussoub, A. Moameni, *Antisymmetric Hamiltonians (II): Variational resolution of Navier-Stokes equations and other nonlinear evolutions*, Ann. Inst. Henri Poincaré, Analyse non-linéaire (in press).
66. N. Ghoussoub, A. Moradifam, *Simultaneous preconditioning and symmetrization of non-symmetric linear systems* (submitted), Preprint available at: http://www.birs.ca/ nassif/, 2008.
67. N. Ghoussoub, L. Tzou, *A variational principle for gradient flows*, Math. Ann., 30(3): 519–549, 2004.
68. N. Ghoussoub, L. Tzou, *Antiself-dual Lagrangians II: Unbounded non self-adjoint operators and evolution equations*, Ann. Mat. Pura Appl., 187:323–352, 2008.
69. N. Ghoussoub, L. Tzou, *Iterations of antiself-dual Lagrangians and applications to Hamiltonian systems and multiparameter gradient flows*, Calculus Variations Partial Differential Equations, 26(4):511–534, 2006.
70. D. Gilbarg, N. S. Trudinger, *Elliptic Partial Differential Equations of Second Order*, 2nd ed., Classics in Mathematics series, Springer-Verlag, New York, 19??.
71. G. Gilardi, U. Stefanelli, *Time-discretization and global solution for a doubly nonlinear Volterra equation*, J. Differential Equations, 228(2):707–736, 2006.

72. M. E. Gurtin, *Variational principles in the linear theory of viscoelasticity*, Arch. Ration. Mech. Anal., 13:179–191, 1963.

73. M. E. Gurtin, *Variational principles for linear elastodynamics*, Arch. Ration. Mech. Anal., 16:34–50, 1964.

74. M. E. Gurtin, *Variational principles for linear initial-value problems*, Q. Appl. Math., 22:252–256, 1964.

75. W. Han, B. D. Reddy, *Convergence of approximations to the primal problem in plasticity under conditions of minimal regularity*, Numer. Math., 87(2):283–315, 2000.

76. N. Hirano, *Existence of periodic solutions for nonlinear evolution equations in Hilbert spaces,* Proc. Am. Math. Soc., 120 (1): 185–192, 1994.

77. N. Hirano, N. Shioji, *Invariant sets for nonlinear evolution equations, Couchy problems and periodic problems*, Abstr. Appl. Anal. (3):183–203, 2004.

78. I. Hlaváček, *Variational principles for parabolic equations,* Appl. Mat., 14:278–297, 1969.

79. R. Jordan, D. Kinderlehrer, F. Otto, *The variational formulation of the Fokker-Planck equation*, SIAM J. Math. Anal., 29:1–17, 1998.

80. J. Jost, *Riemannian Geometry and Geometric Analysis*, Universitytext, Springer, New York, 2002.

81. D. Kinderlehrer, G. Stampachia, *An introduction to variational inequalities and their applications*, Classics in Applied Mathematics, volume 31, SIAM, Philadelphia, 2000.

82. E. Krauss, *A representation of maximal monotone operators by saddle functions*, Rev. Roum. Math. Pures Appl. 309: 823–837, 1985.

83. J. Lemaitre, J.-L. Chaboche, *Mechanics of solid materials*, Cambridge University Press, Cambridge,1990.

84. B. Lemaire, *An asymptotical variational principle associated with the steepest descent method for a convex function*, J. Convex Anal., 3(1):63–70, 1996.

85. J. Leray, *Sur le mouvement d'un liquide visqueux emplissant l'espace*, Acta Math., 63: 193–248, 1934.

86. A. Lew, J. E. Marsden, M. Ortiz, M. West, *Asynchronous variational integrators*, Arch. Ration. Mech. Anal., 167(2):85–146, 2003.

87. A. Lew, J. E. Marsden, M. Ortiz, M. West, *Variational time integrators*, Int. J. Numer. Methods Eng., 60(1):153–212, 2004.

88. J. L. Lions, E. Magenes, *Nonhomogeneous boundary value problems and applications*. Volume 3, Springer-Verlag, Berlin, 1973.

89. M. Mabrouk, *Sur un problème d'évolution à données mesures; approche variationnelle*, C. R. Acad. Sci. Paris Sér. I. Math., 318(1):47–52, 1994.

90. M. Mabrouk, *A variational approach for a semilinear parabolic equation with measure data*, Ann. Fac. Sci. Toulouse Math. Ser. VI, 9(1):91–112, 2000.

91. M. Mabrouk, *Un principe variationnel pour une équation non linéaire du second ordre en temps,* C. R. Acad. Sci. Paris Sér. I, Math., 332(4):381–386, 2001.

92. M. Mabrouk, *A variational principle for a nonlinear differential equation of second order*, Adv. Appl. Math., 31(2):388–419, 2003.

93. A. Mainik and A. Mielke, *Existence results for energetic models for rate-independent systems*, Calc., Variations Partial Differential Equations, 22(1):73–99, 2005.

94. K. Masuda, *Weak solutions of Navier-Stokes equations*, Tohoku Math. J. 36: 623–646, 1984.

95. J. E. Marsden, G. W. Patrick, S. Shkoller, *Multisymplectic geometry, variational integrators, and nonlinear PDEs*, Commun. Math. Phys., 199(2):351–395, 1998.

96. J. Mawhin, M. Willem, *Critical point theory and Hamiltonian systems*. Applied Mathematical Sciences, volume74, Springer Verlag, New York,1989.

97. V. G. Maz'ja, *Sobolev Spaces*, Springer-Verlag, Berlin,1985.

98. A. Mielke, *Finite elastoplasticity Lie groups and geodesics on* SL(d), Geometry, Mechanics, and Dynamics, pages 61–90, Springer, New York, 2002.

99. A. Mielke, *Energetic formulation of multiplicative elasto-plasticity using dissipation distances*, Contin. Mech. Thermodyn., 15(4):351–382, 2003.

100. A. Mielke, *Evolution of rate-independent inelasticity with microstructure using relaxation and Young measures,* Solid Mech. Appl., 108: 33–44, 2003.

101. A. Mielke, *Existence of minimizers in incremental elasto-plasticity with finite strains,* SIAM J. Math. Anal., 36(2):384–404 (electronic), 2004.

102. A. Mielke, *Evolution of rate-independent systems,* C. Dafermos and E. Feireisl, editors, *Handbook of Differential Equations, Evolutionary Equations,* volume 2, pages 461–559. Elsevier, Amsterdam, 2005.

103. A. Mielke, T. Roubíček, *A rate-independent model for inelastic behavior of shape-memory alloys,* Multiscale Model. Simul., 1(4):571–597 (electronic), 2003.

104. A. Mielke, T. Roubíček, *Rate-independent damage processes in nonlinear elasticity,* Math. Models Methods Appl. Sci., 16(2):177–209, 2006.

105. A. Mielke, R. Rossi, *Existence and uniqueness results for general rate-independent hysteresis problems,* Math. Models Methods Appl. Sci., 17:81–123, 2007.

106. A. Mielke, T. Roubíček, U. Stefanelli, *Γ-limits and relaxations for rate-independent evolutionary problems,* Calculus Variations Partial Differential Equations (to appear).

107. A. Mielke, R. Rossi, G. Savaré, *A metric approach to a class of doubly nonlinear evolution equations and applications,* Ann. Scu. Normale. Super. Pisa Cl. Sci. (to appear). Preprint available at: http://www.wias-berlin.de/people/mielke/.

108. A. Mielke, U. Stefanelli, *A discrete variational principle for rate-independent evolutions,* Preprint, 2008.

109. A. Mielke, F. Theil, *A mathematical model for rate-independent phase transformations with hysteresis,* H.-D. Alber, R. Balean, and R. Farwig, editors, *Proceedings of the Workshop on "Models of Continuum Mechanics in Analysis and Engineering,"* pages 117–129, Shaker-Verlag, 1999.

110. A. Mielke, A. M. Timofte, *An energetic material model for time-dependent ferroelectric behavior: existence and uniqueness,* Math. Models Appl. Sci., 29:1393–1410, 2005.

111. A. Mielke, F. Theil, V. I. Levitas, *A variational formulation of rate-independent phase transformations using an extremum principle,* Arch. Ration. Mech. Anal., 162(2):137–177, 2002.

112. A. Mielke, M. Ortiz, *A class of minimum principles for characterizing the trajectories and the relaxation of dissipative systems,* ESAIM Control Optim. Calc. Var. (to appear), Preprint available at: http://www.wias-berlin.de/main/publications/wias-publ/index.cgi.en.

113. S. Müller and M. Ortiz, *On the Γ-convergence of discrete dynamics and variational integrators,* J. Nonlinear Sci., 14(3):279–296, 2004.

114. J.-J. Moreau, *La notion de sur-potentiel et les liaisons unilatérales en élastostatique,* C. R. Acad. Sci. Paris Sér. A-B, 267:A954–A957, 1968.

115. J.-J. Moreau, *Sur les lois de frottement, de viscosité et plasticité,* C. R. Acad. Sci. Paris Sér. II, Méc. Phys. Chim. Sci. Univ. Sci. Terre, 271:608–611, 1970.

116. J.-J. Moreau, *Sur l'évolution d'un système élasto-visco-plastique,* C. R. Acad. Sci. Paris Sér. A-B, 273:A118–A121, 1971.

117. U. Mosco, *Convergence of convex sets and of solutions of variational inequalities,* Adv. Math., 3:510–585, 1969.

118. B. Nayroles, *Deux théorèmes de minimum pour certains systèmes dissipatifs,* C. R. Acad. Sci. Paris Sér. A-B, 282(17):Aiv, A1035–A1038, 1976.

119. B. Nayroles, *Un théorème de minimum pour certains systèmes dissipatifs, Variante hilbertienne,* Travaux Sém. Anal. Convexe, 6(Exp. 2):22, 1976.

120. L. Nirenberg, *Variational Methods in nonlinear problems,* Lecture notes in Mathematics, volume **1365**, pages 100–119, Springer-Verlag, Berlin 1989.

121. R. H. Nochetto, G. Savaré, *Nonlinear evolution governed by accretive operators in Banach spaces: error control and applications,* Math. Models Methods Appl. Sci., 16(3):439–477, 2006.

122. R. Nochetto, G. Savaré, C. Verdi, *Error control of nonlinear evolution equations,* C. R. Acad. Sci. Paris Sér. I, Math., 326(12):1437–1442, 1998.

123. R. Nochetto, G. Savaré, C. Verdi, *A posteriori error estimates for variable time-step discretization of nonlinear evolution equations*, Commun. Pure Appl. Math., 53(5):525–589, 2000.

124. N. Okazawa, T. Yokota, *Monotonicity method applied to the complex Ginzburg-Landau and related equations*, J. Math. Anal. Appl., 267: 247–263, 2002.

125. C. Ortner, *Two variational techniques for the approximation of curves of maximal slope*, Technical Report NA05/10, Oxford University Computing Laboratory, Oxford, 2005.

126. F. Otto, *The geometry of dissipative evolution equations: the porous medium equation*, Comm. Partial Differential Equations, 26: 101–174, 2001.

127. A. Pazy, *Semigroups of Linear Operators and Applications to Partial Differential Equations*, Applied Mathematical Sciences, volume 44, Springer-Verlag, New York, 1983.

128. H. Pecher, W. Von Wahl, *Time dependent nonlinear schrodinger equations,* Manuscripta Math, 27: 125–157, 1979.

129. J.-C. Peralba, *Un problème d'évolution relatif à un opérateur sous-différentiel dépendant du temps*, C. R. Acad. Sci. Paris Sér. A-B, 275:A93–A96, 1972.

130. R.R. Phelps, *Convex functions, monotone operators and differentiability*, 2nd edition, Lecture Notes in Mathematics, 1364, Springer-Verlag, Berlin,1998.

131. H. Rios, *Étude de la question d'existence pour certains problèmes d'évolution par minimisation d'une fonctionnelle convexe*, C. R. Acad. Sci. Paris Sér. A-B, 283(3):Ai, A83–A86, 1976.

132. H. Rios, *La question d'existence de solutions pour certaines équations à opérateurs monotones vue comme problème de minimum ou comme problème de point-selle*, Travaux Sém. Anal. Convexe, 6(Exp. 13):16, 1976.

133. H. Rios, *Étude de certains problèmes paraboliques: existence et approximation des solutions*, Travaux Sém. Anal. Convexe, 8(1) (Exp. No. 1): 96, 1978.

134. H. Rios, *Une étude d'existence sur certains problèmes paraboliques*, Ann. Fac. Sci. Toulouse Math., Ser. V, 1(3):235–255, 1979.

135. R. T. Rockafellar, *Convex analysis*, Princeton Mathematical Series, No. 28, Princeton University Press, Princeton, N.J. 1970

136. R. T. Rockafellar, *Existence and duality theorems for convex problems of Bolza*. Trans. Amer. Math. Soc. (159): 1–40, 1971.

137. R. T. Rockafellar, *Conjugate duality and optimization*, Lectures given at the Johns Hopkins University, Baltimore, Md., June, 1973. Conference Board of the Mathematical Sciences Regional Conference Series in Applied Mathematics, No. 16. Society for Industrial and Applied Mathematics, Philadelphia, Pa., 1974.

138. T. Roubíček, *Direct method for parabolic problems*, Adv. Math. Sci. Appl., 10(1):57–65, 2000.

139. T. Roubíček, *Nonlinear partial differential equations with applications*, International Series of Numerical Mathematics, volume 153, Birkhäuser Verlag, Basel, 2005.

140. G. Rossi, G. Savaré, *Gradient flows of non-convex functionals in Hilbert spaces and applications*, ESAIM Control Optim. Calculus of Variations, 12(3):564–614 (electronic), 2006.

141. G. Savaré, *Error estimates for dissipative evolution problems*, In *Free boundary problems (Trento, 2002)*, International Series of Numerical Mathematics,volume 147, pages 281–291, Birkhäuser, Basel, 2004.

142. T. Senba, *On some nonlinear evolution equationS*, Funkcial. Ekval., 29:243–257, 1986.

143. T. Shigeta, *A characterization of m-accretivity and an application to nonlinear Schrodinger type equations,* Nonlinear Anal. Theor.Methods.App., (10): 823-838, 1986

144. R. E. Showalter, *Monotone operators in Banach Space and nonlinear partial differential equations* Mathematical Survey Monographs, Volume 49, American Mathematical Society, Providence, RI, 1997.

145. U. Stefanelli, *On a class of doubly nonlinear nonlocal evolution equations*, Differential Integral Equations, 15(8):897–922, 2002.

146. U. Stefanelli, *On some nonlocal evolution equations in Banach spaces*, J. Evol. Equations, 4(1):1–26, 2004.

147. U. Stefanelli, *Some quasivariational problems with memory*, Boll. Unione Mat. Ital. Sez. B Artic. Ric. Mat. (8), 7:319–333, 2004.

148. U. Stefanelli, *Nonlocal quasivariational evolution problems*, J. Differential Equations, 229:204–228, 2006.

149. U. Stefanelli, *Some remarks on convergence and approximation for a class of hysteresis problems*, Istit. Lombardo Accad. Sci. Lett. Rend. A: (to appear) 2008. Preprint available at: http://www.imati.cnr.it/ulisse/pubbl.html.

150. U. Stefanelli, *The discrete Brezis-Ekeland principle*, Preprint IMATI - CNR, 14PV07/14/0, 2007.

151. U. Stefanelli, *A variational principle for hardening elasto-plasticity*, preprint IMATI - CNR, 11PV07/11/8, 2007. Available at: http://arxiv.org/abs/0710.2425

152. U. Stefanelli, *A variational principle in non-smooth mechanics*, in *MFO Workshop: Analysis and Numerics for Rate-Independent Processes*, Oberwolfach Reports, 2007.

153. M. Struwe, *Variational Methods and Their Applications to Nonlinear Partial Differential Equations and Hamiltonian Systems*, Springer-Verlag, New York, 1990.

154. B. F. Svaiter, *Fixed points in the family of convex representations of a maximal monotone operator*, Proc. Am. Math. Soc., 131 (12): 3851–3859, 2003.

155. J. J. Telega, *Extremum principles for nonpotential and initial-value problems*, Arch. Mech. (Arch. Mech. Stos.) 54, (5-6): 663–690, 2002.

156. R. Temam, *Navier-Stokes equations. Theory and numerical analysis*, with an appendix by F. Thomasset, 3rd edition, Studies in Mathematics and Its Applications, volume 2, North-Holland, Amsterdam, 1984.

157. R. Temam, *Infinite-dimensional dynamical systems in mechanics and physics*, Applied mathematical sciences, volumw 68, Springer-Verlag, 1997.

158. A. Visintin, *Differential Models of Hysteresis*, Applied Mathematical Sciences, volume 111, Springer, Berlin, 1994.

159. A. Visintin, *A new approach to evolution*, C. R. Acad. Sci. Paris Sér. I, Math., 332(3): 233–238, 2001.

160. I. I. Vrabie, *Compactness Methods for Nonlinear Evolutions*, Pitman Monographs and Surveys in Pure and Applied Mathematics, volume 32, Longman Scientific and Technical, 1987.

161. I. I. Vrabie, *Periodic solutions for nonlinear evolution equations in a Banach space*, Proc. Am. Math. Soc., 109 (3): 653–661, 1990.

162. C. Zanini, *Singular perturbations of finite-dimensional gradient flows*, Discrete Contin. Dyn. Syst., 18(4):657–675, 2007.

163. R. Zarate, PhD dissertation, The University of British Columbia, 2008.

Index

Printed in the United States of America